1401618841

AF606521

FOUNDATIONS OF SOLID MECHANICS

SOLID MECHANICS AND ITS APPLICATIONS
Volume 3

Series Editor: G.M.L. GLADWELL
Solid Mechanics Division, Faculty of Engineering
University of Waterloo
Waterloo, Ontario, Canada N2L 3G1

Aims and Scope of the Series

The fundamental questions arising in mechanics are: *Why?*, *How?*, and *How much?* The aim of this series is to provide lucid accounts written by authoritative researchers giving vision and insight in answering these questions on the subject of mechanics as it relates to solids.

The scope of the series covers the entire spectrum of solid mechanics. Thus it includes the foundation of mechanics; variational formulations; computational mechanics; statics, kinematics and dynamics of rigid and elastic bodies; vibrations of solids and structures; dynamical systems and chaos; the theories of elasticity, plasticity and viscoelasticity; composite materials; rods, beams, shells and membranes; structural control and stability; soils, rocks and geomechanics; fracture; tribology; experimental mechanics; biomechanics and machine design.

The median level of presentation is the first year graduate student. Some texts are monographs defining the current state of the field; others are accessible to final year undergraduates; but essentially the emphasis is on readability and clarity.

For a list of related mechanics titles, see final pages.

Foundations of Solid Mechanics

by

P. KARASUDHI

Asian Institute of Technology, Bangkok, Thailand

KLUWER ACADEMIC PUBLISHERS

DORDRECHT / BOSTON / LONDON

Library of Congress Cataloging-in-Publication Data

Pisidhi Karasudhi.
Foundations of solid mechanics / by Pisidhi Karasudhi.
p. cm. -- (Solid mechanics and its applications ; v. 3)
Includes bibliographical references and indexes.
ISBN 0-7923-0772-0 (acid free paper)
1. Strength of materials. 2. Mechanics, Applied. I. Title.
II. Series.
TA405.P54 1990
620.1'05--dc20 90-48510

ISBN 0-7923-0772-0

Published by Kluwer Academic Publishers,
P.O. Box 17, 3300 AA Dordrecht, The Netherlands.

Kluwer Academic Publishers incorporates
the publishing programmes of
D. Reidel, Martinus Nijhoff, Dr W. Junk and MTP Press.

Sold and distributed in the U.S.A. and Canada
by Kluwer Academic Publishers,
101 Philip Drive, Norwell, MA 02061, U.S.A.

In all other countries, sold and distributed
by Kluwer Academic Publishers Group,
P.O. Box 322, 3300 AH Dordrecht, The Netherlands.

Printed on acid-free paper

Printed in the Netherlands

..... in memory of, and dedicated to

my father.

TABLE OF CONTENTS

X FINITE DEFORMATION

PREFACE

This book has been written with two purposes, as a textbook for engineering courses and as a reference book for engineers and scientists. The book is an outcome of several lecture courses. These include lectures given to graduate students at the Asian Institute of Technology for several years, a course on elasticity for University of Tokyo graduate students in the spring of 1979, and courses on elasticity, viscoelasticity and finite deformation at the National University of Singapore from May to November 1985.

In preparing this book, I kept three objectives in mind: first, to provide sound fundamental knowledge of solid mechanics in the simplest language possible; second, to introduce effective analytical and numerical solution methods; and third, to impress on readers that the subject is beautiful, and is accessible to those with only a standard mathematical background.

In order to meet those objectives, the first chapter of the book is a review of mathematical foundations intended for anyone whose background is an elementary knowledge of differential calculus, scalars and vectors, and Newton's laws of motion. Cartesian tensors are introduced carefully. From then on, only Cartesian tensors in the indicial notation, with subscript as indices, are used to derive and represent all theories. Any combination of indicial and Gibbs notations is avoided except in the sections on curvilinear coordinates in the finite deformation chapter. Conditions under small deformation for cylindrical and spherical coordinates are put in explicit symbols. Most of the pertinent theorems and formulas are compiled, proved and/or verified. The only theorems and formulas which are quoted without proofs are those which can be seen in standard mathematical books. Whenever possible, presentations are made by induction processes, i.e. emerging from the simplest special cases to the most general. New analytical tools and methods such as the Dirac-delta distribution, integral transforms and integral equations are introduced along with their limitations. It is emphasized that an effective solution must have a rational basis. Where deemed appropriate, tables are provided to save mundane though straightforward operations. Approximate formulas and proven numerical algorithms are brought to the attention of the reader. All exercise problems are accompanied by hints and/or answers.

Readers will learn that the major conditions governing the mechanics of a solid domain are the equilibrium equations, strain-displacement relationships, constitutive relationships, and boundary conditions. Chapters II to VIII of this book are concerned with linear solid continua, Chapters IX and X with nonlinear. Both static and dynamic linear problems, and static and quasi-static nonlinear problems are treated.

I have used the book as a textbook in three graduate courses:

1) Introduction to Solid Mechanics – mostly concerned with linear elastostatics. This uses Sections 1.1 to 1.10 and 1.16 and Chapters II to VI.
2) Advanced Solid Mechanics – time dependent constitutive relationships,

material nonlinearity, and geometrical nonlinearity. This uses Sections 1.11 to 1.15 and Chapters VII, IX and X.

3) Wave Propagation in Elastic and Viscoelastic Media – special emphasis on half spaces. This uses Chapter VIII.

A lecturer using this book should find it possible to present it to students equation-by-equation, page-by-page, and section-by-section. The book is also supposed to contain all the essentials of solid mechanics normally expected from a good reference book on the subject. In addition, the arrangement of the contents in this book, together with the Author and Subject Indices and the List of Symbols, should make it simple to use.

My appreciation of solid mechanics was enhanced through many years of association with Professors J.D. Achenbach, J. Dundurs, L.M. Keer, and S.L. Lee of Northwestern University. I would like to acknowledge my indebtedness to the lectures and publications of those professors. I am grateful for the warm friendship, constructive criticism, and excellent hospitality provided by Professor F. Nishino of the University of Tokyo during my sabbatical stay at the University of Tokyo in 1979. I cannot find suitable words to record my heartfelt gratitude to Professor S.L. Lee. He was my professor at Northwestern, my colleague at the Asian Institute of Technology, my host during the sabbatical leave spent at the National University of Singapore in 1985, my close friend for more than two decades, and he has given me many valuable suggestions and constant encouragement.

I am indebted to Professor G.M.L. Gladwell of the University of Waterloo for reviewing the book and suggesting several changes in it.

This book has been written in memory of the person who was the first one who taught me reading, writing and arithmetic, and who was always ready to give me love and care. That was my own late father, to whom this piece of work is humbly dedicated.

P.K.

LIST OF SYMBOLS

Common symbols are defined when they first appear, and when used at other places stand for the following (unless specified differently):

B	bulk creep compliance;
$\bar{B}$	Laplace transform of bulk creep compliance;
C.P.V.	Cauchy's principal value;
c_p	pressure wave speed;
c_s	shear wave speed;
E	Young's modulus, Eqs. 3.36 to 3.38;
$\bar{E}$	Laplace transform of Young's modulus;
e_{ij}	strain deviator, Eq. 2.47;
$H(\ )$	Heaviside step function, Eqs. 1.104 and 1.106;
$H_m^{(1)}(\ )$	a Hankel function or third kind Bessel function of ( );
$H_m^{(2)}(\ )$	another Hankel function or third kind Bessel function of ( );
$\Im(\ )$	imaginary part of ( );
i	imaginary number, $i^2 = -1$;
J	shear creep compliance;
$\bar{J}$	Laplace transform of shear creep compliance;
$J_m(\ )$	first kind Bessel function of ( );
K	bulk modulus;
$\bar{K}$	Laplace transform of bulk modulus;
M_r	a plate bending moment in cylindrical coordinates;
$M_{r\theta}$	plate twisting moment in cylindrical coordinates;
M_x, M_y	plate bending moments in Cartesian coordinates, Eqs. 5.10a and b;
M_{xy}	plate twisting moment in Cartesian coordinates, Eq. 5.10c;
M_θ	another plate bending moment in cylindrical coordinates;
n_i	unit normal vector to a surface;
p	Laplace transform parameter;
Q_r	a transverse shear force in plate bending in cylindrical coordinates;
Q_x, Q_y	transverse shear forces in plate bending in Cartesian coordinates, Eqs. 5.20;

Q_θ	another transverse shear force in plate bending in cylindrical coordinates;
R	radial spherical coordinate, Fig. 2.12;
$\Re(\ )$	real part of ( );
r	radial cylindrical coordinate, Fig. 2.12;
S	surface;
s_{ij}	stress deviator, Eq. 2.34;
T	temperature; or transpose of a matrix when appears as a superscript;
t	time;
u	displacement component in x-direction;
$\mathbf{u}$	displacement vector;
u_i	displacement vector;
V	volume;
V_i	velocity jump, Eq. 8.72;
V_r	a supplemented shear force in plate bending in cylindrical coordinates, Eq. 5.40a;
V_x, V_y	supplemented shear forces in plate bending in Cartesian coordinates; Eqs. 5.26;
V_θ	another supplemented shear force in plate bending in cylindrical coordinates, Eq. 5.40b;
v	displacement component in y-direction;
v_i	Cartesian components of velocity;
w	displacement component in z-direction, plate deflection;
X	body force in x-direction;
$\overline{X}$	surface force in x-direction;
$\mathbf{X}$	body force vector;
X_i	body force vector;
$\overline{\mathbf{X}}$	surface force vector;
$\overline{X}_i$	surface force vector;
x	x_1;
x_i	original position vector, Lagrangian coordinates;
$Y_m(\ )$	second kind Bessel function of ( );
y	x_2;

y_i	position vector after deformation, Eulerian coordinates;
z	x_3;
$\alpha_p(\eta)$	a radical quantity related to pressure wave, Eq. 8.167a;
$\overline{\alpha}_p(\eta)$	$i\alpha_p(\eta)$, Eq. 8.173;
$\alpha_s(\eta)$	a radical quantity related to shear wave, Eq. 8.167b;
$\overline{\alpha}_s(\eta)$	$i\alpha_s(\eta)$, Eq. 8.171;
δ	virtual quantity, variation symbol;
$\delta(\ )$	Dirac-delta function, Eqs. 1.102, 1.103 and 1.106;
δ_{ij}	Kronecker delta, Kronecker symbol, Eq. 1.23;
δ_i^j	δ_{ij};
E_{ij}	Green's strain tensor, Eq. 10.98;
ε_{ij}	Almansi's strain tensor, Eqs. 2.44 and 10.99;
ε_{ijk}	permutation tensor, Eq. 1.33;
ζ	a complex variable, $\eta + i\xi$;
η	transform parameter, $\Re(\zeta)$;
η_L	dimensionless surface Love wavenumber;
η_p	dimensionless pressure wavenumber, Eq. 8.164b;
η_R	dimensionless surface Rayleigh wavenumber;
η_s	dimensionless shear wavenumber, Eq. 8.164c;
θ	angular cylindrical coordinate = an angular spherical coordinate, Fig. 2.12;
θ^λ	curvilinear coordinates, Fig. 1.14;
κ	an elastic constant in plane problems, Eqs. 4.8;
λ	a Lamé's constant;
$\tilde{\lambda}$	Laplace transform of λ;
μ	another Lamé's constant, shear modulus;
$\tilde{\mu}$	Laplace transform of shear modulus;
ν	Poisson's ratio;
$\tilde{\nu}$	Laplace transform of Poisson's ratio;
ξ	$\Im(\zeta)$;
Π	product series, Eq. 7.29g;
ρ	mass density;

Σ_{ij}	Kirchhoff stress tensor, Eq. 10.58;
σ_{ij}	Eulerian stress tensor, Figs. 2.2 and 10.6;
ϕ	another angular spherical coordinate, Fig. 2.12; or stress function, Chapter IV;
ψ	wavefront function, Eqs. 8.49;
ω	angular frequency;
∇^2	Laplace operator, Eq. 1.85d; and
[ ]	jump = discontinuity at wavefront, Eq. 8.39b.

CHAPTER I

MATHEMATICAL FOUNDATIONS

1.1 TENSORS AND CONTINUUM MECHANICS

Continuum mechanics is that part of physical science dealing with the deformation and motion of continuous material media under the influence of external agencies. The foundations of the theory involve laws of motion and constitutive laws that are conveniently stated in terms of *tensors*. As a mathematical entity, a tensor has an existence independent of any coordinate system, yet it may be specified in a particular coordinate system by a certain set of quantities known as its *components*. Components of a tensor in different coordinate systems are related by the *law of transformations*, to be elaborated later.

When attention is restricted to transformations from one rectangular coordinate system to another, the tensors involved are referred to as *Cartesian tensors*. Since much of the theory of continuum mechanics may be developed in terms of Cartesian tensors, the word 'tensor' in this text means 'Cartesian tensor' unless specifically stated otherwise. Physical laws of continuum mechanics are expressed by tensor equations. If such an equation is valid in one (Cartesian) coordinate system, it is valid in any other (Cartesian) coordinate system.

Tensors are classified by *rank* or *order* according to the number of components they possess. In three-dimensional *Euclidean space* such as the ordinary physical space, a tensor of order N has 3^N components. Accordingly, a tensor of *order zero* is specified in any coordinate system by *one* component. Such quantities are also known as *scalars*. A tensor of *order one* has *three* coordinate components in three dimensional Euclidean space and known as a *vector*. A second order tensor is called a *dyadic*. Third order tensors, *triadics*, and fourth order tensors, *tetradics*, also appear in the mathematics of continuum mechanics.

1.2 SCALARS AND VECTORS

Certain physical quantities, such as mass, length, time, temperature, energy, etc. possess magnitude only and are *scalars*, tensors of order zero. Quantities such as force, displacement, velocity, acceleration, etc. which possess both magnitude and direction are *vectors*, first order tensors. In three-dimensional space, a vector may be represented as an arrow pointing in the appropriate direction and having a length proportional to the magnitude of the vector. Two vectors, regardless of their position, are *equal* if they are in the same direction and have the same magnitude. The *negative* of a vector is that vector having the same magnitude but opposite direction. A *unit* vector is a vector of unit magnitude. In the *symbolic* or *Gibbs notation*, scalars are denoted by italic letters such as a, b, c etc., vectors by bold-faced letters such as $\mathbf{a}$, $\mathbf{b}$, etc. Unit vectors are often distinguished by a caret placed over the bold-faced letter, e.g. $\hat{\mathbf{a}}$. In Fig. 1.1, arbitrary vectors $\mathbf{a}$ and $\mathbf{b}$ are shown along with a unit vector $\hat{\mathbf{e}}$ and a pair of equal vectors $\mathbf{c}$ and $\mathbf{d}$. The magnitude of any vector $\mathbf{a}$ is written as a.

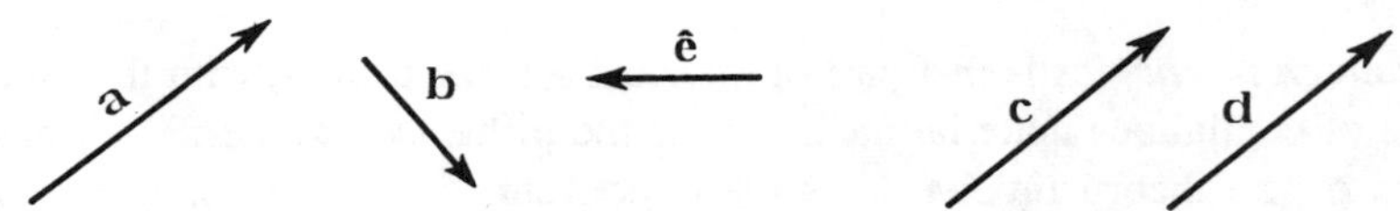

Fig. 1.1 Pictorial views of vectors in three-dimensional space.

Vector addition obeys the *parallelogram law*, which defines a vector sum of two vectors as the diagonal of a parallelogram having the summed vectors as adjacent sides. This law is the same as the *triangle rule* which defines the sum of two vectors as the vector extending from the tail of the first to the head of the second when the summed vectors are adjoined head to tail. The graphical construction for the addition of $\mathbf{a}$ and $\mathbf{b}$ by the parallelogram law is shown in Fig. 1.2(a). Algebraically, the addition process is expressed by the vector equation

$$\mathbf{a}+\mathbf{b}=\mathbf{b}+\mathbf{a}=\mathbf{c} \tag{1.1}$$

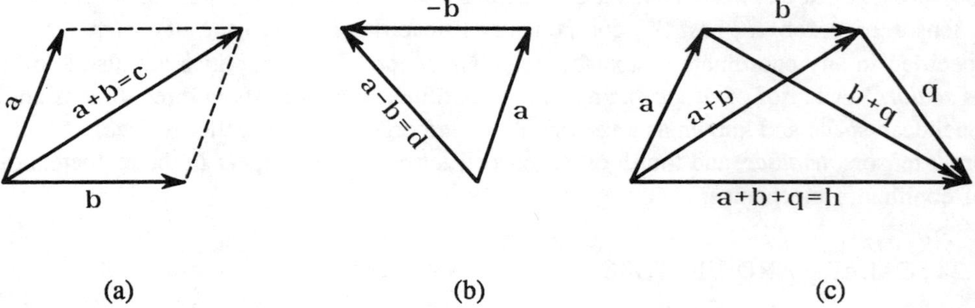

Fig. 1.2 Vector addition and subtraction.

Vector subtraction is accomplished by addition of the negative vector as shown, for example, in Fig. 1.2(b) where the triangular rule is used. Thus

$$\mathbf{a} - \mathbf{b} = -\mathbf{b} + \mathbf{a} = \mathbf{d} \tag{1.2}$$

Thus Eqs. 1.1 and 1.2 state that the operations of vector addition and subtraction are *commutative*. Extensions of such operations to more than two tensors are immediate, since such operations are also *associative*, as illustrated by the case in Fig. 1.2(c) for which the appropriate equations are

$$(\mathbf{a} + \mathbf{b}) + \mathbf{q} = \mathbf{a} + (\mathbf{b} + \mathbf{q}) = \mathbf{h} \tag{1.3}$$

Multiplying a vector by the reciprocal of its magnitude results in a *unit vector* in the direction of the original vector. This relationship is expressed by the equation

$$\hat{\mathbf{b}} = \mathbf{b}/b \tag{1.4}$$

The *dot* or *scalar product* of two vectors $\mathbf{a}$ and $\mathbf{b}$ is the scalar

$$\lambda = \mathbf{a} \cdot \mathbf{b} = \mathbf{b} \cdot \mathbf{a} = ab \cos\theta \tag{1.5}$$

in which θ is the angle between the two vectors. Evidently [Fig. 1.3(a)], $\mathbf{a} \cdot \mathbf{b}$ may be regarded as the quantity obtained by multiplying the magnitude a of $\mathbf{a}$ by the projection $b\cos\theta$ of $\mathbf{b}$ upon $\mathbf{a}$, or as the quantity obtained by multiplying b by the projection of $\mathbf{a}$ upon $\mathbf{b}$.

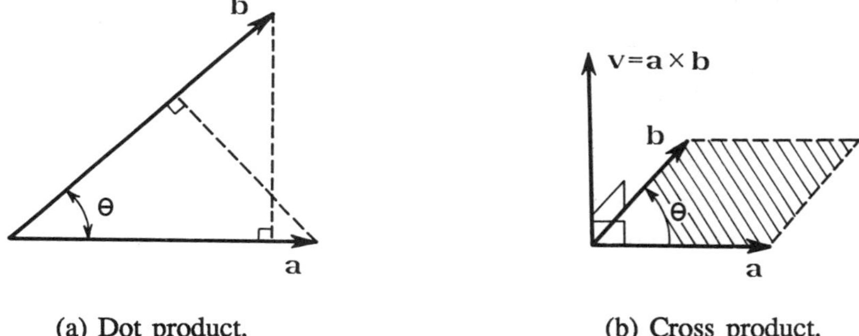

(a) Dot product.

(b) Cross product.

Fig. 1.3 Products of vectors.

The *cross* or *vector product* of $\mathbf{a}$ into $\mathbf{b}$ is the vector $\mathbf{v}$ given by

$$\mathbf{v} = \mathbf{a} \times \mathbf{b} = -\mathbf{b} \times \mathbf{a} = (ab \sin\theta)\hat{\mathbf{v}} \tag{1.6}$$

in which θ is the angle less than π between the vectors $\mathbf{a}$ and $\mathbf{b}$, and $\mathbf{v}$ is perpendicular to the plane of $\mathbf{a}$ and $\mathbf{b}$ so sensed that $\mathbf{a}$, $\mathbf{b}$ and $\mathbf{v}$ form a *right-handed system*, i.e. a right threaded screw rotating through the angle θ from $\mathbf{a}$ to $\mathbf{b}$ will advance in the direction of $\mathbf{v}$. Note that the magnitude of $\mathbf{a} \times \mathbf{b}$ is equal to the area of the parallelogram whose adjacent sides are $\mathbf{a}$ and $\mathbf{b}$ shown shaded in Fig. 1.3(b).

The cross product is not *commutative*. In general, three vectors **a**, **b** and **c** which have coincident initial points are said to form a *right-handed system* or *dextral system* if a right threaded screw rotated through the angle less than π from **a** to **b** will advance in the direction **c** as shown in Fig. 1.4.

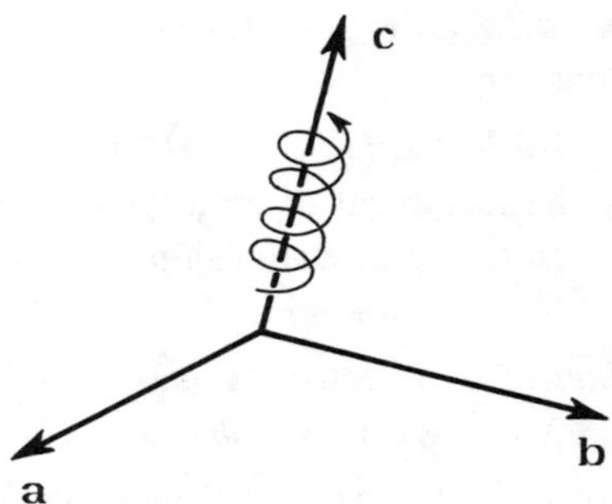

Fig. 1.4 A right-handed or dextral system formed by **a**, **b** and **c**.

The *preferred* coordinate systems for a Euclidean space of any dimension (to be described more in Section 1.13) are *rectangular Cartesian coordinate systems,* or simply *rectangular coordinate systems* for brevity. For a three dimensional Euclidean space, the *well-known* rectangular Cartesian coordinate system is often represented by mutually perpendicular axes x, y, z shown in Fig. 1.5. It should be noted that, in this text, only right-handed coordinate systems are used unless stated otherwise. Any vector **v** in such a system may be expressed as a linear combination of three arbitrary, nonzero, noncoplanar vectors of the system, which are called *base vectors*. The most frequent choice of base vectors for the rectangular Cartesian system is the set of unit vectors $\hat{\mathbf{i}}$, $\hat{\mathbf{j}}$, $\hat{\mathbf{k}}$ along the coordinate axes as shown in Fig. 1.5. These base vectors constitute a right-handed *unit vector triad*, for which

$$\begin{gathered}\hat{\mathbf{i}}\cdot\hat{\mathbf{i}}=\hat{\mathbf{j}}\cdot\hat{\mathbf{j}}=\hat{\mathbf{k}}\cdot\hat{\mathbf{k}}=1,\\ \hat{\mathbf{i}}\cdot\hat{\mathbf{j}}=\hat{\mathbf{j}}\cdot\hat{\mathbf{k}}=\hat{\mathbf{k}}\cdot\hat{\mathbf{i}}=0\end{gathered}\tag{1.7}$$

and

$$\begin{gathered}\hat{\mathbf{i}}\times\hat{\mathbf{j}}=\hat{\mathbf{k}},\qquad \hat{\mathbf{j}}\times\hat{\mathbf{k}}=\hat{\mathbf{i}},\qquad \hat{\mathbf{k}}\times\hat{\mathbf{i}}=\hat{\mathbf{j}},\\ \hat{\mathbf{i}}\times\hat{\mathbf{i}}=\hat{\mathbf{j}}\times\hat{\mathbf{j}}=\hat{\mathbf{k}}\times\hat{\mathbf{k}}=0\end{gathered}\tag{1.8}$$

In terms of unit triads $\hat{\mathbf{i}}$, $\hat{\mathbf{j}}$, $\hat{\mathbf{k}}$, the vector **v** shown in Fig. 1.5 may be expressed by

$$\mathbf{v}=v_x\hat{\mathbf{i}}+v_y\hat{\mathbf{j}}+v_z\hat{\mathbf{k}}\tag{1.9}$$

in which the *Cartesian components*

$$v_x=\mathbf{v}\cdot\hat{\mathbf{i}}=v\cos\alpha\tag{1.10a}$$

$$v_y = \mathbf{v}\cdot\hat{\mathbf{j}} = v\cos\beta \tag{1.10b}$$

$$v_z = \mathbf{v}\cdot\hat{\mathbf{k}} = v\cos\gamma \tag{1.10c}$$

are the projections of $\mathbf{v}$ on to the coordinate axes. The unit vector in the direction of $\mathbf{v}$ is given according to Eq. 1.4 by

$$\hat{\mathbf{v}} = \mathbf{v}/v \tag{1.11}$$

$$= (\cos\alpha)\hat{\mathbf{i}} + (\cos\beta)\hat{\mathbf{j}} + (\cos\gamma)\hat{\mathbf{k}} \tag{1.12}$$

It follows that the unit vector corresponding to a given vector has the *direction cosines* of that vector as its *Cartesian components*.

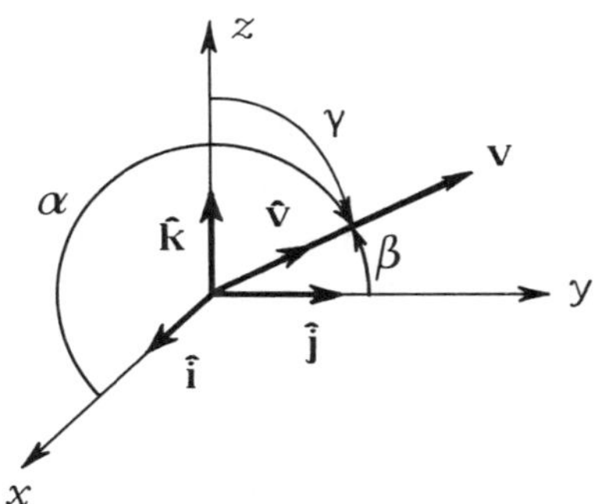

Fig. 1.5 Cartesian unit base vectors and direction cosines of a vector $\mathbf{v}$.

1.3 INDICIAL NOTATION

The components of a tensor of any order, and *indeed the tensor itself*, can be represented clearly and concisely by the use of *indicial notation*. In this notation, letter indices (either subscripts or superscripts) are appended to the *generic* or *kernel* letter representing the tensor quantity of interest. Typical examples illustrating the use of indices are the tensor symbols

$$a_i,\ b^j,\ T_{ij},\ \varepsilon_{ijk},\ R^{pq}.$$

When an index occurs unrepeated in a term, that index is known as a *free* index and can take on one of the values $1, 2, \cdots, N$, where N is a specified integer that determines the *range* of the index. The tensor rank of a given term is equal to the number of free indices appearing in that term. A correctly written tensor equation must have the same letters as free indices in every term.

The repetition of an index in a term denotes a *contraction*, i.e. *summation* with respect to that index over its range. In such so-called *Einstein summation convention*, repeated indices are often referred to as *dummy or umbral indices*, since their replacement by any other letter not appearing as a free index does not change the meaning of the term in which they occur. No index occurs more than twice in a properly written term. First order tensors (i.e. vectors) are denoted by kernel letters bearing *one free index*. Thus a vector **a** is represented by a symbol having a single subscript or superscript, i.e. in one or the other of the two forms,

$$a_i, \ a^i.$$

The following terms, having only one free index, are also recognized as first order tensors,

$$a_{ij}b_j, \ F_{ikk}, \ e_{ijk}u_j v_k.$$

Second order tensors are denoted by symbols having *two* free indices. Thus a dyadic **D** will appear in one of the two possible forms,

$$D_{ij}, \ D^{ij}.$$

By a logical continuation of the above scheme, third order tensors are expressed by symbols with *three* free indices, while symbols such as λ and e_{kk} which have *no* free indices represent scalars or zero order tensors.

When only rectangular Cartesian coordinates are considered, the distinction between *covariance* and *contravariance* (to be discussed later in Section 1.14) is immaterial, and we can choose to write all indices of a tensor as subscripts. In curvilinear coordinate systems, however, such distinction must be recognized by using subscripts for covariance and superscripts for contravariance.

In ordinary physical space, a *basis* is composed of three non-coplanar vectors, so any vector in this space is completely specified by its three components in the directions of those base vectors, and the range of an index is 3. Accordingly, the symbol a_i (where $i = 1, 2, 3$) is understood to represent a vector **a** completely, and also the i th component of the vector, i.e. the component of **a** in the direction of the i th base vector. For such a range of three on both indices, the symbol A_{ij} represents the second order tensor (dyadic) **A**, which has nine components. The tensor A_{ij} is often presented explicitly by the nine components in a square array, i.e.

$$[A_{ij}] = \begin{bmatrix} A_{11} & A_{12} & A_{13} \\ A_{21} & A_{22} & A_{23} \\ A_{31} & A_{32} & A_{33} \end{bmatrix} \tag{1.13}$$

In the same way, the component of a first order tensor (vector) in three-dimensional spaces may be displayed explicitly by a row or column arrangement of the form

$$(a_i) = (a_1 \quad a_2 \quad a_3) \qquad \text{or} \qquad \{a_i\} = \begin{Bmatrix} a_1 \\ a_2 \\ a_3 \end{Bmatrix} \tag{1.14}$$

In general, an n th order tensor with a range N has N^n components.

The usefulness of the indicial notation in presenting systems of equations in compact form is illustrated by the following two examples. For a range of three on both i and j the indicial equation

$$\overline{X}_i = \sigma_{ij} n_j \tag{1.15}$$

represents the three equations

$$\begin{aligned} \overline{X}_1 &= \sigma_{11} n_1 + \sigma_{12} n_2 + \sigma_{13} n_3 \\ \overline{X}_2 &= \sigma_{21} n_1 + \sigma_{22} n_2 + \sigma_{23} n_3 \\ \overline{X}_3 &= \sigma_{31} n_1 + \sigma_{32} n_2 + \sigma_{33} n_3 \end{aligned} \tag{1.16}$$

For a range of two on all indices, the indicial equation

$$A_{ij} = B_{ip} C_{jq} D_{pq} \tag{1.17}$$

represents the four equations

$$\begin{aligned} A_{11} &= B_{11}C_{11}D_{11} + B_{11}C_{12}D_{12} + B_{12}C_{11}D_{21} + B_{12}C_{12}D_{22}, \\ A_{12} &= B_{11}C_{21}D_{11} + B_{11}C_{22}D_{12} + B_{12}C_{21}D_{21} + B_{12}C_{22}D_{22}, \\ A_{21} &= B_{21}C_{11}D_{11} + B_{21}C_{12}D_{12} + B_{22}C_{11}D_{21} + B_{22}C_{12}D_{22}, \\ A_{22} &= B_{21}C_{21}D_{11} + B_{21}C_{22}D_{12} + B_{22}C_{21}D_{21} + B_{22}C_{22}D_{22} \end{aligned} \tag{1.18}$$

1.4 ALGEBRA OF CARTESIAN TENSORS

Let $x_1 x_2 x_3$ and $x_1' x_2' x_3'$ represent two right-handed rectangular Cartesian coordinate systems with a common origin at 0 as shown in Fig. 1.6. If the symbol a_{ij} denotes the cosine of the angle between ith unprimed and jth primed coordinate axes, i.e. $a_{ij} = \cos(x_i, x_j')$, the relative orientation of individual axes of each system with respect to the other is conveniently given by the table inset into Fig. 1.6. An arbitrary vector $\mathbf{v}$ in the same figure has vector components v_i; $i = 1, 2, 3$ in x_1, x_2 and x_3 directions respectively. Alternatively, the same vector can be considered as composed of vector components v_i'; $i = 1, 2, 3$ in x_1', x_2' and x_3' directions respectively. From the definition of a_{ij} and in view of Eqs. 1.10, each component in the primed system can be obtained from those in the unprimed system as

$$\begin{aligned} v_1' &= a_{11} v_1 + a_{21} v_2 + a_{31} v_3, \\ v_2' &= a_{12} v_1 + a_{22} v_2 + a_{32} v_3, \\ v_3' &= a_{13} v_1 + a_{23} v_2 + a_{33} v_3 \end{aligned} \tag{1.19}$$

or, in indicial notation, as

$$v_i' = a_{ji} v_j \tag{1.20}$$

By interchanging the roles of the primed and unprimed vectors in the above develop-

	x'_1	x'_2	x'_3
x_1	a_{11}	a_{12}	a_{13}
x_2	a_{21}	a_{22}	a_{23}
x_3	a_{31}	a_{32}	a_{33}

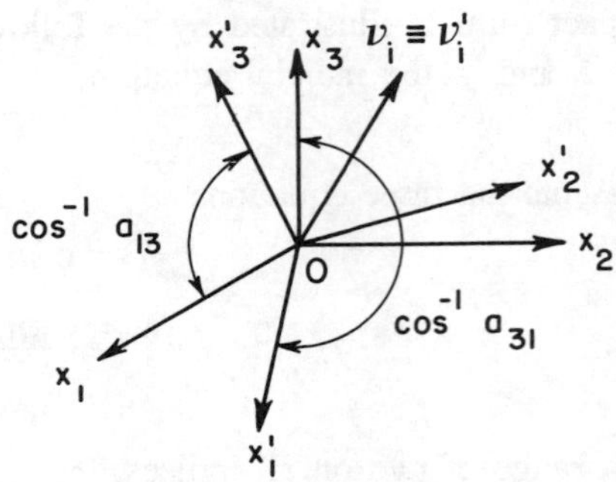

Fig. 1.6 Transformation for first order tensor.

ment, the inverse of Eq. 1.20 is found to be

$$v_i = a_{ij} v'_j \tag{1.21}$$

It is important to note that in Eq. 1.20 the free index on a_{ij} appears as the second index; in Eq. 1.21 the free index appears as the first.

Substitution of Eq. 1.20 into Eq. 1.21 with an appropriate choice of dummy indices yields

$$v_i = a_{ij} a_{kj} v_k \tag{1.22}$$

Since Eq. 1.22 must give the identity $v_i = v_i$, the coefficient $a_{ij}a_{kj}$ must be identical to the *Kronecker delta* δ_{ik}, defined by

$$\delta_{ik} = \delta_{ki} = \begin{cases} 1, & \text{for } i = k \\ 0, & \text{for } i \neq k \end{cases} \tag{1.23}$$

i.e.

$$a_{ij} a_{kj} = \delta_{ik} \tag{1.24}$$

In expanded form, Eq. 1.24 consists of nine equations which are known as the *orthogonality* or *orthonormality conditions* on the direction cosines a_{ij}.

On the other hand, if Eq. 1.21 is substituted into Eq. 1.20 to produce $v'_i = a_{ji} a_{jk} v'_k$, then orthogonality conditions appear in an alternative form $a_{ji} a_{jk} = \delta_{ik}$, thus

$$a_{ij} a_{kj} = a_{ji} a_{jk} = \delta_{ik} \tag{1.25}$$

Equations 1.20 and 1.21 are called the *transformation laws for first order tensors.*

According to Eq. 1.20, a *dyad* $u_i v_j$ has components in the primed coordinate system given by

$$u'_i v'_j = (a_{pi} u_p)(a_{qj} v_q) = a_{pi} a_{qj} u_p v_q \tag{1.26}$$

In an obvious generalization of Eq. 1.26, any second order Cartesian tensor T_{ij} obeys the transformation law

$$T'_{ij} = a_{pi}a_{qj}T_{pq} \tag{1.27}$$

With the help of the orthogonality conditions (Eq. 1.25) it is a simple calculation to invert Eq. 1.27, thereby obtaining the transformation rule from primed components to unprimed components as

$$T_{ij} = a_{ip}a_{jq}T'_{pq} \tag{1.28}$$

The transformation can be generalized for Nth order tensor as

$$T'_{ijk\ldots} = a_{pi}a_{qj}a_{rk}\ldots T_{pqr\ldots} \tag{1.29}$$

Cartesian tensors of the same order may be added (or subtracted) component by component in accordance with the rule

$$A_{ijk\ldots} \pm B_{ijk\ldots} = T_{ijk\ldots} \tag{1.30}$$

The sum is a tensor of the same order as those added.

Multiplication of every component of a tensor by a given scalar produces a new tensor of the same order. For the scalar multiplier λ, typical examples written in both indicial and symbolic notations are

$$b_i = \lambda a_i, \qquad \text{or} \qquad \mathbf{b} = \lambda\mathbf{a};$$

$$B_{ij} = \lambda A_{ij}, \qquad \text{or} \qquad \mathbf{B} = \lambda\mathbf{A}.$$

The *outer product* of two tensors of arbitrary order is the tensor formed by simply setting down the factor tensors in juxtaposition, for examples:

(a) $a_ib_j = T_{ij}.$ (c) $D_{ij}T_{km} = \Phi_{ijkm}.$

(b) $v_iF_{jk} = \alpha_{ijk}.$ (d) $\varepsilon_{ijk}v_m = \Theta_{ijkm}.$

This process produces a tensor having an order which is the sum of the orders of the factor tensors.

The *inner product* of two tensors of first order is the result of a *contraction,* involving one index from each tensor, performed on the outer product of the two tensors. Such a scalar product written in both indicial and symbolic notations is

$$a_ib_i \equiv \mathbf{a}\cdot\mathbf{b} \tag{1.31}$$

The *cross product* $\mathbf{a}\times\mathbf{b}$ is expressed in indicial notation by

$$\varepsilon_{ijk}a_jb_k = \mathbf{a}\times\mathbf{b} \tag{1.32}$$

where ε_{ijk} is the *permutation tensor,* i.e.

$$\varepsilon_{ijk} = \begin{cases} 1; & \text{if } i,j,k \text{ permute like, 1, 2, 3, etc.} \\ -1; & \text{if } i,j,k \text{ permute like, 3, 2, 1, etc.} \\ 0; & \text{if two or more of the indices, } i,j,k \text{ have the same value.} \end{cases} \tag{1.33}$$

1.5 MATRICES AND DETERMINANTS

A rectangular array of *elements*, enclosed by square brackets and subject to certain laws of combination is called a *matrix*. An $M \times N$ matrix is one having M (horizontal) rows and N (vertical) columns of elements. In the symbol A_{ij}, used to represent the typical element of a matrix, the first subscript denotes the row, the second subscript the column occupied by the element. The matrix itself is designated by enclosing the elements in square brackets [ ]. For example, a 3×3 matrix $[A_{ij}]$ is the array given by

$$[A_{ij}] = \begin{bmatrix} A_{11} & A_{12} & A_{13} \\ A_{21} & A_{22} & A_{23} \\ A_{31} & A_{32} & A_{33} \end{bmatrix}$$

A matrix for which $M = N$ is called a *square* matrix. A $1 \times N$ matrix is called a *row matrix* and is written by enclosing the elements in parentheses ( ) as follows

$$(v_i) = (v_1 \quad v_2 \quad \ldots \quad v_N) \tag{1.34}$$

while a *column matrix* is an $M \times 1$ matrix and is written by enclosing the elements in braces { }, i.e.

$$\{v_i\} = \begin{Bmatrix} v_1 \\ v_2 \\ \cdot \\ \cdot \\ \cdot \\ v_M \end{Bmatrix} \tag{1.35}$$

Thus every tensor of order two or less can be represented by a matrix; a tensor of second order by a square matrix, a vector by a row or a column matrix, and a scalar by a single term. A matrix having only zeros as elements is called the *zero matrix*. A square matrix with zeros everywhere except on the main diagonal $(A_{11}, A_{22}, A_{33}, \ldots, A_{MM})$ is called a *diagonal matrix*. If the nonzero elements of a diagonal matrix are all unity, the matrix is called the *unity* or *identity* matrix. The $N \times M$ matrix $[A_{ij}]^T$, formed by interchanging rows and columns of the $M \times N$ matrix $[A_{ij}]$, is called the *transpose matrix* of $[A_{ij}]$. A square matrix $[A_{ij}]$ is called *symmetric* if $[A_{ij}]^T = [A_{ij}]$.

Matrices having the same number of rows and columns may be added (or subtracted) *element by element*. Multiplication of the matrix $[A_{ij}]$ by a scalar λ results in the matrix $[\lambda A_{ij}]$. The product of two matrices is defined only if the matrices are *conformable*, i.e. if the *prefactor* matrix has the same number of columns as the *postfactor* matrix has rows. The product of an $M \times P$ matrix multiplied into a $P \times N$ matrix is an $M \times N$ matrix.

Matrix multiplication is usually denoted by simply setting down the matrix symbols in juxtaposition as follows:

$$[A_{ij}]\,[B_{jk}] = [C_{ik}] \tag{1.36}$$

in which the element $C_{ik} = A_{ij}B_{jk}$ and the summation convention applies from 1 to P

for the repeating index j. It follows from the definition of matrix multiplication, but is left to the reader to prove, that the transpose of a product of matrices equals the product of the transposed matrices taken in the reverse order, e.g.

$$[[A_{ij}]\,[B_{jk}]\,[C_{kl}]]^T = [C_{kl}]^T\,[B_{jk}]^T\,[A_{ij}]^T \tag{1.37}$$

Matrix multiplication is not commutative in general. In three-dimensional Euclidean space, the transformation of Cartesian vectors and second order tensors in Eqs. 1.20, 1.21, 1.27 and 1.28 can be rewritten respectively in matrix form as follows;

$$\{v_i'\} = [a_{ji}]^T\,\{v_j\} \tag{1.38}$$

$$\{v_i\} = [a_{ij}]\,\{v_j'\} \tag{1.39}$$

$$[T_{ij}'] = [a_{pi}]^T\,[T_{pq}]\,[a_{qj}] \tag{1.40}$$

$$[T_{ij}] = [a_{ip}]\,[T_{pq}']\,[a_{jq}]^T \tag{1.41}$$

The *determinant* of a square matrix $[A_{ij}]$ is normally denoted by the symbol $|A_{ij}|$ or $\det[A_{ij}]$, and can be determined by a standard method, which involves the concept of cofactors and minors. The *cofactor* of an element A_{ij} of a square matrix $[A_{ij}]$, denoted by A_{ij}^*, is defined by

$$A_{ij}^* = (-1)^{i+j} M_{ij} \tag{1.42}$$

in which M_{ij} is the *minor* of A_{ij}, i.e. the determinant of the square array remaining after the row and column of A_{ij} are deleted. The value of a determinant is then defined as sum of the products of the elements in any row (or column) by their corresponding cofactors, i.e. *Laplace expansion* of the form

$$\det[A_{ij}] = \sum_{j=1}^{N} A_{ij}A_{ij}^*, \quad (i \text{ is optional and not summed}) \tag{1.43a}$$

$$= \sum_{i=1}^{N} A_{ij}A_{ij}^*, \quad (j \text{ is optional and not summed}) \tag{1.43b}$$

where N is the number of rows (and columns as well) of the matrix $[A_{ij}]$.

Here are some important theorems on determinants.

Theorem 1.1. The value of a determinant remains the same if rows and columns are interchanged. In symbols,

$$\det[A_{ij}] = \det[A_{ij}]^T \tag{1.44}$$

Theorem 1.2. An interchange of any two rows (or columns) changes the sign of the determinant.

Theorem 1.3. If any two rows (or columns) are the same or proportional, the determinant is zero.

Theorem 1.4. If $[A_{ij}]$ and $[B_{ij}]$ are square matrices of the same order, then

$$\det[[A_{ij}]\,[B_{jk}]] = \det[A_{ij}]\,\det[B_{ij}] \tag{1.45}$$

A square matrix $[A_{ij}]$ whose determinant is zero is called a *singular* matrix.

The determinant of a 3×3 matrix $[T_{ij}]$ can be put in indicial notation in various ways, as follows

$$\det[T_{ij}] = \varepsilon_{ijk}T_{i1}T_{j2}T_{k3} \tag{1.46}$$

$$= \varepsilon_{ijk}T_{1i}T_{2j}T_{3k} \tag{1.47}$$

$$\varepsilon_{ijk}\det[T_{ij}] = \varepsilon_{lmn}T_{li}T_{mj}T_{nk} \tag{1.48}$$

$$= \varepsilon_{lmn}T_{il}T_{jm}T_{kn} \tag{1.49}$$

$$\det[T_{ij}] = \frac{1}{6}\varepsilon_{ijk}\varepsilon_{lmn}T_{li}T_{mj}T_{nk} \tag{1.50}$$

$$= \frac{1}{6}\varepsilon_{ijk}\varepsilon_{lmn}T_{il}T_{jm}T_{kn} \tag{1.51}$$

where ε_{ijk} is the permutation tensor in three dimensional space as defined in Eq. 1.33.

The *adjoint* matrix of $[A_{ij}]$ is obtained by replacing each element by its cofactor and then interchanging rows and columns. If a square matrix $[A_{ij}]$ is non-singular, it possesses a unique *inverse matrix* $[A_{ij}]^{-1}$ which is defined as the adjoint matrix of $[A_{ij}]$ divided by the determinant of $[A_{ij}]$. Thus

$$[A_{ij}]^{-1} = \frac{[A^*_{ji}]}{|A_{ij}|} \tag{1.52}$$

From the inverse matrix definition (Eq. 1.52), it may be shown that

$$[A_{ij}]^{-1}[A_{ij}] = [A_{ij}][A_{ij}]^{-1} = I \tag{1.53}$$

where I is the *identity matrix.*

1.6 LINEAR EQUATIONS AND EIGENVALUE PROBLEMS

Let $[\sigma_{ij}]$ be an $n\times n$ matrix, and $\{x_i\}$ and $\{b_i\}$ column matrices each with n elements. A set of equations having the form

$$[\sigma_{ij}]\{x_j\} = \{b_i\} \tag{1.54a}$$

or, in indicial notation,

$$\sigma_{ij}x_j = b_i \tag{1.54b}$$

is called *a system of n linear equations in the n unknowns* $x_1, x_2, \ldots, x_n$. If $b_1, b_2, \ldots, b_n$ are all zero the system is called *homogeneous.* If they are not all zero it is called *nonhomogeneous.* Any set of $x_1, x_2, \ldots, x_n$, which satisfies Eqs. 1.54 is called a *solution* of the system and may be put in the form

$$\{x_i\} = [\sigma_{ij}]^{-1}\{b_j\} \tag{1.55}$$

More explicitly, such a solution can be put according to the *Cramer's rule* as follows

$$\{x_i\} = \frac{\{\Delta_i\}}{\det[\sigma_{ij}]} \tag{1.56a}$$

or, in indicial notation,

$$x_i = \frac{\Delta_i}{\det[\sigma_{ij}]} \tag{1.56b}$$

where Δ_i, $i = 1,2......,n$ is the determinant obtained from $[\sigma_{ij}]$ by removing the i th column and replacing it by the column matrix $\{b_i\}$. The following four cases can arise.

Case 1, $\det[\sigma_{ij}] \neq 0$, $\{b_i\} \neq \{0\}$. In this case there will be a unique solution where not all x_i will be zero.

Case 2, $\det[\sigma_{ij}] \neq 0$, $\{b_i\} = \{0\}$. In this case the only solution is the *trivial solution*, i.e. $\{x_i\} = \{0\}$.

Case 3, $\det[\sigma_{ij}] = 0$, $\{b_i\} = \{0\}$. In this case there will be infinitely many solution other than the trivial solution. This means that at least one of the equations can be obtained from the others, i.e. the equations are linearly dependent. To be discussed more later is this case which is normally called an *eigenvalue problem.*

Case 4, $\det[\sigma_{ij}] = 0$, $\{b_i\} \neq \{0\}$. In this case infinitely many solutions will exists if and only if all Δ_i in Eqs. 1.56 vanish. Otherwise there will be no solution.

Next, consider a homogeneous system of n equations of the form

$$[\sigma_{ij}]\,\{x_j\} - \lambda\{x_i\} = 0 \tag{1.57a}$$

or, in indicial notation,

$$(\sigma_{ij} - \lambda\delta_{ij})x_j = 0 \tag{1.57b}$$

where λ is a number, and δ_{ij} denotes the *Kronecker symbol* as defined in Eq. 1.23, or in matrix form $[\delta_{ij}]$ is an identity matrix, i.e.

$$[\delta_{ij}] = \begin{bmatrix} 1 & 0 & 0 & . & . & 0 \\ 0 & 1 & 0 & . & . & 0 \\ . & . & & . & & . \\ . & . & & & & . \\ 0 & 0 & . & . & . & 1 \end{bmatrix} \tag{1.58}$$

As discussed previously, the system of linear equations, Eq. 1.57a or b will have non-trivial solutions if and only if

$$\det[\sigma_{ij} - \lambda\delta_{ij}] = 0 \tag{1.59}$$

which is a polynomial equation of degree n in λ. The problem characterized by Eqs. 1.57 is normally called an *eigenvalue problem.* The polynomial equation in λ, Eq. 1.59, is the *characteristic equation,* and its roots are *eigenvalues* or *characteristic values* of the matrix $[\sigma_{ij}]$. Corresponding to each eigenvalue there will be a solution $\{x_i\} \neq \{0\}$, i.e. a non-trivial solution, which is called an *eigenvector* or *characteristic vector*. Writing Eq. 1.57a for the case where the eigenvalue is equal to λ_k, we have

$$[\sigma_{ij}]\,\{x_j^{(k)}\} = \lambda_k\{x_i^{(k)}\}\,, \qquad (k \text{ not summed}) \tag{1.60a}$$

or, in indicial notation,

$$\sigma_{ij}x_j^{(k)} = \lambda_k x_i^{(k)}\,, \qquad (k \text{ not summed}) \tag{1.60b}$$

Note that the equations in Eq. 1.60a or b are linearly dependent, thus $x_1^{(k)}, x_2^{(k)}, \ldots, x_n^{(k)}$ cannot be obtained explicitly but rather as known multiples of one of them, say $x_l^{(k)}$ which must be non-zero.

It is proved in standard texts of mathematics (e.g. Spiegel, 1971) that *a symmetric real matrix has real eigenvalues and orthogonal eigenvectors.* The latter property for a symmetric real $[\sigma_{ij}]$ can be put as

$$x_j^{(k)}x_j^{(l)} = 0\,, \qquad (k \neq l) \tag{1.61}$$

It will be found useful to transform each of these eigenvectors $x_j^{(k)}$ into a unit eigenvector $\{\chi_{ik}\}$ by the following *normalization*

$$\chi_{ik} = \frac{x_i^{(k)}}{\sqrt{x_j^{(k)}x_j^{(k)}}}\,, \qquad (k \text{ not summed}) \tag{1.62a}$$

or, in matrix form,

$$\{\chi_{ik}\} = \frac{\{x_i^{(k)}\}}{\sqrt{\{x_j^{(k)}\}^T\{x_j^{(k)}\}}}\,, \qquad (k \text{ not summed}) \tag{1.62b}$$

Since

$$\chi_{ik}\chi_{ik} = 1\,, \qquad (k \text{ not summed}) \tag{1.63}$$

we find, on combining the last equation with Eq. 1.61, that we can write the *orthogonality condition for the unit eigenvectors* in the form

$$\chi_{il}\chi_{ik} = \delta_{kl} \tag{1.64a}$$

or, in matrix form,

$$\{\chi_{il}\}^T\{\chi_{ik}\} = 1\,, \qquad (k = l) \tag{1.64b}$$

$$= 0\,, \qquad (k \neq l) \tag{1.64c}$$

In terms of these unit eigenvectors, Eq. 1.60b can be put in the form

$$\sigma_{ij}\chi_{jk} = \lambda_k\chi_{ik}\,, \qquad (k \text{ not summed}) \tag{1.65}$$

Multiplying the last equation by χ_{il}, and using Eq. 1.64a, we find

$$\sigma_{ij}\chi_{il}\chi_{jk} = \lambda_k\delta_{lk}\,, \qquad (k \text{ not summed}) \tag{1.66}$$

or, in matrix form,

$$\{\chi_{il}\}^T[\sigma_{ij}]\,\{\chi_{jk}\} = \lambda_k\,, \qquad (k = l) \tag{1.67a}$$

$$= 0\,, \qquad (k \neq l) \tag{1.67b}$$

Equation 1.66 can be rewritten in matrix form as

$$[\chi_{il}]^T [\sigma_{ij}] [\chi_{jk}] = \begin{bmatrix} \lambda_1 & 0 & 0 & . & . & 0 \\ 0 & \lambda_2 & & . & . & 0 \\ . & . & & & & . \\ . & . & & & & . \\ 0 & 0 & . & . & . & \lambda_n \end{bmatrix} \tag{1.68}$$

Here $[\chi_{ik}]$ is an $n \times n$ matrix with unit eigenvectors assembled as columns in the matrix, i.e. χ_{ik} denotes the value of x_i for the eigenvalue λ_k. Equation 1.68 shows how the symmetric real matrix $[\sigma_{ij}]$ may be transformed into the diagonal matrix constructed from the eigenvalues of $[\sigma_{ij}]$. Thus introducing the linear transformation

$$\{x_j\} = [\chi_{jk}] \{y_k\} \tag{1.69a}$$

or

$$x_j = \chi_{jk} y_k \tag{1.69b}$$

into Eq. 1.54a, premultiplying the result by $[\chi_{il}]^T$, and using Eq. 1.66 or 1.68, we find

$$y_l = \frac{\{\chi_{il}\}^T \{b_i\}}{\lambda_l}, \qquad (\,l \text{ not summed}\,) \tag{1.70a}$$

or, in indicial form,

$$y_l = \frac{b_i \chi_{il}}{\lambda_l}, \qquad (\,l \text{ not summed}\,) \tag{1.70b}$$

This enables us to say that the system of linear equations of Eqs. 1.54 may be *uncoupled* by the transformation defined in Eq. 1.69. For this reason, $\{y_i\}$ are frequently referred to as the *normal coordinates*, to distinguish them from the *generalized coordinates* x_i.

Next, consider the *quadratic form* of generalized coordinates

$$V = \{x_i\}^T [\sigma_{ij}] \{x_j\} \tag{1.71a}$$

$$= \sigma_{ij} x_i x_j \tag{1.71b}$$

The transformation defined in Eq. 1.69 will readily change this quadratic form into a new form with no cross product terms, i.e.

$$V = \sum_{k=1}^{n} \lambda_k y_k y_k$$

$$= \lambda_1 y_1^2 + \lambda_2 y_2^2 + \ldots + \lambda_n y_n^2 \tag{1.72}$$

which is called the *canonical form*. A symmetric real matrix $[\sigma_{ij}]$ and its quadratic form as defined in Eqs. 1.71 are said to be *positive definite* if $V > 0$ for all real $\{x_i\} \neq \{0\}$. If $[\sigma_{ij}]$ is positive definite, then the terms on the main diagonal of $[\sigma_{ij}]$ must all be positive; for if one, say σ_{22}, were negative or zero, then V would be negative or zero when x_2 is the only non-zero coordinate. This condition is insufficient to ensure the positive definiteness. A necessary and sufficient condition is that (Frazer, Duncan and Collar, 1938)

$$\sigma_{11} > 0\,, \quad \begin{vmatrix} \sigma_{11} & \sigma_{12} \\ \sigma_{21} & \sigma_{22} \end{vmatrix} > 0\,, \quad \begin{vmatrix} \sigma_{11} & \sigma_{12} & \sigma_{13} \\ \sigma_{21} & \sigma_{22} & \sigma_{23} \\ \sigma_{31} & \sigma_{32} & \sigma_{33} \end{vmatrix} > 0\,, \quad \ldots, \quad \det[\sigma_{ij}] > 0 \tag{1.73}$$

Alternatively, the canonical form defined in Eq. 1.72 gives another necessary and sufficient condition for this purpose, i.e. the positive definiteness is ensured provided that all eigenvalues are positive.

Before the advent of digital computers, the solution to the eigenvalue problem for a square matrix with a large dimension (n) had to be done manually using an appropriate iterative scheme. The use of digital computers and appropriate softwares for this purpose has greatly simplified the task.

For $n = 2$, the characteristic equation, Eq. 1.59, becomes a quadratic equation of the form

$$\lambda^2 - (\sigma_{11} + \sigma_{22})\lambda + (\sigma_{11}\sigma_{22} - \sigma_{12}^2) = 0 \tag{1.74}$$

the roots of which are

$$\lambda_1,\ \lambda_2 = \frac{1}{2}\left\{\sigma_{11} + \sigma_{22} \pm [(\sigma_{11} - \sigma_{22})^2 + 4\sigma_{12}^2]^{1/2}\right\} \tag{1.75a,b}$$

When $n = 3$, the characteristic equation is

$$\lambda^3 - I_1\lambda^2 + I_2\lambda - I_3 = 0 \tag{1.76}$$

where

$$I_1 = \sigma_{ii} \tag{1.77a}$$

$$I_2 = \begin{vmatrix} \sigma_{11} & \sigma_{12} \\ \sigma_{12} & \sigma_{22} \end{vmatrix} + \begin{vmatrix} \sigma_{22} & \sigma_{23} \\ \sigma_{32} & \sigma_{33} \end{vmatrix} + \begin{vmatrix} \sigma_{11} & \sigma_{13} \\ \sigma_{31} & \sigma_{33} \end{vmatrix} \tag{1.77b}$$

$$I_3 = \det[\sigma_{ij}] \tag{1.77c}$$

New symbols may be introduced as follows

$$a = \frac{1}{3}(I_1^2 - 3I_2) \tag{1.78a}$$

$$b = \frac{1}{27}(2I_1^3 - 9I_1I_2 + 27I_3) \tag{1.78b}$$

and we may show that $27b^2 - 4a^3 \leq 0$. If $27b^2 - 4a^3 = 0$, Eq. 1.76 will have three real roots, of which two at least are equal, i.e.

$$\lambda_1 = (4b)^{1/3} + \frac{I_1}{3} \tag{1.79a}$$

$$\lambda_2 = \lambda_3 = -\left(\frac{b}{2}\right)^{1/3} + \frac{I_1}{3} \tag{1.79b,c}$$

On the other hand, if $27b^2 - 4a^3 < 0$, there will be three real and unequal roots, i.e.

$$\lambda_1 = 2\left(\frac{a}{3}\right)^{1/2} \cos\left(\frac{\phi}{3}\right) + \frac{I_1}{3} \tag{1.80a}$$

$$\lambda_2 = 2\left(\frac{a}{3}\right)^{1/2} \cos\left(\frac{2\pi + \phi}{3}\right) + \frac{I_1}{3} \tag{1.80b}$$

$$\lambda_3 = 2\left(\frac{a}{3}\right)^{1/2} \cos\left(\frac{4\pi + \phi}{3}\right) + \frac{I_1}{3} \tag{1.80c}$$

where

$$\cos\phi = \frac{3\sqrt{3}\,b}{2a^{3/2}} \tag{1.81}$$

1.7 THEOREMS ON TENSOR FIELDS

A *tensor field* assigns a tensor $\mathbf{T}(\mathbf{x}, t)$ to every pair $(\mathbf{x}, t)$ where the *position vector* $\mathbf{x}(\equiv x_i)$ varies over a particular region of space and t varies over a particular interval of time. The tensor field is said to be a continuous (or differentiable) function of x_i and t if the components of the field are continuous (or differentiable) functions of x_i and t. If the components are functions of x_i only, the tensor field is said to be *steady*.

With respect to a rectangular Cartesian coordinate system, for which the *position vector* of an arbitrary point is x_i, tensor fields of various orders are represented in indicial and symbolic notations as follows:

(a) scalar field; $\phi = \phi(x_i, t)$ or $\phi = \phi(\mathbf{x}, t)$.

(b) vector field; $v_i = v_i(x_j, t)$ or $\mathbf{v} = \mathbf{v}(\mathbf{x}, t)$.

(c) second order tensor field; $T_{ij} = T_{ij}(x_k, t)$ or $\mathbf{T} = \mathbf{T}(\mathbf{x}, t)$.

Note the following identity

$$\frac{\partial(\ )}{\partial x_i} = \frac{\partial(\ )}{\partial x_j'}\frac{\partial x_j'}{\partial x_i},$$

where x_j' is the position vector with reference to another rectangular coordinate system (Fig. 1.6). Applying the transformation law, Eq. 1.20, to x_j', we find that the equation above becomes

$$\frac{\partial(\ )}{\partial x_i} = \frac{\partial(\ )}{\partial x_j'} a_{kj} \frac{\partial x_k}{\partial x_i} \tag{1.82}$$

Since x_1, x_2 and x_3 are independent variables, a derivative of one of them with respect to another must take the form

$$\frac{\partial x_k}{\partial x_i} = \delta_{ik} \tag{1.83}$$

where δ_{ik} is the Kronecker symbol defined by Eq. 1.23. Thus Eq. 1.82 becomes

$$\frac{\partial(\ )}{\partial x_i} = a_{ij}\frac{\partial(\ )}{\partial x_j'} \tag{1.84}$$

Comparing this equation with Eq. 1.21, which is the transformation law for a first-order tensor, we see that the operator $\partial/\partial x_i$ is transformed in the same way a first-order tensor. For this reason we represent such differentiation by a *comma-subscript* as illustrated by the following examples:

$$(a)\quad \frac{\partial\phi}{\partial x_i} = \phi_{,i}. \qquad (d)\quad \frac{\partial^2 v_i}{\partial x_j \partial x_k} = v_{i,jk}.$$

$$(b)\quad \frac{\partial v_i}{\partial x_i} = v_{i,i}. \qquad (e)\quad \frac{\partial T_{ij}}{\partial x_k} = T_{ij,k}.$$

$$(c)\quad \frac{\partial v_i}{\partial x_j} = v_{i,j}. \qquad (f)\quad \frac{\partial^2 T_{ij}}{\partial x_k \partial x_m} = T_{ij,km}.$$

We note that the operator $\partial/\partial x_i$ produces a tensor of one order higher if i remains a free index, and a tensor of one order lower if i becomes a dummy index in the derivative.

Differential operators which appear often in continuum mechanics are given here for reference.

$$\begin{aligned} \text{grad}\,\phi &= \nabla\phi && \text{or} \quad \phi_{,i}, \\ \text{div}\,\mathbf{v} &= \nabla\cdot\mathbf{v} && \text{or} \quad v_{i,i}, \\ \text{curl}\,\mathbf{v} &= \nabla\times\mathbf{v} && \text{or} \quad \varepsilon_{ijk}v_{k,j}, \\ \text{Laplacian}\,\phi &= \nabla^2\phi = \nabla\cdot\nabla\phi && \text{or} \quad \phi_{,ii} \end{aligned} \tag{1.85}$$

Gauss's divergence theorem relates a volume integral to a surface integral. In its traditional form the theorem says that for the vector field $\mathbf{v} = \mathbf{v}(\mathbf{x})$,

$$\int_V \text{div}\mathbf{v}\text{d}V = \int_S \mathbf{n}\cdot\mathbf{v}\text{d}S \tag{1.86}$$

where $\mathbf{n}$ is the outward unit normal to the bounding surface S, of the volume V in which the vector field is defined. In the indicial notation, Eq. 1.86 is written as

$$\int_V v_{i,i}\text{d}V = \int_S v_i n_i \text{d}S \tag{1.87}$$

For an arbitrary tensor field $T_{ijk\ldots}$ the theorem is

$$\int_V T_{ijk\ldots,p}\text{d}V = \int_S T_{ijk\ldots}n_p\text{d}S \tag{1.88}$$

Green's theorem in the plane. If R is a closed region of the x_1x_2 plane bounded by a *simple closed curve* (a curve which does not intersect itself anywhere) C and if

M and N are continuous functions of x_1 and x_2 having continuous derivative in R, then

$$\oint_c [M\mathrm{d}x_1 + N\mathrm{d}x_2] = \int_R \left(\frac{\partial N}{\partial x_1} - \frac{\partial M}{\partial x_2} \right) \mathrm{d}x_1 \mathrm{d}x_2 \tag{1.89}$$

This theorem also holds for a multiply-connected region, i.e. region bounded by a finite number of simple closed curves which do not intersect.

1.8 DIFFERENTIAL GEOMETRY

Differential geometry is the study of *space curves* and *surfaces*. If in ordinary physical space a *position vector* x_i is a function of a single scalar variable u, i.e.

$$x_1 = x_1(u), \qquad x_2 = x_2(u), \qquad x_3 = x_3(u),$$

the terminal point of x_i describes a *space curve* as u changes. As shown in Fig. 1.7, if

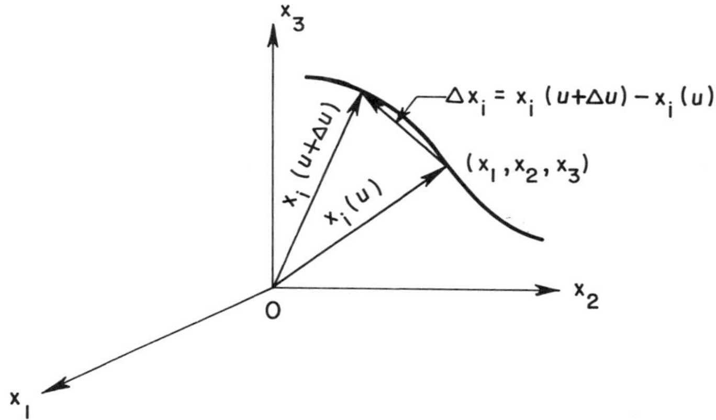

Fig. 1.7 Space curve.

$$\lim_{\Delta u \to 0} \frac{\Delta x_i}{\Delta u} = \frac{\mathrm{d}x_i}{\mathrm{d}u}$$

exists, the limit will be a vector in the direction of the tangent to the space curve at point (x_1, x_2, x_3) and will have Cartesian components $(\mathrm{d}x_1/\mathrm{d}u$, $\mathrm{d}x_2/\mathrm{d}u$, $\mathrm{d}x_3/\mathrm{d}u)$. If u is the time t, $\mathrm{d}x_i/\mathrm{d}t$ represents the *velocity* v_i with which the terminal of vector x_i describes the curve. Similarly, $\mathrm{d}v_i/\mathrm{d}t = \mathrm{d}^2x_i/\mathrm{d}t^2$ represents its *acceleration* along the curve.

If the scalar u is taken as the arc length s measured from some fixed point on the space curve, then $\mathrm{d}x_i/\mathrm{d}s$ is a unit tangent vector to the curve and denoted by T_i (see Fig. 1.8). If N_i and B_i are the unit *principal normal* and *binormal* vectors to the

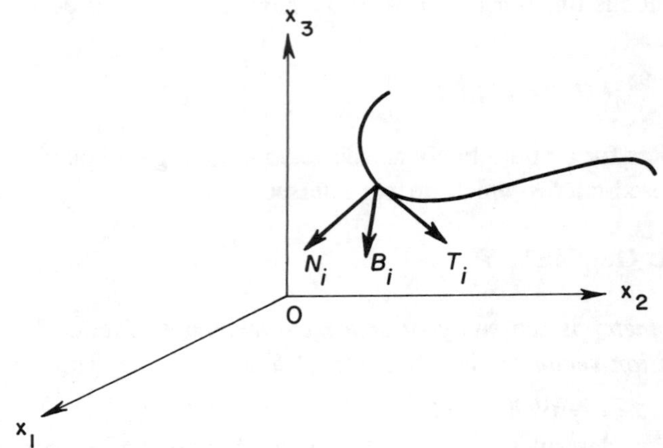

Fig. 1.8 Tangent and normal unit vectors to a space curve.

curve respectively, the three unit vectors T_i, N_i and B_i are mutually orthogonal such that

$$B_i = \varepsilon_{ijk} T_j N_k \tag{1.90}$$

Other relations among them are called *Frenet-Serret formulas*, i.e.

$$\frac{\mathrm{d}T_i}{\mathrm{d}s} = KN_i\,, \quad \frac{\mathrm{d}B_i}{\mathrm{d}s} = -\tau N_i \tag{1.91}$$

and

$$\frac{\mathrm{d}N_i}{\mathrm{d}s} = \tau B_i - KT_i \tag{1.92}$$

where K and τ denote the *curvature* and *torsion* respectively, and the reciprocal of these, $\rho = K^{-1}$ and $\sigma = \tau^{-1}$ are called the *radius of curvature* and *radius of torsion* respectively.

The *parametric equations* which ensure that point (x_1, x_2, x_3) is on the straight line joining points (y_1, y_2, y_3) and (z_1, z_2, z_3) are

$$\frac{x_1 - y_1}{z_1 - y_1} = \frac{x_2 - y_2}{z_2 - y_2} = \frac{x_3 - y_3}{z_3 - y_3} \tag{1.93}$$

See the illustration in Fig. 1.9 (a) in which numbers 1, 2, 3 denote the coordinate axes. In Fig. 1.9 (b), another position vector x_i which is equal to the vector running from (y_1, y_2, y_3) to (z_1, z_2, z_3) is illustrated, thus

$$x_i = z_i - y_i \tag{1.94}$$

The equation of the plane passing through three points $(\alpha_1, \alpha_2, \alpha_3)$, $(\beta_2, \beta_2, \beta_3)$ and $(\gamma_1, \gamma_2, \gamma_3)$, which are not on the same straight line, has the form

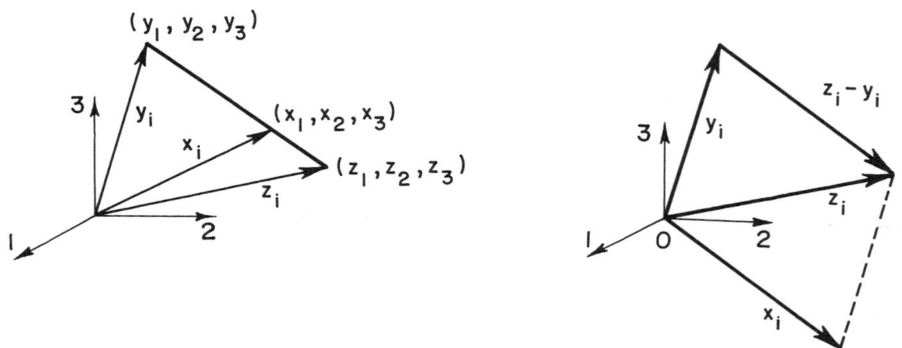

(a) End of x_i on the straight line joining end of y_i with end of z_i.

(b) Position vector $x_i = z_i - y_i$.

Fig. 1.9 Equations of a straight line.

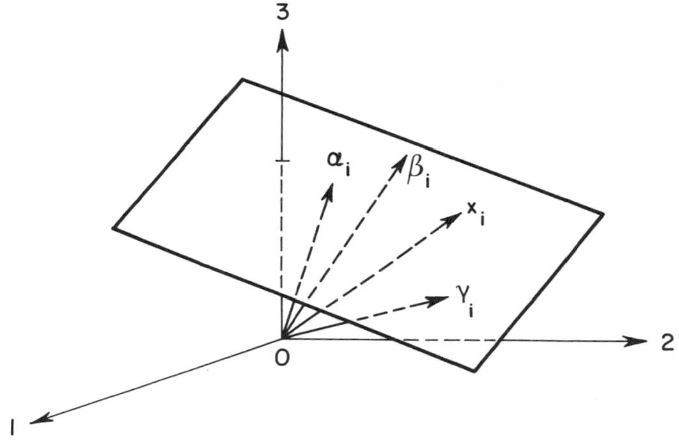

Fig. 1.10 A plane passing through terminal points of α_i, β_i and γ_i.

$$ax_1 + bx_2 + cx_3 + 1 = 0 \tag{1.95}$$

in which a, b and c are constants to be determined from the condition that all three prescribed points be on the plane. This condition gives the equation of the plane in the form

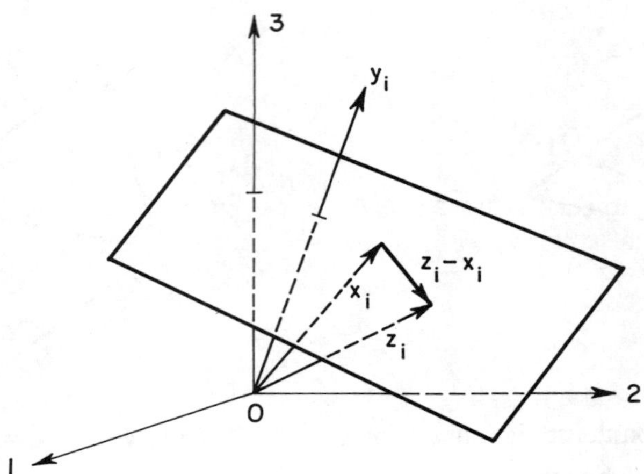

Fig. 1.11 Plane perpendicular to y_i and through terminal point of z_i.

$$\begin{vmatrix} x_1-\alpha_1 & x_2-\alpha_2 & x_3-\alpha_3 \\ \beta_1-\alpha_1 & \beta_2-\alpha_2 & \beta_3-\alpha_3 \\ \gamma_1-\alpha_1 & \gamma_2-\alpha_2 & \gamma_3-\alpha_3 \end{vmatrix} = 0 \tag{1.96}$$

See the illustration in Fig. 1.10. On the other hand, an equation for the plane perpendicular to a vector y_i and passing through the terminal point of another position vector z_i (see Fig. 1.11) is

$$(z_i - x_i)y_i = 0,$$

or

$$x_i y_i = y_i z_i \tag{1.97}$$

or

$$x_1 y_1 + x_2 y_2 + x_3 y_3 = y_1 z_1 + y_2 z_2 + y_3 z_3.$$

Another way of writing the equation of a plane is

$$x_i n_i = d \tag{1.98}$$

where n_i is the unit vector normal to the plane and d is the shortest distance from the origin 0 to the plane.

On a surface $\phi(x_i) = constant$, its total differential at a position defined by the vector x_i is

$$\mathrm{d}\phi = \frac{\partial \phi}{\partial x_i}\mathrm{d}x_i = \phi_{,i}\mathrm{d}x_i = 0.$$

This shows that the vector $\phi_{,i}$ is perpendicular to the vector $\mathrm{d}x_i$ and therefore to the surface. In other words, the *gradient* of a surface is a *normal vector to the surface.* Moreover, a unit vector normal to the surface $\phi(x_i) =$ *constant* can be obtained by the usual normalization

$$n_i = \pm \frac{\phi_{,i}}{\sqrt{\phi_{,j}\phi_{,j}}} \tag{1.99}$$

The sign in Eq. 1.99 is often chosen so that the vector n_i is directed 'outward' from the region on one particular side of the surface $\phi(x_i) =$ *constant.* Often we can express the position vector of a point on a surface in the form

$$x_i = x_i(\alpha, \beta) \tag{1.100}$$

where α and β are two *independent parameters of the surface.* Noting that $\partial x_i/\partial\alpha$ and $\partial x_i/\partial\beta$ are vectors lying in the tangent plane to the surface, we see that the direction of the normal to the surface is the same as the following non-zero vector,

$$\varepsilon_{ijk} \frac{\partial x_j}{\partial \alpha} \frac{\partial x_k}{\partial \beta} \neq 0 \tag{1.101}$$

1.9 DIRAC-DELTA AND HEAVISIDE STEP FUNCTIONS

It has been found advantageous in some problems, such as those involving concentrated forces and impulsive forces, to introduce the *Dirac-delta function*, $\delta(t-a)$, which is *singular* but *integrable* as follows

$$\delta(t-a) = \infty, \quad (t = a) \tag{1.102a}$$

$$= 0, \quad (t \neq a) \tag{1.102b}$$

$$\int_b^c \delta(t-a)\mathrm{d}t = 1, \quad (b < a < c) \tag{1.102c}$$

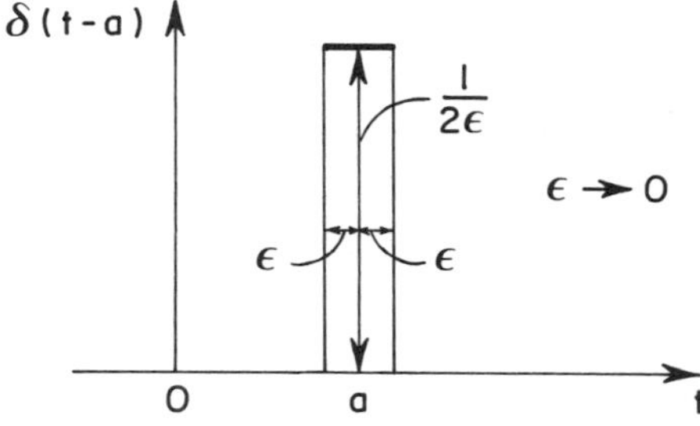

Fig. 1.12 Graphical representation of Dirac-delta function.

If $\phi(t)$ is a continuous function of t in the neighborhood of $t=a$, then

$$\int_b^c \phi(t)\delta(t-a)\mathrm{d}t = \phi(a), \qquad (b<a<c) \tag{1.103}$$

The Dirac-delta function is not an ordinary function, but rather a limit of a special sequence of ordinary functions (Stakgold, 1967). In Fig. 1.12 we show the Dirac-delta function as the limit of a *pulse*. In characterizing a concentrated force by $\delta(t-a)$, t is a spatial variable. On other occasions, when $\delta(t-a)$ characterizes an impulsive force, t is the time variable.

Since all time functions involved are normally in the *state of complete quietude* before the time t is reckoned as zero, it is appropriate to study the *Heaviside step function* which has the properties

$$H(t)=1, \qquad (t>0) \tag{1.104a}$$

$$=0, \qquad (t<0) \tag{1.104b}$$

Graphically, we may put $H(t-a)$ as shown in Fig. 1.13.

To study the derivatives of $H(t)$ and $\delta(t)$, let us consider the following integral

$$\int_b^c \phi(t)\frac{\mathrm{d}H(t-a)}{\mathrm{d}t}\mathrm{d}t,$$

where $b<a<c$ and $\phi(t)$ is continuous at $t=a$. Performing integration by parts, we find

$$\int_b^c \phi(t)\frac{\mathrm{d}H(t-a)}{\mathrm{d}t}\mathrm{d}t = \phi(c)(1)-\phi(b)(0)-\int_b^c H(t-a)\frac{\mathrm{d}\phi(t)}{\mathrm{d}t}\mathrm{d}t$$

$$=\phi(c)-\int_a^c \mathrm{d}\phi(t)$$

$$=\phi(c)-\phi(c)+\phi(a)$$

$$=\phi(a) \tag{1.105}$$

Comparing Eq. 1.105 with Eq. 1.103 leads to the 'in effect' identity as follows

$$\frac{\mathrm{d}H(t-a)}{\mathrm{d}t}=\delta(t-a) \tag{1.106}$$

Another 'in effect' identity induced from Eqs. 1.102 and 1.103 is

$$\phi(t)\delta(t-a)=\phi(a)\delta(t-a) \tag{1.107}$$

For a function $\phi(t)$ which is *differentiable*, we can show that

$$\int_b^c \phi(t)\delta^{(1)}(t-a)\mathrm{d}t = [-\phi^{(1)}(t)]_{t=a}, \qquad (b<a<c) \tag{1.108}$$

in which a superscript in parentheses denotes a derivative with respect to t. As a more general identity, we can show by mathematical induction for a function $\phi(t)$ which is differentiable n times that

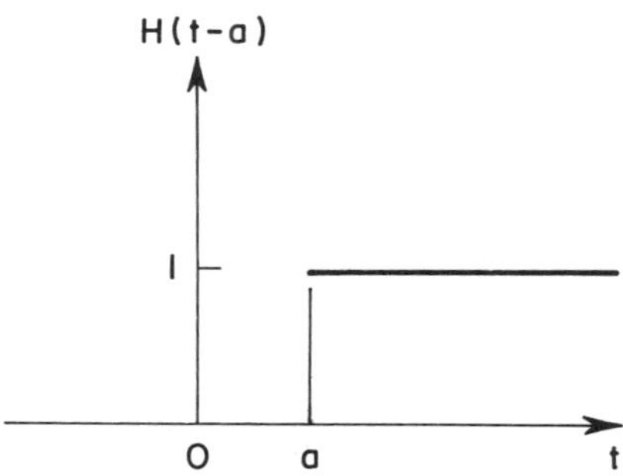

Fig. 1.13 Graphical representation of Heaviside step function.

$$\int_b^c \phi(t)\delta^{(n)}(t-a)\mathrm{d}t = [(-1)^n\phi^{(n)}(t)]_{t=a}, \qquad (b<a<c) \tag{1.109}$$

More 'in effect' identities should be noted here, i.e.

$$t^n\delta^{(n)}(t) = (-1)^n n(n-1)(n-2)\ldots(1)\delta(t) \tag{1.110a}$$

$$t^m\delta^{(n)}(t) = 0, \qquad (m>n) \tag{1.110b}$$

$$\frac{\mathrm{d}}{\mathrm{d}t}[\phi(t)H(t)] = \phi(0)\delta(t) + \phi^{(1)}(t)H(t) \tag{1.110c}$$

$$\frac{\mathrm{d}^n}{\mathrm{d}t^n}[\phi(t)H(t)] = \sum_{i=0}^{n-1}\phi^{(n-1-i)}(0)\delta^{(i)}(t) + \phi^{(n)}(t)H(t) \tag{1.110d}$$

1.10 BESSEL FUNCTIONS

We present a brief review of *Bessel functions*. Further details may be found in the books by Abramowitz and Stegun (1964), and Watson (1966). The *Bessel equation* is the following

$$z^2\frac{\mathrm{d}^2C_m}{\mathrm{d}z^2} + z\frac{\mathrm{d}C_m}{\mathrm{d}z} + (z^2 - m^2)C_m = 0 \tag{1.111}$$

In general, z is a complex variable, and m is a complex constant. Solutions to the equation above are the *Bessel functions of the first kind* $J_{\pm m}(z)$, of the *second kind* (also called *Weber's function*) $Y_m(z)$, and of the *third kind* (also called *Hankel functions*) $H_m^{(1)}(z)$, $H_m^{(2)}(z)$. Each is a *regular* (*analytic* or *holomorphic*) *function* of z throughout the z-plane cut along the negative real axis, and for fixed $z\ (\neq 0)$ each is an *entire* (*integral*) function of m. When m is a positive or negative integer, $J_m(z)$ has no branch point at $(z=0)$ and is an entire function of z. See Churchill (1960) for the definitions of the terms analytic, entire, branch point.

Listed below are *recurrence formulas*,

$$C_{m-1}(z)+C_{m+1}(z)=\frac{2m}{z}C_m(z) \tag{1.112a}$$

$$C_{m-1}(z)-C_{m+1}(z)=2\frac{\mathrm{d}C_m(z)}{\mathrm{d}z} \tag{1.112b}$$

$$\frac{\mathrm{d}C_m(z)}{\mathrm{d}z}=C_{m-1}(z)-\frac{m}{z}C_m(z) \tag{1.112c}$$

$$\frac{\mathrm{d}C_m(z)}{\mathrm{d}z}=-C_{m+1}(z)+\frac{m}{z}C_m(z) \tag{1.112d}$$

in which C denotes J, Y, $H^{(1)}$, $H^{(2)}$ or any linear combination of these functions, the coefficients in which are independent of z and m. These functions of the same *order* m are related as the following

$$H_m^{(1)}(z)=J_m(z)+\mathrm{i}Y_m(z) \tag{1.113a}$$

$$H_m^{(2)}(z)=J_m(z)-\mathrm{i}Y_m(z) \tag{1.113b}$$

where i is the imaginary number ($\mathrm{i}^2=-1$).

To follow are *analytic continuation* formulas,

$$J_m(z\mathrm{e}^{\mathrm{i}n\pi})=\mathrm{e}^{\mathrm{i}mn\pi}J_m(z) \tag{1.114a}$$

$$Y_m(z\mathrm{e}^{\mathrm{i}n\pi})=\mathrm{e}^{-\mathrm{i}mn\pi}Y_m(z)+2\mathrm{i}\sin(mn\pi)\cot(m\pi)J_m(z) \tag{1.114b}$$

$$\sin(m\pi)H_m^{(1)}(z\mathrm{e}^{\mathrm{i}n\pi})=-\sin[(n-1)m\pi]H_m^{(1)}(z)-\mathrm{e}^{-\mathrm{i}m\pi}\sin(mn\pi)H_m^{(2)}(z) \tag{1.114c}$$

$$\sin(m\pi)H_m^{(2)}(z\mathrm{e}^{\mathrm{i}n\pi})=\sin[(n+1)m\pi]H_m^{(2)}(z)+\mathrm{e}^{\mathrm{i}m\pi}\sin(mn\pi)H_m^{(1)}(z) \tag{1.114d}$$

$$H_m^{(1)}(z\mathrm{e}^{\mathrm{i}\pi})=-\mathrm{e}^{-\mathrm{i}m\pi}H_m^{(2)}(z) \tag{1.114e}$$

$$H_m^{(2)}(z\mathrm{e}^{\mathrm{i}\pi})=-\mathrm{e}^{\mathrm{i}m\pi}H_m^{(1)}(z) \tag{1.114f}$$

In Eqs. 1.114a to d, m is an integer.

When m is fixed and $|z|\to 0$,

$$J_m(z)\cong\frac{(z/2)^m}{\Gamma(m+1)},\qquad (m\neq\text{negative integer}) \tag{1.115a}$$

$$Y_0(z)\cong-\mathrm{i}H_0^{(1)}(z)\cong\mathrm{i}H_0^{(2)}(z)\cong\frac{2}{\pi}\log z \tag{1.115b}$$

$$Y_m(z) \cong -iH_m^{(1)}(z) \cong iH_m^{(2)}(z) \cong -\frac{\Gamma(m)}{\pi}\left(\frac{z}{2}\right)^{-m}, \qquad [\Re(m) > 0] \qquad (1.115c)$$

In these equations, the symbol $\cong$ means 'is approximately equal to'. On the other hand, if m is fixed and $|z| \to \infty$,

$$J_m(z) = \left(\frac{2}{\pi z}\right)^{1/2}\left[\cos\left(z - \frac{m\pi}{2} - \frac{\pi}{4}\right) + e^{|\Im(z)|}\,O(|z|^{-1})\right], \qquad (|\arg z| < \pi) \qquad (1.116a)$$

$$Y_m(z) = \left(\frac{2}{\pi z}\right)^{1/2}\left[\sin\left(z - \frac{m\pi}{2} - \frac{\pi}{4}\right) + e^{|\Im(z)|}\,O(|z|^{-1})\right], \qquad (|\arg z| < \pi) \qquad (1.116b)$$

$$H_m^{(1)}(z) \cong [2/(\pi z)]^{1/2} e^{i\left(z - \frac{m\pi}{2} - \frac{\pi}{4}\right)}, \qquad (-\pi < \arg z < 2\pi) \qquad (1.116c)$$

$$H_m^{(2)}(z) \cong [2/(\pi z)]^{1/2} e^{-i\left(z - \frac{m\pi}{2} - \frac{\pi}{4}\right)}, \qquad (-2\pi < \arg z < \pi) \qquad (1.116d)$$

In Eqs. 1.115 and 1.116, $\Gamma(z)$ is a *gamma function* (discussed in the next section for real and non-negative z), $\Re(z)$ and $\Im(z)$ are *real* and *imaginary parts* of z respectively, $\arg z = \tan^{-1}[\Im(z)/\Re(z)]$, and O( ) denotes the *order of the truncation error*.

Closely related to Bessel functions are *modified Bessel functions, Kelvin functions,* and *spherical Bessel functions.* For the modified Bessel functions, the pertinent relationship is

$$I_m(z) = e^{-im\pi/2} J_m(z e^{i\pi/2}), \qquad (-\pi < \arg z \le \pi/2) \qquad (1.117a)$$

$$= e^{3im\pi/2} J_m(z e^{-3i\pi/2}), \qquad (\pi/2 < \arg z \le \pi) \qquad (1.117b)$$

$$K_m(z) = \frac{i\pi}{2} e^{im\pi/2} H_m^{(1)}(z e^{i\pi/2}), \qquad (-\pi < \arg z \le \pi/2) \qquad (1.118a)$$

$$= -\frac{i\pi}{2} e^{-im\pi/2} H_m^{(2)}(z e^{-i\pi/2}), \qquad (-\pi/2 < \arg z < \pi) \qquad (1.118b)$$

The Kelvin functions, of a real and non-negative argument z and a real order m, can be expressed in terms of Bessel functions as

$$ber_m(z) + i\,bei_m(z) = J_m(z e^{3i\pi/4}) \qquad (1.119a)$$

$$ker_m(z) + i\,kei_m(z) = \frac{i\pi}{2} H_m^{(1)}(z e^{3i\pi/4}) \qquad (1.119b)$$

Spherical Bessel functions are defined for $m = 0, \pm 1, \pm 2, \ldots$ by the equations

$$j_m(z) = [\pi/(2z)]^{1/2} J_{m+\frac{1}{2}}(z) \qquad (1.120a)$$

$$y_m(z) = [\pi/(2z)]^{1/2} Y_{m+\frac{1}{2}}(z) \qquad (1.120b)$$

$$h_m^{(1)}(z) = j_m(z) + i y_m(z) = [\pi/(2z)]^{1/2} H_{m+\frac{1}{2}}^{(1)}(z) \qquad (1.120c,d)$$

$$h_m^{(2)}(z) = j_m(z) - \mathrm{i}y_m(z) = [\pi/(2z)]^{1/2} H^{(2)}_{m+\frac{1}{2}}(z) \tag{1.120e,f}$$

Here j_m, y_m and h_m are *spherical Bessel functions of the first kind, second kind,* and *third kind,* respectively. In addition, $h_m^{(1)}$ and $h_m^{(2)}$ are also known as *spherical Hankel functions.*

With the help of the recurrence formulas (Eqs. 1.112) and the asymptotic form when m is fixed and $|z| \to \infty$ (Eqs. 1.116), we can derive the following

$$C_m(z)/z^k \cong 0 \tag{1.121a}$$

$$C_{m+1}(z) + C_{m-1}(z) \cong 0 \tag{1.121b}$$

$$\frac{\mathrm{d}C_m(z)}{\mathrm{d}z} \cong C_{m-1}(z) \tag{1.121c}$$

where $k \geq 1$, and C denotes J, Y, $H^{(1)}$ or $H^{(2)}$, provided the value of $\arg z$ falls within the respective ranges stipulated in Eqs. 1.116. Due to Eqs. 1.120, the same conditions also hold for j, y, $h^{(1)}$ or $h^{(2)}$.

Computation of these special functions has been simplified since the establishment of their *polynomial approximations.* Such approximations are presented in the handbook by Abramowitz and Stegun (1964).

1.11 LAPLACE TRANSFORMS

If the function $v(t)$ is defined for the time interval $0 < t < \infty$, then its *Laplace transform* $\mathrm{L}\{v(t)\}$ or $\bar{v}(p)$ is defined as

$$\mathrm{L}\{v(t)\} = \bar{v}(p) = \int_{0^-}^{\infty} v(t)\mathrm{e}^{-pt}\mathrm{d}t \tag{1.122}$$

Note that the lower limit in the integral above is not a plane 0 but rather 0^-, which is a quantity infinitesimally less than 0. This is used to accommodate the *state of complete quietude* of $v(t)$ before the time t is reckoned as zero, i.e.

$$v(0^-) = v^{(1)}(0^-) = v^{(2)}(0^-) = \ldots. = 0 \tag{1.123}$$

in which a superscript n in parentheses denotes an nth derivative with respect to t. Laplace transforms of some elementary functions are listed below:

$$\mathrm{L}\{H(t)\} = \frac{1}{p} \tag{1.124a}$$

$$\mathrm{L}\{\delta(t)\} = 1 \tag{1.124b}$$

$$\mathrm{L}\{\delta^{(n)}(t)\} = p^n \tag{1.124c}$$

$$\mathrm{L}\{\mathrm{e}^{at}\} = (p-a)^{-1}, \qquad (\text{p-a}>0) \tag{1.124d}$$

$$\mathrm{L}\{\sin at\} = a/(p^2+a^2) \tag{1.124e}$$

$$\mathrm{L}\{\cos at\} = p/(p^2+a^2) \tag{1.124f}$$

$$\mathrm{L}\{\sinh at\} = a/(p^2-a^2), \quad (p>(|a|) \tag{1.124g}$$

$$\mathrm{L}\{\cosh at\} = p/(p^2-a^2), \quad (p>(|a|) \tag{1.124h}$$

$$\mathrm{L}\{t^a\} = \Gamma(a+1)/p^{a+1}, \quad (a>0) \tag{1.124i}$$

In the last equation above $\Gamma(y)$ denotes a *gamma function*, i.e.

$$\Gamma(y) = \int_0^\infty x^{y-1} \mathrm{e}^{-x} \mathrm{d}x, \quad (y>0) \tag{1.125}$$

Accordingly, it can be shown that

$$\Gamma(y+1) = y\Gamma(y), \quad y \neq 0,-1,-2,-3,\ldots \tag{1.126}$$

If y is equal to a positive integer n,

$$\Gamma(n) = (n-1)! \tag{1.127a}$$

where $n!$ denotes the *factorial* n, i.e.

$$n! = n(n-1)(n-2)\ldots(1) \tag{1.127b}$$

Note that

$$1! = 0! = 1 \tag{1.128}$$

Here are some important theorems on Laplace transforms:

Theorem 1.5. *Linearity property*. If c_1 and c_2 are any constants then

$$\mathrm{L}\{c_1 v_1(t) + c_2 v_2(t)\} = c_1 \bar{v}_1(p) + c_2 \bar{v}_2(p) \tag{1.129}$$

Theorem 1.6. *First translation* or *shifting property*. If $\mathrm{L}\{v(t)\} = \bar{v}(p)$ then

$$\mathrm{L}\{\mathrm{e}^{at} v(t)\} = \bar{v}(p-a) \tag{1.130}$$

Theorem 1.7. *Second translation* or *shifting property*. If $\mathrm{L}\{v(t)\} = \bar{v}(p)$ and $f(t) = v(t-a)H(t-a)$, where $H(t)$ denotes the Heaviside step function, then

$$\mathrm{L}\{f(t)\} = \mathrm{e}^{-pa} \bar{v}(p) \tag{1.131}$$

Theorem 1.8. *Laplace transform of derivatives*. If $\mathrm{L}\{v(t)\} = \bar{v}(p)$, then

$$\mathrm{L}\left\{\frac{\mathrm{d}^n v}{\mathrm{d}t^n}\right\} = p^n \bar{v} \tag{1.132}$$

in which the state of complete quietude (Eqs. 1.123) has been incorporated.

Theorem 1.9. *Laplace transform of integrals*. If $\mathrm{L}\{v(t)\} = \bar{v}(p)$, then

$$\mathrm{L}\left\{\int_{0^-}^t v(s)\mathrm{d}s\right\} = \frac{\bar{v}(p)}{p} \tag{1.133}$$

Theorem 1.10. *Laplace transform of convolution integrals*. If $\mathrm{L}\{v(t)\} = \bar{v}(p)$ and a convolution integral is symbolized by

$$v_1 * v_2 = \int_{0^-}^{t} v_1(s) v_2(t-s) \mathrm{d}s \tag{1.134}$$

then

$$\mathrm{L}\{v_1 * v_2\} = \bar{v}_1(p) \bar{v}_2(p) \tag{1.135}$$

Since one often wants to determine the derivative of a convolution integral, it is appropriate to record here that

$$\frac{\mathrm{d}}{\mathrm{d}t} \int_{0^-}^{t} v_1(s) v_2(t-s) \mathrm{d}s = \int_{0^-}^{t} v_1(s) \frac{\partial v_2(t-s)}{\partial t} \mathrm{d}s + v_1(t) v_2(0^+) \tag{1.136}$$

where 0^+ denotes a quantity infinitesimally greater than zero. A more general formula known as *Leibnitz's rule for differentiating an integral* is the following

$$\frac{\mathrm{d}}{\mathrm{d}t} \int_{a(t)}^{b(t)} f(t,s) \mathrm{d}s = \int_{a(t)}^{b(t)} \frac{\partial}{\partial t} f(t,s) \mathrm{d}s + f(t,b) \frac{\mathrm{d}b}{\mathrm{d}t} - f(t,a) \frac{\mathrm{d}a}{\mathrm{d}t} \tag{1.137}$$

Note that such convolution integrals obey the *commutative law*, i.e.

$$v_1 * v_2 = v_2 * v_1 \tag{1.138a}$$

the *associative law*, i.e.

$$v_1 * (v_2 * v_3) = (v_1 * v_2) * v_3 \tag{1.138b}$$

and the *distributive law*, i.e.

$$v_1 * (v_2 + v_3) = v_1 * v_2 + v_1 * v_3 \tag{1.138c}$$

Theorem 1.11. *Lerch's uniqueness theorem.* If $v(t)$ is sectionally continuous in every finite interval $0 < t < N$ and of exponential order for $t > N$, then the *inverse Laplace transform* of $v(t)$, i.e. $\mathrm{L}^{-1}\{\bar{v}(p)\} = v(t)$ is unique.

Theorem 1.12. *Initial-value theorem.* If the indicated limits exist, then

$$\lim_{t \to 0^+} v(t) = \lim_{p \to \infty} p \bar{v}(p) \tag{1.139}$$

Theorem 1.13. *Final-value theorem.* If the indicated limits exist, then

$$\lim_{t \to \infty} v(t) = \lim_{p \to 0^+} p \bar{v}(p) \tag{1.140}$$

Theorem 1.14. *Fourier transform.* For a function $v(t)$ defined for the interval $0 < t < \infty$, its Fourier transform $\mathrm{F}\{v(t)\}$ or $v_*(\omega)$ is defined as

$$\mathrm{F}\{v(t)\} = v_*(\omega) = \int_{0^-}^{\infty} v(t) \mathrm{e}^{-i\omega t} \mathrm{d}t \tag{1.141}$$

A sufficient condition for the existence of $v_*(\omega)$ is that the function $v(t)$ must be *absolutely integrable*, i.e.

$$\int_{0^-}^{\infty} |v(t)| \, \mathrm{d}t < \infty \tag{1.142}$$

Comparing Eq. 1.141 with Eq. 1.122, we get

$$v_*(\omega) = \bar{v}(\mathrm{i}\omega) \tag{1.143}$$

It should be noted that Eq. 1.142 is a sufficient but not necessary condition for the existence of $v_*(\omega)$. Functions which do not satisfy Eq. 1.142 may have Fourier transforms. Here are examples of such functions (Hsu, 1970);

$$\mathrm{F}\{H(t)\} = \int_{0^-}^{\infty} H(t)\mathrm{e}^{-\mathrm{i}\omega t}\mathrm{d}t = \frac{1}{\mathrm{i}\omega} + \pi\delta(\omega) \tag{1.144a}$$

$$\mathrm{F}\{t^n H(t)\} = \int_{0^-}^{\infty} t^n H(t)\mathrm{e}^{-\mathrm{i}\omega t}\mathrm{d}t = \frac{n!}{(\mathrm{i}\omega)^{n+1}} + \pi(\mathrm{i})^n\delta^{(n)}(\omega) \tag{1.144b}$$

Note that the second term on the right-hand side of each of Eqs. 1.144 is an addition to that readily given by Eq. 1.143. In fact, Eq. 1.144b can be obtained by differentiating Eq. 1.144a with respect to ω.

Equating the real parts in Eq. 1.144a gives

$$\delta(\omega) = \frac{1}{\pi}\int_0^{\infty} \cos\omega t\,\mathrm{d}t \tag{1.145}$$

while doing the same for the imaginary parts gives

$$\frac{1}{\omega} = \int_0^{\infty} \sin\omega t\,\mathrm{d}t \tag{1.146}$$

Taking the inverse cosine transform of Eq. 1.145 leads to

$$1 = 2\int_0^{\infty} \delta(\omega)\cos\omega t\,\mathrm{d}\omega, \qquad (t > 0),$$

or

$$\int_0^{\infty} \delta(\omega)\cos\omega t\,\mathrm{d}\omega = \frac{1}{2} \tag{1.147}$$

On the other hand, the inverse sine transform of Eq. 1.146 gives

$$1 = \frac{2}{\pi}\int_0^{\infty} \frac{\sin\omega t}{\omega}\mathrm{d}\omega, \qquad (t > 0),$$

or

$$H(t) = \frac{1}{2} + \frac{1}{\pi}\int_0^{\infty} \frac{\sin\omega t}{\omega}\mathrm{d}\omega \tag{1.148}$$

or

$$\int_0^{\infty} \frac{\sin\omega t}{\omega}\mathrm{d}\omega = \frac{\pi}{2}\operatorname{sgn} t \tag{1.149}$$

where $\operatorname{sgn} t$ (read as *signum* t) is defined as

$$\operatorname{sgn} t = 1, \quad (t > 0) \tag{1.150a}$$

$$= -1, \quad (t < 0) \tag{1.150b}$$

1.12 INVERSE LAPLACE TRANSFORMS

For a given Laplace transform $\bar{v}(p)$, its inverse symbolized by

$$L^{-1}\{\bar{v}(p)\} = v(t) \tag{1.151}$$

may be found from standard mathematical textbooks, e.g. by Spiegel (1965). Otherwise, one may use *Bromwich's integral formula*, i.e.

$$v(t) = \frac{1}{2\pi i}\int_{\gamma - i\infty}^{\gamma + i\infty} e^{pt}\bar{v}(p)dp \tag{1.152}$$

The integration in the equation above is to be performed in the complex plane where $p = q + ir$ along a line $q = \gamma$. The real number γ is chosen so that $q = \gamma$ lies to the right of all singularities (poles, branch points, or essential singularities) but is otherwise arbitrary. The formula provides direct means for calculating $v(t)$ for a given $\bar{v}(p)$.

When the methods mentioned above are not possible or practical, approximate methods may be adopted, but their accuracy and efficiency must be checked or tested case by case. Cost (1964) started his investigation on some approximate methods with the n th derivative with respect to p of the Laplace transform, Eq. 1.122, i.e.

$$\frac{d^n\bar{v}(p)}{dp^n} = (-1)^n\int_{0^-}^{\infty} v(t)t^n e^{-pt}dt,$$

or

$$(-1)^n\frac{p^{n+1}}{n!}\frac{d^n\bar{v}(p)}{dp^n} = \int_{0^-}^{\infty} v(t)\left[\frac{p^{n+1}t^n e^{-pt}}{n!}\right]dt \tag{1.153}$$

Considering the term in the brackets on the right-hand side of the equation above, i.e.

$$\frac{p^{n+1}t^n e^{-pt}}{n!},$$

we note that it has a single peak at $t = n/p$, and due to the definitions of the gamma functions and factorials, Eqs. 1.125 and 1.127, we may write

$$\frac{p^{n+1}}{n!}\int_{0^-}^{\infty} t^n e^{-pt}dt = 1 \tag{1.154}$$

Moreover, the *Stirling's formula*, (Spiegel, 1971) which says that

$$n! \cong \sqrt{2\pi}e^{-n}n^{n+1/2} \tag{1.155}$$

implies the following

$$\lim_{n\to\infty}\frac{p^{n+1}t^n e^{-pt}}{n!} = \infty, \quad (t = n/p) \tag{1.156a}$$

$$= 0, \quad (t \neq n/p) \tag{1.156b}$$

Comparing these results with the properties of the Dirac-delta function given in Eqs. 1.102, we can write

$$\operatorname*{Lim}_{n\to\infty}\frac{p^{n+1}t^{n}e^{-pt}}{n!}=\delta\left(t-\frac{n}{p}\right) \tag{1.157}$$

Substituting Eq. 1.157 into Eq. 1.153, in view of Eq. 1.103, leads to

$$v(t)=\operatorname*{Lim}_{n\to\infty}\left[\frac{(-1)^{n}p^{n+1}}{n!}\frac{d^{n}\bar{v}(p)}{dp^{n}}\right]_{p=n/t} \tag{1.158}$$

which is known as the *Widder's general inversion formula* (Widder, 1946).

Alfrey's approximation (Alfrey, 1944). Putting $n=1$ in Eq. 1.158 yields

$$v(t)\cong\left[-p^{2}\frac{d\bar{v}(p)}{dp}\right]_{p=t^{-1}} \tag{1.159}$$

Haar's approximation (ter Haar, 1951). Rewrite Eq. 1.122 in the form

$$\frac{p^{m+1}\bar{v}(p)}{m!}=\int_{0^-}^{\infty}\frac{v(t)}{t^{m}}\left[\frac{p^{m+1}t^{m}e^{-pt}}{m!}\right]dt \tag{1.160}$$

where m is a positive integer, and put $n=m$ in Eq. 1.157 to get

$$\frac{p^{m+1}t^{m}e^{-pt}}{m!}\cong\delta\left(t-\frac{m}{p}\right) \tag{1.161}$$

Substituting the equation above into Eq. 1.160 yields

$$\frac{p^{m+1}\bar{v}(p)}{m!}\cong\left[\frac{v(t)}{t^{m}}\right]_{t=m/p},$$

or

$$v(t)\cong\left[\frac{m^{m}p\bar{v}(p)}{m!}\right]_{p=m/t} \tag{1.162}$$

For $m=1$, we get

$$v(t)\cong p\bar{v}(p)|_{p=t^{-1}} \tag{1.163}$$

Schapery's approximation. In 1962, R.A. Schapery proposed the following approximate formula

$$v(t)\cong p\bar{v}(p)|_{p=0.5/t} \tag{1.164}$$

Collocation method. Suppose the given Laplace transform has an approximate inverse of the form

$$v(t)\cong v^{*}(t)=\sum_{i=1}^{N}g_{i}e^{-\alpha_{i}t} \tag{1.165}$$

where N is a positive integer and $\alpha_i (i=1$ to $N)$ are positive constants, g_i are unknown constants. The values of N and α_i are specified basing on the nature of the problem involved, while g_i are to be determined by the techniques of the minimum square error, i.e. from the condition

$$\frac{\partial}{\partial g_i}\int_{0^-}^{\infty}[v(t)-v^*(t)]^2\mathrm{d}t=0,$$

or

$$\int_{0^-}^{\infty}[v(t)-v^*(t)]\frac{\partial v^*(t)}{\partial g_i}\mathrm{d}t=0,$$

or, due to Eq. 1.165,

$$\int_{0^-}^{\infty}[v(t)-v^*(t)]\mathrm{e}^{-\alpha_i t}\mathrm{d}t=0,$$

or

$$\int_{0^-}^{\infty}v(t)\mathrm{e}^{-\alpha_i t}\mathrm{d}t=\int_{0^-}^{\infty}v^*(t)\mathrm{e}^{-\alpha_i t}\mathrm{d}t \tag{1.166}$$

Substituting Eq. 1.165 again into the right-hand side of the equation above, while noting its left-hand side is $\bar{v}(\alpha_i)$, we can write

$$\bar{v}(\alpha_i)=\sum_{j=1}^{N}\frac{g_j}{\alpha_i+\alpha_j},\quad (i=1,2,\ldots,N) \tag{1.167}$$

Thus we have N equations in the form as above to determine $g_i(i=1 \text{ to } N)$.

The approximation methods presented above have been found quite successful for *quasi-static* problems. Beside the collocation method, all of them automatically give exact initial and final values, since Eqs. 1.139 and 1.140 are satisfied exactly. The same favorable characteristics can be achieved in the collocation method also by a proper selection of $v^*(t)$ in Eq. 1.165 and assigning appropriate constraint conditions among g_i before carrying out the procedure of the minimum square error. For problems of wave propagation in elastic solids, Swanson (1980) proposed to use the numerical inversion techniques developed by Bellman, Kalaba and Lockett (1966).

1.13 ONE-TO-ONE MAPPINGS

A set of three independent variables x_1, x_2, x_3 may be thought of as specifying the coordinates of a point in a three dimensional space. A transformation from x_1, x_2, x_3 to a set of new variables y_1, y_2, y_3 through the equations

$$y_i=y_i(x_1,x_2,x_3),\quad (i=1,2,3) \tag{1.168a}$$

specifies a *transformation of coordinates*, and may be thought as specifying the coordinates of a point in another three dimensional space. The points specified by y_i and x_i may not belong to the same space, and if they do belong to the same space they may not define the same point. The inverse transformation

$$x_i=x_i(y_1,y_2,y_3) \tag{1.168b}$$

proceeds in the reverse direction. By the standard chain rule of differentiation, we can derive the following identities

$$\frac{\partial A(y_i)}{\partial x_j} = \frac{\partial A}{\partial y_k}\frac{\partial y_k}{\partial x_j}, \quad \frac{\partial B(x_i)}{\partial y_j} = \frac{\partial B}{\partial x_k}\frac{\partial x_k}{\partial y_j} \tag{1.168c,d}$$

$$\frac{\partial y_i}{\partial x_k}\frac{\partial x_k}{\partial y_j} = \delta_{ij} = \frac{\partial x_i}{\partial y_k}\frac{\partial y_k}{\partial x_j} \tag{1.168e,f}$$

When an infinitesimal quantity $\mathrm{d}x_i$ in the domain of the variables x_i is uniquely mapped into $\mathrm{d}y_i$ in the domain of the variables y_i we can write

$$\mathrm{d}y_i = \frac{\partial y_i}{\partial x_j}\mathrm{d}x_j \tag{1.169a}$$

or, reversely

$$\mathrm{d}x_i = \frac{\partial x_i}{\partial y_j}\mathrm{d}y_j \tag{1.169b}$$

Each of the last two equations can be rewritten as a system of three linear equations (Sections 1.6) as the following

$$\{\mathrm{d}y_i\} = \left[\frac{\partial y_i}{\partial x_j}\right]\{\mathrm{d}x_j\}, \quad \{\mathrm{d}x_i\} = \left[\frac{\partial x_i}{\partial y_j}\right]\{\mathrm{d}y_j\} \tag{1.169c,d}$$

where

$$\left[\frac{\partial y_i}{\partial x_j}\right] = \begin{bmatrix} \frac{\partial y_1}{\partial x_1} & \frac{\partial y_1}{\partial x_2} & \frac{\partial y_1}{\partial x_3} \\ \frac{\partial y_2}{\partial x_1} & \frac{\partial y_2}{\partial x_2} & \frac{\partial y_2}{\partial x_3} \\ \frac{\partial y_3}{\partial x_1} & \frac{\partial y_3}{\partial x_2} & \frac{\partial y_3}{\partial x_3} \end{bmatrix} \tag{1.170a}$$

$$\left[\frac{\partial x_i}{\partial y_j}\right] = \begin{bmatrix} \frac{\partial x_1}{\partial y_1} & \frac{\partial x_1}{\partial y_2} & \frac{\partial x_1}{\partial y_3} \\ \frac{\partial x_2}{\partial y_1} & \frac{\partial x_2}{\partial y_2} & \frac{\partial x_2}{\partial y_3} \\ \frac{\partial x_3}{\partial y_1} & \frac{\partial x_3}{\partial y_2} & \frac{\partial x_3}{\partial y_3} \end{bmatrix} \tag{1.170b}$$

$$\left[\frac{\partial y_i}{\partial x_j}\right] = \left[\frac{\partial x_i}{\partial y_j}\right]^{-1} \tag{1.170c}$$

Thus in order to insure that the transformation is *reversible* and in *one-to-one correspondence* in a certain domain V of the variables x_i, i.e. in order that each x_i defines a unique y_i for x_i in V and vice versa, it is sufficient that :

(a) The functions y_i are single-valued, continuous, and possess continuous first partial derivatives in V, and

(b) The *Jacobian determinant* $J = |\partial y_i/\partial x_j|$ does not vanish at any point in V. Naturally, since the equations of the inverse transformation, Eq. 1.168b, are to be uniquely solvable for x_i, the Jacobian determinant of the inverse transformation, i.e. $|\partial x_i/\partial y_j|$, must also be different from zero throughout V.

Coordinate transformation with the properties (a) and (b) named above are called *admissible transformations*. If the Jacobian is positive everywhere, then a right-hand set of coordinates is transformed into another right-hand set, and the transformation is said to be *proper*. If the Jacobian is negative everywhere, a right-hand set of coordinates is transformed into a left-hand one, and the transformation is said to be *improper*. *In this book, we shall tacitly assume admissible and proper transformations.* Note that the transformation principle described above can be readily generalized for a set of any number of independent variables.

1.14 CURVILINEAR COORDINATES

Suppose that any point P^0 in a three dimensional space (see Fig. 1.14) can be located by two different sets of coordinates, the usual rectangular Cartesian coordinates (x^1, x^2, x^3) and the *curvilinear coordinates* $(\theta^1, \theta^2, \theta^3)$, so that

$$x^1 = x^1(\theta^1, \theta^2, \theta^3),$$
$$x^2 = x^2(\theta^1, \theta^2, \theta^3),$$
$$x^3 = x^3(\theta^1, \theta^2, \theta^3),$$

or

$$x^i = x^i(\theta^1, \theta^2, \theta^3) \tag{1.171}$$

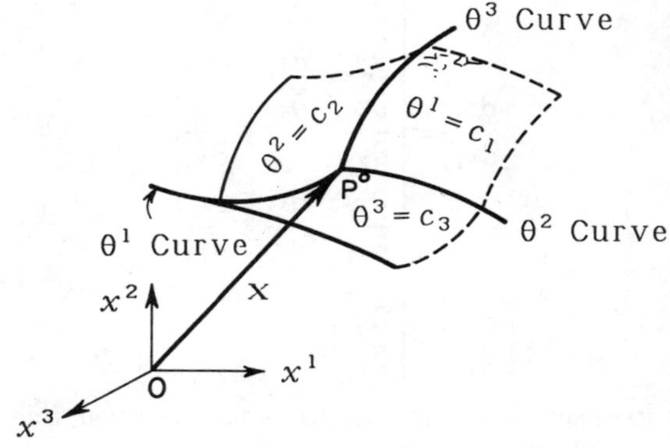

Fig. 1.14 Coordinate surfaces and coordinate curves.

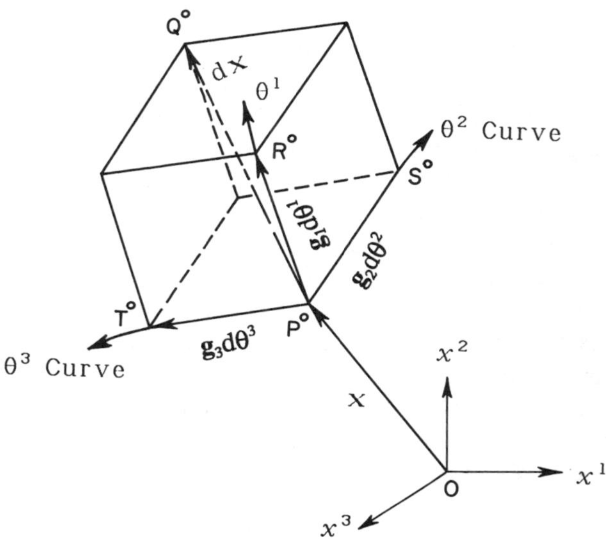

Fig. 1.15 Vector **dx** with components in directions of three coordinate curves.

and, in reverse,

$$\theta^i = \theta^i(x^1, x^2, x^3) \tag{1.172}$$

The sets of equations above (Eqs. 1.171 and 1.172) define a *transformation of coordinates*, when the functions involved make the transformation admissible and proper. In practice, this assumption may not apply at certain points and special consideration is required.

In Fig. 1.14, the surfaces $\theta^1 = c_1$, $\theta^2 = c_2$ and $\theta^3 = c_3$ where c_1, c_2 and c_3 are constants, are called *coordinate surfaces*. The intersections of these surfaces are known as *coordinate curves* or *lines*. If a neighboring point Q^0 is located by the position vector $\mathbf{x} + \mathbf{dx}$, the vector $\mathbf{dx}$ drawn from P^0 to Q^0 must have a mathematical form as follows

$$\mathbf{dx} = \frac{\partial \mathbf{x}}{\partial \theta^1} d\theta^1 + \frac{\partial \mathbf{x}}{\partial \theta^2} d\theta^2 + \frac{\partial \mathbf{x}}{\partial \theta^3} d\theta^3,$$

or

$$\mathbf{dx} = \mathbf{g}_\lambda d\theta^\lambda \tag{1.173}$$

where $\mathbf{g}_\lambda$ is known as the *covariant base vector*, i.e.

$$\mathbf{g}_\lambda = \frac{\partial \mathbf{x}}{\partial \theta^\lambda} \tag{1.174}$$

Thus, as described in Section 1.8, $\mathbf{g}_\lambda$ is a vector tangent to the θ^λ curve at point P^0. Accordingly, an infinitesimal parallelepiped with P^0Q^0 as a diagonal is constructed and shown in Fig. 1.15.

At this stage, we may notice that some notation rules are different from those already used in rectangular Cartesian coordinate systems. We now have to show the difference between *covariance* and *contravariance* by using *subscripts for covariant tensors and superscript for contravariant tensors.* Equating and summing two covariant indices or two contravariant indices do not yield a tensor of lower order and not a proper contraction. An equation does not have a correct tensor form if having a free index as a subscript on one side but as a superscript on the other side. However, the Einstein summation convention is still in use, while vectors and tensors are written in symbolic (Gibbs) notation together with indexed scripts. An example of this point is Eq. 1.173 where the notation is essentially symbolic but with the added feature of the summation convention. In such a 'combination' style of notations, which seems to be the most efficient way when dealing with curvilinear coordinates, tensor character is not given by the free index rule prevailing in a true indicial notation.

The position vector $\mathbf{x}$ can be put explicitly in terms of three Cartesian components as follows

$$\mathbf{x} = x^1\hat{\mathbf{i}} + x^2\hat{\mathbf{j}} + x^3\hat{\mathbf{k}} \tag{1.175}$$

where $\hat{\mathbf{i}}$, $\hat{\mathbf{j}}$ and $\hat{\mathbf{k}}$ are unit base vectors of axes x^1, x^2 and x^3, respectively, and have the product properties given in Eqs. 1.7 and 1.8. Accordingly, Eq. 1.174 becomes

$$\mathbf{g}_\lambda = \frac{\partial x^1}{\partial \theta^\lambda}\hat{\mathbf{i}} + \frac{\partial x^2}{\partial \theta^\lambda}\hat{\mathbf{j}} + \frac{\partial x^3}{\partial \theta^\lambda}\hat{\mathbf{k}} \tag{1.176}$$

In general, $\mathbf{g}_1$, $\mathbf{g}_2$ and $\mathbf{g}_3$ are not mutually orthogonal nor unit vectors. When they are mutually orthogonal, they form an *orthogonal curvilinear coordinate system.* The length $\mathrm{d}s$ of the infinitesimal line element P^0Q^0 in Fig. 1.15 is given by

$$(\mathrm{d}s)^2 = \mathrm{d}\mathbf{x} \cdot \mathrm{d}\mathbf{x} = \mathrm{d}x^i \mathrm{d}x^i$$

or, due to Eq. 1.173,

$$(\mathrm{d}s)^2 = g_{\lambda\mu}\mathrm{d}\theta^\lambda \mathrm{d}\theta^\mu \tag{1.177}$$

where $g_{\lambda\mu}$ is the *covariant metric tensor* and defined as

$$g_{\lambda\mu} = \mathbf{g}_\lambda \cdot \mathbf{g}_\mu \tag{1.178}$$

The covariant metric tensor is a symmetric tensor, i.e.

$$g_{\lambda\mu} = g_{\mu\lambda} \tag{1.179}$$

Substituting Eq. 1.176 into Eq. 1.178 yields

$$g_{\lambda\mu} = \frac{\partial x^i}{\partial \theta^\lambda}\frac{\partial x^i}{\partial \theta^\mu} \tag{1.180}$$

or, in matrix form,

$$[g_{ij}] = [\chi_{ki}\chi_{kj}] = [\chi_{ki}]^T[\chi_{kj}] \tag{1.181}$$

where χ_{ij} denotes the ith Cartesian components of $\mathbf{g}_j$ (given in Eq. 1.176), i.e.

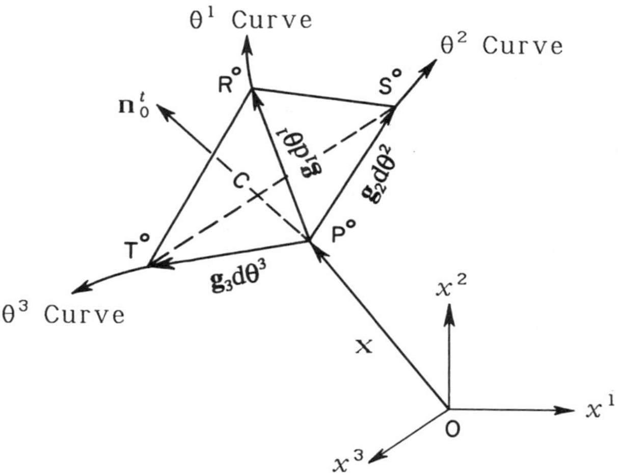

Fig. 1.16 Infinitesimal tetrahedron formed by three line elements along coordinate curves.

$$\chi_{ij} = \frac{\partial x^i}{\partial \theta^j} \tag{1.182}$$

Moreover, we can show that

$$\mathbf{g}_\lambda \cdot (\mathbf{g}_\lambda \times \mathbf{g}_\mu) = 0, \quad (\lambda \text{ not summed}) \tag{1.183a}$$

$$\sqrt{g} = \mathbf{g}_1 \cdot (\mathbf{g}_2 \times \mathbf{g}_3) = \mathbf{g}_2 \cdot (\mathbf{g}_3 \times \mathbf{g}_1) = \mathbf{g}_3 \cdot (\mathbf{g}_1 \times \mathbf{g}_2) \tag{1.183b}$$

$$= \det[\chi_{ij}] \tag{1.183c}$$

and

$$g = \det[g_{ij}] = \det[\chi_{ki}\chi_{kj}] \tag{1.183d}$$

In a general N dimensional space, a curvilinear coordinate system involved must have N coordinates $(\theta^1, \theta^2, \ldots, \theta^N)$, and an infinitesimal line element with a length $\mathrm{d}s$ is given by the *metric form*

$$(\mathrm{d}s)^2 = g_{\lambda\mu} \mathrm{d}\theta^\lambda \mathrm{d}\theta^\mu, \quad (\lambda, \mu = 1, 2, \ldots, N) \tag{1.184}$$

If there exists a transformation into a rectangular coordinate system $(x^1, x^2, \ldots, x^N)$ such that the metric form becomes

$$(\mathrm{d}s)^2 = \mathrm{d}x^i \mathrm{d}x^i, \quad (i = 1, 2, \ldots, N) \tag{1.185}$$

the space is called an N dimensional *Euclidean space*, otherwise the space is *Riemannian.*

In three dimensional Euclidean space like the ordinary physical space, the volume of the infinitesimal parallelepiped in Fig. 1.15 is given by

$$dV^0 = \sqrt{g}\, d\theta^1 d\theta^2 d\theta^3 \tag{1.186}$$

where g is as defined in Eqs. 1.183b to 1.183d. Further, passing a plane through corners R^0, S^0 and T^0 of the same parallelepiped, we obtain an infinitesimal tetrahedron as shown in Fig. 1.16, and the following relation can be noted

$$\begin{aligned} 2\mathbf{n}_0^t dS_0^t &= (\mathbf{g}_2 d\theta^2 - \mathbf{g}_1 d\theta^1) \times (\mathbf{g}_3 d\theta^3 - \mathbf{g}_1 d\theta^1) \\ &= \mathbf{g}_1 \times \mathbf{g}_2 d\theta^1 d\theta^2 + \mathbf{g}_2 \times \mathbf{g}_3 d\theta^2 d\theta^3 + \mathbf{g}_3 \times \mathbf{g}_1 d\theta^3 d\theta^1 \end{aligned} \tag{1.187}$$

where dS_0^t is the area of the triangle $R^0S^0T^0$ and $\mathbf{n}_0^t$ is the unit vector normal to dS_0^t. Performing a dot product of Eq. 1.187 with $\mathbf{g}_1$, in view of Eqs. 1.183a and b, yields

$$2dS_0^t \mathbf{g}_1 \cdot \mathbf{n}_0^t = \sqrt{g}\, d\theta^2 d\theta^3 \tag{1.188a}$$

and, in a similar manner,

$$2dS_0^t \mathbf{g}_2 \cdot \mathbf{n}_0^t = \sqrt{g}\, d\theta^3 d\theta^1 \tag{1.188b}$$

$$2dS_0^t \mathbf{g}_3 \cdot \mathbf{n}_0^t = \sqrt{g}\, d\theta^1 d\theta^2 \tag{1.188c}$$

At this stage, it is appropriate to introduce two very important contravariant tensors. Firstly, the *contravariant metric tensor* $g^{\lambda\mu}$ may be introduced as follows

$$g^{\lambda\mu} = \frac{\overset{*}{g}_{\lambda\mu}}{g} \tag{1.189}$$

where $\overset{*}{g}_{\lambda\mu}$ is, as defined in Eq. 1.42, the cofactor of an element $g_{\lambda\mu}$ of the square matrix $[g_{\lambda\mu}]$. Thus it can be shown that

$$g^{\lambda\mu} = g^{\mu\lambda} \tag{1.190a}$$

$$g^{\lambda\kappa} g_{\kappa\mu} = \delta_\mu^\lambda \tag{1.190b}$$

where δ_μ^λ is the Kronecker symbol. Secondly, the *contravariant base vector* $\mathbf{g}^\lambda$ may be introduced as follows

$$\mathbf{g}^\lambda = g^{\lambda\mu} \mathbf{g}_\mu \tag{1.191}$$

There are interesting identities, namely

$$g^{\lambda\mu} = \mathbf{g}^\lambda \cdot \mathbf{g}^\mu \tag{1.192a}$$

$$\mathbf{g}_\lambda = g_{\lambda\mu} \mathbf{g}^\mu \tag{1.192b}$$

$$\mathbf{g}_\mu \cdot \mathbf{g}^\lambda = \mathbf{g}^\lambda \cdot \mathbf{g}_\mu = \delta_\mu^\lambda \tag{1.192c}$$

The last equation means physically that a contravariant base vector $\mathbf{g}^\lambda$ is orthogonal to the coordinate surface $\theta^\lambda = \text{constant}$. Using this property for $\lambda = 1$, we can obtain the area of the side $P^0S^0T^0$ of the infinitesimal tetrahedron in Fig. 1.16 as

$$dS_0^1 = \frac{1}{2} g^1 \mathbf{g}_1 \cdot (\mathbf{g}_2 \times \mathbf{g}_3) d\theta^2 d\theta^3,$$

where g^1 is the magnitude of $\mathbf{g}^1$. Due to Eqs. 1.183b and 1.192a, the equation above can be written in the form

$$dS_0^1 = \frac{1}{2}\sqrt{g g^{11}}\, d\theta^2 d\theta^3.$$

Likewise, the area of any side, where θ^λ = constant, of the same tetrahedron is

$$dS_0^\lambda = \frac{1}{2}\sqrt{g g^{\lambda\lambda}}\, d\theta^\mu d\theta^\kappa, \quad (\lambda \text{ not summed}, \ \lambda \neq \mu \neq \kappa) \tag{1.193}$$

1.15 DERIVATIVES WITH RESPECT TO CURVILINEAR COORDINATES

Henceforth, curvilinear coordinates will be indexed by Greek scripts and a partial differentiation with respect to θ^λ will follow the comma-subscript convention. The derivative of the covariant base vector $\mathbf{g}_\mu$ with respect to θ^ν is again a vector, and can be put in the form

$$\mathbf{g}_{\mu,\nu} = \begin{Bmatrix} \lambda \\ \mu\nu \end{Bmatrix} \mathbf{g}_\lambda \tag{1.194}$$

where $\begin{Bmatrix} \lambda \\ \mu\nu \end{Bmatrix}$ is the *Christoffel symbol of the second kind* relating the component of $\mathbf{g}_{\mu,\nu}$ in the direction of $\mathbf{g}_\lambda$. Recalling the definition of $\mathbf{g}_\mu$ in Eq. 1.174, we can write

$$\mathbf{g}_{\mu,\nu} = \mathbf{g}_{\nu,\mu} \tag{1.195}$$

thus

$$\begin{Bmatrix} \lambda \\ \mu\nu \end{Bmatrix} = \begin{Bmatrix} \lambda \\ \nu\mu \end{Bmatrix} \tag{1.196}$$

In addition, the Christoffel symbol can be expressed in terms of metric tensors as

$$\begin{Bmatrix} \lambda \\ \mu\nu \end{Bmatrix} = \frac{1}{2} g^{\lambda\kappa}(g_{\kappa\mu,\nu} + g_{\kappa\nu,\mu} - g_{\mu\nu,\kappa}) \tag{1.197}$$

and the derivative of the contravariant base vector $\mathbf{g}^\mu$ with respect to θ^ν can be obtained as

$$\mathbf{g}^\mu_{,\nu} = -\begin{Bmatrix} \mu \\ \lambda\nu \end{Bmatrix} \mathbf{g}^\lambda \tag{1.198}$$

A vector $\mathbf{u}$ can be resolved into three vectors in the directions of $\mathbf{g}_\lambda$ as followsy

$$\mathbf{u} = v^\lambda \mathbf{g}_\lambda \tag{1.199}$$

where v^λ is called the *contravariant component* of the vector $\mathbf{u}$. Differentiating $\mathbf{u}$ with respect to θ^ν and using Eq. 1.194 lead to

$$\mathbf{u}_{,\nu} = v^\lambda_{;\nu} \mathbf{g}_\lambda \tag{1.200}$$

where $v^\lambda_{;\nu}$ is the *covariant derivative of the contravariant component* v^λ and given by

$$v^{\lambda}_{;\nu} = v^{\lambda}_{,\nu} + \begin{Bmatrix} \lambda \\ \rho\nu \end{Bmatrix} v^{\rho} \tag{1.201}$$

On the other hand, the same vector $\mathbf{u}$ can be resolved into three vectors in the directions of $\mathbf{g}^{\lambda}$ as follows

$$\mathbf{u} = v_{\lambda}\mathbf{g}^{\lambda} \tag{1.202}$$

where v_{λ} is the *covariant component* of $\mathbf{u}$ and related to v^{λ} by the relation

$$v_{\lambda} = g_{\lambda\mu}v^{\mu} \tag{1.203}$$

Differentiating Eq. 1.202 with respect to θ^{ν} yields

$$\mathbf{u}_{,\nu} = v_{\lambda;\nu}\mathbf{g}^{\lambda} \tag{1.204}$$

where $v_{\lambda;\nu}$ is the *covariant derivative of the covariant component* v_{λ} and given by

$$v_{\lambda;\nu} = v_{\lambda,\nu} - \begin{Bmatrix} \rho \\ \lambda\nu \end{Bmatrix} v_{\rho} \tag{1.205}$$

For a general tensor $T^{\lambda_1 \ldots \lambda_r}_{\mu_1 \ldots \mu_s}$, its covariant derivative can be put as

$$T^{\lambda_1 \ldots \lambda_r}_{\mu_1 \ldots \mu_s;\nu} = T^{\lambda_1 \ldots \lambda_r}_{\mu_1 \ldots \mu_s,\nu} + \sum_{i=1}^{r} T^{\lambda_1 \ldots \lambda_{i-1}\rho\lambda_{i+1} \ldots \lambda_r}_{\mu_1 \ldots\ldots\ldots\ldots\ldots\ldots \mu_s} \begin{Bmatrix} \lambda_i \\ \rho\nu \end{Bmatrix}$$

$$- \sum_{j=1}^{s} T^{\lambda_1 \ldots\ldots\ldots\ldots\ldots\ldots \lambda_r}_{\mu_1 \ldots \mu_{j-1}\rho\mu_{j+1} \ldots \mu_s} \begin{Bmatrix} \rho \\ \mu_j\nu \end{Bmatrix} \tag{1.206}$$

Two applications of the last equation are illustrated in Eqs. 1.201 and 1.205. Another is that the covariant derivatives of the covariant and contravariant metric tensors vanish, i.e.,

$$g_{\lambda\mu;\nu} = 0\,, \qquad g^{\lambda\mu}_{;\nu} = 0 \tag{1.207}$$

Moreover, the covariant derivative of a product of two tensors $S^{\lambda\ldots}_{\mu\ldots}T^{\kappa\ldots}_{\rho\ldots}$ follows a simple formula,

$$(S^{\lambda\ldots}_{\mu\ldots}T^{\kappa\ldots}_{\rho\ldots})_{;\nu} = S^{\lambda\ldots}_{\mu\ldots;\nu}T^{\kappa\ldots}_{\rho\ldots} + S^{\lambda\ldots}_{\mu\ldots}T^{\kappa\ldots}_{\rho\ldots;\nu} \tag{1.208}$$

The covariant derivative of Eq. 1.203, due to Eqs. 1.207 and 1.208 takes the form

$$v_{\lambda;\nu} = g_{\lambda\mu}v^{\mu}_{;\nu} \tag{1.209}$$

For an orthogonal curvilinear coordinate system, $\mathbf{g}^{\lambda}$ and $\mathbf{g}_{\lambda}$ are in the same direction, and the following conditions hold:

$$g_{\lambda\mu} \equiv g^{\lambda\mu} = 0\,, \qquad (\lambda \neq \mu) \tag{1.210a}$$

$$g_{\lambda\lambda}g^{\lambda\lambda} = 1 \tag{1.210b}$$

$$\begin{Bmatrix} \lambda \\ \lambda\lambda \end{Bmatrix} = \frac{1}{\sqrt{g_{\lambda\lambda}}} \frac{\partial\sqrt{g_{\lambda\lambda}}}{\partial\theta^{\lambda}} \tag{1.210c}$$

$$\begin{Bmatrix} \lambda \\ \lambda\mu \end{Bmatrix} = \frac{1}{\sqrt{g_{\lambda\lambda}}} \frac{\partial \sqrt{g_{\lambda\lambda}}}{\partial \theta^{\mu}} \tag{1.210d}$$

$$\begin{Bmatrix} \lambda \\ \mu\mu \end{Bmatrix} = -\frac{\sqrt{g_{\mu\mu}}}{g_{\lambda\lambda}} \frac{\partial \sqrt{g_{\mu\mu}}}{\partial \theta^{\lambda}} \tag{1.210e}$$

$$\begin{Bmatrix} \lambda \\ \mu\kappa \end{Bmatrix} = 0 , \qquad (\lambda \neq \mu \neq \kappa) \tag{1.210f}$$

In Eqs. 1.210, a repetition of an index script does not imply a summation.

A rectangular Cartesian coordinate system may be viewed as a special case of an orthogonal curvilinear system. In this special case, $\mathbf{g}^{\lambda}$ and $\mathbf{g}_{\lambda}$ become an identical unit vector, i.e.

$$\mathbf{g}^{\lambda} \equiv \mathbf{g}_{\lambda} \tag{1.211a}$$

$$|\mathbf{g}^{\lambda}| \equiv |\mathbf{g}_{\lambda}| \equiv g^{\lambda} \equiv g_{\lambda} = 1 \tag{1.211b}$$

Consequently, the contravariant and covariant metric tensors become identical and equal to Kronecker symbol, g defined in Eqs. 1.183b to d becomes unity, and the Christoffel symbol of the second kind given in Eq. 1.197 vanishes, i.e.

$$g^{\lambda\mu} \equiv g_{\lambda\mu} \equiv \delta_{\lambda\mu} \tag{1.212a}$$

$$g = 1 \tag{1.212b}$$

$$\begin{Bmatrix} \lambda \\ \mu\nu \end{Bmatrix} = 0 \tag{1.212c}$$

Moreover, the contravariant and covariant components of a vector such as in Eqs. 1.199 and 1.202 become identical, and a covariant partial differentiation such as in Eqs. 1.201, 1.205 and 1.206 is reduced into a usual partial differentiation, i.e.

$$v^{\lambda} = v_{\lambda} \tag{1.213a}$$

$$v^{\lambda}_{;\nu} = v^{\lambda}_{,\nu} \tag{1.213b}$$

$$v_{\lambda;\nu} = v_{\lambda,\nu} \tag{1.213c}$$

$$T^{\lambda_1 \cdots \lambda_r}_{\mu_1 \cdots \mu_s;\nu} = T^{\lambda_1 \cdots \lambda_r}_{\mu_1 \cdots \mu_s,\nu} \tag{1.213d}$$

Strictly speaking, equations such as Eqs. 1.211a, 1.212a and 1.213a are inadmissible, since the free index scripts are not of the same type.

A special type of curvilinear coordinate systems is a system of *skew coordinates*, in which every coordinate (θ^{λ}) has a unit of length and has a fixed orientation with respect to the global (x^{i}) coordinate system. Hence, in such a system: a covariant base vector $\mathbf{g}_{\mu}$ is independent of any skew coordinate θ^{λ}; a Christoffel symbol of the second kind due to Eq. 1.194 vanishes as put in Eq. 1.212c; a covariant partial differentiation as in Eqs. 1.201, 1.205 and 1.206 is reduced into a usual partial differentiation as in Eqs. 1.213; and we can put each and every Cartesian coordinate as a linear function of the skew coordinates as follows

$$x^i = \chi_{ij}\theta^j + c^i \tag{1.214}$$

where c^i is a constant Cartesian component of the position vector of the origin of the skew coordinate system, and χ_{ij} is the cosine of the angle between axes x^i and θ^j. Thus

$$\chi_{ij}\chi_{ij} = 1, \quad (j \text{ not summed}) \tag{1.215}$$

In fact, χ_{ij} is as defined in Eq. 1.182 for any curvilinear coordinate systems.

1.16 EXERCISE PROBLEMS

1.16.1 Prove the equivalence between Gibbs and indicial notations as in Eqs. 1.31, 1.32 and 1.85.

1.16.2 Prove the identities that involve the permutation tensor and Kronecker delta as

$$\varepsilon_{ijk} = \delta_{i1}\delta_{j2}\delta_{k3} + \delta_{j1}\delta_{k2}\delta_{i3} + \delta_{k1}\delta_{i2}\delta_{j3}$$
$$-\delta_{j1}\delta_{i2}\delta_{k3} - \delta_{k1}\delta_{j2}\delta_{i3} - \delta_{i1}\delta_{k2}\delta_{j3} \tag{1.216a}$$

$$\varepsilon_{ijk}\varepsilon_{ilm} = \delta_{jl}\delta_{km} - \delta_{jm}\delta_{kl} \tag{1.216b}$$

$$\varepsilon_{ijk}\varepsilon_{klj} = 2\delta_{il} \tag{1.216c}$$

and Eqs. 1.46 to 1.51.

1.16.3 (a) If $[C_{ij}] = [A_{ik}]\,[B_{kj}]$, where the ranges of indices i, j and k are 2, 3 and 4, respectively; find each and every element C_{ij} in terms of elements A_{ik} and B_{kj}. (b) Write the set of simultaneous equations

$$3a_1 - 2a_2 + 2a_3 = 10,$$

$$a_1 + 2a_2 - 3a_3 = -1,$$

$$4a_1 + a_2 + 2a_3 = 3.$$

in matrix form, and in indicial notation. Find the determinant and the inverse of the square matrix involved, and solve the equations for a_1, a_2 and a_3.

Answer: Determinant $= 35$, $a_1 = 2, a_2 = -3, a_3 = -1$.

1.16.4 Find eigenvalues and eigenvectors of the following matrix

$$[\sigma_{ij}] = \begin{bmatrix} 10 & -6 & 0 \\ -6 & 10 & 0 \\ 0 & 0 & 1 \end{bmatrix}.$$

Rewrite the generalized coordinates in terms of the normal coordinates. Show that $[\sigma_{ij}]$ is positive definite by Eq. 1.72 and by Eq. 1.73.
Answer: $\lambda_1 = 16$, $\lambda_2 = 4$, $\lambda_3 = 1$; $\{x_j\} = [\chi_{jk}]\{y_k\}$, where

$$[\chi_{jk}] = \frac{1}{\sqrt{2}} \begin{bmatrix} 1 & 1 & 0 \\ -1 & 1 & 0 \\ 0 & 0 & \sqrt{2} \end{bmatrix}.$$

1.16.5 (a) Use Gauss's divergence theorem to show that

$$\int_S x_i n_j \mathrm{d}S = V\delta_{ij}.$$

(b) Verify Green's theorem in the plane for

$$\oint_C [(x_1 x_2 + x_2^2)\mathrm{d}x_1 + x_1^2 \mathrm{d}x_2],$$

where C is the closed curve of the region bounded by $x_1 = x_2$ and $x_2 = x_1^2$ as shown in Fig. 1.17.

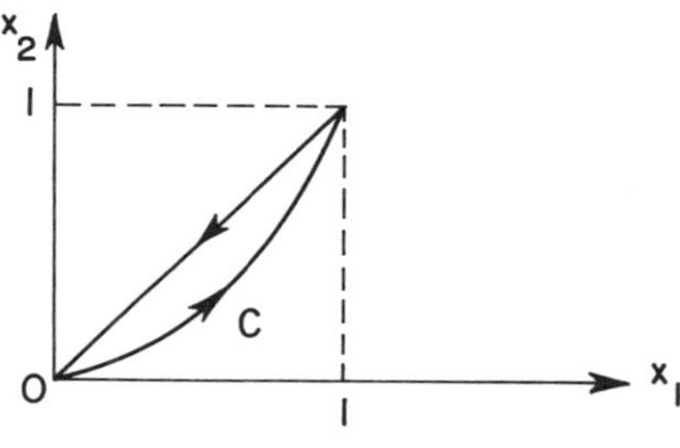

Fig. 1.17.

1.16.6 Find a unit normal and an equation for the tangent plane to the surface $2x_1x_3^2 - 3x_1x_2 - 4x_1 = 7$ at the point $(1, -1, 2)$.
Answer: A unit normal $(n_1, n_2, n_3) = \pm(7, -3, 8)/\sqrt{122}$, and an equation for the tangent plane is

$$7(x_1 - 1) - 3(x_2 + 1) + 8(x_3 - 2) = 0.$$

1.16.7 Find an equation for the plane perpendicular to the position vector $(y_1, y_2, y_3) = (2, 3, 6)$ and passing through the terminal point of the position vector $(z_1, z_2, z_3) = (1, 5, 3)$. Find also the distance from the origin to the plane.

Answer: An equation for the plane is $2x_1+3x_2+6x_3=35$, and the plane is 5 from the origin.

1.16.8 By means of Gauss's divergence theorem show that

$$\int_S \mathbf{n}\times(\mathbf{a}\times\mathbf{x})dS = 2\mathbf{a}V,$$

where V is the volume enclosed by the surface S having the outward unit normal $\mathbf{n}$. The position vector to any point in V is $\mathbf{x}$, and $\mathbf{a}$ is an arbitrary constant vector.
Hint: Write the expression in indicial notation.

1.16.9 Find the tangent and normal vectors, and verify the Frenet-Serret formulas of the following space curves: (a) a plane circle defined by $x_1=a\cos\theta$, $x_2=a\sin\theta$, $x_3=b$ where a and b are constants while θ varies. (b) a cylindrical helix defined by $x_1=a\cos\theta$, $x_2=a\sin\theta$, $x_3=b\theta$.

CHAPTER II

STRESS AND STRAIN TENSORS

2.1 INTRODUCTION

The present book is restricted to *small deformation* of solid continua until Chapter X, where the finite deformation theory will start. Indicial notation rules, using subscripts to denote Cartesian components of tensors, will be followed until Section 10.6. Unless specified otherwise, subscripts 1, 2 and 3 denote Cartesian components in x_1, x_2 and x_3-directions respectively. Many textbooks denote x_1, x_2, x_3 by x, y, z respectively. Figure 2.1 shows the equivalence between the notations.

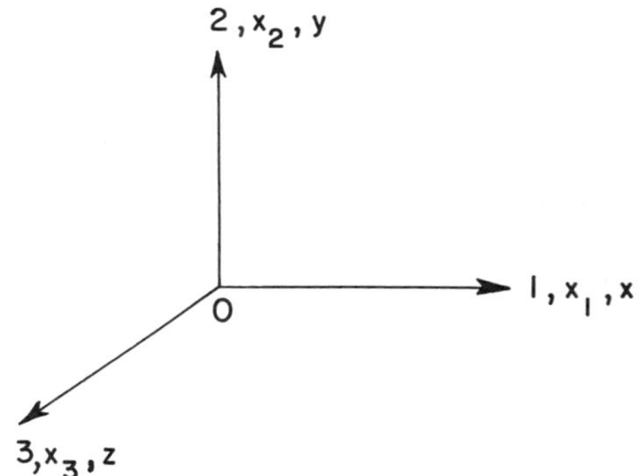

Fig. 2.1 Various notations of Cartesian coordinate axes.

2.2 FORCE DISTRIBUTION AND STRESSES

One of the main concerns in the study of continuous media is how forces are transmitted through the media. These forces can be classified into two classes, i.e. *body forces* and *surface tractions*. A body force acts directly on the distribution of the matter in the domain of interest, and thus has the dimensions of force per unit volume of material. (In finite deformation theory, Chapter X, we will specify a body-force dis-

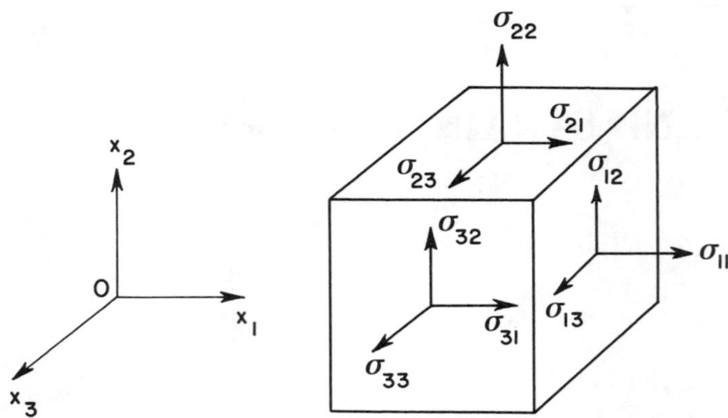

Fig. 2.2 Cartesian components of stress tensor.

tribution per unit *mass*, since mass is the most basic characteristic of the material, and must be *conserved.*)

For a continuum there may be some actual boundary that encloses the domain of interest, or we may specify a domain of interest being enclosed by an artificial boundary. There can be physical forces acting on an actual boundary, while forces on an artificial boundary can be generated by the material immediately outside the domain of interest. In either case, there will be a *surface force distribution* or a *surface traction* which has the dimensions of force per unit area.

Consider an infinitesimal rectangular parallelepiped with its edges parallel to x_1, x_2, x_3 axes, as in Fig. 2.2. The surface traction on each rectangular boundary plane can be decomposed into three Cartesian components. Each of these components can be represented by the symbol σ_{ij} $(i,j=1,2,3)$, where the first subscript denotes the direction of the normal to the plane, while the second subscript denotes the direction of the force. The *stress component* σ_{ij} has the dimensions of force per unit area. As an example, σ_{12} is the x_2-component of the *stress tensor* on a plane normal to the x_1-direction. We will call the stress components with repeated numeral subscripts *normal stresses*, since the force corresponding to each of these stresses is normal to the surface upon which the force acts; and we call the remaining stress components *shear stresses*.

A positive normal stress is directed outward from the material domain of interest, i.e. a *tensile stress*. A shear stress is positive, if it is directed in a positive coordinate direction when its associated normal stress (on the same rectangular boundary plane) is also positive in another positive coordinate direction. A better understanding may be obtained from Fig. 2.3, which is an illustration of the convention for a two-dimensional case. Moreover it should be a good exercise for a reader to complete Fig. 2.2 with positive stress components on the remaining three faces of the infinitesimal parallelepiped.

We may display the stress components in the matrix

$$[\sigma_{ij}] = \begin{bmatrix} \sigma_{11} & \sigma_{12} & \sigma_{13} \\ \sigma_{21} & \sigma_{22} & \sigma_{23} \\ \sigma_{31} & \sigma_{32} & \sigma_{33} \end{bmatrix} = \begin{bmatrix} \sigma_{xx} & \sigma_{xy} & \sigma_{xz} \\ \sigma_{yx} & \sigma_{yy} & \sigma_{yz} \\ \sigma_{zx} & \sigma_{zy} & \sigma_{zz} \end{bmatrix} \tag{2.1}$$

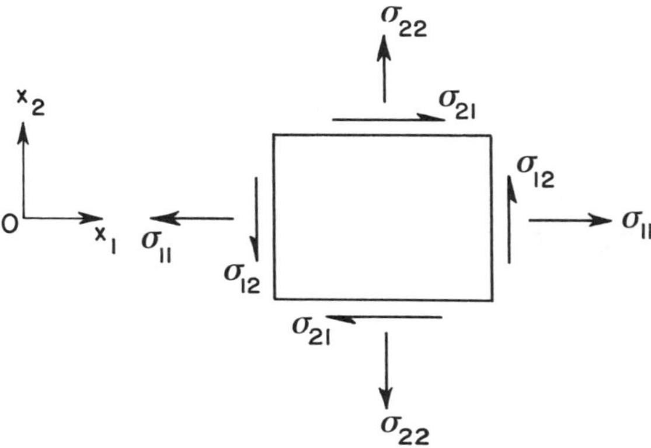

Fig. 2.3 Positive stresses for two-dimensional case.

where the terms in the main diagonal are the normal stresses, while the off-diagonal terms are the shear stresses.

2.3 STRESS VECTOR AND EQUATIONS OF MOTION

First consider the two-dimensional situation with x_1 and x_2 as coordinates of the frame of reference, as in Fig. 2.4. The plane AB is infinitesimally close to the point P and is parallel to the x_3-axis. Let n_1 and n_2 denote the direction cosines of the unit vector n_i normal to the plane AB, and let $\overline{X}_1$ and $\overline{X}_2$ be Cartesian components of the surface traction across the plane AB. If σ_{31} and σ_{32} vanish, the *equilibrium of forces* in the x_1-direction on the element PAB requires

$$\overline{X}_1(AB) - \sigma_{11}\mathrm{d}x_2 - \sigma_{21}\mathrm{d}x_1 = 0,$$

or

$$\overline{X}_1(AB) - \sigma_{11}n_1(AB) - \sigma_{21}n_2(AB) = 0,$$

or

$$\overline{X}_1 = \sigma_{11}n_1 + \sigma_{21}n_2 \tag{2.2}$$

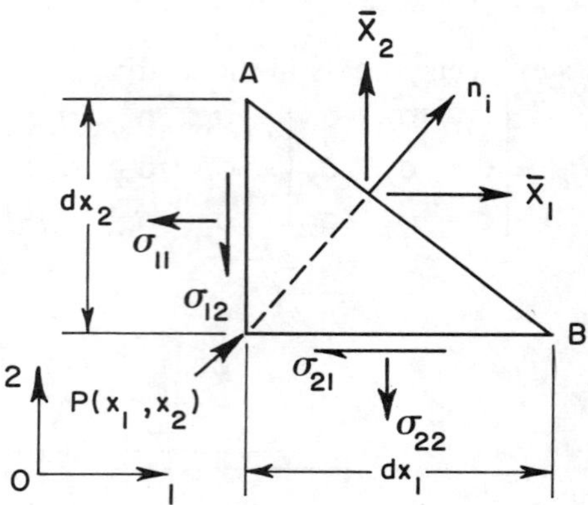

Fig. 2.4 Stress vector for a two-dimensional case.

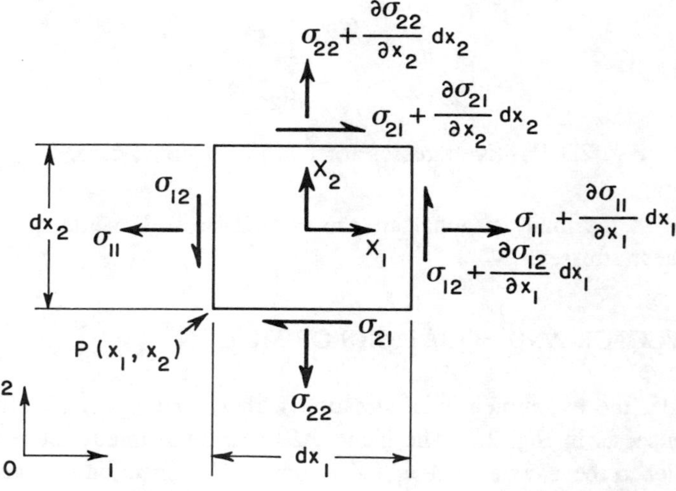

Fig. 2.5 Two-dimensional neighborhood of point P in equilibrium.

Similarly, equilibrium in the x_2-direction gives

$$\overline{X}_2 = \sigma_{12} n_1 + \sigma_{22} n_2 \tag{2.3}$$

Consequently, the surface traction $\overline{X}_i$, being a vector but having the same dimensions as a stress component σ_{ij}, is called a *stress vector*.

Each stress component can be a function of position so that the same stress components on any two opposing boundary plane need not be equal. Next, consider the equilibrium of forces acting on an infinitesimally small neighborhood of the point $P(x_1, x_2)$ subjected to a body force X_i and surface tractions σ_{ji} as shown in Fig. 2.5. The equilibrium of forces in x_1 and x_2-directions yields, respectively, the following

$$\frac{\partial \sigma_{11}}{\partial x_1} + \frac{\partial \sigma_{21}}{\partial x_2} + X_1 = 0 \tag{2.4a}$$

$$\frac{\partial \sigma_{12}}{\partial x_1} + \frac{\partial \sigma_{22}}{\partial x_2} + X_2 = 0 \tag{2.4b}$$

On the other hand, the *equilibrium of moments* yields the *stress symmetry*, i.e.

$$\sigma_{12} = \sigma_{21} \tag{2.5}$$

After understanding the two-dimensional case, we can easily obtain the corresponding three dimensional conditions at any point P located by (x_1, x_2, x_3). For the stress vector, the illustration is depicted in Fig. 2.6 and the formulas can be shown to be

$$\overline{X}_1 = \sigma_{11}n_1 + \sigma_{21}n_2 + \sigma_{31}n_3,$$

$$\overline{X}_2 = \sigma_{12}n_1 + \sigma_{22}n_2 + \sigma_{32}n_3,$$

$$\overline{X}_3 = \sigma_{13}n_1 + \sigma_{23}n_2 + \sigma_{33}n_3,$$

or, in indicial notation,

$$\overline{X}_i = \sigma_{ji}n_j \tag{2.6}$$

where n_i is the unit vector normal to the plane where $\overline{X}_i$ acts, thus

$$n_i n_i \equiv n_1^2 + n_2^2 + n_3^2 = 1 \tag{2.7}$$

Moreover, n_i must be 'outward', i.e. pointing away from the material domain of interest. For static equilibrium, the equation is

$$\sigma_{ji,j} + X_i = 0 \tag{2.8}$$

and the *symmetry of stresses* is

$$\sigma_{ij} = \sigma_{ji} \tag{2.9}$$

The last equation implies that the matrix $[\sigma_{ij}]$ shown in Eq. 2.1 is a symmetric matrix, i.e. $[\sigma_{ij}] = [\sigma_{ij}]^T$, and that there are six (*not* nine) distinct stress components. For dynamic equilibrium, *Newton's second law of motion* (I. Newton, 1687) gives, in place of Eq. 2.8, the following equation

$$\sigma_{ji,j} + X_i = \rho \ddot{u}_i \tag{2.10}$$

where X_i is the body force, u_i the *displacement vector* (see Section 2.8), ρ the *mass density* (mass per unit volume), and a super dot $(\dot{\ })$ a derivative with respect to the *time* t. Equation 2.10 can be rewritten in the form of Eq. 2.8 if the reversed mass acceleration density $-\rho \ddot{u}_i$, the so-called *D'Alembert's force* (J. le R. D'Alembert, 1743), is added to the actual body force.

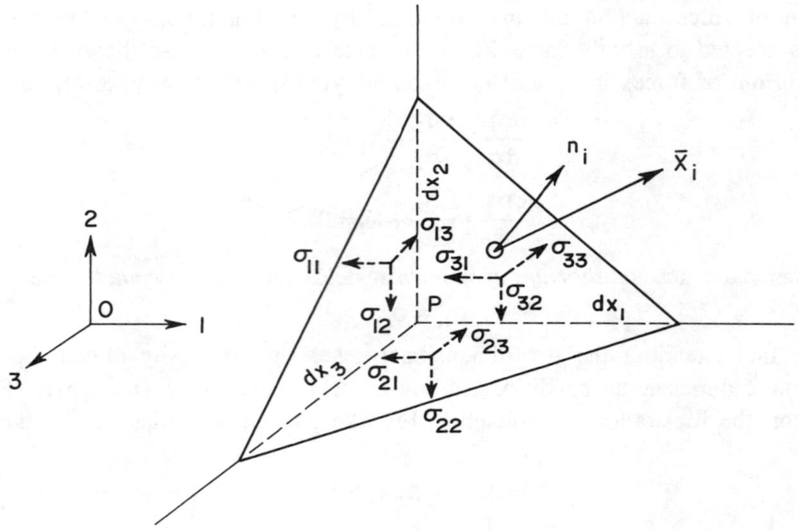

Fig 2.6 Stress vector for a three-dimensional case.

2.4 EULER'S LAWS OF MOTION

The laws that state that *the time rate of the linear momentum is equal to the applied force*, and that *the time rate of the moment of momentum is equal to the applied torque*, are known as *Euler's first and second laws of motion*, respectively (L. Euler, 1752). In fact, Euler's first law of motion is identical to Newton's second law. For an infinitesimally small domain of volume dV which is completely enclosed by a surface dS, the particular law can be expressed as

$$X_i dV + \bar{X}_i dS = \frac{d}{dt}(\rho v_i dV) \tag{2.11}$$

where v_i is the velocity, i.e. dx_i/dt. For small deformation, it can be shown that

$$v_i = \frac{\partial u_i}{\partial t} \tag{2.12}$$

and that the changes of ρ and dV with time are negligible. Thus Eq. 2.11 becomes

$$X_i dV + \bar{X}_i dS = \rho \ddot{u}_i dV,$$

or, if considering X_i to include both actual and D'Alembert's forces,

$$X_i dV + \bar{X}_i dS = 0 \tag{2.13}$$

Substituting the condition on the surface S, Eq. 2.6, into Eq. 2.13 yields

$$X_i \mathrm{d}V + \sigma_{ji} n_j \mathrm{d}S = 0.$$

Applying Gauss's divergence theorem, Eq. 1.88, to the second term in the equation above, we get

$$X_i \mathrm{d}V + \sigma_{ji,j} \mathrm{d}V = 0,$$

which holds for any $\mathrm{d}V$ at a particular point $P(x_1, x_2, x_3)$. Thus the common factor $\mathrm{d}V$ can be removed from the equation above, and the latter becomes

$$\sigma_{ji,j} + X_i = 0 \tag{2.14}$$

which is the equation of motion (Eq. 2.8) derived before. If the equation of motion is used in Eq. 2.13, we shall obtain

$$-\sigma_{ji,j} \mathrm{d}V + \overline{X}_i \mathrm{d}S = 0,$$

or, due to Gauss's divergence theorem,

$$-\sigma_{ji} n_j \mathrm{d}S + \overline{X}_i \mathrm{d}S = 0,$$

or

$$\overline{X}_i = \sigma_{ji} n_j,$$

which is the formula of the stress vector (Eq. 2.6). Hence, we can see that for any domain V enclosed completely by a surface S, the equation of motion for any point in V can be derived from the condition of surface traction on S, and in the reverse the condition of surface traction at any point on S can be derived from the equation of motion in V.

Euler's second law of motion may be expressed as

$$\varepsilon_{ijk} x_j (X_k \mathrm{d}V + \overline{X}_k \mathrm{d}S) = \frac{\mathrm{d}}{\mathrm{d}t}(\varepsilon_{ijk} x_j \rho v_k \mathrm{d}V)$$

$$= \varepsilon_{ijk} x_j \rho \ddot{u}_k \mathrm{d}V + \rho \varepsilon_{ijk} v_j v_k \mathrm{d}V.$$

However, the second term on the right-hand side of the equation above must vanish, since $\varepsilon_{ijk} v_j v_k$ denotes a cross product of two identical vectors. Thus, we have

$$\varepsilon_{ijk} x_j (X_k \mathrm{d}V + \overline{X}_k \mathrm{d}S) = 0 \tag{2.15}$$

in which X_i includes both actual and D'Alembert's forces. In fact, Eq. 2.15 can be obtained directly from Eq. 2.13 by taking the moment of the latter. Proceeding in the same way as Eq. 2.14 has been obtained, one should arrive at the following

$$\varepsilon_{ijk}(x_j \sigma_{lk})_{,l} + \varepsilon_{ijk} x_j X_k = 0,$$

or

$$\varepsilon_{ijk} x_j (\sigma_{lk,l} + X_k) + \varepsilon_{ijk} x_{j,l} \sigma_{lk} = 0.$$

The first term vanishes due to the equation of motion, $x_{j,l}$ in the second term is δ_{jl}, so that

$$\varepsilon_{ijk} \sigma_{jk} = 0,$$

This gives for $i = 1$,

$$\sigma_{23} = \sigma_{32};$$

for $i = 2$,

$$\sigma_{31} = \sigma_{13};$$

and for $i = 3$,

$$\sigma_{12} = \sigma_{21}.$$

Thus the symmetry of stresses, Eq. 2.9, is proved once more.

2.5 STRESS TENSOR

At this stage, it is appropriate to demonstrate that σ_{ij} is, in fact, a tensor. We start with the stress vector formula

$$\sigma_{ji} n_j = \overline{X}_i.$$

The right-hand side is a vector, so that, upon applying the transformation law, Eq. 1.21, we can rewrite the equation above as

$$\sigma_{ji} n_j = a_{ik} \overline{X}'_k,$$

where a prime ($'$) denotes a Cartesian component in the rectangular coordinate system (x'_1, x'_2, x'_3). Applying the stress vector formula to $\overline{X}'_i$ in the last equation yields

$$\sigma_{ji} n_j = a_{ik} \sigma'_{lk} n'_l.$$

Applying the transformation law (Eq. 1.20) for the vector n'_l changes the equation into

$$\sigma_{ji} n_j = a_{ik} a_{jl} \sigma'_{lk} n_j.$$

But n_j is an arbitrary unit vector, so that

$$\sigma_{ji} = a_{jl} a_{ik} \sigma'_{lk} \tag{2.16a}$$

The inverse to Eq. 2.16a can be obtained by the same procedure, i.e.

$$\sigma'_{ij} = a_{ki} a_{lj} \sigma_{kl} \tag{2.16b}$$

Thus we have shown that σ_{ij} is a tensor of second order, since it obeys the transformation law for such a tensor, Eqs. 1.27 and 1.28. Note that the transformed stress is symmetric as well, i.e. $\sigma'_{ij} = \sigma'_{ji}$.

Letting $i = j$ in Eq. 2.16a or 2.16b and using the orthogonality conditions, Eq. 1.25, we find

$$\sigma'_{ii} = \sigma_{kk},$$

which shows that the scalar quantity σ_{ii} does not change with the reference coordinate system. Any term which has such property is normally referred to as an *invariant*. The following are also *stress invariants*;

$$\sigma_{ij}\sigma_{ji}, \quad \sigma_{ij}\sigma_{jk}\sigma_{ki}, \quad \sigma_{ij}\sigma_{jk}\sigma_{kl}\sigma_{li}, \text{ etc.}$$

The three quantities, I_1, I_2, I_3 given in Eqs. 1.17, and known as the *first*, *second* and *third stress invariants*, are combinations of these invariants; in fact

$$I_1 = \sigma_{ii} \tag{2.17a}$$

$$I_2 = \frac{1}{2}(I_1^2 - \sigma_{ij}\sigma_{ji}) \tag{2.17b}$$

$$I_3 = \frac{1}{6}(2\sigma_{ij}\sigma_{jk}\sigma_{ki} - 3I_1\sigma_{ij}\sigma_{ji} + I_1^3) \tag{2.17c}$$

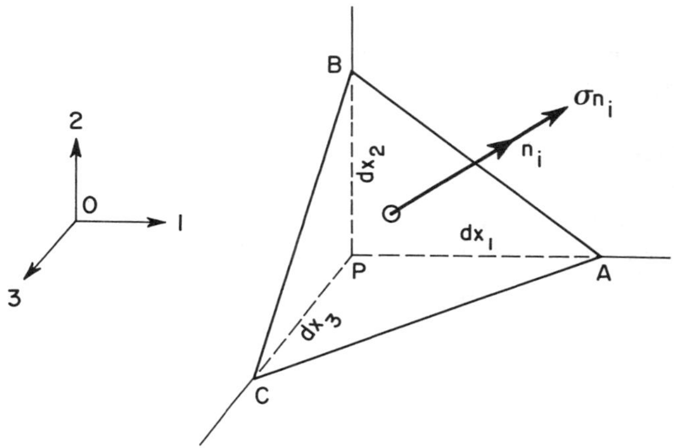

Fig. 2.7 A principal plane and corresponding principal stress.

Among all the planes passing through a point P, there can be one that has no shear traction. Such a plane is called a *principal plane*, the normal traction on the plane is known as a *principal stress*, and the direction of the outward normal vector is called a *principal direction* or *principal axis*. In Fig. 2.7, ABC is a principal plane with n_i as the unit outward normal vector, and σ is the value (scalar) of the principal stress. Following the stress vector formula, Eq. 2.6, we can write

$$\sigma n_i = \sigma_{ij} n_j,$$

or

$$(\sigma_{ij} - \sigma\delta_{ij})n_j = 0.$$

Since the matrix $[\sigma_{ij}]$ is symmetric, the analysis of Section 1.6 shows that there are three real values of σ, i.e. three principal stresses, denoted by σ_1, σ_2, σ_3, for which this equation has a non-trivial solution for n_j. These values, the eigenvalues, are the roots of the equation

$$\det[\sigma_{ij} - \sigma\delta_{ij}] = 0,$$

or

$$\sigma^3 - I_1\sigma^2 + I_2\sigma - I_3 = 0,$$

as given in Eq. 1.76. Stress invariants given by Eqs. 1.77 or Eqs. 2.17 can be expressed in terms of principal stresses as

$$I_1 = \sigma_1 + \sigma_2 + \sigma_3 \tag{2.18a}$$

$$I_2 = \sigma_1\sigma_2 + \sigma_2\sigma_3 + \sigma_3\sigma_1 \tag{2.18b}$$

$$I_3 = \sigma_1\sigma_2\sigma_3 \tag{2.18c}$$

If the eigenvalues are distinct, each eigenvalue defines a unique eigenvector n_j, and the three eigenvectors are mutually orthogonal. In other words, when all three principal stresses are distinct, there is a unique triad of mutually orthogonal principal planes, and a corresponding unique triad of mutually orthogonal principal axes.

On the other hand, if two principal stresses are equal, only the principal axis in the direction of the distinct principal stress is unique. Using principal axes as reference Cartesian coordinate axes we have

$$\sigma_{ij} = \sigma_1\delta_{i1}\delta_{j1} + \sigma_2(\delta_{i2}\delta_{j2} + \delta_{i3}\delta_{j3}) \tag{2.19}$$

or, in matrix form,

$$[\sigma_{ij}] = \begin{bmatrix} \sigma_1 & 0 & 0 \\ 0 & \sigma_2 & 0 \\ 0 & 0 & \sigma_2 \end{bmatrix},$$

where we suppose $\sigma_3 = \sigma_2 \neq \sigma_1$. In the primed coordinate system, the stress components are given by Eq. 2.16b, i.e.

$$\sigma'_{ij} = a_{1i}a_{1j}\sigma_1 + (a_{2i}a_{2j} + a_{3i}a_{3j})\sigma_2 \tag{2.20}$$

For x_1 and x'_1 being identical, and x'_2 and x'_3 being any pair of mutually orthogonal axes, the transformation tensor a_{ij} is

$$[a_{ij}] = \begin{bmatrix} 1 & 0 & 0 \\ 0 & a_{22} & a_{23} \\ 0 & a_{32} & a_{33} \end{bmatrix}.$$

Then, in view of the orthogonal property of a_{ij}, Eq. 1.25, we can show that Eq. 2.20 gives

$$[\sigma'_{ij}] = \begin{bmatrix} \sigma_1 & 0 & 0 \\ 0 & \sigma_2 & 0 \\ 0 & 0 & \sigma_2 \end{bmatrix},$$

so that the primed axes are also principal.

In the hydrostatic case, when all three principal stresses are equal, then, referred to principal axes

$$\sigma_{ij} = \sigma\delta_{ij} \tag{2.21}$$

or, in matrix form,

$$[\sigma_{ij}] = \begin{bmatrix} \sigma & 0 & 0 \\ 0 & \sigma & 0 \\ 0 & 0 & \sigma \end{bmatrix}.$$

For any other set of axes, Eq. 2.16b gives

$$\sigma'_{ij} = \sigma\delta_{ij}.$$

Thus any three mutually orthogonal axes form a set of principal axes.

2.6 STATIONARY SHEAR STRESSES

Using principal axes as the reference Cartesian coordinate axes, we can write the stress tensor σ_{ij} in the form

$$\sigma_{ij} = \sigma_1\delta_{i1}\delta_{j1} + \sigma_2\delta_{i2}\delta_{j2} + \sigma_3\delta_{i3}\delta_{j3} \tag{2.22}$$

or, in matrix form,

$$[\sigma_{ij}] = \begin{bmatrix} \sigma_1 & 0 & 0 \\ 0 & \sigma_2 & 0 \\ 0 & 0 & \sigma_3 \end{bmatrix} \tag{2.23}$$

The stress vector on a plane with the unit normal vector n_i, as shown in Fig. 2.8, is given by Eq. 2.6, and its components normal to and tangential to the plane are respectively

$$N_i = \overline{X}_k n_k n_i \tag{2.24}$$

$$S_i = \overline{X}_i - N_i \tag{2.25}$$

Substituting Eqs. 2.6 and 2.24 into Eq. 2.25, and using Eq. 2.22, we find

$$S_i = \sigma_1\delta_{i1}n_1 + \sigma_2\delta_{i2}n_2 + \sigma_3\delta_{i3}n_3$$

$$- (\sigma_1 n_1^2 + \sigma_2 n_2^2 + \sigma_3 n_3^2)n_i \tag{2.26}$$

The square of the magnitude of the shear traction is

$$S_i S_i = (\sigma_1^2 n_1^2 + \sigma_2^2 n_2^2 + \sigma_3^2 n_3^2)$$

$$- (\sigma_1 n_1^2 + \sigma_2 n_2^2 + \sigma_3 n_3^2)^2 \tag{2.27}$$

The components of the unit vector n_i satisfy

$$n_3^2 = 1 - (n_1^2 + n_2^2) \tag{2.28}$$

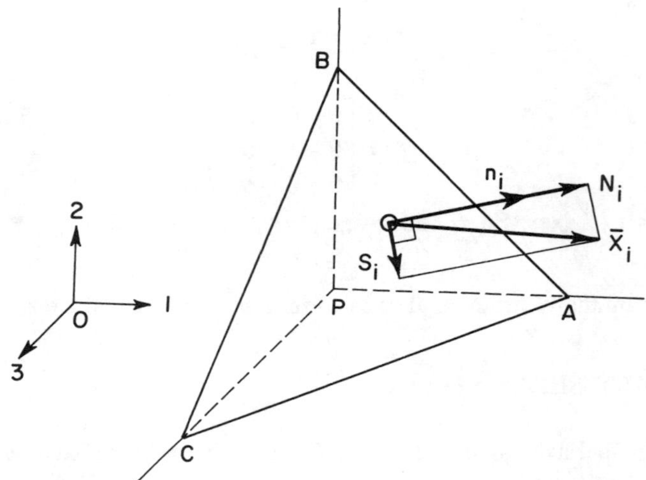

Fig. 2.8 Principal axes as reference coordinate axes for plane *ABC* with normal vector n_i.

Eliminating n_3^2 from Eq. 2.27, taking derivatives of the result S_iS_i with respect to n_1 and n_2, then equating these derivatives to zero; we obtain

$$\begin{bmatrix} (\sigma_1-\sigma_3)n_1 & (\sigma_2-\sigma_3)n_1 \\ (\sigma_1-\sigma_3)n_2 & (\sigma_2-\sigma_3)n_2 \end{bmatrix} \begin{Bmatrix} n_1^2 \\ n_2^2 \end{Bmatrix} = \frac{1}{2} \begin{Bmatrix} n_1(\sigma_{11}-\sigma_{33}) \\ n_2(\sigma_{22}-\sigma_{33}) \end{Bmatrix} \tag{2.29}$$

The direction n_i are specified by n_1 and n_2, which are governed by the equation above. We seek n_i for which S_iS_i is a *maximum* or *minimum*, or more generally, *stationary*. One obvious solution to Eq. 2.29 is that $n_1 = n_2 = 0$, which is the case of a principal stress σ_3. Another possible solution is that $n_1 = 0$, $n_2 = \pm 1/\sqrt{2}$ and $n_3 \pm 1/\sqrt{2}$, for which Eq. 2.27 gives

$$(S_iS_i)^{1/2} = \pm\frac{1}{2}(\sigma_2 - \sigma_3) \tag{2.30}$$

and Eq. 2.24 gives

$$N_i = \frac{1}{2}(\sigma_2 + \sigma_3) \tag{2.31}$$

Investigating for all possible solutions, we obtain, as compiled in Table 2.1, the three cases of principal stresses, and three cases of stationary shear stresses. Note that a stationary shear stress occurs on a plane bisecting the angle between two principal directions and has a value equal to half the difference between those two principal stresses. In fact, each stationary shear stress may be obtained individually from a

Table 2.1

n_1 n_2 n_3	± 1 0 0	0 ± 1 0	0 0 ± 1	0 $\pm 1/\sqrt{2}$ $\pm 1/\sqrt{2}$	$\pm 1/\sqrt{2}$ 0 $\pm 1/\sqrt{2}$	$\pm 1/\sqrt{2}$ $\pm 1/\sqrt{2}$ 0
				Stationary shear stresses		
Shear Stress	0	0	0	$\pm\dfrac{\sigma_2-\sigma_3}{2}$	$\pm\dfrac{\sigma_3-\sigma_1}{2}$	$\pm\dfrac{\sigma_1-\sigma_2}{2}$
	Principal stresses					
Normal Stress	σ_1	σ_2	σ_3	$\dfrac{\sigma_2+\sigma_3}{2}$	$\dfrac{\sigma_3+\sigma_1}{2}$	$\dfrac{\sigma_1+\sigma_2}{2}$

two-dimensional problem with two principal axes forming a plane of reference. As an example, consider the fifth column in Table 2.1, and take axes $1'$ and $2'$ as shown in Fig. 2.9 and the $3'$-axis along the 1-axis. The equilibrium of forces acting on the element *PAB* in $1'$ and $2'$-directions give, respectively, the normal stress σ'_{11} and the stationary shear stress σ'_{12} the listed values in the fifth column, i.e. $\sigma'_{11}=(\sigma_2+\sigma_3)/2$ and $\sigma'_{12}=(\sigma_3-\sigma_2)/2$.

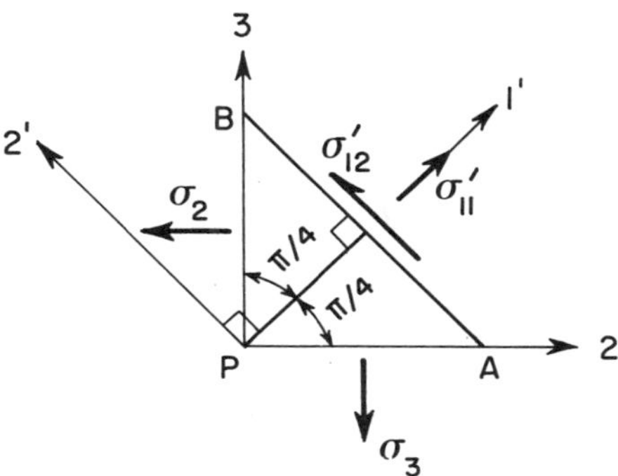

Fig. 2.9 A stationary shear stress treated as a two-dimensional problem.

2.7 OCTAHEDRAL SHEAR STRESS AND STRESS DEVIATOR

There can be eight planes making an equal angle with each principal direction as shown in Fig. 2.10. Each such plane is called an *octahedral plane*, and has direction cosines, with respective to the principal axes, given by

$$n_1^2 = n_2^2 = n_3^2 = \frac{1}{3} \tag{2.32}$$

Equation 2.27 shows that the magnitude of the shear stress on each octahedral plane, called the *octahedral shear stress* τ_0, is given by

$$9\tau_0^2 = (\sigma_1 - \sigma_2)^2 + (\sigma_2 - \sigma_3)^2 + (\sigma_3 - \sigma_1)^2 \tag{2.33}$$

We shall now show that τ_0^2 is a stress invariant.

We introduce the *stress deviator*

$$s_{ij} = \sigma_{ij} - \frac{1}{3}\sigma_{kk}\delta_{ij} \tag{2.34}$$

This equation, which may be rewritten as

$$\sigma_{ij} = s_{ij} + \frac{1}{3}\sigma_{kk}\delta_{ij} \tag{2.35}$$

shows the stress tensor σ_{ij} decomposed into a *deviatoric stress tensor* s_{ij} and a *hydrostatic* or *spherical stress tensor* $\sigma_{kk}\delta_{ij}/3$. We note that σ_{kk} is identical to the first stress invariant I_1. Equation 2.34 may be rewritten in matrix form

$$[s_{ij}] = \begin{bmatrix} \sigma_{11} - \frac{\sigma_{kk}}{3} & \sigma_{12} & \sigma_{13} \\ \sigma_{21} & \sigma_{22} - \frac{\sigma_{kk}}{3} & \sigma_{23} \\ \sigma_{31} & \sigma_{32} & \sigma_{33} - \frac{\sigma_{kk}}{3} \end{bmatrix} \tag{2.36}$$

Note that s_{ij} is a symmetric second order tensor i.e.

$$s_{ij} = s_{ji} \tag{2.37}$$

Now the *first*, *second* and *third deviatoric stress invariants* are, respectively, defined as the following

$$J_1 \equiv s_{ii} = 0 \tag{2.38}$$

$$J_2 = \frac{1}{2}s_{ij}s_{ij} \tag{2.39}$$

$$J_3 = \frac{1}{3}s_{ij}s_{jk}s_{ki} \tag{2.40}$$

While J_1 is identically zero, J_2 and J_3 can be expressed in terms of other stress invariants as

$$J_2 = \frac{I_1^2}{3} - I_2 \tag{2.41a}$$

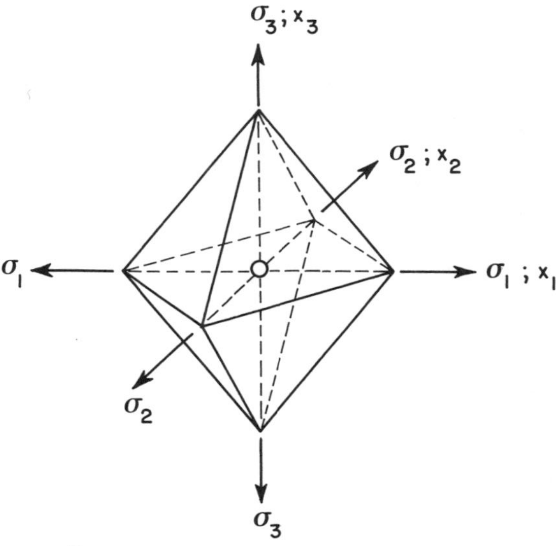

Fig. 2.10 Octahedral planes.

$$J_3 = I_3 - \frac{I_1 I_2}{3} + \frac{2I_1^3}{27} \tag{2.41b}$$

Thus we obtain the relationship between the octahedral shear stress and the second deviatoric stress invariant in the form

$$J_2 = \frac{3}{2}\tau_0^2 \tag{2.42}$$

In 1913, Richard von Mises proposed that yielding of some materials occurs at a constant value of J_2, then Nadai (1915) apparently because of Eq. 2.42 interpreted that the yielding occurs at a constant value of the octahedral shear stress.

2.8 STRAIN TENSOR

As usual, let us begin with a two-dimensional case with x_1 and x_2 as reference axes, and consider the situation that a point P located by (x_1, x_2) at a reference time moves to a point P' located by (x_1+u_1, x_2+u_2) at any time t. In other words u_1 and u_2 are *displacements* of the point P in x_1 and x_2-directions respectively. Similarly, other points Q and R infinitesimally close to P move to points Q' and R' respectively as shown in Fig. 2.11.

From the figure, we may notice 'small deformation' as follows; the change of the line element PQ per unit original length is approximately $\partial u_1/\partial x_1$, the change of the

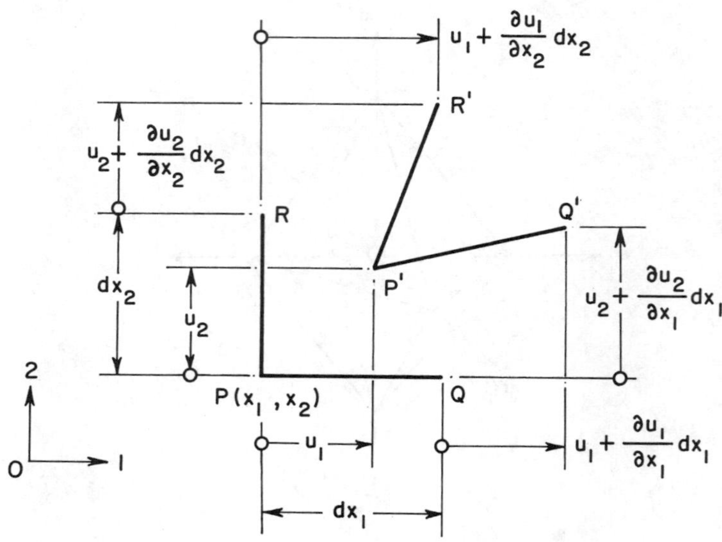

Fig. 2.11 Displacement and deformation in a two-dimensional case.

line element PR per unit original length is approximately $\partial u_2/\partial x_2$ and the change of the angle QPR is approximately $\partial u_1/\partial x_2 + \partial u_2/\partial x_1$. Denote these changes by new symbols as follows

$$\varepsilon_{11} = \frac{\partial u_1}{\partial x_1} \tag{2.43a}$$

$$\varepsilon_{22} = \frac{\partial u_2}{\partial x_2} \tag{2.43b}$$

$$\varepsilon_{12} = \frac{1}{2}\left(\frac{\partial u_1}{\partial x_2} + \frac{\partial u_2}{\partial x_1}\right) \tag{2.43c}$$

In their common names, ε_{11} is the *normal strain* in x_1-direction, ε_{22} the *normal strain* in x_2-direction, and ε_{12} the *shear strain* in x_1x_2-plane. Note, however, the shear strain defined above is only half of the whole angle change. The latter (whole angle change) is normally called an *engineering shear strain.*

For three-dimensional cases, strains can be put in terms of displacement components u_i as follows

$$\varepsilon_{ij} = \frac{1}{2}(u_{i,j} + u_{j,i}) \tag{2.44}$$

and *engineering shear strains*

$$\gamma_{12} = u_{1,2} + u_{2,1} \tag{2.45a}$$

$$\gamma_{23} = u_{2,3} + u_{3,2} \tag{2.45b}$$

$$\gamma_{31} = u_{3,1} + u_{1,3} \tag{2.45c}$$

Thus, we have symmetry in strains, i.e.

$$\varepsilon_{ij} = \varepsilon_{ji} \tag{2.46a}$$

and

$$\gamma_{ij} = \gamma_{ji} \tag{2.46b}$$

which means that there are six (*not* nine) distinct strain components. It was shown in Section 1.7 that differentiation of a tensor with respect to x_i yields tensor of one order higher if the index i remains a free index. The quantity ε_{ij}, consisting of such a derivative of the displacement vector, is therefore a tensor of second order. So all ideas developed for σ_{ij}, e.g. stress invariants, principal stresses, etc., have counterparts for ε_{ij}, i.e. strain invariants, principal strains, etc.

In particular, the first strain invariant ε_{kk} is called the *dilatation* since it represents the proportional volume change of an infinitesimal element, and the *strain deviator* is defined by

$$e_{ij} = \varepsilon_{ij} - \frac{1}{3}\varepsilon_{kk}\delta_{ij} \tag{2.47}$$

The tensor defined by

$$\omega_{ij} = \frac{1}{2}(u_{i,j} - u_{j,i}) \tag{2.48}$$

is the *rotation tensor*, since each component of it represents an average angle of rotation about a coordinate axis. In Fig. 2.11, the line PQ rotates through an angle $\partial u_2/\partial x_1$ about the x_3-axis to become $P'Q'$ and the line PR rotates through an angle $-\partial u_1/\partial x_2$ about the same axis to become $P'R'$. Hence the average rotation of the material body framed by lines PQ and PR about the x_3-axis is $(\partial u_2/\partial x_1 - \partial u_1/\partial x_2)/2 = -\omega_{12}$. The rotation tensor is *antisymmetric*, i.e.

$$\omega_{ij} = -\omega_{ji} \tag{2.49}$$

In a matrix form, $[\omega_{ij}]$ has zero elements on its main diagonal, since

$$\omega_{11} = \omega_{22} = \omega_{33} = 0 \tag{2.50}$$

Thus ω_{ij} has only three distinct components corresponding to the angles of rotation about x_1, x_2 and x_3 axes. The *rotation vector* ω_i having these angles as its Cartesian components can be written as

$$\omega_i = \frac{1}{2}\varepsilon_{ijk}\omega_{kj} \tag{2.51a}$$

$$= \frac{1}{4}\varepsilon_{ijk}(u_{k,j} - u_{j,k}) \tag{2.51b}$$

or, more explicitly,

$$\omega_1 = \omega_{32} = \frac{1}{2}(u_{3,2} - u_{2,3}) \tag{2.52a}$$

$$\omega_2 = \omega_{13} = \frac{1}{2}(u_{1,3} - u_{3,1}) \tag{2.52b}$$

$$\omega_3 = \omega_{21} = \frac{1}{2}(u_{2,1} - u_{1,2}) \tag{2.52c}$$

When there is displacement, but there is zero strain and constant rotation, the material body is displaced as a rigid body. The Cartesian components of the *rigid-body displacement* vector will be given by

$$u_i = C_i + \varepsilon_{ijk} B_j x_k \tag{2.53}$$

where B_i and C_i are arbitrary constants. Thus

$$u_{i,j} = \varepsilon_{imj} B_m, \qquad u_{j,i} = \varepsilon_{jmi} B_m,$$

so that

$$\varepsilon_{ij} = 0, \qquad \omega_i = B_i.$$

2.9 COMPATIBILITY CONDITIONS

If we take the strain components as continuous functions of coordinates of material points, the deformation will be single-valued and continuous only if the following conditions are satisfied

$$\varepsilon_{ij,kl} + \varepsilon_{kl,ij} - \varepsilon_{ik,jl} - \varepsilon_{jl,ik} = 0 \tag{2.54}$$

which are known as *compatibility conditions*. The proof that these conditions are necessary can be made simply by differentiating Eq. 2.44 which is the definition of the strain, but the proof of their sufficiency is more involved, as will be shown next, following Cesaro (1906) and Sokolnikoff (1956).

Let $P^0(x_1^0, x_2^0, x_3^0)$ be a point at which displacements u_i and ε_{ij} are known. Then the displacement of any other point P^a in the domain can be obtained by a line integral along a continuous curve C from P^0 to P^a as follows

$$u_i^a = u_i^0 + \int_{P^0}^{P^a} \mathrm{d}u_i$$

$$= u_i^0 + \int_{P^0}^{P^a} u_{i,j}\mathrm{d}x_j$$

$$= u_i^0 + \int_{P^0}^{P^a} \varepsilon_{ij}\mathrm{d}x_j + \int_{P^0}^{P^a} \omega_{ij}\mathrm{d}x_j \tag{2.55}$$

The derivative of the rotation tensor ω_{ij}, defined in Eq. 2.48, with respect to x_k is

$$\omega_{ij,k} = \frac{1}{2}(u_{i,jk} - u_{j,ik})$$

$$= \frac{1}{2}(u_{i,kj} - u_{k,ij}) - \frac{1}{2}(u_{j,ki} - u_{k,ji})$$

$$= \varepsilon_{ik,j} - \varepsilon_{jk,i} \tag{2.56}$$

In the last integral in Eq. 2.55, we can replace dx_j by $d(x_j - x_j^a)$, then integrate by parts, to obtain

$$\int_{P^0}^{P^a} \omega_{ij} dx_j = \omega_{ij}^0 (x_j^a - x_j^0) - \int_{P^0}^{P^a} (x_j - x_j^a) \omega_{ij,k} dx_k,$$

or, due to Eq. 2.56,

$$\int_{P^0}^{P^a} \omega_{ij} dx_j = \omega_{ij}^0 (x_j^a - x_j^0) + \int_{P^0}^{P^a} (x_j^a - x_j)(\varepsilon_{ik,j} - \varepsilon_{jk,i}) dx_k.$$

Substituting the last equation into Eq. 2.55 yields

$$u_i^a = u_i^0 + \omega_{ij}^0 (x_j^a - x_j^0) + \int_{P^0}^{P^a} U_{ik} dx_k \tag{2.57}$$

where

$$U_{ik} = \varepsilon_{ik} + (x_j^a - x_j)(\varepsilon_{ik,j} - \varepsilon_{jk,i}) \tag{2.58}$$

For singlevaluedness and continuity of u_i, the integral in Eq. 2.57 must be independent of the integrating path C. This is possible only if $U_{ik} dx_k$ is an exact differential, say

$$U_{ik} dx_k = dV_i$$

$$= V_{i,k} dx_k$$

or, by expansion,

$$U_{i1} dx_1 + U_{i2} dx_2 + U_{i3} dx_3 = V_{i,1} dx_1 + V_{i,2} dx_2 + V_{i,3} dx_3.$$

Then, since x_1, x_2 and x_3 are independent variables, the last equation implies that

$$U_{i1} = V_{i,1},$$

$$U_{i2} = V_{i,2},$$

$$U_{i3} = V_{i,3},$$

or, in a compact form,

$$U_{ij} = V_{i,j} \tag{2.59}$$

Differentiating the equation above with respect to x_k yields

$$U_{ij,k} = V_{i,jk} \tag{2.60}$$

but $V_{i,jk}$ in the last equation is identical to $V_{i,kj}$, hence we have

$$U_{ij,k} = U_{ik,j} \tag{2.61}$$

Substituting Eq. 2.58 into Eq. 2.61 yields

$$\varepsilon_{ik,l} - \delta_{jl}(\varepsilon_{ik,j} - \varepsilon_{jk,i}) + (x_j^a - x_j)(\varepsilon_{ik,jl} - \varepsilon_{jk,il})$$

$$= \varepsilon_{il,k} - \delta_{jk}(\varepsilon_{il,j} - \varepsilon_{jl,i}) + (x_j^a - x_j)(\varepsilon_{il,jk} - \varepsilon_{jl,ik}),$$

or

$$(x_j^a - x_j)(\varepsilon_{ik,jl} - \varepsilon_{jk,il} - \varepsilon_{il,jk} + \varepsilon_{jl,ik}) = 0.$$

Since $x_j^a - x_j$ is in general not zero, its factor in the equation above must vanish leading to the sufficiency (and necessity as well) of the compatibility conditions as defined in Eq. 2.54.

Actually, only six of the total eighty-one compatibility conditions are independent, while the rest are identities or repetitions due to the symmetry of the strain tensor. These six distinct conditions can be easily identified by contracting Eq. 2.54 with respect to two indices, say k and l, upon which we get more compact form of compatibility conditions as

$$\varepsilon_{ij,kk} + \varepsilon_{kk,ij} - \varepsilon_{ik,jk} - \varepsilon_{jk,ik} = 0,$$

which is symmetric with respect to the remaining two free indices i and j. Moreover, writing the equation above explicitly as a set of six independent equations, we can say that the six independent equations can be obtained from the equation above even with k not summed. In conclusion, the most compact form for the compatibility conditions is

$$\varepsilon_{ij,kk} + \varepsilon_{kk,ij} - \varepsilon_{ik,jk} - \varepsilon_{jk,ik} = 0, \qquad (k \neq i, k \neq j, k \text{ not summed}) \tag{2.62}$$

A more formal derivation of Eq. 2.62 from Eq. 2.54 can start by defining R_{ijkl} as the absolute value of the left-hand side of Eq. 2.54, i.e.

$$R_{ijkl} = |\,\varepsilon_{ij,kl} + \varepsilon_{kl,ij} - \varepsilon_{ik,jl} - \varepsilon_{jl,ik}|\,.$$

The tensor R_{ijkl} has following symmetry properties

$$R_{ijkl} = R_{klij} = R_{jikl} = R_{ijlk} = R_{ikjl} = R_{ljki}.$$

Moreover, R_{ijkl} vanishes if more than two of its subscripts are equal, i.e.

$$R_{1111}, R_{2223}, \ldots \equiv 0.$$

Hence Eq. 2.54 is equivalent to the following distinct equations,

$$R_{1122} = 0, \quad R_{2233} = 0, \quad R_{3311} = 0,$$

$$R_{2311} = 0, \quad R_{3122} = 0, \quad R_{1233} = 0.$$

which can be put collectively as Eq. 2.62. The tensor R_{ijkl} is the *Riemann-Christoffel covariant curvature tensor* to be described in Section 10.11.

2.10 CYLINDRICAL AND SPHERICAL COORDINATES

The most frequently used curvilinear coordinate systems are the cylindrical coordinate (r, θ, z) and the spherical coordinate (R, ϕ, θ) systems as shown in Fig. 2.12. Both of them are orthogonal curvilinear coordinate systems. Although the governing equations for these systems may be found using the general tensor analysis for orthogonal curvilinear systems, it is convenient to derive them in explicit (not indicial) notation.

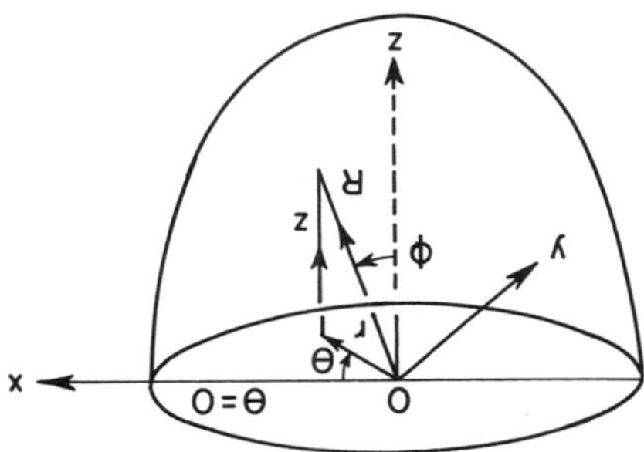

Fig. 2.12 Cartesian (x, y, z), cylindrical (r, θ, z) and spherical (R, ϕ, θ) coordinates.

Cartesian and cylindrical coordinate systems are related by

$$x = r\cos\theta, \quad y = r\sin\theta, \quad z = z \tag{2.63a,b,c}$$

which may be reverted to give

$$r = +(x^2 + y^2)^{1/2}, \quad \theta = \tan^{-1}\frac{y}{x}, \quad z = z \tag{2.64a,b,c}$$

The positive sign in front of the square root in Eq. 2.64a denotes that r always assumes a non-negative value. The derivatives of cylindrical coordinates with respect to Cartesian coordinates are

$$\frac{\partial r}{\partial x} = \frac{x}{r} = \cos\theta, \quad \frac{\partial r}{\partial y} = \frac{y}{r} = \sin\theta \tag{2.65a,b}$$

$$\frac{\partial \theta}{\partial x} = -\frac{y}{r^2} = -\frac{\sin\theta}{r}, \quad \frac{\partial \theta}{\partial y} = \frac{x}{r^2} = \frac{\cos\theta}{r} \tag{2.65c,d}$$

Any derivative in Cartesian equations can be put in terms of cylindrical coordinates as

$$\frac{\partial}{\partial x} = \frac{\partial}{\partial r}\frac{\partial r}{\partial x} + \frac{\partial}{\partial \theta}\frac{\partial \theta}{\partial x} = \cos\theta\frac{\partial}{\partial r} - \frac{\sin\theta}{r}\frac{\partial}{\partial \theta} \tag{2.66a}$$

$$\frac{\partial}{\partial y}=\frac{\partial}{\partial r}\frac{\partial r}{\partial y}+\frac{\partial}{\partial \theta}\frac{\partial \theta}{\partial y}=\sin\theta\frac{\partial}{\partial r}+\frac{\cos\theta}{r}\frac{\partial}{\partial \theta} \tag{2.66b}$$

Since r, θ and z form a right-hand orthogonal system, the rules for transformation among Cartesian tensor components and cylindrical tensor components described in Section 1.4 hold. As an example, the relationship among displacement components with reference to these two different coordinate systems can be written as

$$\{u_i\} = [a_{ij}]\,\{u_j'\} \tag{2.67a}$$

or

$$\{u_i'\} = [a_{ji}]^T\,\{u_j\} \tag{2.67b}$$

where

$$\{x_i\} = \begin{Bmatrix} x \\ y \\ z \end{Bmatrix}, \quad \{x_i'\} = \begin{Bmatrix} r \\ \theta \\ z \end{Bmatrix} \tag{2.68a,b}$$

$$\{u_i\} = \begin{Bmatrix} u \\ v \\ w \end{Bmatrix}, \quad \{u_i'\} = \begin{Bmatrix} u_r \\ u_\theta \\ u_z \end{Bmatrix} \tag{2.68c,d}$$

$$[a_{ij}] = \begin{bmatrix} \cos\theta & -\sin\theta & 0 \\ \sin\theta & \cos\theta & 0 \\ 0 & 0 & 1 \end{bmatrix} \tag{2.68e}$$

Figure 2.13 shows the decomposition of a vector **u** into alternative orthogonal components u, v and u_r, u_θ, thus reconfirming Eqs. 2.67. For tensors of second order, e.g. the strain tensor, we have

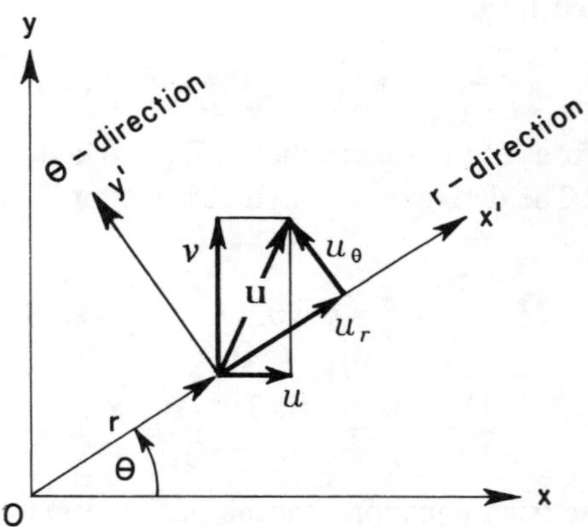

Fig. 2.13 Orthogonal components of a vector **u** in xy or $r\theta$ plane.

$$[\varepsilon'_{ij}] = [a_{ki}]^T [\varepsilon_{kl}] [a_{lj}] \tag{2.69}$$

where

$$[\varepsilon'_{ij}] = \begin{bmatrix} \varepsilon_{rr} & \varepsilon_{r\theta} & \varepsilon_{rz} \\ & \varepsilon_{\theta\theta} & \varepsilon_{\theta z} \\ \text{Symm.} & & \varepsilon_{zz} \end{bmatrix} \tag{2.70}$$

Substituting Eq. 2.44, in view of Eqs. 2.66 to 2.68 and Eqs. 2.70, into Eq. 2.69 yields the strain-displacement relationship in cylindrical coordinates as

$$\varepsilon_{rr} = \frac{\partial u_r}{\partial r}, \quad \varepsilon_{\theta\theta} = \frac{1}{r}\left(u_r + \frac{\partial u_\theta}{\partial \theta}\right), \quad \varepsilon_{zz} = \frac{\partial u_z}{\partial z} \tag{2.71a,b,c}$$

$$\varepsilon_{r\theta} = \varepsilon_{\theta r} = \frac{1}{2}\left[\frac{1}{r}\left(\frac{\partial u_r}{\partial \theta} - u_\theta\right) + \frac{\partial u_\theta}{\partial r}\right] \tag{2.71d}$$

$$\varepsilon_{\theta z} = \varepsilon_{z\theta} = \frac{1}{2}\left(\frac{1}{r}\frac{\partial u_z}{\partial \theta} + \frac{\partial u_\theta}{\partial z}\right) \tag{2.71e}$$

$$\varepsilon_{zr} = \varepsilon_{rz} = \frac{1}{2}\left(\frac{\partial u_r}{\partial z} + \frac{\partial u_z}{\partial r}\right) \tag{2.71f}$$

The dilatation, which is a strain invariant, can be written

$$\varepsilon_{kk} = \varepsilon_{rr} + \varepsilon_{\theta\theta} + \varepsilon_{zz} = \frac{1}{r}\left[\frac{\partial}{\partial r}(ru_r) + \frac{\partial u_\theta}{\partial \theta}\right] + \frac{\partial u_z}{\partial z} \tag{2.72}$$

Substituting the Cartesian rigid-body displacement components, Eq. 2.53, into Eq. 2.67b leads to the cylindrical rigid-body displacement components

$$u_r = C_1 \cos\theta + C_2 \sin\theta + z(-B_1 \sin\theta + B_2 \cos\theta) \tag{2.73a}$$

$$u_\theta = -C_1 \sin\theta + C_2 \cos\theta + B_3 r - z(B_1 \cos\theta + B_2 \sin\theta) \tag{2.73b}$$

$$u_z = C_3 + r(-B_2 \cos\theta + B_1 \sin\theta) \tag{2.73c}$$

By similar procedures, the Cartesian equation of motion, Eq. 2.14, can be transformed to obtain those in r, θ and z directions, respectively as

$$\frac{\partial \sigma_{rr}}{\partial r} + \frac{1}{r}\frac{\partial \sigma_{r\theta}}{\partial \theta} + \frac{\partial \sigma_{rz}}{\partial z} + \frac{\sigma_{rr} - \sigma_{\theta\theta}}{r} + X_r = 0 \tag{2.74a}$$

$$\frac{\partial \sigma_{r\theta}}{\partial r} + \frac{1}{r}\frac{\partial \sigma_{\theta\theta}}{\partial \theta} + \frac{\partial \sigma_{\theta z}}{\partial z} + \frac{2}{r}\sigma_{r\theta} + X_\theta = 0 \tag{2.74b}$$

$$\frac{\partial \sigma_{rz}}{\partial r} + \frac{1}{r}\frac{\partial \sigma_{\theta z}}{\partial \theta} + \frac{\partial \sigma_{zz}}{\partial z} + \frac{1}{r}\sigma_{rz} + X_z = 0 \tag{2.74c}$$

in which X_r, X_θ and X_z are cylindrical components of the body force, thus related to the Cartesian components by

$$\{X_i'\} = [a_{ji}]^T \{X_j\} \tag{2.75}$$

where

$$\{X_j'\} = (X_r \quad X_\theta \quad X_z)^T \tag{2.76}$$

Figure 2.14 depicts positive cylindrical components of the stress tensor.

Cylindrical and spherical coordinate systems are related by

$$r = R\sin\phi\,, \quad \theta = \theta\,, \quad z = R\cos\phi \qquad \text{(2.77a,b,c)}$$

In reverse,

$$R = +(r^2 + z^2)^{1/2} = +(x^2 + y^2 + z^2)^{1/2} \tag{2.78a}$$

$$\phi = \tan^{-1}\frac{r}{z}\,, \quad \theta = \theta \qquad \text{(2.78b,c)}$$

Note that R and ϕ are non-negative, and the maximum value of ϕ is π.

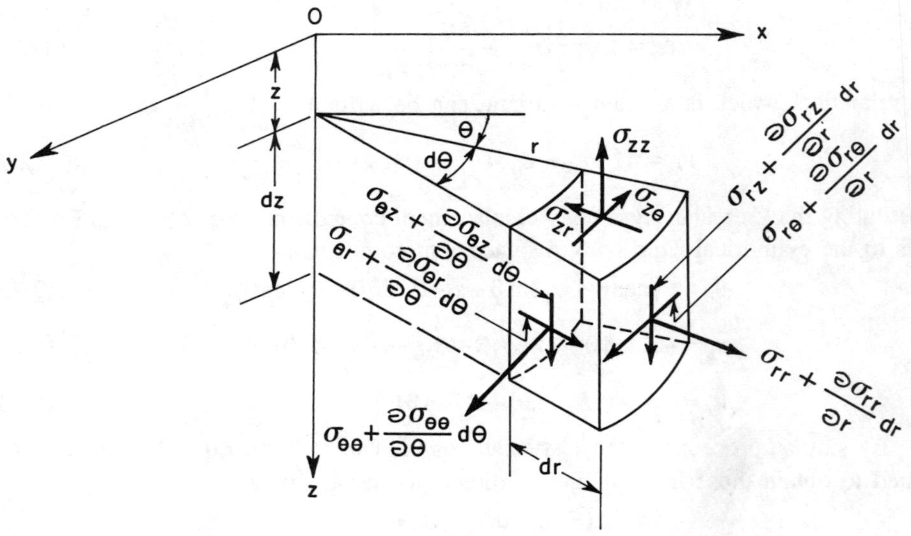

Fig. 2.14 Positive cylindrical components of stress tensor.

Spherical and cylindrical components of displacements and body forces can be related by

$$u_R = u_r \sin\phi + u_z \cos\phi \tag{2.79a}$$

$$u_\phi = u_r \cos\phi - u_z \sin\phi \tag{2.79b}$$

$$u_\theta = u_\theta \tag{2.79c}$$

$$X_R = X_r \sin\phi + X_z \cos\phi \tag{2.80a}$$

$$X_\phi = X_r \cos\phi - X_z \sin\phi \tag{2.80b}$$

$$X_\theta = X_\theta \tag{2.80c}$$

Figure 2.15 shows the decomposition of a vector **u** into its cylindrical and spherical components, thus reconfirming Eqs. 2.79 and 2.80. The strain-displacement relationship can be written

$$\varepsilon_{RR} = \frac{\partial u_R}{\partial R}, \quad \varepsilon_{\phi\phi} = \frac{1}{R}\left(u_R + \frac{\partial u_\phi}{\partial \phi}\right) \tag{2.81a,b}$$

$$\varepsilon_{\theta\theta} = \frac{1}{R}\left(\operatorname{cosec}\phi \frac{\partial u_\theta}{\partial \theta} + u_R + u_\phi \cot\phi\right) \tag{2.81c}$$

$$\varepsilon_{R\phi} = \varepsilon_{\phi R} = \frac{1}{2}\left[\frac{1}{R}\left(\frac{\partial u_R}{\partial \phi} - u_\phi\right) + \frac{\partial u_\phi}{\partial R}\right] \tag{2.81d}$$

$$\varepsilon_{\phi\theta} = \varepsilon_{\theta\phi} = \frac{1}{2R}\left(\frac{\partial u_\theta}{\partial \phi} - u_\theta \cot\phi + \operatorname{cosec}\phi \frac{\partial u_\phi}{\partial \theta}\right) \tag{2.81e}$$

$$\varepsilon_{\theta R} = \varepsilon_{R\theta} = \frac{1}{2}\left[\frac{1}{R}\left(\operatorname{cosec}\phi \frac{\partial u_R}{\partial \theta} - u_\theta\right) + \frac{\partial u_\theta}{\partial R}\right] \tag{2.81f}$$

The dilatation is

$$\begin{aligned}\varepsilon_{kk} &= \varepsilon_{RR} + \varepsilon_{\phi\phi} + \varepsilon_{\theta\theta} \\ &= \frac{1}{R^2}\frac{\partial}{\partial R}(R^2 u_R) + \frac{\operatorname{cosec}\phi}{R}\left[\frac{\partial}{\partial \phi}(u_\phi \sin\phi) + \frac{\partial u_\theta}{\partial \theta}\right]\end{aligned} \tag{2.82}$$

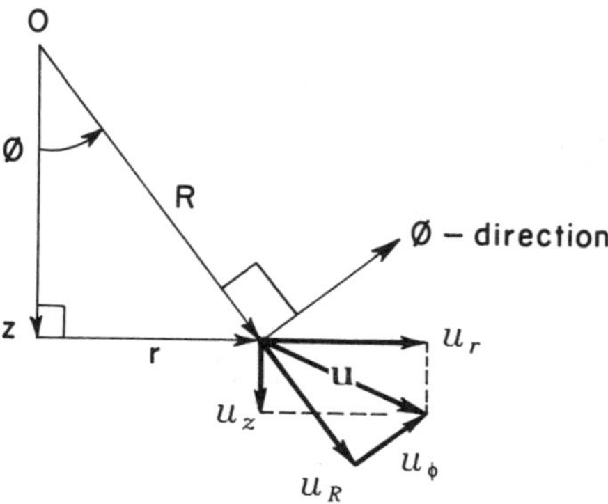

Fig. 2.15 Orthogonal components of a vector **u** in rz or $R\phi$ plane.

The spherical rigid-body displacement components are

$$u_R = \sin\phi(C_1\cos\theta + C_2\sin\theta) + C_3\cos\phi \tag{2.83a}$$

$$u_\phi = \cos\phi(C_1\cos\theta + C_2\sin\theta) - C_3\sin\phi - B_1 R\sin\theta + B_2 R\cos\theta \tag{2.83b}$$

$$u_\theta = -C_1\sin\theta + C_2\cos\theta - R\cos\phi(B_1\cos\theta + B_2\sin\theta) + B_3 R\sin\phi \tag{2.83c}$$

where B_i and C_i are given in Eq. 2.53. The equations of equilibrium in R, ϕ and θ directions are, respectively

$$\frac{\partial\sigma_{RR}}{\partial R} + \frac{1}{R\sin\phi}\frac{\partial\sigma_{R\theta}}{\partial\theta} + \frac{1}{R}\frac{\partial\sigma_{R\phi}}{\partial\phi} + \frac{2\sigma_{RR} - \sigma_{\theta\theta} - \sigma_{\phi\phi} - \sigma_{R\phi}\cot\phi}{R} + X_R = 0 \tag{2.84a}$$

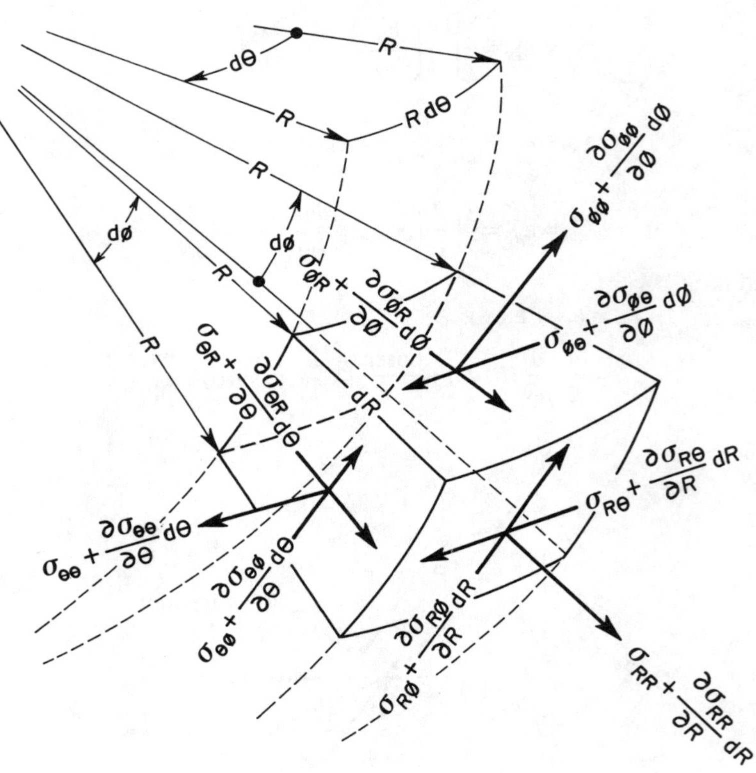

Fig. 2.16 Positive spherical components of stress tensor.

$$\frac{\partial \sigma_{R\phi}}{\partial R}+\frac{1}{R\sin\phi}\frac{\partial \sigma_{\theta\phi}}{\partial \theta}+\frac{1}{R}\frac{\partial \sigma_{\phi\phi}}{\partial \phi}+\frac{3\sigma_{R\phi}+(\sigma_{\phi\phi}-\sigma_{\theta\theta})\cot\phi}{R}+X_{\phi}=0 \tag{2.84b}$$

$$\frac{\partial \sigma_{R\theta}}{\partial R}+\frac{1}{R\sin\phi}\frac{\partial \sigma_{\theta\theta}}{\partial \theta}+\frac{1}{R}\frac{\partial \sigma_{\phi\theta}}{\partial \phi}+\frac{3\sigma_{R\theta}+2\sigma_{\theta\phi}\cot\phi}{R}+X_{\theta}=0 \tag{2.84c}$$

Figure 2.16 depicts positive spherical components of the stress tensor.

2.11 PROBLEMS

2.11.1 *Stress or Strain Computation from Three Different In-Plane Normal Components*

When the axes x_1 and x_2 are rotated about the x_3 axis through an angle θ to become axes x_1' and x_2' as shown in Fig 2.17, the 'in-plane' strain components of the primed coordinate system can be shown to be in terms of those of the nonprimed one as

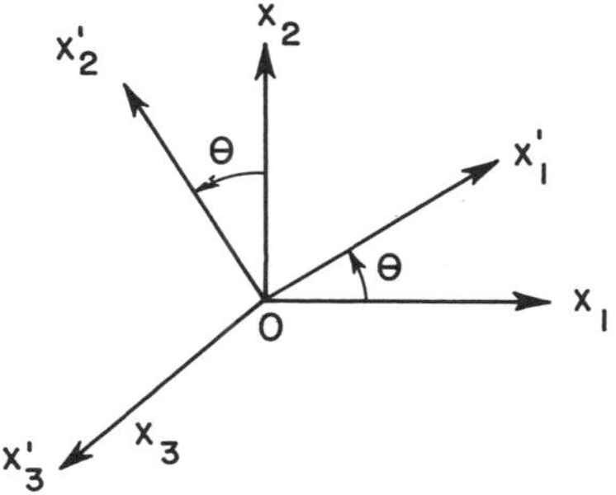

Fig. 2.17

$$\varepsilon_{11}' = \cos^2\theta\varepsilon_{11}+\sin 2\theta\varepsilon_{12}+\sin^2\theta\varepsilon_{22} \tag{2.85a}$$

$$\varepsilon_{22}' = \sin^2\theta\varepsilon_{11}-\sin 2\theta\varepsilon_{12}+\cos^2\theta\varepsilon_{22} \tag{2.85b}$$

$$\varepsilon_{12}' = \frac{1}{2}\sin 2\theta(\varepsilon_{22}-\varepsilon_{11})+\cos 2\theta\varepsilon_{12} \tag{2.85c}$$

while the 'out-of-plane' components are

$$\varepsilon_{33}' = \varepsilon_{33} \tag{2.85d}$$

$$\varepsilon_{23}' = \cos\theta\varepsilon_{23}-\sin\theta\varepsilon_{31} \tag{2.85e}$$

$$\varepsilon_{31}' = \sin\theta\varepsilon_{23}+\cos\theta\varepsilon_{31} \tag{2.85f}$$

The equations above hold also for stress components, i.e. if ε is replaced by σ in those equations. In practice, it is not easy to measure a shear component directly. Instead, normal strain components are measured in a plane in three different directions. Then Eq. 2.85a is applied for each measured component to establish a set of three linear equations for determining ε_{11}, ε_{22} and ε_{12}, which are in-plane strain components aligned with designated coordinate axes x_1 and x_2.

2.11.2 *Cylindrical and Spherical Rotation Components*

Being mutually orthogonal components of a vector, cylindrical rotation components can be obtained from Cartesian ones by means of the usual transformation formula for vectors, i.e.

$$\{\omega^c\} = [a_{ij}]^T \{\omega_i\} \tag{2.86}$$

where $[a_{ij}]$ is the transformation matrix given in Eq. 2.68e, and

$$\{\omega^c\} = (\omega_r \quad \omega_\theta \quad \omega_z)^T, \quad \{\omega_i\} = (\omega_x \quad \omega_y \quad \omega_z)^T \tag{2.87a,b}$$

e.g. ω_θ is the rotation angle about the θ-direction. Substituting Eqs. 2.52, 2.66 and 2.68e into Eq. 2.86 yields

$$\omega_r = \frac{1}{2}\left(\frac{1}{r}\frac{\partial u_z}{\partial \theta} - \frac{\partial u_\theta}{\partial z}\right) \tag{2.88a}$$

$$\omega_\theta = \frac{1}{2}\left(\frac{\partial u_r}{\partial z} - \frac{\partial u_z}{\partial r}\right) \tag{2.88b}$$

$$\omega_z = \frac{1}{2}\left[\frac{1}{r}\frac{\partial (r u_\theta)}{\partial r} - \frac{1}{r}\frac{\partial u_r}{\partial \theta}\right] \tag{2.88c}$$

Spherical rotation components can be derived from the cylindrical components in a similar way, thus

$$\omega_R = \frac{1}{2R\sin\phi}\left[\frac{\partial(u_\theta \sin\phi)}{\partial \phi} - \frac{\partial u_\phi}{\partial \theta}\right] \tag{2.89a}$$

$$\omega_\phi = \frac{1}{2R\sin\phi}\left[\frac{\partial u_R}{\partial \theta} - \sin\phi \frac{\partial (R u_\theta)}{\partial R}\right] \tag{2.89b}$$

$$\omega_\theta = \frac{1}{2R}\left[\frac{\partial (R u_\phi)}{\partial R} - \frac{\partial u_R}{\partial \phi}\right] \tag{2.89c}$$

2.11.3 *Rigid-Body Rotation and Translation Components*

When a continuum rotates as a rigid body about the x-axis, following conditions arise

$$\omega_x^* = \frac{\partial w}{\partial y} = -\frac{\partial v}{\partial z}, \quad \varepsilon_{yz} = 0 \tag{2.90a,b}$$

in which a super-asterisk (*) denotes a rigid body component. For rigid-body rotation about the y-axis,

$$\omega_y^* = \frac{\partial u}{\partial z} = -\frac{\partial w}{\partial x}, \quad \varepsilon_{zx} = 0 \tag{2.91a,b}$$

and about the z-axis,

$$\omega_z^* = \frac{\partial v}{\partial x} = -\frac{\partial u}{\partial y}, \quad \varepsilon_{xy} = 0 \tag{2.91c,d}$$

In cylindrical coordinates; for rigid-body rotation about the r-direction,

$$\omega_r^* = \frac{\partial u_z}{r\partial\theta} = -\frac{\partial u_\theta}{\partial z}, \quad \varepsilon_{\theta z} = 0 \tag{2.92a,b}$$

about the θ-direction,

$$\omega_\theta^* = \frac{\partial u_r}{\partial z} = -\frac{\partial u_z}{\partial r}, \quad \varepsilon_{zr} = 0 \tag{2.92c,d}$$

and about the z-axis,

$$\omega_z^* = \frac{\partial u_\theta}{\partial r} = \frac{u_\theta}{r} - \frac{\partial u_r}{r\partial\theta}, \quad \varepsilon_{r\theta} = 0 \tag{2.92e,f}$$

In spherical coordinates; for rigid-body rotation about the R-direction,

$$\omega_R^* = \frac{\partial u_\theta}{R\partial\phi} = \frac{1}{R\sin\phi}\left(\cos\phi\, u_\theta - \frac{\partial u_\phi}{\partial\theta}\right), \quad \varepsilon_{\phi\theta} = 0 \tag{2.93a,b}$$

about the ϕ-direction,

$$\omega_\phi^* = \frac{1}{R\sin\phi}\left(\frac{\partial u_R}{\partial\theta} - \sin\phi\, u_\theta\right) = -\frac{\partial u_\theta}{\partial R}, \quad \varepsilon_{\theta R} = 0 \tag{2.93c,d}$$

and about the θ-direction,

$$\omega_\theta^* = \frac{\partial u_\phi}{\partial R} = -\frac{1}{R}\left(\frac{\partial u_R}{\partial\phi} - u_\phi\right), \quad \varepsilon_{R\phi} = 0 \tag{2.93e,f}$$

These formulas of rigid-body rotation components in terms of displacement components are useful in the derivation of governing equations of bending of plate or shell structures which have small thickness in one direction. In such problems, it is very common to assume that a normal in the same direction as the small thickness remains normal after deformation. This is equivalent to the assumption that the thin cross section rotates as a rigid body. We adopt this assumption in the thin plate problems in Chapter V.

We showed in Section 2.8 that if the body is displaced as a rigid body, according to Eq. 2.53, then the Cartesian rigid-body rotation components are

$$\omega_x^* = B_1, \quad \omega_y^* = B_2, \quad \omega_z^* = B_3 \tag{2.94a,b,c}$$

If it is merely translated (i.e. if $B_i = 0$), then its Cartesian rigid-body translation components can be put as

$$u^* = C_1, \quad v^* = C_2, \quad w^* = C_3 \tag{2.95a,b,c}$$

The corresponding equations in cylindrical coordinates are

$$u_r^* = C_1 \cos\theta + C_2 \sin\theta \tag{2.96a}$$

$$u_\theta^* = -C_1 \sin\theta + C_2 \cos\theta \tag{2.96b}$$

$$u_z^* = C_3 \tag{2.96c}$$

$$\omega_r^* = B_1 \cos\theta + B_2 \sin\theta \tag{2.97a}$$

$$\omega_\theta^* = -B_1 \sin\theta + B_2 \cos\theta \tag{2.97b}$$

$$\omega_z^* = B_3 \tag{2.97c}$$

and in spherical coordinates

$$u_R^* = \sin\phi(C_1 \cos\theta + C_2 \sin\theta) + C_3 \cos\phi \tag{2.98a}$$

$$u_\phi^* = \cos\phi(C_1 \cos\theta + C_2 \sin\theta) - C_3 \sin\phi \tag{2.98b}$$

$$u_\theta^* = -C_1 \sin\theta + C_2 \cos\theta \tag{2.98c}$$

$$\omega_R^* = \sin\phi(B_1 \cos\theta + B_2 \sin\theta) + B_3 \cos\phi \tag{2.99a}$$

$$\omega_\phi^* = \cos\phi(B_1 \cos\theta + B_2 \sin\theta) - B_3 \sin\phi \tag{2.99b}$$

$$\omega_\theta^* = -B_1 \sin\theta + B_2 \cos\theta \tag{2.99c}$$

2.12 EXERCISE PROBLEMS

With permission of the publisher, McGraw-Hill Book Company, Problems 2.12.1 to 2.12.27 were taken from the book by Mase (1970) and Problems 2.12.28 to 2.12.30 from the book by Wang (1953).

2.12.1 The state of stress at a point is given by the stress tensor

$$[\sigma_{ij}] = \begin{bmatrix} \sigma & a\sigma & b\sigma \\ a\sigma & \sigma & c\sigma \\ b\sigma & c\sigma & \sigma \end{bmatrix},$$

where a, b, c are constants and σ is some stress value. Determine the constants a, b, c so that the stress vector vanishes on the plane which makes equal angles with the reference Cartesian axes.
Answer : $a = b = c = -1/2$.

2.12.2 The stress tensor at a point P is given by the array

$$[\sigma_{ij}] = \begin{bmatrix} 7 & -5 & 0 \\ -5 & 3 & 1 \\ 0 & 1 & 2 \end{bmatrix}.$$

Determine the stress vector on the plane passing through P and parallel to the plane ABC shown in Fig 2.18.
Answer : $(\overline{X}_1, \overline{X}_2, \overline{X}_3) = (-9, 5, 10)/7$.

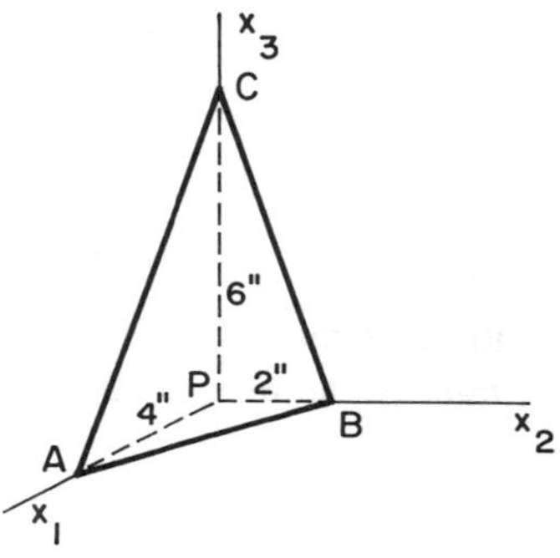

Fig. 2.18

2.12.3 The state of stress throughout a continuum is given with respect to the Cartesian axes by the array

$$[\sigma_{ij}] = \begin{bmatrix} 3x_1x_2 & 5x_2^2 & 0 \\ 5x_2^2 & 0 & 2x_2 \\ 0 & 2x_3 & 0 \end{bmatrix}.$$

Determine the stress vector acting at the point $P(2, 1, \sqrt{3})$ of the plane that is tangent to the cylindrical surface $x_2^2 + x_3^2 = 4$ at P (Fig. 2.19).

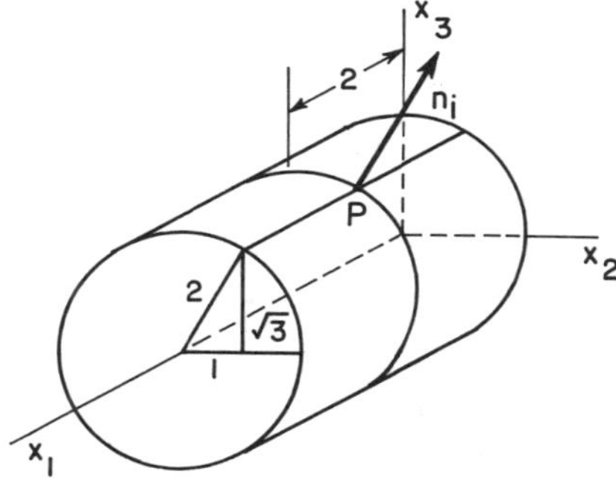

Fig. 2.19

Answer : $(\overline{X}_1, \overline{X}_2, \overline{X}_3) = (5/2, 3, \sqrt{3})$.

2.12.4 For the distribution of the state of stress given in Problem 2.12.3, what form must the body force components have if the equilibrium equations are to be satisfied everywhere.
Answer : $X_1 = -13x_2$, $X_2 = -2$, $X_3 = 0$.

2.12.5 The state of stress at a point is given with respect to the Cartesian axes by the array

$$[\sigma_{ij}] = \begin{bmatrix} 2 & -2 & 0 \\ -2 & \sqrt{2} & 0 \\ 0 & 0 & -\sqrt{2} \end{bmatrix}.$$

Determine the stress tensor σ'_{ij} for the rotated axes related to the unprimed axes by the transformation tensor

$$[a_{ij}] = \begin{bmatrix} 0 & 1/\sqrt{2} & -1/\sqrt{2} \\ 1/\sqrt{2} & 1/2 & 1/2 \\ 1/\sqrt{2} & -1/2 & -1/2 \end{bmatrix}.$$

Answer :

$$[\sigma'_{ij}] = \begin{bmatrix} 0 & 0 & 2 \\ 0 & 1-\sqrt{2} & -1 \\ 2 & -1 & 1+\sqrt{2} \end{bmatrix}.$$

2.12.6 The stress tensor at a point P is given as

$$[\sigma_{ij}] = \begin{bmatrix} 3 & 1 & 1 \\ 1 & 0 & 2 \\ 1 & 2 & 0 \end{bmatrix}.$$

Determine the principal stress values and the principal stress directions.
Answer :

$$(\sigma_1, \sigma_2, \sigma_3) = (-2, 1, 4),$$

$$(n_1^{(1)}, n_2^{(1)}, n_3^{(1)}) = (0, 1, -1)/\sqrt{2},$$

$$(n_1^{(2)}, n_2^{(2)}, n_3^{(2)}) = (1, -1, -1)/\sqrt{3},$$

$$(n_1^{(3)}, n_2^{(3)}, n_3^{(3)}) = (-2, -1, -1)/\sqrt{6}.$$

2.12.7 The stress tensor at a point is given by

$$[\sigma_{ij}] = \begin{bmatrix} 5 & 0 & 0 \\ 0 & -6 & -12 \\ 0 & -12 & 1 \end{bmatrix}.$$

Determine the principal stresses and the maximum shear stress at the point.
Answer :

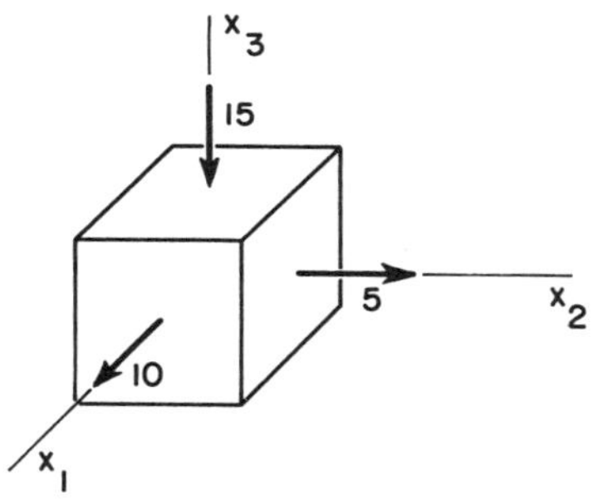

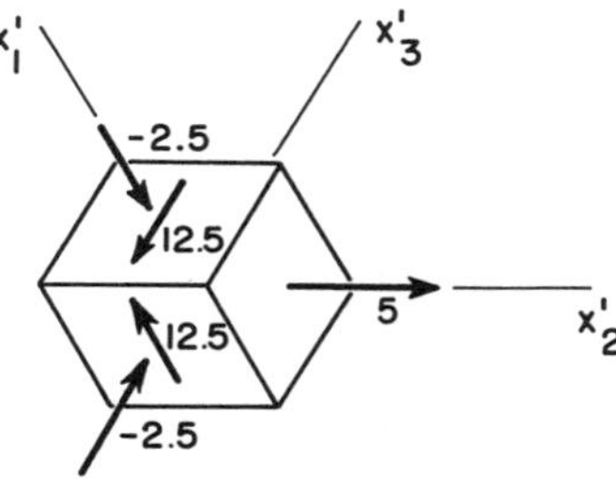

Fig. 2.20

2.12.8 The state of stress throughout a body is given by the stress tensor

$$[\sigma_{ij}] = \begin{bmatrix} 0 & Cx_3 & 0 \\ Cx_3 & 0 & -Cx_1 \\ 0 & -Cx_1 & 0 \end{bmatrix},$$

where C is an arbitrary constant. (a) Show that the equilibrium equations are satisfied if body forces are zero. (b) At the point $P(4, -4, 7)$ calculate the stress vector on the plane

$$2x_1 + 2x_2 - x_3 = -7,$$

and on the sphere

$$x_1^2 + x_2^2 + x_3^2 = 81.$$

(c) Determine the principal stresses, and maximum shear stresses at P.
Answer : (b) For the plane, $(\overline{X}_1, \overline{X}_2, \overline{X}_3) = (14, 18, -8)C/3$; for the sphere, $(-28, 0, 16)C/9$. (c) $(\sigma_1, \sigma_2, \sigma_3) = (1, 0, -1)\sqrt{65}\,C$, $\sigma_s = \pm\sqrt{65}\,C$.

2.12.9 At point P stress tensor is

$$[\sigma_{ij}] = \begin{bmatrix} 14 & 7 & -7 \\ 7 & 21 & 0 \\ -7 & 0 & 35 \end{bmatrix}.$$

Determine the stress vector on the plane at P parallel to planes (a) BGE, (b) $BGFC$ of the small parallelepiped shown in Fig. 2.21.

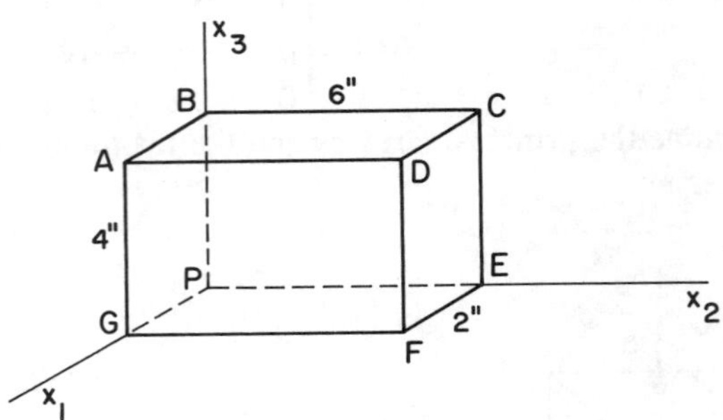

Fig. 2.21

Answer : (a) $(\overline{X}_1,\overline{X}_2,\overline{X}_3)=(11,12,9)$, (b) $(21,14,21)/\sqrt{5}$.

2.12.10 Determine the normal and shear stress components on the plane $BGFC$ of Problem 2.12.9.
Answer : $\sigma_N = 63/5$, $\sigma_S = 7\sqrt{29}/5$.

2.12.11 The principal stresses at point P are $\sigma_1 = 12$, $\sigma_2 = 3$, $\sigma_3 = -6$. Determine the stress vector and its normal component on the octahedral plane at P.
Answer : $(\overline{X}_1,\overline{X}_2,\overline{X}_3)=(12,3,-6)/\sqrt{3}$; $\sigma_N = 3$.

2.12.12 Determine the principal stress values for

$$\text{(a)}\ [\sigma_{ij}] = \begin{bmatrix} 0 & 1 & 1 \\ 1 & 0 & 1 \\ 1 & 1 & 0 \end{bmatrix} \quad \text{and} \quad \text{(b)}\ [\sigma_{ij}] = \begin{bmatrix} 2 & 1 & 1 \\ 1 & 2 & 1 \\ 1 & 1 & 2 \end{bmatrix}$$

and show that both have the same principal directions.
Answer : (a) $\sigma_1 = 2$, $\sigma_2 = \sigma_3 = -1$, (b) $\sigma_1 = 4$, $\sigma_2 = \sigma_3 = 1$.

2.12.13 Show that the normal component of the stress vector on the octahedral plane is equal to one third the first invariant of the stress tensor.

2.12.14 The stress tensor at a point is given as

$$[\sigma_{ij}] = \begin{bmatrix} 0 & 1 & 2 \\ 1 & \sigma_{22} & 1 \\ 2 & 1 & 0 \end{bmatrix},$$

with σ_{22} unspecified. Determine σ_{22} so that the stress vector on some plane at the point will be zero. Give the unit normal for this traction-free plane.
Answer : $\sigma_{22} = 1$; $(n_1,n_2,n_3)=(1,-2,1)/\sqrt{6}$.

2.12.15 In a continuum, the stress field is given by the tensor

$$[\sigma_{ij}] = \begin{bmatrix} x_1^2 x_2 & (1-x_2^2)x_1 & 0 \\ (1-x_2^2)x_1 & (x_2^3-3x_2)/3 & 0 \\ 0 & 0 & 2x_3^2 \end{bmatrix}.$$

Determine (a) the body force distribution if the equilibrium equations are to be satisfied throughout the field, (b) the principal stress values at the point $P\,(a, 0, 2\sqrt{a}\,)$, (c) the maximum shear stress at P.
Answer : (a) $X_3 = -4x_3$, (b) $a, -a, 8a$, (c) $\pm 4.5a$.

2.12.16 For a displacement field given by

$$(u_1, u_2, u_3) = ((x_1 - x_2)^2, (x_2 + x_3)^2, -x_1 x_2).$$

Determine at point $P(0, 2, -1)$, (a) the linear strain tensor, and (b) the change in length per unit length in the direction

$$(n_1, n_2, n_3) = (8, -1, 4)/9.$$

Answer :

$$\text{(a)} \begin{bmatrix} 2 & 0 & -2 \\ 0 & 2 & 1 \\ -2 & 1 & 0 \end{bmatrix}, \qquad \text{(b)} \ -2/27.$$

2.12.17 Use the results of Problem 2.12.16 to compute the change in the angle between two unit line elements $(n_1, n_2, n_3) = (8, -1, 4)/9$ and $(m_1, m_2, m_3) = (4, 4, -7)/9$ at the point P.
Answer : 106/27 radians.

2.12.18 A linear (small strain) deformation is specified by

$$u_1 = 4x_1 - x_2 + 3x_3, \quad u_2 = x_1 + 7x_2, \quad u_3 = -3x_1 + 4x_2 + 4x_3.$$

Determine the principal strains.
Answer : 8, 4, 3.

2.12.19 A 45° strain-rosette measures longitudinal strain along the axes shown in Fig. 2.22. At a point P,

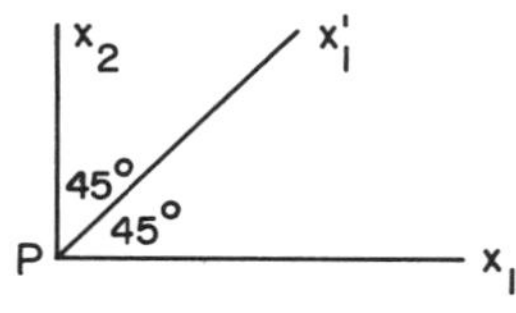

Fig 2.22

$$\varepsilon_{11} = 5 \times 10^{-4}, \quad \varepsilon'_{11} = 4 \times 10^{-4}, \quad \varepsilon_{22} = 7 \times 10^{-4} \quad \text{in/in.}$$

Determine the shear strain ε_{12} at the point.
Answer: $\varepsilon_{12} = -2 \times 10^{-4}$.

2.12.20 The state of strain throughout a continuum is specified by

$$[\varepsilon_{ij}] = \begin{bmatrix} x_1^2 & x_2^2 & x_1 x_3 \\ x_2^2 & x_3 & x_3^2 \\ x_1 x_3 & x_3^2 & x_1^2 \end{bmatrix}.$$

Are the compatibility equations for strain satisfied?
Answer: Yes.

2.12.21 For a homogeneous deformation the small strain tensor is given by

$$[\varepsilon_{ij}] = \begin{bmatrix} 0.01 & -0.005 & 0 \\ -0.005 & 0.02 & 0.01 \\ 0 & 0.01 & -0.03 \end{bmatrix}.$$

What is the change in the 90° angle ADC depicted by the small tetrahedron $OABC$ in Fig. 2.23 if $OA = OB = OC$, and D is the midpoint of AB?
Answer: $-0.01/\sqrt{3}$.

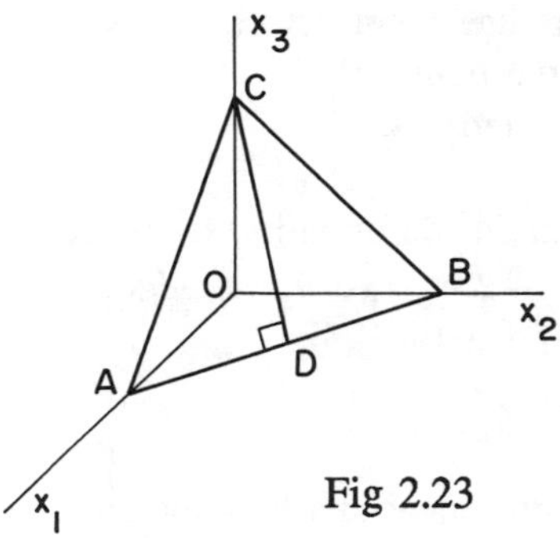

Fig 2.23

2.12.22 At a point the strain tensor is given by

$$[\varepsilon_{ij}] = \begin{bmatrix} 5 & -1 & -1 \\ -1 & 4 & 0 \\ -1 & 0 & 4 \end{bmatrix}.$$

Show that it is in principal form as

$$[\varepsilon^*_{ij}] = \begin{bmatrix} 6 & 0 & 0 \\ 0 & 4 & 0 \\ 0 & 0 & 3 \end{bmatrix}.$$

Calculate the strain invariants for each of these tensors and show their equivalence.
Answer: $I_1 = 13, \quad I_2 = 54, \quad I_3 = 72.$

2.12.23 Show that the displacement field

$$u_1 = Ax_1 + 3x_2, \quad u_2 = 3x_1 - Bx_2, \quad u_3 = 5,$$

gives a state of *plane strain* ($\varepsilon_{33} = \varepsilon_{31} = \varepsilon_{32} = 0$) and determine the relationship between A and B for which the deformation is *isochoric* (constant volume deformation).
Answer: $A = B$.

2.12.24 A so-called delta-rosette for measuring longitudinal surface strains has the shape of the equilateral triangle Δ and records normal strains ε_{11}, ε_{11}', ε_{11}'' in the directions shown in Fig. 2.24. If $\varepsilon_{11} = a$, $\varepsilon_{11}' = b$, $\varepsilon_{11}'' = c$, determine ε_{12} and ε_{22} at the point.
Answer: $\varepsilon_{12} = (b - c)/\sqrt{3}$ and $\varepsilon_{22} = (-a + 2b + 2c)/3$.

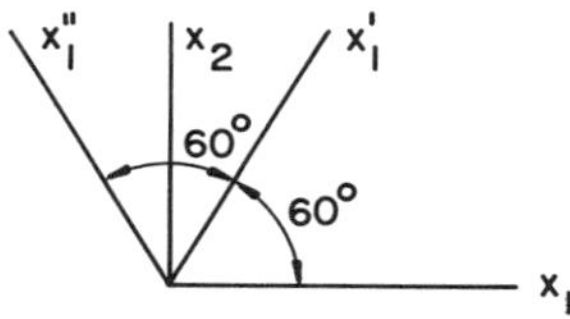

Fig. 2.24

2.12.25 The strain tensor at a point is given by

$$[\varepsilon_{ij}] = \begin{bmatrix} 1 & -3 & \sqrt{2} \\ -3 & 1 & -\sqrt{2} \\ \sqrt{2} & -\sqrt{2} & 4 \end{bmatrix}.$$

Determine: (a) the normal strain in the direction of a unit vector n_i, of which $(n_1, n_2, n_3) = (1, -1, \sqrt{2})/2$; (b) the shear strain between n_i and another unit vector m_i, of which $(m_1, m_2, m_3) = (-1, 1, \sqrt{2})/2$; (c) determine the principal-axes form of ε_{ij} and note that n_i and m_i are principal directions; (b) calculate the three strain invariants.
Answer: (a) 6, (b) 0, (c) $(\varepsilon_1, \varepsilon_2, \varepsilon_3) = (6, 2, -2)$, (d) $(I_1, I_2, I_3) = (6, -4, -24)$.

2.12.26 A displacement field is given by

$$u_1 = 3x_1x_2^2, \quad u_2 = 2x_3x_1, \quad u_3 = x_3^2 - x_1x_2.$$

Determine the strain tensor ε_{ij} and check whether the compatibility conditions are satisfied.
Answer:

$$[\varepsilon_{ij}] = \begin{bmatrix} 3x_2^2 & 3x_1x_2+x_3 & -x_2/2 \\ 3x_1x_2+x_3 & 0 & x_1/2 \\ -x_2/2 & x_1/2 & 2x_3 \end{bmatrix}.$$

Of course, the compatibility conditions are assured because the strains arise from a displacement field which is given.

2.12.27 For a delta-strain-rosette the normal strains are found to be those shown in Fig. 2.25. Determine ε_{12} and ε_{22} at the location.
Answer: $\varepsilon_{22} = 1 \times 10^{-4}$, $\varepsilon_{12} = -0.2885 \times 10^{-4}$.

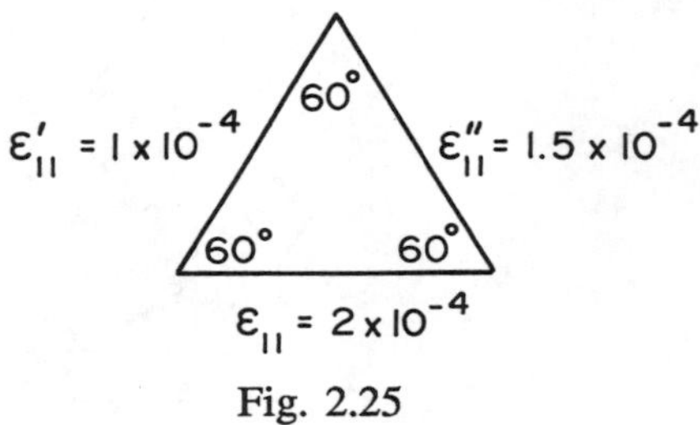

Fig. 2.25

2.12.28 Show that

$$\varepsilon_{xx} = k(x^2+y^2), \quad \varepsilon_{yy} = ky^2, \quad \gamma_{xy} = 2kxy, \quad \varepsilon_{zz} = \gamma_{yz} = \gamma_{zx} = 0,$$

where k is a small constant, is a possible state of strain; while

$$\varepsilon_{xx} = kz(x^2+y^2), \quad \varepsilon_{yy} = ky^2z, \quad \gamma_{xy} = 2kxyz, \quad \varepsilon_{zz} = \gamma_{yz} = \gamma_{zx} = 0,$$

is not a possible one.

2.12.29 An elastic solid is heated nonuniformly to a temperature distribution $T(x,y,z)$, where $T(x,y,z)$ is a function of x,y,z. If each element in the body has unrestrained thermal expansion, the strain components will be

$$\varepsilon_{xx} = \varepsilon_{yy} = \varepsilon_{zz} = \alpha T, \qquad \gamma_{xy} = \gamma_{yz} = \gamma_{zx} = 0,$$

where α is the coefficient of thermal expansion and is a constant. Prove that this can occur only when T is a linear function of x,y,z.

2.12.30 Determine the relations among the constants A_0, A_1, B_0, B_1, C_0, C_1, C_2, so that the following is a possible system of strains:

$$\varepsilon_{xx} = A_0 + A_1(x^2+y^2) + (x^4+y^4),$$

$$\varepsilon_{yy} = B_0 + B_1(x^2 + y^2) + (x^4 + y^4),$$

$$\gamma_{xy} = C_0 + C_1 xy(x^2 + y^2 + C_2),$$

$$\varepsilon_{zz} = \gamma_{yz} = \gamma_{zx} = 0.$$

Answer: $A_1 + B_1 - 2C_2 = 0, \quad C_1 = 4$.

CHAPTER III

LINEAR ELASTICITY

3.1 STRAIN ENERGY FUNCTION

So far we have discussed some general concepts of mechanics of solids under small deformation. A solid is classified as *elastic*, if it possesses a homogeneous stress-free *natural state*, and if in an appropriately defined finite neighborhood of this state there exists a one-to-one correspondence between the stress tensor σ_{ij} and the strain tensor ε_{ij}. This definition is based on Cauchy's approach (A. L. Cauchy, 1827). It is understood here that the natural state is a state of thermodynamic equilibrium in which all components of stress and strain are zero throughout the body. In other words, the definition incorporates the idea that the body returns to the natural state when all loads are removed.

Consider the case where the body force X_i acts on an infinitesimal body of volume $\mathrm{d}V$ at point P, and the body is deformed in such a way that point P moves through the distances $\mathrm{d}u_i$. Then the increment of work done by the body force X_i is

$$\mathrm{d}W_b = X_i \mathrm{d}u_i \mathrm{d}V \tag{3.1}$$

At the same time, the work done by the surface force $\overline{X}_i$ on the infinitesimal surface $\mathrm{d}S$ (enclosing $\mathrm{d}V$) is

$$\mathrm{d}W_s = \overline{X}_i \mathrm{d}u_i \mathrm{d}S,$$

or, due to Eq. 2.6,

$$\mathrm{d}W_s = \sigma_{ij} n_i \mathrm{d}u_j \mathrm{d}S.$$

By Gauss's divergence theorem, we can write

$$\mathrm{d}W_s = (\sigma_{ij} \mathrm{d}u_j)_{,i} \mathrm{d}V,$$

or, after performing the indicated differentiation,

$$\mathrm{d}W_s = (\sigma_{ij,i} \mathrm{d}u_j + \sigma_{ij} \mathrm{d}u_{j,i}) \mathrm{d}V \tag{3.2}$$

If the process is *adiabatic*, i.e. without loss or gain of heat, the whole work done will be stored up inside the material body as *potential energy*, or *strain energy* as it is called in the mechanics of deformable bodies. Denoting such energy per unit volume of the material by U, we have

$$\mathrm{d}U\mathrm{d}V = \mathrm{d}W_b + \mathrm{d}W_s \tag{3.3}$$

or, due to Eqs. 3.1 and 3.2,

$$\mathrm{d}U\mathrm{d}V = [(X_j + \sigma_{ij,i})\mathrm{d}u_j + \sigma_{ij}\mathrm{d}u_{j,i}]\mathrm{d}V,$$

or, due to the equilibrium (Eq. 2.14),

$$\mathrm{d}U = \sigma_{ij}\mathrm{d}u_{j,i} \tag{3.4}$$

It should be noted that Eq. 3.3 is valid for *conservation of energy* in dynamics if the body force X_i in Eq. 3.1 is taken as including the D'Alembert's force $-\rho\ddot{u}_i$. On the other hand, if X_i is taken strictly as the actual body force only, Eq. 3.3 should be written

$$\mathrm{d}U\mathrm{d}V + \mathrm{d}T\mathrm{d}V = \mathrm{d}W_b + \mathrm{d}W_s \tag{3.5}$$

where T is the *kinetic energy* per unit volume of the material, i.e.

$$T = \frac{1}{2}\rho\dot{u}_i^2 \tag{3.6}$$

Thus

$$\begin{aligned} \mathrm{d}T &= \rho\dot{u}_i\mathrm{d}\dot{u}_i \\ &= \rho\dot{u}_i\ddot{u}_i\mathrm{d}t \\ &= \rho\mathrm{d}u_i\ddot{u}_i. \end{aligned}$$

Substituting the equation above together with Eqs. 3.1 and 3.2 into Eq. 3.5, in view of Eq. 2.10, yields the same result as before, i.e. Eq. 3.4. Rewrite the latter as

$$\mathrm{d}U = \frac{1}{2}(\sigma_{ij}\mathrm{d}u_{j,i} + \sigma_{ji}\mathrm{d}u_{i,j}),$$

or, due to the symmetry of σ_{ij},

$$\mathrm{d}U = \sigma_{ij}\mathrm{d}\varepsilon_{ij} \tag{3.7}$$

If the stress components are unique homogeneous functions of the strain components, there exists the strain energy function U as a scalar homogeneous function of the strain components, i.e.

$$U = U(\varepsilon_{ij}) \tag{3.8}$$

Accordingly, its total differential is

$$\mathrm{d}U = \frac{\partial U}{\partial \varepsilon_{ij}}\mathrm{d}\varepsilon_{ij} \tag{3.9}$$

In Eqs. 3.7 and 3.9, each of the nine components of the strain tensor are considered as an independent variable in the function U. Comparing these two equations, we have

$$\sigma_{ij} = \frac{\partial U}{\partial \varepsilon_{ij}} \tag{3.10}$$

The existence of a strain energy function such that the stress components are unique homogeneous functions of the strain components is a basic property of elastic solids. For one-dimensional cases, the stress-strain relationship for elastic solids are shown in Fig. 3.1, where the loading and unloading paths coincide as implied by the theory.

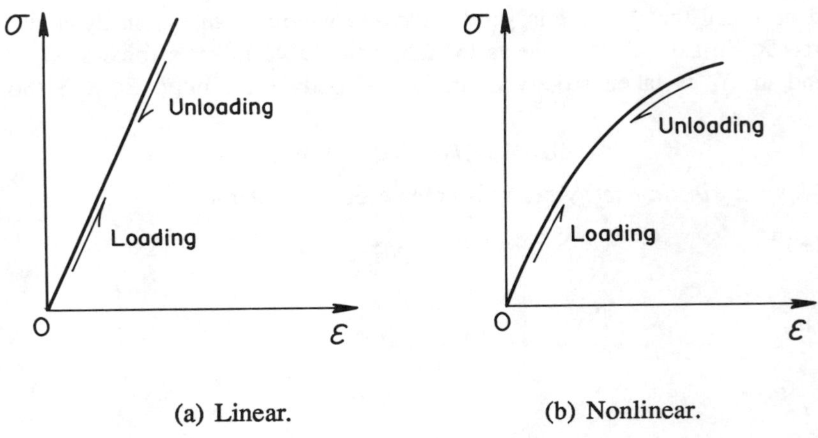

(a) Linear. (b) Nonlinear.

Fig. 3.1 One-dimensional stress-strain relationship for elastic solids.

When *Hooke's law* (R. Hooke, 1678) is obeyed the material is classified as linear elastic. This law states: *Stresses are linear and homogeneous functions of strains*, i.e.

$$\sigma_{ij} = C_{ijkl}\varepsilon_{kl} \tag{3.11}$$

We shall now show that the symmetry of the stress and strain tensors implies that there are only 21 independent components among the 81 C_{ijkl}.

$$C_{ijkl}\varepsilon_{kl} = \frac{\partial U}{\partial \varepsilon_{ij}},$$

the derivative of which with respect to ε_{mn} is

$$C_{ijkl}\delta_{km}\delta_{ln} = \frac{\partial^2 U}{\partial \varepsilon_{ij}\partial \varepsilon_{mn}},$$

or

$$C_{ijmn} = \frac{\partial^2 U}{\partial \varepsilon_{ij}\partial \varepsilon_{mn}} \tag{3.12}$$

Interchanging i with m and j with n in the equation above gives

$$C_{mnij} = \frac{\partial^2 U}{\partial \varepsilon_{mn}\partial \varepsilon_{ij}} \tag{3.13}$$

which shows that $C_{ijkl}=C_{klij}$. Inasmuch as the components σ_{ij} are symmetric, an interchange of the idices i and j in Eq. 3.11 does not alter the formula, thus $C_{ijkl}=C_{jikl}$ and consequently,

$$C_{ijkl}=C_{klij}=C_{jikl}=C_{ijlk} \tag{3.14}$$

Substituting Eq. 3.11 into Eq. 3.7 yields

$$\begin{aligned} \mathrm{d}U &= C_{ijkl}\varepsilon_{kl}\mathrm{d}\varepsilon_{ij} \\ &= \frac{1}{2}(C_{ijkl}\varepsilon_{kl}\mathrm{d}\varepsilon_{ij}+C_{ijkl}\varepsilon_{kl}\mathrm{d}\varepsilon_{ij}) \\ &= \frac{1}{2}(C_{ijkl}\varepsilon_{kl}\mathrm{d}\varepsilon_{ij}+C_{klij}\varepsilon_{ij}\mathrm{d}\varepsilon_{kl}), \end{aligned}$$

or, due to Eq. 3.14,

$$\begin{aligned} \mathrm{d}U &= \frac{1}{2}(C_{ijkl}\varepsilon_{kl}\mathrm{d}\varepsilon_{ij}+C_{ijkl}\varepsilon_{ij}\mathrm{d}\varepsilon_{kl}) \\ &= \frac{1}{2}C_{ijkl}\mathrm{d}(\varepsilon_{ij}\varepsilon_{kl}). \end{aligned}$$

Integrating the equation above from the natural state, where all stresses and strains are identically zero, leads to

$$U=\frac{1}{2}C_{ijkl}\varepsilon_{ij}\varepsilon_{kl} \tag{3.15a}$$

$$=\frac{1}{2}\sigma_{ij}\varepsilon_{ij} \tag{3.15b}$$

To verify Eq. 3.10, let us take the derivative of Eq. 3.15a with respect to ε_{mn}, i.e.

$$\begin{aligned} \frac{\partial U}{\partial \varepsilon_{mn}} &= \frac{1}{2}C_{ijkl}\left(\frac{\partial \varepsilon_{kl}}{\partial \varepsilon_{mn}}\varepsilon_{ij}+\frac{\partial \varepsilon_{ij}}{\partial \varepsilon_{mn}}\varepsilon_{kl}\right) \\ &= \frac{1}{2}C_{ijkl}(\delta_{km}\delta_{ln}\varepsilon_{ij}+\delta_{im}\delta_{jn}\varepsilon_{kl}) \\ &= \frac{1}{2}(C_{ijmn}\varepsilon_{ij}+C_{mnkl}\varepsilon_{kl}), \end{aligned}$$

or, due to Eq. 3.14,

$$\begin{aligned} \frac{\partial U}{\partial \varepsilon_{mn}} &= C_{mnij}\varepsilon_{ij} \\ &= \sigma_{mn}, \end{aligned}$$

according to Eq. 3.11.

On the other hand, we can take the advantage of the symmetry of the stress and strain tensors and consider each of these tensors to be tensors of the first order but with a range 6 in the form of matrices as follows

$$\{\sigma_i\} = \begin{Bmatrix} \sigma_{11} \\ \sigma_{22} \\ \sigma_{33} \\ \sigma_{12} \\ \sigma_{23} \\ \sigma_{31} \end{Bmatrix}, \qquad \{\varepsilon_i\} = \begin{Bmatrix} \varepsilon_{11} \\ \varepsilon_{22} \\ \varepsilon_{33} \\ \gamma_{12} \\ \gamma_{23} \\ \gamma_{31} \end{Bmatrix} \tag{3.16a,b}$$

By expansion, we can show that Eq. 3.7 can be rewritten in the form

$$dU = \sigma_i d\varepsilon_i, \quad (i \text{ summed from 1 to 6}) \tag{3.17}$$

Note that the engineering (*not* the tensorial) shear strain components must be used. Assuming the existence of the strain energy function U as a function of the six independent strain components as listed in Eq. 3.16b, we can write

$$U = U(\varepsilon_i), \quad (i = 1 \text{ to } 6) \tag{3.18a}$$

and

$$dU = \frac{\partial U}{\partial \varepsilon_i} d\varepsilon_i \tag{3.18b}$$

Comparing Eq. 3.18b with Eq. 3.17 leads to

$$\sigma_i = \frac{\partial U}{\partial \varepsilon_i} \tag{3.19}$$

which is essentially identical to Eq. 3.10 obtained earlier.

For *linear elastic* solids, the stress-strain relationship, Eq. 3.11, can be put in the form

$$\sigma_i = C_{ij}\varepsilon_j \tag{3.20}$$

where C_{ij} is the *stiffness tensor*. Substituting Eq. 3.20 into Eq. 3.19 leads to

$$C_{ij}\varepsilon_j = \frac{\partial U}{\partial \varepsilon_i},$$

the derivative of which with respect to ε_m is

$$C_{ij}\delta_{jm} = \frac{\partial^2 U}{\partial \varepsilon_i \partial \varepsilon_m},$$

or

$$C_{im} = \frac{\partial^2 U}{\partial \varepsilon_i \partial \varepsilon_m} \tag{3.21}$$

Interchanging i and m in the equation above, we can show the symmetry of C_{ij}, i.e. $C_{ij} = C_{ji}$. Thus Hooke's law for the most general kind of *anisotropy* has the matrix form

$$\begin{Bmatrix}\sigma_{11}\\ \sigma_{22}\\ \sigma_{33}\\ \sigma_{12}\\ \sigma_{23}\\ \sigma_{31}\end{Bmatrix}=\begin{bmatrix}C_{11} & C_{12} & C_{13} & C_{14} & C_{15} & C_{16}\\ & C_{22} & C_{23} & C_{24} & C_{25} & C_{26}\\ & & C_{33} & C_{34} & C_{35} & C_{36}\\ & & & C_{44} & C_{45} & C_{46}\\ & \text{Symm.} & & & C_{55} & C_{56}\\ & & & & & C_{66}\end{bmatrix}\begin{Bmatrix}\varepsilon_{11}\\ \varepsilon_{22}\\ \varepsilon_{33}\\ \gamma_{12}\\ \gamma_{23}\\ \gamma_{31}\end{Bmatrix} \tag{3.22}$$

The matrix $[C_{ij}]$, being symmetric, has 21 independent elastic constants. Since the strain energy function is

$$U=\frac{1}{2}C_{ij}\varepsilon_i\varepsilon_j \tag{3.23a}$$

$$=\frac{1}{2}\sigma_i\varepsilon_i \tag{3.23b}$$

and is positive for any non-trivial set of strain components, the matrix $[C_{ij}]$ must be positive definite as described in Section 1.6 and elaborated in Section 3.6.5 for various types of linear elastic solids.

3.2 ORTHOTROPIC AND ISOTROPIC ELASTIC SOLIDS

Almost all of the materials used for engineering purposes exhibit certain degrees of elastic symmetry, and require less than 21 independent elastic constants. First let the 1, 2, 3 reference frame in Fig. 3.2 be such that the $1-2$ plane is a plane of symmetry. Let another frame $1'$, $2'$, $3'$ be generated by a 180° rotation about the 3-axis. The transformation tensor a_{ij} linking the two systems is

$$[a_{ij}]=\begin{bmatrix}-1 & 0 & 0\\ 0 & -1 & 0\\ 0 & 0 & 1\end{bmatrix}$$

Transforming stresses according to Eq. 1.27 yields

$$\begin{aligned}\sigma'_{11}&=\sigma_{11}, & \sigma'_{12}&=\sigma_{12}\\ \sigma'_{22}&=\sigma_{22}, & \sigma'_{23}&=-\sigma_{23}\\ \sigma'_{33}&=\sigma_{33}, & \sigma'_{31}&=-\sigma_{31}\end{aligned} \tag{3.24}$$

Similarly, the transformation of strains gives

$$\begin{aligned}\varepsilon'_{11}&=\varepsilon_{11}, & \gamma'_{12}&=\gamma_{12}\\ \varepsilon'_{22}&=\varepsilon_{22}, & \gamma'_{23}&=-\gamma_{23}\\ \varepsilon'_{33}&=\varepsilon_{33}, & \gamma'_{31}&=-\gamma_{31}\end{aligned} \tag{3.25}$$

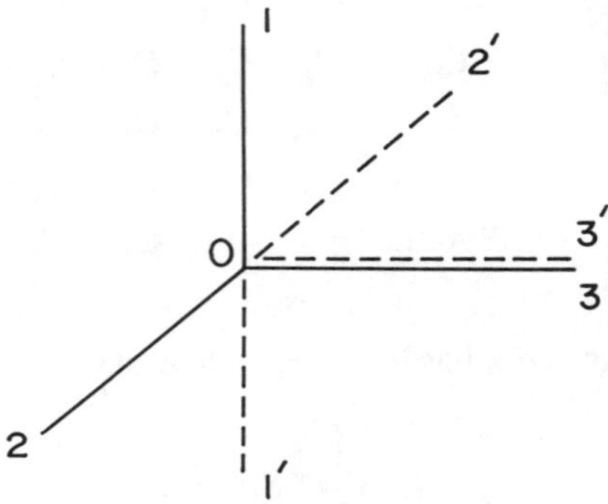

Fig. 3.2 Transformation for orthotropic symmetry with respect to plane $1-2$.

Substituting Eqs. 3.24 and 3.25 into Eq. 3.22 leads to

$$\begin{Bmatrix} \sigma'_{11} \\ \sigma'_{22} \\ \sigma'_{33} \\ \sigma'_{12} \\ \sigma'_{23} \\ \sigma'_{31} \end{Bmatrix} = \begin{bmatrix} C_{11} & C_{12} & C_{13} & C_{14} & -C_{15} & -C_{16} \\ & C_{22} & C_{23} & C_{24} & -C_{25} & -C_{26} \\ & & C_{33} & C_{34} & -C_{35} & -C_{36} \\ & & & C_{44} & -C_{45} & -C_{46} \\ & \text{Symm.} & & & C_{55} & C_{56} \\ & & & & & C_{66} \end{bmatrix} \begin{Bmatrix} \varepsilon'_{11} \\ \varepsilon'_{22} \\ \varepsilon'_{33} \\ \gamma'_{12} \\ \gamma'_{23} \\ \gamma'_{31} \end{Bmatrix} \tag{3.26}$$

Since the stress and strain relationships of $1 \quad 2 \quad 3$ and $1' \quad 2' \quad 3'$ systems must be identical, the stiffness matrix $[C_{ij}]$ in Eq. 3.26 must be equal to the one in Eq. 3.22. This implies that the negative elements C_{ij} in Eq. 3.26 must be zero, i.e.

$$[C_{ij}] = \begin{bmatrix} C_{11} & C_{12} & C_{13} & C_{14} & 0 & 0 \\ & C_{22} & C_{23} & C_{24} & 0 & 0 \\ & & C_{33} & C_{34} & 0 & 0 \\ & & & C_{44} & 0 & 0 \\ & \text{Symm.} & & & C_{55} & C_{56} \\ & & & & & C_{66} \end{bmatrix} \tag{3.27}$$

Note that there remain 13 essential elastic constants for the case of a single plane of symmetry.

An additional plane of symmetry, say the plane $2-3$, will further reduce the stiffness matrix to

$$[C_{ij}] = \begin{bmatrix} C_{11} & C_{12} & C_{13} & 0 & 0 & 0 \\ & C_{22} & C_{23} & 0 & 0 & 0 \\ & & C_{33} & 0 & 0 & 0 \\ & & & C_{44} & 0 & 0 \\ & \text{Symm.} & & & C_{55} & 0 \\ & & & & & C_{66} \end{bmatrix} \tag{3.28}$$

If the plane $3-1$ is also the plane of symmetry, no further reduction in the number of elastic constants can be made. Such an elastic solid, which has three mutually orthogonal planes of symmetry, is said to be *orthotropic*; and it has nine elastic constants. Wood and fiber reinforced plastics are examples of orthotropic elastic solids.

Materials, which behave mechanically in the same manner for all directions are called *isotropic* materials. An isotropic elastic solid has more elastic symmetry than an orthotropic one, and therefore has fewer elastic constants. We must have, for an isotropic elastic solid,

$$C_{11} = C_{22} = C_{33},$$

$$C_{12} = C_{23} = C_{31},$$

and

$$C_{44} = C_{55} = C_{66} \tag{3.29}$$

Equations 3.29 show that there are three essential elastic constants, but they are not independent, as we now show. Rotate the coordinates $1, 2, 3$ about the 3-axis through an angle $45°$, to give $1', 2', 3'$ as shown in Fig. 3.3. The transformation tensor is

$$[a_{ij}] = \begin{bmatrix} 1/\sqrt{2} & -1/\sqrt{2} & 0 \\ 1/\sqrt{2} & 1/\sqrt{2} & 0 \\ 0 & 0 & 1 \end{bmatrix}.$$

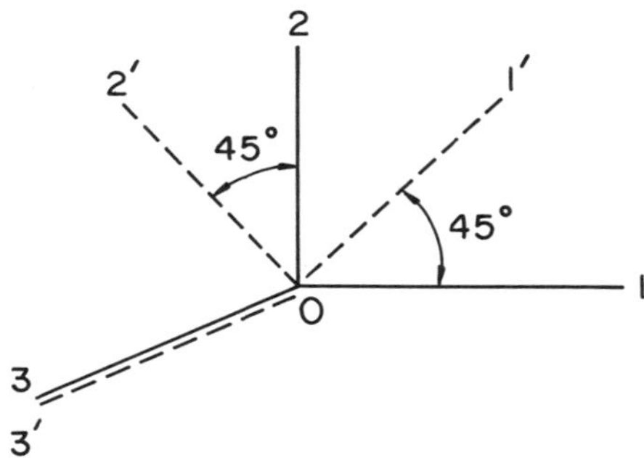

Fig. 3.3 A transformation for isotropic solid.

Repeating the analysis already used for orthotropy yields the relation

$$2C_{44} = C_{11} - C_{12} \tag{3.30}$$

Equations 3.29 and 3.30 show that the stiffness matrix for isotropic elastic solids has the form

$$[C_{ij}] = \begin{bmatrix} \lambda+2\mu & \lambda & \lambda & 0 & 0 & 0 \\ & \lambda+2\mu & \lambda & 0 & 0 & 0 \\ & & \lambda+2\mu & 0 & 0 & 0 \\ & & & \mu & 0 & 0 \\ & \text{Symm.} & & & \mu & 0 \\ & & & & & \mu \end{bmatrix} \tag{3.31}$$

where only two constants λ and μ, commonly known as *Lamé's constants* (G. Lamé, 1852), remain. Since μ relates a shear stress with a shear strain, it is called the shear modulus, and it is sometimes replaced by the symbol G. In indicial notation, Hooke's law (Eq. 3.11) for an isotropic solid can be written as

$$\sigma_{ij} = \lambda \varepsilon_{kk} \delta_{ij} + 2\mu \varepsilon_{ij} \tag{3.32}$$

In fact, Eq. 3.11 can be reduced to Eq. 3.32 upon setting

$$C_{ijkl} = \lambda \delta_{ij} \delta_{kl} + \mu(\delta_{ik}\delta_{jl} + \delta_{il}\delta_{jk}) \tag{3.33}$$

With the introduction of the *stress deviator* s_{ij} and the *strain deviator* e_{ij} as in Eqs. 2.34 and 2.47, Hooke's law for isotropic elastic solids separates into two parts, governing shear and bulk behavior, i.e.

$$s_{ij} = 2\mu e_{ij} \tag{3.34}$$

$$\sigma_{kk} = 3K \varepsilon_{kk} \tag{3.35}$$

where K, known as the *bulk modulus*, is equal to $\lambda + 2\mu/3$.

For the state of uniaxial stress in the x_1-direction (only the stress σ_{11} is non-zero), *Young's modulus* E (T. Young, 1807) and *Poisson's ratio* ν (S. D. Poisson, 1829) can be introduced through the relationship

$$\sigma_{11} = E \varepsilon_{11} \tag{3.36a}$$

and

$$\varepsilon_{22} = \varepsilon_{33} = -\nu \varepsilon_{11} \tag{3.36b}$$

In terms of E and ν the constitutive law for isotropic elastic solids becomes

$$\sigma_{ij} = \frac{E}{1+\nu}\left(\varepsilon_{ij} + \frac{\nu}{1-2\nu}\delta_{ij}\varepsilon_{kk}\right) \tag{3.37}$$

Table 3.1 shows a number of useful relations among the isotropic constants. In the state of a uniaxial stress where the non-zero stress and strains are as defined in Eqs. 3.36, real observations show that the signs of σ_{11} and ε_{11} are the same but opposite to that of ε_{22} and ε_{33}. Hence we may state that

$$E > 0, \qquad \nu > 0.$$

Table 3.1

	λ,μ	E,ν	μ,K	μ,ν
λ	λ	$\frac{E\nu}{(1+\nu)(1-2\nu)}$	$\frac{3K-2\mu}{3}$	$\frac{2\mu\nu}{1-2\nu}$
μ	μ	$\frac{E}{2(1+\nu)}$	μ	μ
K	$\lambda+\frac{2}{3}\mu$	$\frac{E}{3(1-2\nu)}$	K	$\frac{2\mu(1+\nu)}{3(1-2\nu)}$
E	$\frac{\mu(3\lambda+2\mu)}{\lambda+\mu}$	E	$\frac{9K\mu}{3K+\mu}$	$2\mu(1+\nu)$
ν	$\frac{\lambda}{2(\lambda+\mu)}$	ν	$\frac{3K-2\mu}{2(3K+\mu)}$	ν

Another special state of stress, known as *hydrostatic pressure*, is defined by $\sigma_{ij}=-p\delta_{ij}$. By employing Eq. 3.35 we find $p=-K\varepsilon_{kk}$, where $K=E/[3(1-2\nu)]$. Since a hydrostatic pressure p should give a negative dilatation, we should have $K>0$, i.e.

$$1-2\nu>0.$$

Hence the ranges of values of the engineering elastic constants E and ν are

$$E>0, \qquad 0\le\nu\le 1/2 \tag{3.38}$$

When $\nu=1/2$, the material is called *incompressible* since the dilatation remains zero irrespective of the stress components.

3.3 YOUNG'S MODULI AND POISSON'S RATIOS FOR ORTHOTROPIC ELASTIC SOLIDS

The inverse of Eq. 3.22, for $[C_{ij}]$ of orthotropy as defined in Eq. 3.28, is

$$\begin{Bmatrix}\varepsilon_{xx}\\ \varepsilon_{yy}\\ \varepsilon_{zz}\\ \gamma_{xy}\\ \gamma_{yz}\\ \gamma_{zx}\end{Bmatrix}=\begin{bmatrix}S_{11}&S_{12}&S_{13}&0&0&0\\ S_{21}&S_{22}&S_{23}&0&0&0\\ S_{31}&S_{32}&S_{33}&0&0&0\\ 0&0&0&S_{44}&0&0\\ 0&0&0&0&S_{55}&0\\ 0&0&0&0&0&S_{66}\end{bmatrix}\begin{Bmatrix}\sigma_{xx}\\ \sigma_{yy}\\ \sigma_{zz}\\ \sigma_{xy}\\ \sigma_{yz}\\ \sigma_{zx}\end{Bmatrix} \tag{3.39}$$

where

$$S_{ij}=S_{ji} \tag{3.40a}$$

$$[S_{ij}]=[C_{ij}]^{-1}; \quad i,j=1,2,3 \tag{3.40b}$$

and

$$S_{44} = C_{44}^{-1}, \quad S_{55} = C_{55}^{-1}, \quad S_{66} = C_{66}^{-1} \tag{3.40c}$$

An elastic solid in which

$$\sigma_{zz} = \sigma_{yz} = \sigma_{zx} = 0 \tag{3.41}$$

is said to be in a state of *plane stress*. Its essential stress-strain relations become

$$\begin{Bmatrix} \varepsilon_{xx} \\ \varepsilon_{yy} \\ \gamma_{xy} \end{Bmatrix} = \begin{bmatrix} S_{11} & S_{12} & 0 \\ S_{21} & S_{22} & 0 \\ 0 & 0 & S_{44} \end{bmatrix} \begin{Bmatrix} \sigma_{xx} \\ \sigma_{yy} \\ \sigma_{xy} \end{Bmatrix} \tag{3.42}$$

Taking the inverse of Eq. 3.42 yields,

$$\begin{Bmatrix} \sigma_{xx} \\ \sigma_{yy} \end{Bmatrix} = \begin{bmatrix} E_{11} & E_{12} \\ E_{21} & E_{22} \end{bmatrix} \begin{Bmatrix} \varepsilon_{xx} \\ \varepsilon_{yy} \end{Bmatrix} \tag{3.43a}$$

$$\gamma_{xy} = C_{44} \gamma_{xy} \tag{3.43b}$$

where

$$\begin{bmatrix} E_{11} & E_{12} \\ E_{21} & E_{22} \end{bmatrix} = \begin{bmatrix} S_{11} & S_{12} \\ S_{21} & S_{22} \end{bmatrix}^{-1} \tag{3.44a}$$

and

$$E_{ij} = E_{ji}; \quad i, j = 1, 2 \tag{3.44b}$$

Suppose it is found in an experiment that σ_{xx} is the only non-zero stress. Equation 3.43a gives

$$\sigma_{xx} = E_{11}\varepsilon_{xx} + E_{12}\varepsilon_{yy} \tag{3.45a}$$

$$0 = E_{12}\varepsilon_{xx} + E_{22}\varepsilon_{yy} \tag{3.45b}$$

or

$$\varepsilon_{xx} = \frac{\sigma_{xx} E_{22}}{E_{11}E_{22} - E_{12}^2} \tag{3.46a}$$

$$\varepsilon_{yy} = -\frac{\sigma_{xx} E_{12}}{E_{11}E_{22} - E_{12}^2} \tag{3.46b}$$

From this experiment we may determine the Young's modulus E_x and the Poisson's ratio ν_{xy} by using the equations

$$E_x = \frac{\sigma_{xx}}{\varepsilon_{xx}} \tag{3.47a}$$

$$\nu_{xy} = -\frac{\varepsilon_{yy}}{\varepsilon_{xx}} \tag{3.47b}$$

Substituting Eq. 3.47a into Eq. 3.46a yields

$$E_x = E_{11} - \frac{E_{12}^2}{E_{22}} \tag{3.48a}$$

Equation 3.45b gives

$$\nu_{xy} = \frac{E_{12}}{E_{22}} \tag{3.48b}$$

A repetition of this operation with σ_{yy} being the only non-zero stress leads to

$$E_y = E_{22} - \frac{E_{12}^2}{E_{11}} \tag{3.49a}$$

$$\nu_{yx} = \frac{E_{12}}{E_{11}} \tag{3.49b}$$

It should be noted that

$$\frac{E_x}{E_y} = \frac{\nu_{xy}}{\nu_{yx}} \tag{3.50}$$

In addition, it is customary to name C_{44} a shear modulus G_{xy}, and the 4 distinct elastic constants for the case of plane stress of orthotropic elastic materials (see Eqs. 3.43) are

$$E_{11} = \frac{E_x}{1 - \nu_{xy}\nu_{yx}} \tag{3.51a}$$

$$E_{12} = E_{21} = \frac{E_y \nu_{xy}}{1 - \nu_{xy}\nu_{yx}} = \frac{E_x \nu_{yx}}{1 - \nu_{xy}\nu_{yx}} \tag{3.51b}$$

$$E_{22} = \frac{E_y}{1 - \nu_{xy}\nu_{yx}} \tag{3.51c}$$

$$C_{44} = G_{xy} \tag{3.51d}$$

Moreover, S_{ij} in Eq. 3.42 can be obtained as

$$S_{11} = \frac{1}{E_x} \tag{3.52a}$$

$$S_{12} = S_{21} = -\frac{\nu_{xy}}{E_x} = -\frac{\nu_{yx}}{E_y} \tag{3.52b}$$

$$S_{22} = \frac{1}{E_y} \tag{3.52c}$$

$$S_{44} = \frac{1}{G_{xy}} \tag{3.52d}$$

Finally, we can write the remaining S_{ij} 's in Eq. 3.39 as

$$S_{33}=\frac{1}{E_z} \tag{3.52e}$$

$$S_{23}=S_{32}=-\frac{\nu_{yz}}{E_y}=-\frac{\nu_{zy}}{E_z} \tag{3.52f}$$

$$S_{55}=\frac{1}{G_{yz}} \tag{3.52g}$$

$$S_{13}=S_{31}=-\frac{\nu_{xz}}{E_x}=-\frac{\nu_{zx}}{E_z} \tag{3.52h}$$

$$S_{66}=\frac{1}{G_{zx}} \tag{3.52i}$$

In total, there are 9 elastic constants for an orthotropic elastic material.

The constitutive law for an orthotropic elastic material can be specialized for an isotropic elastic material by setting

$$E_x=E_y=E_z=E \tag{3.53a}$$

$$\nu_{xy}=\nu_{yz}=\nu_{zx}=\nu_{yx}=\nu_{zy}=\nu_{xz}=\nu \tag{3.53b}$$

and

$$G_{xy}=G_{yz}=G_{zx}=\mu=\frac{E}{2(1+\nu)} \tag{3.54}$$

3.4 SOLUTION SCHEMES

It was stated in Section 3.1 that if the process in an elastic system is adiabatic, the work done by the forces must be equal to the strain energy which is stored up in the system. This fact can be expressed in another form known as the *principle of virtual work*, i.e.

$$W_E=W_I \tag{3.55}$$

where W_E is external virtual work, i.e.

$$W_E=\int_V X_i\delta u_i dV+\int_S \overline{X}_i\delta u_i dS \tag{3.56}$$

and W_I is the internal virtual work or virtual strain energy (Eq. 3.7), i.e.

$$W_I=\int_V \sigma_{ij}\delta\varepsilon_{ij}dV \tag{3.57}$$

In the equations above, V denotes the material domain, S the surface completely enclosing V, and δ the *variational symbol* signifying a *virtual quantity* (Spiegel, 1971).

Due to Eq. 2.44, Eq. 3.57 can be written as

$$W_I = \int_V \sigma_{ij}\delta(u_{i,j})\mathrm{d}V,$$

or, upon applying the divergence theorem (Eq. 1.88),

$$W_I = \int_S \sigma_{ji}n_j\delta u_i \mathrm{d}S - \int_V \sigma_{ji,j}\delta u_i \mathrm{d}V \tag{3.58}$$

Substituting Eqs. 3.56 and 3.58 into Eq. 3.55 yields

$$\int_V (\sigma_{ji,j} + X_i)\delta u_i \mathrm{d}V + \int_S (\overline{X}_i - \sigma_{ji}n_j)\delta u_i \mathrm{d}S = 0,$$

which can be satisfied for arbitrary δu_i only if

$$\sigma_{ji,j} + X_i = 0, \quad \text{in } V \tag{3.59}$$

and either

$$\delta u_i = 0, \text{ on } S \tag{3.60a}$$

or

$$\overline{X}_i = \sigma_{ji}n_j, \text{ on } S \tag{3.60b}$$

Note that Eq. 3.59 is the equilibrium equation, Eq. 2.8, and Eq. 3.60b is the stress vector formula, Eq. 2.6. Equation 3.59 is a *field equation.* On the other hand, Eqs. 3.60 specify *boundary conditions* in a proper way, i.e. at a certain point on the boundary S either

$$u_i = \overline{u}_i \text{ or } \sigma_{ji}n_j = \overline{X}_i \tag{3.61}$$

in which a bar $(^{-})$ denotes a prescribed quantity. These *well-posed boundary conditions* state that, at a certain point on the material boundary, either $\overline{u}_1$ or $\overline{X}_1$, either $\overline{u}_2$ or $\overline{X}_2$, and either $\overline{u}_3$ or $\overline{X}_3$ must be prescribed. If, as an example, the boundary conditions are the prescription of $\overline{u}_1$, $\overline{u}_2$ and $\overline{X}_2$, the solution to the problem may not be possible since the boundary conditions are not well posed. In other words, boundary conditions are improper and the solution may not exist if a displacement and a traction in the same direction are prescribed at a certain point on the boundary of the material domain.

The principle of virtual work and its consequences derived so far hold for nonlinear as well as linear elastic materials. Each *boundary value problem* is governed by fifteen field equations to be solved for fifteen unknown variables. The fifteen field equations are three equations of motion (Eq. 3.59), six equations for the strain-displacement relationship (Eq. 2.44), and six equations for the constitutive relationship (e.g. Eq. 3.11 for linear elastic solids). On the other hand, the fifteen unknown variables are three displacements, six stresses and six strains. The boundary conditions (Eqs. 3.61) are needed for determining uniquely the constants arising from integrating the equations of motion and the strain-displacement relations, which are differential equations.

For linear elastic solids, solution methods may be classified into two main schemes, one in which the basic unknowns are displacements, one in which they are stresses or strains. In the first scheme, the fifteen field equations can be reduced, by simple substitutions, to three field equations for the three displacements. For an isotropic elastic solid the equations are

$$\mu u_{i,kk}+(\lambda+\mu)\varepsilon_{kk,i}+X_i=0.$$

as described in the next section. In the second scheme, since strains (or stresses) are chosen to be the basic unknowns, the compatibility conditions (Eq. 2.62) must be incorporated in the approach to guarantee the uniqueness and continuity of the solution. In general, the first scheme is more straight-forward and should be more efficient than the second scheme, since it involves fewer unknowns. The second scheme is effective for elastostatics of isotropic plane problems, to be discussed in Chapter IV. In Chapter V a 'mixed' scheme developed for elastic bending of thin plates is presented. In Chapter VI and VIII the first scheme is described for elastostatics and elastodynamics of isotropic solids.

Sometimes it is convenient to replace one boundary value problem by using the *Maxwell-Betti reciprocal theorem* (J. C. Maxwell, 1870 and E. Betti, 1872) which states that:

'In a linear elastic solid, the work done by a set of forces acting through the corresponding displacements produced by a second set of forces is equal to the work done by the second set of forces acting through the corresponding displacements produced by the first set of forces.'

Denoting the first and second sets by superscripts 1 and 2 respectively, we can state the theorem as

$$\int_V X_i^{(1)}u_i^{(2)}\mathrm{d}V+\int_S \overline{X}_i^{(1)}u_i^{(2)}\mathrm{d}S=\int_V X_i^{(2)}u_i^{(1)}\mathrm{d}V+\int_S \overline{X}_i^{(2)}u_i^{(1)}\mathrm{d}S \tag{3.62}$$

The proof of this theorem starts with rewriting the left-hand side of the equation above by using the procedure for obtaining Eq. 3.7. The result is

$$\int_V \sigma_{ij}^{(1)}\varepsilon_{ij}^{(2)}\mathrm{d}V,$$

or, due to Hooke's law (Eq. 3.11),

$$\int_V C_{ijkl}\varepsilon_{kl}^{(1)}\varepsilon_{ij}^{(2)}\mathrm{d}V.$$

Similarly, the right-hand side of Eq. 3.62 can be rewritten as

$$\int_V C_{ijkl}\varepsilon_{kl}^{(2)}\varepsilon_{ij}^{(1)}\mathrm{d}V.$$

The last two expressions are identical since $C_{ijkl}=C_{klij}$ (Eq. 3.14), and thus the theorem is proved.

3.5 FIELD EQUATIONS IN TERMS OF DISPLACEMENTS

In this section, displacement field equations will be presented for the Cartesian, cylindrical and spherical coordinates. The presentation will be limited to linear isotropic elastic solids, of which the constitutive laws are: for the Cartesian system (Eq. 3.32),

$$\sigma_{ij} = \lambda\varepsilon_{kk}\delta_{ij} + 2\mu\varepsilon_{ij} \tag{3.63}$$

for the cylindrical system,

$$\begin{Bmatrix}\sigma_{rr}\\ \sigma_{\theta\theta}\\ \sigma_{zz}\\ \sigma_{r\theta}\\ \sigma_{\theta z}\\ \sigma_{zr}\end{Bmatrix} = \begin{bmatrix}\lambda+2\mu & \lambda & \lambda & 0 & 0 & 0\\ & \lambda+2\mu & \lambda & 0 & 0 & 0\\ & & \lambda+2\mu & 0 & 0 & 0\\ & & & 2\mu & 0 & 0\\ & \text{Symm.} & & & 2\mu & 0\\ & & & & & 2\mu\end{bmatrix}\begin{Bmatrix}\varepsilon_{rr}\\ \varepsilon_{\theta\theta}\\ \varepsilon_{zz}\\ \varepsilon_{r\theta}\\ \varepsilon_{\theta z}\\ \varepsilon_{zr}\end{Bmatrix} \tag{3.64}$$

and for the spherical system,

$$\begin{Bmatrix}\sigma_{RR}\\ \sigma_{\phi\phi}\\ \sigma_{\theta\theta}\\ \sigma_{R\phi}\\ \sigma_{\phi\theta}\\ \sigma_{\theta R}\end{Bmatrix} = \begin{bmatrix}\lambda+2\mu & \lambda & \lambda & 0 & 0 & 0\\ & \lambda+2\mu & \lambda & 0 & 0 & 0\\ & & \lambda+2\mu & 0 & 0 & 0\\ & & & 2\mu & 0 & 0\\ & \text{Symm.} & & & 2\mu & 0\\ & & & & & 2\mu\end{bmatrix}\begin{Bmatrix}\varepsilon_{RR}\\ \varepsilon_{\phi\phi}\\ \varepsilon_{\theta\theta}\\ \varepsilon_{R\phi}\\ \varepsilon_{\phi\theta}\\ \varepsilon_{\theta R}\end{Bmatrix} \tag{3.65}$$

Substituting Eq. 3.63 into the equations of motion (Eq. 2.14), and using the strain-displacement relationship (Eq. 2.44), we obtain the Cartesian displacement field equations known as *Navier equations* (C. L. M. H. Navier, 1823 and 1827), i.e.

$$\mu u_{i,kk} + (\lambda+\mu)\varepsilon_{kk,i} + X_i = 0 \tag{3.66}$$

in which it has been assumed that the material is homogeneous, i.e. λ and μ are constants. The cylindrical Navier equations can be obtained, by using Eqs. 2.71, 2.74 and 3.64, as

$$(\lambda+\mu)\frac{\partial}{\partial r}(\varepsilon_{kk}) + \mu\left(\nabla^2 u_r - \frac{u_r}{r^2} - \frac{2}{r^2}\frac{\partial u_\theta}{\partial\theta}\right) + X_r = 0 \tag{3.67a}$$

$$(\lambda+\mu)\frac{1}{r}\frac{\partial}{\partial\theta}(\varepsilon_{kk}) + \mu\left(\nabla^2 u_\theta + \frac{2}{r^2}\frac{\partial u_r}{\partial\theta} - \frac{u_\theta}{r^2}\right) + X_\theta = 0 \tag{3.67b}$$

$$(\lambda+\mu)\frac{\partial}{\partial z}(\varepsilon_{kk}) + \mu(\nabla^2 u_z) + X_z = 0 \tag{3.67c}$$

The spherical Navier equations can be obtained, by using Eqs. 2.81, 2.84 and 3.65, as

$$(\lambda+\mu)\frac{\partial}{\partial R}(\varepsilon_{kk}) + \mu\left(\nabla^2 u_R - \frac{2u_R}{R^2} - \frac{2}{R^2}\frac{\partial u_\phi}{\partial\phi} - \frac{2u_\phi\cot\phi}{R^2} - \frac{2}{R^2\sin\phi}\frac{\partial u_\theta}{\partial\theta}\right) + X_R = 0,$$

$$(\lambda+\mu)\frac{1}{R}\frac{\partial}{\partial\phi}(\varepsilon_{kk})+\mu\left(\nabla^2 u_\phi+\frac{2}{R^2}\frac{\partial u_R}{\partial\phi}-\frac{u_\phi}{R^2\sin^2\phi}-\frac{2\cos\phi}{R^2\sin^2\phi}\frac{\partial u_\theta}{\partial\theta}\right)+X_\phi=0,$$

$$(\lambda+\mu)\frac{1}{R\sin\phi}\frac{\partial}{\partial\theta}(\varepsilon_{kk})+\mu\left(\nabla^2 u_\theta-\frac{u_\theta}{R^2\sin^2\phi}+\frac{2}{R^2\sin^2\phi}\frac{\partial u_R}{\partial\theta}+\frac{2\cos\phi}{R^2\sin^2\phi}\frac{\partial u_\phi}{\partial\theta}\right)+X_\theta=0$$

(3.68a,b,c)

In these equations, ∇^2 is an invariant operator as defined in Eq. 1.85d, i.e.

$$\nabla^2=\frac{\partial^2}{\partial x_k\partial x_k} \tag{3.69a}$$

$$=\frac{1}{r}\frac{\partial}{\partial r}\left(r\frac{\partial}{\partial r}\right)+\frac{1}{r^2}\frac{\partial^2}{\partial\theta^2}+\frac{\partial^2}{\partial z^2} \tag{3.69b}$$

$$=\frac{1}{R^2}\frac{\partial}{\partial R}\left(R^2\frac{\partial}{\partial R}\right)+\frac{1}{R^2\sin\phi}\frac{\partial}{\partial\phi}\left(\sin\phi\frac{\partial}{\partial\phi}\right)+\frac{1}{R^2\sin^2\phi}\frac{\partial^2}{\partial\theta^2} \tag{3.69c}$$

and ε_{kk} is the dilatation defined in Eq. 2.72 for cylindrical coordinates and in Eq. 2.82 for spherical coordinates.

3.6 PROBLEMS

3.6.1 *Beltrami-Michell Compatibility Conditions*

The stress-strain relationship for an isotropic elastic solid, Eq. 3.37, has the inverse

$$E\varepsilon_{ij}=(1+\nu)\sigma_{ij}-\nu\delta_{ij}\sigma_{kk} \tag{3.70}$$

Thus, for a homogeneous material,

$$E\varepsilon_{ij,kk}=(1+\nu)\sigma_{ij,kk}-\nu\delta_{ij}\sigma_{kk,mm} \tag{3.71a}$$

$$E\varepsilon_{kk,ij}=(1-2\nu)\sigma_{kk,ij} \tag{3.71b}$$

$$E\varepsilon_{ik,jk}=(1+\nu)\sigma_{ik,jk}-\nu\sigma_{kk,ij} \tag{3.72a}$$

$$E\varepsilon_{jk,ik}=(1+\nu)\sigma_{jk,ik}-\nu\sigma_{kk,ij} \tag{3.72b}$$

A spatial derivative of the equation of equilibrium, Eq. 2.8, can be put as

$$\sigma_{ik,jk}=-X_{i,j} \tag{3.73}$$

Due to Eqs. 3.72 and 3.73, we can write

$$E(\varepsilon_{ik,jk}+\varepsilon_{jk,ik})=-(1+\nu)(X_{i,j}+X_{j,i})-2\nu\sigma_{kk,ij} \tag{3.74}$$

Substituting Eqs. 3.71 and 3.74 into the compatibility conditions, Eq. 2.62, with k summed, we obtain

$$(1+\nu)\sigma_{ij,kk}+\sigma_{kk,ij}-\nu\delta_{ij}\sigma_{kk,mm}+(1+\nu)(X_{i,j}+X_{j,i})=0 \tag{3.75}$$

Summing the equation above for $i=j$, we find

$$\nabla^2\sigma_{kk} = -\frac{1+\nu}{1-\nu}X_{k,k} \tag{3.76}$$

Substituting Eq. 3.76 into Eq. 3.75 yields the *Beltrami-Michell Compatibility conditions* (E. Beltrami, 1892 and J. H. Michell, 1900), i.e.

$$\nabla^2\sigma_{ij} + \frac{1}{1+\nu}\sigma_{kk,ij} + \frac{\nu}{1-\nu}\delta_{ij}X_{k,k} + X_{i,j} + X_{j,i} = 0 \tag{3.77}$$

If the body force X_i is zero or constant, Eqs. 3.76 and 3.77 show that σ_{kk} is *harmonic*, in the sense that it satisfies *Laplace* (*harmonic*) *equation*,

$$\nabla^2\sigma_{kk} = 0 \tag{3.78}$$

and Eq. 3.77 becomes

$$\nabla^2\sigma_{ij} + \frac{1}{1+\nu}\sigma_{kk,ij} = 0 \tag{3.79}$$

Applying the Laplace operator ∇^2 on the equation above, and using Eq. 3.78, we obtain the *biharmonic equation*

$$\nabla^4\sigma_{ij} = 0 \tag{3.80}$$

where ∇^4 is the *biharmonic operator*, i.e.

$$\nabla^4 = \nabla^2\nabla^2 \tag{3.81}$$

Equation 3.80 is a set of six distinct uncoupled partial differential equations. It can be shown that any *biharmonic function* Φ, i.e. a solution of

$$\nabla^4\Phi = 0 \tag{3.82}$$

can be written

$$\Phi = x_k x_k \Psi + x_i \Psi_i + \Psi + C_{ijk} x_i x_j x_k \tag{3.83}$$

where C is an arbitrary constant tensor, Ψ a harmonic scalar field, and Ψ_i a harmonic vector field, i.e.

$$\nabla^2\Psi = 0, \qquad \nabla^2\Psi_i = 0 \tag{3.84a,b}$$

It is interesting to note that any function χ, defined by

$$\chi = x_i x_j \Psi_{,ij} \tag{3.85}$$

is harmonic if Ψ is harmonic.

3.6.2 *Conservative Forces and Potentials*

If (and only if) the body force field is *conservative*, then it has a potential V such that

$$X_i = -V_{,i} \tag{3.86}$$

So that a simple check for X_i to be conservative is the equation

$$X_{i,j} = X_{j,i} \tag{3.87}$$

and the compatibility conditions (Eq. 3.77) become

$$\nabla^2 \sigma_{ij} + \frac{1}{1+\nu} \sigma_{kk,ij} - \frac{\nu}{1-\nu} \delta_{ij} \nabla^2 V - 2V_{,ij} = 0 \tag{3.88}$$

The static equilibrium equation, Eq. 2.8, is satisfied exactly if the stress tensor is expressed in terms of a stress potential Φ as the following

$$\sigma_{ij} = \delta_{ij}(\nabla^2 \Phi + V) - \Phi_{,ij} \tag{3.89}$$

Note if X_i vanishes, we can set V to zero without loss of generality. Substituting Eq. 3.89 into Eq. 3.88, we find

$$\delta_{ij}\left(\nabla^4 \Phi + \frac{1-2\nu}{1-\nu} \nabla^2 V\right) + \frac{1-\nu}{1+\nu}\left(\nabla^2 \Phi + \frac{1-2\nu}{1-\nu} V\right)_{,ij} = 0 \tag{3.90}$$

Summing the equation above with $i = j$ leads to

$$\nabla^4 \Phi + \frac{1-2\nu}{1-\nu} \nabla^2 V = 0 \tag{3.91}$$

Incorporating Eq. 3.91 in Eq. 3.90 yields

$$\nabla^2 \Phi_{,ij} = -\frac{1-2\nu}{1-\nu} V_{,ij} \tag{3.92}$$

Thus for an elastostatic, isotropic, homogeneous, conservative system, the solution via stresses leads to the search for a single scalar function Φ linked to the stresses by Eq. 3.89 and satisfying Eq. 3.92. The solution to Eq. 3.92 can be taken as a sum of a *particular solution* and a *complementary solution.* While such a particular solution satisfies the governing equation exactly without involving any arbitrary constants, any complementary solution satisfies the homogeneous governing equation, i.e.

$$\nabla^2 \Phi_{,ij} = 0 \tag{3.93}$$

The solution to which can be shown to be

$$\Phi = \Psi + C_{ijk} x_i x_j x_k + C_{ij} x_i x_j + C_i x_i x_k x_k + C x_k x_k \tag{3.94}$$

where Ψ is any harmonic function, and C's are constant tensors. Note that polynomials of x_i of lower degrees than two are disregarded since they give trivial stress components. The same solution scheme is described in more detail in Chapter IV, particularly for two-dimensional cases.

3.6.3 *Elastostatic Displacement Potentials*

We shall now describe solutions which start from the displacements rather than the stresses. If the displacement components are given by

$$u_i = \Phi_{,i} \tag{3.95}$$

then the dilatation ε_{kk} becomes

$$\varepsilon_{kk} = \nabla^2\Phi \tag{3.96}$$

and the static equilibrium equations for homogeneous isotropic solids, Eq. 3.66, become

$$(\lambda + 2\mu)(\nabla^2\Phi)_{,i} + X_i = 0 \tag{3.97}$$

where X_i is the actual body force vector. The function Φ is called *Lamé's strain potential.*

Similarly, if

$$u_i = \varepsilon_{ijk}\Psi_{k,j} \tag{3.98}$$

the dilatation becomes

$$\begin{aligned}
\varepsilon_{ii} &= \varepsilon_{ijk}\Psi_{k,ij} \\
&= \frac{1}{2}(\varepsilon_{ijk}\Psi_{k,ij} + \varepsilon_{jik}\Psi_{k,ji}) \\
&= \frac{1}{2}(\varepsilon_{ijk}\Psi_{k,ij} - \varepsilon_{ijk}\Psi_{k,ji}) \\
&= 0
\end{aligned} \tag{3.99}$$

and Eq. 3.66 is specialized into

$$\mu\varepsilon_{ijk}(\nabla^2\Psi_k)_{,j} + X_i = 0 \tag{3.100}$$

Lastly, if we set (Galerkin, 1930)

$$u_i = \frac{\lambda + 2\mu}{\lambda + \mu}\nabla^2\chi_i - \chi_{k,ki} \tag{3.101}$$

we can show that the dilatation and the equilibrium equations become, respectively,

$$\varepsilon_{kk} = \frac{\mu}{\lambda + \mu}(\nabla^2\chi_k)_{,k} \tag{3.102}$$

and

$$\mu\left(\frac{\lambda + 2\mu}{\lambda + \mu}\right)\nabla^4\chi_i + X_i = 0 \tag{3.103}$$

The vector χ_i is called the *Galerkin's vector.*

The functions Φ, Ψ_i and χ_i as introduced above are known as displacement potentials. If the actual body force X_i is conservative (Eq. 3.86) and each displacement component is put as a linear combination of Eqs. 3.95, 3.98 and 3.101, i.e.

$$u_i = \Phi_{,i} + \varepsilon_{ijk}\Psi_{k,j} + \frac{\lambda + 2\mu}{\lambda + \mu}\nabla^2\chi_i - \chi_{k,ki} \tag{3.104}$$

the equilibrium equations, Eq. 3.66, can be satisfied exactly, provided that Φ satisfies the equation

$$(\lambda + 2\mu)\nabla^2\Phi = V \tag{3.105}$$

that Ψ_i is a harmonic function, i.e.

$$\nabla^2 \Psi_i = 0 \tag{3.106}$$

and that χ_i is a biharmonic function, i.e.

$$\nabla^4 \chi_i = 0 \tag{3.107}$$

Thus Navier equations can be uncoupled by means of the potentials as defined by Eqs. 3.104 to 3.107.

In Section 6.3, elastostatic displacement potentials appropriate for a homogeneous isotropic domain infinitely extended radially ($0 \le r < \infty$) are described. Other elastostatic displacement potentials such as those proposed by Papkovich (1932) and Neuber (1934) appear in several books on solid mechanics, e.g. Sokolnikoff (1956).

3.6.4 *Elastodynamic Displacement Potentials*

The introduction of the potential Φ as in Eq. 3.95 yields the dilatation as in Eq. 3.96, and the equations of motion (Eq. 3.66) for the case of zero actual body force as

$$(\lambda + 2\mu)(\nabla^2 \Phi)_{,i} = \rho \left(\frac{\partial^2 \Phi}{\partial t^2} \right)_{,i} \tag{3.108}$$

Similarly, the potential Ψ_i introduced as in Eq. 3.98 gives the dilatation as in Eq. 3.99, and the equations of motion as

$$\mu \varepsilon_{ijk} (\nabla^2 \Psi_k)_{,j} = \rho \varepsilon_{ijk} \left(\frac{\partial^2 \Psi_k}{\partial t^2} \right)_{,j} \tag{3.109}$$

Incorporating the results above, we can state that the displacement vector

$$u_i = \Phi_{,i} + \varepsilon_{ijk} \Psi_{k,j} \tag{3.110}$$

satisfies the equations of motion exactly, provided that Φ and Ψ_i are solutions to the *wave equations*

$$c_p^2 \nabla^2 \Phi = \frac{\partial^2 \Phi}{\partial t^2} \tag{3.111}$$

and

$$c_s^2 \nabla^2 \Psi_i = \frac{\partial^2 \Psi_i}{\partial t^2} \tag{3.112}$$

Here

$$c_p = [(\lambda + 2\mu)/\rho]^{1/2} \tag{3.113a}$$

is the *pressure wave speed*, and

$$c_s = (\mu/\rho)^{1/2} \tag{3.113b}$$

the *shear wave speed*. They are discussed further in Chapter VIII. So we can say that the equations of motion of a homogeneous isotropic elastic domain under zero actual body force can be uncoupled into two wave equations by means of the potentials as defined by Eqs. 3.110 to 3.112.

In Section 8.8, we describe elastodynamic displacement potentials appropriate for a homogeneous isotropic domain infinitely extended radially ($0 \le r < \infty$). More general forms of cylindrical and spherical displacement components in terms of appropriate potentials have been listed by Achenbach (1973).

3.6.5 *Positive Definiteness of the Strain Energy Function*

Since the strain energy function must be positive for any non-trivial set of the strain components, the stiffness tensor C_{ij} in Eq. 3.22 must satisfy the conditions stipulated by Eq. 1.73, i.e.

$$C_{11} > 0, \quad \begin{vmatrix} C_{11} & C_{12} \\ C_{21} & C_{22} \end{vmatrix} > 0, \quad \begin{vmatrix} C_{11} & C_{12} & C_{13} \\ C_{21} & C_{22} & C_{23} \\ C_{31} & C_{32} & C_{33} \end{vmatrix} > 0, \ldots,$$

$$\begin{vmatrix} C_{11} & C_{12} & C_{13} & C_{14} & C_{15} & C_{16} \\ & C_{22} & C_{23} & C_{24} & C_{25} & C_{26} \\ & & C_{33} & C_{34} & C_{35} & C_{36} \\ & & & C_{44} & C_{45} & C_{46} \\ & \text{Symm.} & & & C_{55} & C_{56} \\ & & & & & C_{66} \end{vmatrix} > 0 \qquad (3.114)$$

For orthotropic solids (Eq. 3.28), these conditions can be specialized into

$$C_{11} > 0, \quad \begin{vmatrix} C_{11} & C_{12} \\ C_{21} & C_{22} \end{vmatrix} > 0, \quad \begin{vmatrix} C_{11} & C_{12} & C_{13} \\ C_{21} & C_{22} & C_{23} \\ C_{31} & C_{32} & C_{33} \end{vmatrix} > 0 \qquad (3.115a,b,c)$$

$$C_{44} > 0, \quad C_{55} > 0, \quad C_{66} > 0 \qquad (3.115d,e,f)$$

Simplest among all, are the conditions for isotropic solids (Eq. 3.31), i.e.

$$\mu > 0, \quad 3\lambda + 2\mu > 0 \qquad (3.116a,b)$$

Equation 3.116b states that Poisson's ratio ν is less than $1/2$.

3.6.6 *Stress and Strain Computation from Measured Results*

At a point 0 on a surface enclosing an isotropic elastic domain with Poisson's ratio $\nu = 0.3$, the axes x and y are drawn tangent to the surface and the axis z normal outwardly to the domain. A surface traction of a magnitude 10 is applied at point 0

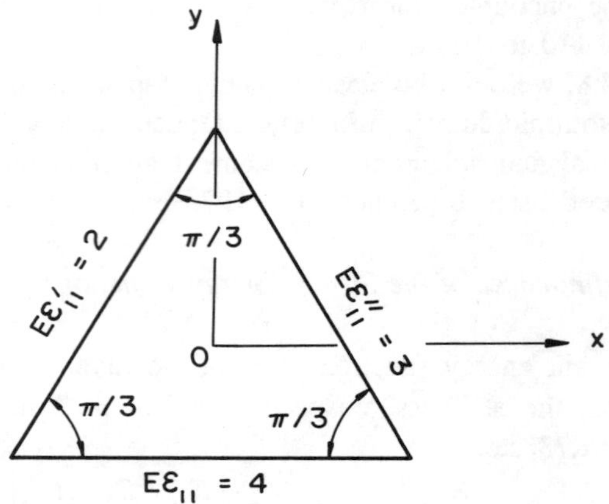

Fig. 3.4

in the direction of a vector N_i with $(N_1, N_2, N_3) = (1, -1, 1)$. In addition, three strain gages placed at point 0 give normal strains as shown in Fig. 3.4. Determine the stresses and strains at point 0.

Answer:

$$[\sigma_{ij}] = \begin{bmatrix} 7.530 & -0.444 & 5.774 \\ & 5.991 & -5.774 \\ \text{Symm.} & & 5.774 \end{bmatrix},$$

$$E[\varepsilon_{ij}] = \begin{bmatrix} 4 & -0.577 & 7.506 \\ & 2 & -7.506 \\ \text{Symm.} & & 1.718 \end{bmatrix}.$$

3.6.7 *Saturated Porous Elastic Media*

Biot (1941) proposed the following theory for *quasi-statics* of porous isotropic elastic solids completely saturated with fluids. In the absence of body forces, the equilibrium conditions can be written

$$\sigma_{ij,j} + \delta_{ij} p_{f,j} = 0 \tag{3.117}$$

where σ_{ij} is the *effective stress* tensor, and p_f the *excess pore pressure* (positive if in tension). The stress-strain relationship is in the usual form

$$\sigma_{ij} = 2\mu\varepsilon_{ij} + \delta_{ij}\lambda\varepsilon_{kk} \tag{3.118}$$

where λ and μ are Lamé's constants of the bulk material. The strains and the dilatation also take the common forms, i.e.

$$\varepsilon_{ij} = \frac{1}{2}(u_{i,j} + u_{j,i}),$$

$$\varepsilon_{kk} = u_{k,k},$$

where u_i is a displacement component of the solid matrix. The fluid flow is governed by *Darcy's law* in the form

$$\frac{\partial w_i}{\partial t} = k p_{f,i} \tag{3.119}$$

where k is the *coefficient of permeability*, and

$$w_i = f(U_i - u_i) \tag{3.120}$$

in which U_i is a displacement component of the fluid, and f the *porosity*, i.e. the volume of the voids per unit total volume. The condition of perfect saturation with an incompressible fluid implies that

$$\varepsilon_{kk} = -w_{i,i} \tag{3.121}$$

So that Darcy's law becomes

$$(\lambda + 2\mu) k \nabla^2 (w_{i,i}) = \frac{\partial w_{i,i}}{\partial t} \tag{3.122}$$

Substituting Eq. 3.118 into Eq. 3.117 yields the Navier equations for the saturated porous solid as

$$\mu \nabla^2 u_i + (\lambda + \mu)\varepsilon_{kk,i} + p_{f,i} = 0 \tag{3.123}$$

while Darcy's law (Eq. 3.122), due to Eq. 3.121, becomes

$$c \nabla^2 \varepsilon_{kk} = \frac{\partial \varepsilon_{kk}}{\partial t} \tag{3.124}$$

where c is the *coefficient of consolidation*, i.e.

$$c = (\lambda + 2\mu) k \tag{3.125}$$

Note that the Navier equations for solids (Eq. 3.66) and for saturated solids (Eq. 3.123) become identical if the body force X_i in the former is replaced by the *excess pore pressure gradient* $p_{f,i}$. Darcy's law can be put in another form after substituting Eq. 3.121 into Eq. 3.124 and then using Eq. 3.119, i.e.

$$c \nabla^2 \varepsilon_{kk} = -k \nabla^2 p_f \tag{3.126}$$

The well-posed boundary conditions for the present case can be obtained by means of a variational principle. They state that, at a certain point on the boundary, either

$$u_i = \overline{u}_i \quad \text{or} \quad n_j(\sigma_{ij} + \delta_{ij} p_f) = \overline{X}_i \tag{3.127a}$$

and, either

$$p_f = \overline{p}_f \quad \text{or} \quad k p_{f,i} n_i = \overline{Q} \tag{3.127b}$$

Here the bar ($^{-}$) denotes a prescribed quantity, and $\overline{Q}$ the prescribed rate of the outflow.

It can be seen that there are four basic unknown functions u_1, u_2, u_3 and p_f, and four governing field equations Eqs. 3.123 and 3.124. The problem becomes identical to that of pure elastic solids if p_f tends to zero, which happens if the system is subjected to constant forces for a sufficiently large time. In Section 6.8, we discuss problems of saturated porous elastic domains infinitely extended radially ($0 \leq r < \infty$).

CHAPTER IV

ELASTOSTATIC PLANE PROBLEMS

4.1 PLANE PROBLEMS OF ORTHOTROPIC ELASTIC MATERIALS

For orthotropic elastic case of *plane stress* in the xy-plane, the three stresses on the plane normal to the z-axis vanish as in Eq. 3.41. The remaining stresses are related to strains as in Eqs. 3.42, with the coefficients S_{ij} derived in Eqs. 3.52a to d. Altogether there are eight field equations: two equilibrium equations, three stress-strain relations, and three strain-displacement relations. The equilibrium equations are Eqs. 2.4 rewritten in explicit symbols as

$$\frac{\partial \sigma_{xx}}{\partial x}+\frac{\partial \sigma_{xy}}{\partial y}+X=0 \tag{4.1a}$$

$$\frac{\partial \sigma_{xy}}{\partial x}+\frac{\partial \sigma_{yy}}{\partial y}+Y=0 \tag{4.1b}$$

where X and Y are the body forces in the x and y directions, respectively. The stress-strain relations are Eqs. 3.42. The strain-displacement relations are Eqs. 2.43, i.e.

$$\varepsilon_{xx}=\frac{\partial u}{\partial x} \tag{4.2a}$$

$$\varepsilon_{yy}=\frac{\partial v}{\partial y} \tag{4.2b}$$

$$\gamma_{xy}=\frac{\partial v}{\partial x}+\frac{\partial u}{\partial y} \tag{4.2c}$$

Equally, there are eight unknowns: 3 stresses $(\sigma_{xx},\sigma_{yy},\sigma_{xy})$, 3 strains $(\varepsilon_{xx},\varepsilon_{yy},\gamma_{xy})$, and 2 displacements (u,v). In addition, since we will encounter some constants of integration in solving the differential equations, Eqs. 4.1 and 4.2, we will have to prescribe some boundary conditions.

The boundary conditions, Eqs. 3.61, may be written

$$\overline{X}=\sigma_{xx}n_x+\sigma_{xy}n_y \quad \text{or} \quad \overline{u}=u \tag{4.3a}$$

$$\overline{Y}=\sigma_{xy}n_x+\sigma_{yy}n_y \quad \text{or} \quad \overline{v}=v \tag{4.3b}$$

where a bar $(^-)$ denotes a prescribed quantity on the boundary, and n_x and n_y are directions cosines of the outward drawn normal to the boundary, thus

$$n_x^2+n_y^2=1 \tag{4.3c}$$

Rigid-body displacements (Eq. 2.53 for 3-dimensional cases) for plane problems are

$$u=Ay+B \tag{4.3d}$$

$$v=-Ax+C \tag{4.3e}$$

where A, B and C are arbitrary constants.

A material body is said to be in state of *plane strain* in the xy-plane if the displacement w vanishes and u and v are independent of z. It is found that all the governing conditions derived already for plane stress problems hold, with the exception that the stress-strain relations for plane strain should be derived from the general relationship, Eqs. 3.39 with

$$\varepsilon_{zz}=\gamma_{yz}=\gamma_{zx}=0 \tag{4.4}$$

This leads to the following relationship among the pertinent stresses and strains.

$$\begin{Bmatrix}\varepsilon_{xx}\\ \varepsilon_{yy}\\ \gamma_{xy}\end{Bmatrix}=\begin{bmatrix} S_{11}-\dfrac{S_{13}S_{31}}{S_{33}} & S_{12}-\dfrac{S_{13}S_{32}}{S_{33}} & 0\\ S_{21}-\dfrac{S_{23}S_{31}}{S_{33}} & S_{22}-\dfrac{S_{23}S_{32}}{S_{33}} & 0\\ 0 & 0 & S_{44}\end{bmatrix}\begin{Bmatrix}\sigma_{xx}\\ \sigma_{yy}\\ \sigma_{xy}\end{Bmatrix} \tag{4.5}$$

where S_{ij}'s are given in Eqs. 3.52. It should be noted that σ_{zz} in a plane strain problem does not vanish.

For an isotropic elastic material, the stress-strain relations for both plane stress and plane strain problems can be written in a form known as the *unified Hooke's law for isotropic plane problems*, i.e.

$$\varepsilon_{xx}=\frac{1}{2\mu}\left[\sigma_{xx}-\frac{3-\kappa}{4}(\sigma_{xx}+\sigma_{yy})\right] \tag{4.6a}$$

$$\varepsilon_{yy}=\frac{1}{2\mu}\left[\sigma_{yy}-\frac{3-\kappa}{4}(\sigma_{xx}+\sigma_{yy})\right] \tag{4.6b}$$

$$\gamma_{xy}=\frac{1}{\mu}\sigma_{xy} \tag{4.6c}$$

or, in reverse,

$$\sigma_{xx}=\frac{\mu}{\kappa-1}[(\kappa+1)\varepsilon_{xx}+(3-\kappa)\varepsilon_{yy}] \tag{4.7a}$$

$$\sigma_{yy}=\frac{\mu}{\kappa-1}[(\kappa+1)\varepsilon_{yy}+(3-\kappa)\varepsilon_{xx}] \tag{4.7b}$$

$$\sigma_{xy}=\mu\gamma_{xy} \tag{4.7c}$$

where

$$\kappa = \begin{cases} (3-\nu)/(1+\nu), & \text{for plane stress} \\ 3-4\nu, & \text{for plane strain} \end{cases} \tag{4.8a,b}$$

Thus a solution to a plane strain problem with a Poisson's ratio ν is equivalent to that of a plane stress with a Poisson's ratio $= \nu/(1-\nu)$.

4.2 AIRY FUNCTION FOR ISOTROPIC PLANE PROBLEMS

This is an approach using stresses as basic unknowns. Equilibrium is governed by Eqs. 4.1, stress-strain relationship by Eqs. 4.6 or 4.7, strain-displacement relationship by Eqs. 4.2, and boundary conditions are in the form of Eqs. 4.3. Among the 6 compatibility conditions stated in Eq. 2.62, we have to consider only the condition

$$\frac{\partial^2 \varepsilon_{xx}}{\partial y^2} + \frac{\partial^2 \varepsilon_{yy}}{\partial x^2} - \frac{\partial^2 \gamma_{xy}}{\partial x \partial y} = 0 \tag{4.9}$$

Other 5 compatibility conditions are satisfied identically in plane strain problems. For plane stress problems, the other 5 compatibility conditions will be satisfied if

$$\frac{\partial^2 \varepsilon_{zz}}{\partial x^2} = \frac{\partial^2 \varepsilon_{zz}}{\partial y^2} = \frac{\partial^2 \varepsilon_{zz}}{\partial x \partial y} = 0 \tag{4.10}$$

or

$$\varepsilon_{zz} = Ax + By + C \tag{4.11}$$

where A, B and C are constants of integration. Such a solution in Eq. 4.11 is rather an exception than a general solution. Thus the solution to a plane stress problem satisfying Eq. 4.9 is not exact, but should be a good approximation to a thin plate problem.

Substituting Eqs. 4.6 into Eq. 4.9 yields

$$\frac{(1+\kappa)}{4}\nabla^2(\sigma_{xx} + \sigma_{yy}) - 2\frac{\partial^2 \sigma_{xy}}{\partial x \partial y} - \frac{\partial^2 \sigma_{xx}}{\partial x^2} - \frac{\partial^2 \sigma_{yy}}{\partial y^2} = 0 \tag{4.12}$$

For the present two-dimensional case, the *Laplace operator* (*harmonic operator*) ∇^2 is

$$\nabla^2 = \frac{\partial^2}{\partial x^2} + \frac{\partial^2}{\partial y^2} \tag{4.13}$$

Determining $\partial^2 \sigma_{xy}/\partial x \partial y$ from Eqs. 4.1, we get

$$-2\frac{\partial^2 \sigma_{xy}}{\partial x \partial y} = \frac{\partial^2 \sigma_{xx}}{\partial x^2} + \frac{\partial^2 \sigma_{yy}}{\partial y^2} + \frac{\partial X}{\partial x} + \frac{\partial Y}{\partial y} \tag{4.14}$$

Substituting Eq. 4.14 into Eq. 4.12 leads to the following compatibility condition for the normal stresses

$$\nabla^2(\sigma_{xx} + \sigma_{yy}) = -\frac{4}{1+\kappa}\left(\frac{\partial X}{\partial x} + \frac{\partial Y}{\partial y}\right) \tag{4.15}$$

If the body force is *conservative*, then it has a potential V, and

$$X = -\frac{\partial V}{\partial x} \tag{4.16a}$$

$$Y = -\frac{\partial V}{\partial y} \tag{4.16b}$$

In fact, the condition for a conservative body force is

$$\frac{\partial X}{\partial y} = \frac{\partial Y}{\partial x} \tag{4.17}$$

Equation 4.15 becomes

$$\nabla^2(\sigma_{xx} + \sigma_{yy}) = \frac{4}{1+\kappa}\nabla^2 V \tag{4.18}$$

In 1862, G.B. Airy proposed that the stresses be expressed in terms of another potential $\phi(x,y)$ by

$$\sigma_{xx} = \frac{\partial^2 \phi}{\partial y^2} + V \tag{4.19a}$$

$$\sigma_{yy} = \frac{\partial^2 \phi}{\partial x^2} + V \tag{4.19b}$$

$$\sigma_{xy} = -\frac{\partial^2 \phi}{\partial x \partial y} \tag{4.19c}$$

These satisfy the equilibrium equations, Eqs. 4.1, automatically, and change the compatibility condition, Eq. 4.18, into

$$\nabla^4 \phi = -2\frac{\kappa - 1}{\kappa + 1}\nabla^2 V \tag{4.20}$$

where ∇^4 is the *plane biharmonic operator* and defined by

$$\nabla^4 = \nabla^2 \nabla^2 \tag{4.21}$$

We note that ϕ in terms of polynomials of x and y of lower degrees than two can be neglected, since it gives zero stress components. If both body force components vanish, then V can be set to zero without of loss of generality. Lastly, the three-dimensional form of Eq. 4.20 is Eq. 3.91.

4.3 ISOTROPIC ELASTIC PLANE PROBLEMS IN CYLINDRICAL COORDINATES

Various conditions in cylindrical coordinates are derived in Sections 2.10 and 3.5. For plane problems in the $r\theta$-plane, equations of equilibrium in the r and θ directions can be found from Eqs. 2.74a and 2.74b, respectively, as

$$\frac{\partial \sigma_{rr}}{\partial r} + \frac{1}{r}\frac{\partial \sigma_{r\theta}}{\partial \theta} + \frac{\sigma_{rr} - \sigma_{\theta\theta}}{r} + X_r = 0 \tag{4.22a}$$

$$\frac{\partial \sigma_{r\theta}}{\partial r}+\frac{1}{r}\frac{\partial \sigma_{\theta\theta}}{\partial \theta}+\frac{2}{r}\sigma_{r\theta}+X_\theta=0 \tag{4.22b}$$

The strain-displacement relations are Eqs. 2.71, while the unified Hooke's law is the same as in Eqs. 4.6 or 4.7 with r and θ replacing x and y, respectively, i.e.

$$\varepsilon_{rr}=\frac{1}{2\mu}\left[\sigma_{rr}-\frac{3-\kappa}{4}(\sigma_{rr}+\sigma_{\theta\theta})\right] \tag{4.23a}$$

$$\varepsilon_{\theta\theta}=\frac{1}{2\mu}\left[\sigma_{\theta\theta}-\frac{3-\kappa}{4}(\sigma_{rr}+\sigma_{\theta\theta})\right] \tag{4.23b}$$

$$\varepsilon_{r\theta}=\frac{1}{2\mu}\sigma_{r\theta} \tag{4.23c}$$

$$\sigma_{rr}=\frac{\mu}{\kappa-1}[(\kappa+1)\varepsilon_{rr}+(3-\kappa)\varepsilon_{\theta\theta}] \tag{4.24a}$$

$$\sigma_{\theta\theta}=\frac{\mu}{\kappa-1}[(\kappa+1)\varepsilon_{\theta\theta}+(3-\kappa)\varepsilon_{rr}] \tag{4.24b}$$

$$\sigma_{r\theta}=2\mu\varepsilon_{r\theta} \tag{4.24c}$$

The compatibility condition in terms of normal stresses is

$$\nabla^2(\sigma_{rr}+\sigma_{\theta\theta})=-\frac{4}{1+\kappa}\left(\frac{\partial X_r}{\partial r}+\frac{X_r}{r}+\frac{1}{r}\frac{\partial X_\theta}{\partial \theta}\right) \tag{4.25}$$

In *polar (plane cylindrical) coordinates*, the Laplace operator ∇^2 is

$$\nabla^2=\frac{1}{r}\frac{\partial}{\partial r}\left(r\frac{\partial}{\partial r}\right)+\frac{1}{r^2}\frac{\partial^2}{\partial \theta^2} \tag{4.26}$$

Equation 4.25 is equivalent to Eq. 4.15 in Cartesian coordinates. For a conservative body force, X_r and X_θ can be expressed in terms of a potential V as

$$X_r=-\frac{\partial V}{\partial r} \tag{4.27a}$$

$$X_\theta=-\frac{1}{r}\frac{\partial V}{\partial \theta} \tag{4.27b}$$

The condition to check whether the body force is conservative is

$$\frac{\partial X_r}{\partial \theta}=\frac{\partial (rX_\theta)}{\partial r} \tag{4.28}$$

Substituting Eqs. 4.27 into Eq. 4.25 leads to

$$\nabla^2(\sigma_{rr}+\sigma_{\theta\theta})=\frac{4}{\kappa+1}\nabla^2 V \tag{4.29}$$

The basic unknown stresses can be expressed in the following form

$$\sigma_{rr} = \frac{1}{r}\frac{\partial\phi}{\partial r} + \frac{1}{r^2}\frac{\partial^2\phi}{\partial\theta^2} + V \tag{4.30a}$$

$$\sigma_{\theta\theta} = \frac{\partial^2\phi}{\partial r^2} + V \tag{4.30b}$$

$$\sigma_{r\theta} = -\frac{\partial}{\partial r}\left(\frac{1}{r}\frac{\partial\phi}{\partial\theta}\right) \tag{4.30c}$$

which satisfy the equilibrium equations, Eqs. 4.22, exactly and change the compatibility condition, Eq. 4.29, into

$$\nabla^4\phi = -2\frac{\kappa-1}{\kappa+1}\nabla^2 V \tag{4.31}$$

Note that Eqs. 4.29 and 4.31 are identical to Eqs. 4.18 and 4.20, respectively.

Rigid-body displacements for plane problems in cylindrical coordinates can be written

$$u_r = C_1\cos\theta + C_2\sin\theta \tag{4.32a}$$

$$u_\theta = -C_1\sin\theta + C_2\cos\theta + C_3 r \tag{4.32b}$$

4.4 DISPLACEMENT FOR A GIVEN BIHARMONIC FUNCTION

In the absence of body forces, the *Airy function* ϕ is governed by the *plane biharmonic equation*

$$\nabla^4\phi = 0 \tag{4.33}$$

We seek a solution in the form of a Fourier series

$$\phi = \sum_{m=0}^{\infty} R_m\cos m\theta + \sum_{m=1}^{\infty} R_m'\sin m\theta \tag{4.34}$$

in which R_m and R_m' are functions of the radial coordinate r only. Substituting Eq. 4.34 into Eq. 4.33, we obtain, for each m, an ordinary differential equation

$$\nabla_m^4 R_m = 0$$

where $\nabla_m^4 = \nabla_m^2\nabla_m^2$, and

$$\nabla_m^2 = \frac{1}{r}\frac{\mathrm{d}}{\mathrm{d}r}\left(r\frac{\mathrm{d}}{\mathrm{d}r}\right) - \frac{m^2}{r^2}$$

The general solution of R_m is, for $m=0$,

$$R_0 = A_0 + B_0 r^2 + C_0\log r + D_0 r^2\log r \tag{4.35a}$$

for $m=1$,

$$R_1 = A_1 r + B_1 r^3 + C_1 r^{-1} + D_1 r\log r \tag{4.35b}$$

and for $m>1$,

$$R_m = A_m r^m + B_m r^{-m} + C_m r^{m+2} + D_m r^{-m+2} \tag{4.35c}$$

where the constants A_m, B_m, C_m, and D_m are arbitrary. Similar expressions can be written for R_m'. However, we should not have the impression that only the functions in Eq. 4.34 are *plane biharmonic functions*. There are several other functions which are also biharmonic, since they happen to satisfy Eq. 4.33.

If ϕ is biharmonic, and

$$\nabla^2\phi = P(x,y) \tag{4.36}$$

then

$$\nabla^2 P(x,y) = 0 \tag{4.37}$$

which means that P is a *plane harmonic function*. Consider an *analytic function* (Churchill, 1960) of the form

$$f(z) = P(x,y) + \mathrm{i}Q(x,y) \tag{4.38}$$

where Q is another harmonic function, i is the imaginary number ($\mathrm{i}^2 = -1$), and z in this particular section is the *complex variable*, i.e.

$$z = x + \mathrm{i}y \tag{4.39}$$

or, in polar coordinates,

$$z = r\mathrm{e}^{\mathrm{i}\theta} \tag{4.40}$$

We may introduce another analytic function $\psi(z)$, the derivative of which is

$$\frac{\mathrm{d}\psi}{\mathrm{d}z} = \frac{f(z)}{4} \tag{4.41}$$

Let ψ be composed of the real and imaginary parts as follows

$$\psi = p(x,y) + \mathrm{i}q(x,y) \tag{4.42}$$

Thus p and q are also harmonic functions, and $\mathrm{d}\psi/\mathrm{d}z$ can be put in the form

$$\frac{\mathrm{d}\psi}{\mathrm{d}z} = \frac{\partial p}{\partial x} + \mathrm{i}\frac{\partial q}{\partial x} = \frac{\partial q}{\partial y} - \mathrm{i}\frac{\partial p}{\partial y} \tag{4.43}$$

Note that we have used the *Cauchy-Riemann conditions*

$$\frac{\partial p}{\partial x} = \frac{\partial q}{\partial y}, \qquad \frac{\partial p}{\partial y} = -\frac{\partial q}{\partial x} \tag{4.44a,b}$$

in obtaining Eq. 4.43. The Cauchy-Riemann conditions in polar coordinates are

$$\frac{\partial p}{\partial r} = \frac{\partial q}{r\partial\theta}, \qquad \frac{\partial q}{\partial r} = -\frac{\partial p}{r\partial\theta} \tag{4.45a,b}$$

These functions p and q are also called *conjugate harmonic functions*. Substituting Eqs. 4.38 and 4.43 into Eq. 4.41 leads to

$$\frac{\partial p}{\partial x} = \frac{\partial q}{\partial y} = \frac{P}{4} \tag{4.46a}$$

$$\frac{\partial q}{\partial x} = -\frac{\partial p}{\partial y} = \frac{Q}{4} \tag{4.46b}$$

The stress components for the case of ϕ governed by Eq. 4.33 can be obtained from Eqs. 4.19 with $V = 0$, i.e.

$$\sigma_{xx} = \frac{\partial^2 \phi}{\partial y^2} \tag{4.47a}$$

$$\sigma_{yy} = \frac{\partial^2 \phi}{\partial x^2} \tag{4.47b}$$

$$\sigma_{xy} = -\frac{\partial^2 \phi}{\partial x \partial y} \tag{4.47c}$$

Substituting strain-displacement relationship and Eqs. 4.47 into the stress-strain relationship (Eqs. 4.6), in view of Eqs. 4.36 and 4.46a, leads to

$$2\mu \frac{\partial u}{\partial x} = -\frac{\partial^2 \phi}{\partial x^2} + (1+\kappa)\frac{\partial p}{\partial x} \tag{4.48a}$$

$$2\mu \frac{\partial v}{\partial y} = -\frac{\partial^2 \phi}{\partial y^2} + (1+\kappa)\frac{\partial q}{\partial y} \tag{4.48b}$$

$$\mu\left(\frac{\partial u}{\partial y} + \frac{\partial v}{\partial x}\right) = -\frac{\partial^2 \phi}{\partial x \partial y} \tag{4.48c}$$

Integrating Eq. 4.48a with respect to x, and Eq. 4.48b with respect to y leads to, respectively

$$2\mu u = -\frac{\partial \phi}{\partial x} + (1+\kappa)p + g_1(y) \tag{4.49a}$$

$$2\mu v = -\frac{\partial \phi}{\partial y} + (1+\kappa)q + g_2(x) \tag{4.49b}$$

where $g_1(y)$ and $g_2(x)$ result from the integration. Upon substituting the preceding equations into Eq. 4.48c, we find

$$\frac{\mathrm{d}g_1}{\mathrm{d}y} = -\frac{\mathrm{d}g_2}{\mathrm{d}x} = A \tag{4.50}$$

where A is an arbitrary constant. Integrating the above equation leads to

$$g_1(y) = Ay + B \tag{4.51a}$$

$$g_2(x) = -Ax + C \tag{4.51b}$$

where B and C are also arbitrary constants. Recalling Eqs. 4.3d and e, we notice that Eqs. 4.51 are rigid-body displacement components. Ignoring such components for the time being, Eqs. 4.49 become

$$2\mu u = -\frac{\partial\phi}{\partial x} + (1+\kappa)p \tag{4.52a}$$

$$2\mu v = -\frac{\partial\phi}{\partial y} + (1+\kappa)q \tag{4.52b}$$

With the relationship among cylindrical and Cartesian coordinates as listed in Eqs. 2.65 to 2.68, we may write displacement components in the cylindrical coordinate system as

$$2\mu u_r = -\frac{\partial\phi}{\partial r} + (1+\kappa)(p\cos\theta + q\sin\theta) \tag{4.53a}$$

$$2\mu u_\theta = -\frac{\partial\phi}{r\partial\theta} + (1+\kappa)(-p\sin\theta + q\cos\theta) \tag{4.53b}$$

Let a real function p_1 be denoted by

$$p_1 = \phi - \Re[\bar{z}\psi(z)] \tag{4.54}$$

where $\Re$ denotes the real part and $\bar{z}$ the *conjugate* of z, i.e.

$$\bar{z} = x - \mathrm{i}y \tag{4.55}$$

Applying ∇^2 on Eq. 4.54, in view of Eqs. 4.36, 4.42 and 4.55, yields

$$\nabla^2 p_1 = P - \left[x\nabla^2 p + y\nabla^2 q + 2\left(\frac{\partial p}{\partial x} + \frac{\partial q}{\partial y}\right)\right],$$

or, due to Eqs. 4.46,

$$\nabla^2 p_1 = -x\nabla^2 p - y\nabla^2 q.$$

Moreover, the fact that $\nabla^2 p = \nabla^2 q = 0$ reduces the last equation into

$$\nabla^2 p_1 = 0 \tag{4.56}$$

Thus p_1 as introduced is a harmonic function.

Then Eq. 4.54 can be rewritten in the form

$$\phi = \Re[\bar{z}\psi(z)] + p_1$$

$$= \Re[(\bar{z} + z)\psi(z)] - \Re[z\psi(z)] + p_1,$$

or, due to Eqs. 4.39, 4.42 and 4.55,

$$\phi = 2xp + p_2 \tag{4.57}$$

where $p_2 = -\Re[(z\psi(z)] + p_1$ thus is a harmonic function. Alternatively, we can write Eq. 4.54 in the form

$$\phi = \Re[(\bar{z} - z)\psi(z)] + \Re[z\psi(z)] + p_1$$

$$= 2yq + p_2 \tag{4.58}$$

where $p_2 = \Re[(z\psi(z)] + p_1$ thus is also a harmonic function. Equation 4.58 is another general form of biharmonic functions. On the other hand, if we rewrite Eq. 4.54 as follows

$$\phi = \Re[\bar{z}z\psi(z)/z] + p_1,$$

we would arrive at

$$\phi = r^2 p_3 + p_1 \tag{4.59}$$

where

$$\begin{aligned} p_3 &= \Re[\psi(z)/z] \\ &= \Re[(px+qy)+\mathrm{i}(-py+qx)]/r^2 \\ &= \Re[(p\cos\theta + q\sin\theta)+\mathrm{i}(-p\sin\theta+q\cos\theta)]/r \\ &= \Re(p_3+\mathrm{i}q_3), \end{aligned}$$

where

$$p_3 = (p\cos\theta + q\sin\theta)/r \tag{4.60a}$$

$$q_3 = (-p\sin\theta + q\cos\theta)/r \tag{4.60b}$$

or,

$$p = p_3 x - q_3 y \tag{4.61a}$$

$$q = p_3 y + q_3 x \tag{4.61b}$$

Since p_3 and q_3 are real and imaginary parts, respectively, of the same analytic function $[\psi(z)/z]$, they are conjugate harmonic functions and related by the Cauchy-Riemann conditions. Incorporating Eqs. 4.61 and 4.60 into Eqs. 4.52 and 4.53 respectively, we have

$$2\mu u = -\frac{\partial\phi}{\partial x} + (1+\kappa)(p_3 x - q_3 y) \tag{4.62a}$$

$$2\mu v = -\frac{\partial\phi}{\partial y} + (1+\kappa)(p_3 y + q_3 x) \tag{4.62b}$$

$$2\mu u_r = -\frac{\partial\phi}{\partial r} + (1+\kappa) r p_3 \tag{4.63a}$$

$$2\mu u_\theta = -\frac{\partial\phi}{r\partial\theta} + (1+\kappa) r q_3 \tag{4.63b}$$

Thus when a harmonic function takes the form of Eq. 4.59, q_3 can be determined from p_3 by simple integration of the Cauchy-Riemann conditions (Eqs. 4.44 or 4.45), then the displacement components can be simply obtained by Eqs. 4.62 or 4.63. For the case of biharmonic functions in the form of Eq. 4.57, q should be determined by the Cauchy-Riemann conditions, and the displacement components by Eqs. 4.52 or 4.53. The case of biharmonic functions in the form of Eq. 4.58 can be treated in the same way as the case of Eq. 4.57 with the exception that p must be determined from the known function q using the Cauchy-Riemann conditions. Arbitrary constants arising from the integration of the Cauchy-Riemann conditions correspond to rigid-body displacement components, Eqs. 4.3d and e for Cartesian components, and Eqs. 4.32 for cylindrical components.

Table 4.1 at the end of the present chapter, contains some common harmonic and biharmonic functions and their corresponding stresses and displacements (but not rigid-body components). When these functions are combined properly, we have new harmonic functions. Here are some harmonic functions which can be used in some *half plane* problems.

$$r(\log r \sin\theta + \theta\cos\theta) \equiv y\log r + x\theta \tag{4.64a}$$

$$r(\log r \cos\theta - \theta\sin\theta) \equiv x\log r - y\theta \tag{4.64b}$$

$$r^2(\log r \sin 2\theta + \theta\cos 2\theta) \equiv 2xy\log r + (x^2 - y^2)\theta \tag{4.64c}$$

$$r^2(\log r \cos 2\theta - \theta\sin 2\theta) \equiv (x^2 - y^2)\log r - 2xy\theta \tag{4.64d}$$

One may notice that each term in the equation above is a product of two harmonic functions, hence it may be found expedient to note that a Laplacian of a product of harmonic functions $f_1(x,y)$ and $f_2(x,y)$ can be put as

$$\nabla^2(f_1 f_2) = 2\left[\frac{\partial f_1}{\partial x}\frac{\partial f_2}{\partial x} + \frac{\partial f_1}{\partial y}\frac{\partial f_2}{\partial y}\right] \tag{4.65}$$

The total solution to Eq. 4.20 or 4.31 can be a linear combination of *complementary solutions* and a *particular solution.* A complementary solution is a biharmonic function; while a particular solution is one that satisfies the governing equations exactly without involving any arbitrary constants. In addition to complementary solutions, Table 4.1 also contains particular solutions and their corresponding stresses and displacements for various common potentials V. The arbitrary constants, some being attached to complementary solutions while the rest associated with rigid-body displacement components, are to be determined from the condition that the total solution must satisfy all pertinent boundary conditions.

4.5 EXAMPLES OF INFINITE PLANE PROBLEMS

The first problem to be considered is an infinite plane under uniform tension in the x-direction as shown in Fig. 4.1. Since there is no body force $(X = Y = 0)$, we can set

$$V = 0.$$

The stress field everywhere is

$$\sigma_{xx} = \sigma_0, \quad \sigma_{yy} = 0, \quad \sigma_{xy} = 0.$$

So that the Airy stress function ϕ of Eq. 4.19 satisfies

$$\frac{\partial^2\phi}{\partial y^2} = \sigma_0, \quad \frac{\partial^2\phi}{\partial x^2} = 0, \quad \frac{\partial^2\phi}{\partial x\partial y} = 0,$$

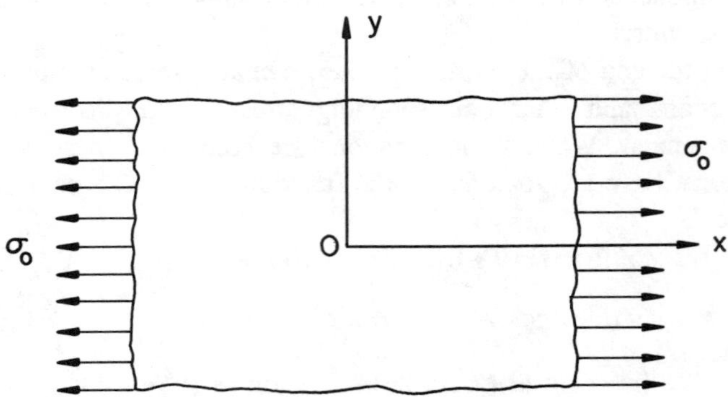

Fig. 4.1 Infinite plane under uniform tension.

the solution to which can be shown to be

$$\phi = \sigma_0 \frac{y^2}{2} + C_1 x + C_2 y + C_3,$$

where C_1 to C_3 are constants of integration, and can be set to zero since they yield zero stresses. Thus

$$\phi = \sigma_0 \frac{y^2}{2} \tag{4.66a}$$

In polar coordinates,

$$\phi = \frac{\sigma_0}{4}(r^2 - r^2 \cos 2\theta) \tag{4.66b}$$

and the stresses (see Table 4.1) are

$$\sigma_{rr} = \frac{\sigma_0}{2}(1 + \cos 2\theta) \tag{4.67a}$$

$$\sigma_{r\theta} = -\frac{\sigma_0}{2} \sin 2\theta \tag{4.67b}$$

$$\sigma_{\theta\theta} = \frac{\sigma_0}{2}(1 - \cos 2\theta) \tag{4.67c}$$

The second example is an infinite plane with a circular hole and under uniform tension far away from the hole as shown in Fig. 4.2. The stresses at the far field ($r \to \infty$) are identical to those in the previous example. Thus, the complementary solution may be written

$$\phi = \phi_1 + \phi_2,$$

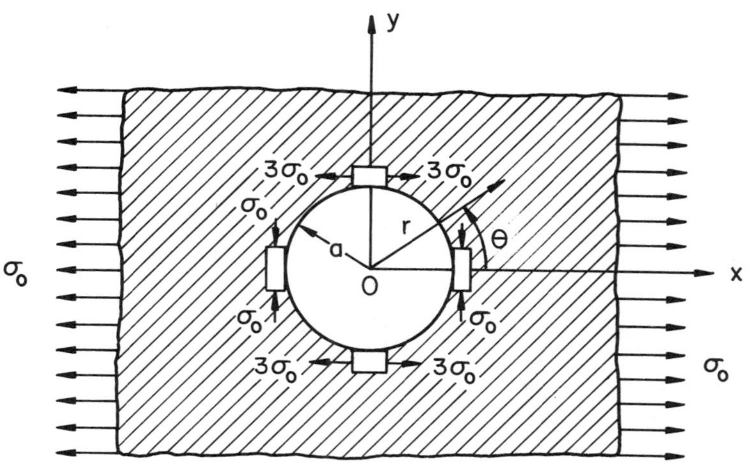

Fig 4.2 Infinite plane with a hole under uniform tension at infinity.

where ϕ_1 is the function ϕ of the previous example, i.e.

$$\phi_1 = \frac{\sigma_0}{4}(r^2 - r^2 \cos 2\theta),$$

and ϕ_2 should be such that it gives zero stresses as r tends to infinity, and leads to the satisfaction of the stress free boundary conditions at $r = a$, i.e.

$$\sigma_{rr}(a, \theta) = 0 \tag{4.68a}$$

$$\sigma_{r\theta}(a, \theta) = 0 \tag{4.68b}$$

Considering all these requirements and after consulting Table 4.1, we find

$$\phi = \frac{\sigma_0}{4}\left(r^2 - r^2 \cos 2\theta + A a^2 \log r + B a^4 \frac{\cos 2\theta}{r^2} + C a^2 \cos 2\theta\right) \tag{4.69}$$

where A, B and C are constants to be determined by substituting Eq. 4.69 into Eqs. 4.68, i.e.

$$2 + 2\cos 2\theta + A - 6B \cos 2\theta - 4C \cos 2\theta = 0,$$

$$-2 \sin 2\theta + 0 - 6B \sin 2\theta - 2C \sin 2\theta = 0.$$

Since the two equations hold for every value of θ, they imply the following three independent conditions

$$2 + A = 0,$$

$$2 - 6B - 4C = 0,$$

$$-2 - 6B - 2C = 0.$$

Finally, we find

$$\phi = \frac{\sigma_0}{4}\left[r^2 - r^2\cos 2\theta - 2a^2\log r - \frac{a^4}{r^2}\cos 2\theta + 2a^2\cos 2\theta\right] \tag{4.70}$$

$$\sigma_{rr} = \frac{\sigma_0}{2}\left[1 - \frac{a^2}{r^2} + \left(1 - 4\frac{a^2}{r^2} + 3\frac{a^4}{r^4}\right)\cos 2\theta\right] \tag{4.71a}$$

$$\sigma_{r\theta} = \frac{\sigma_0}{2}\left(-1 - 2\frac{a^2}{r^2} + 3\frac{a^4}{r^4}\right)\sin 2\theta \tag{4.71b}$$

$$\sigma_{\theta\theta} = \frac{\sigma_0}{2}\left[1 + \frac{a^2}{r^2} - \left(1 + 3\frac{a^4}{r^4}\right)\cos 2\theta\right] \tag{4.71c}$$

The *hoop stress*, which is $\sigma_{\theta\theta}$ at $r = a$, is

$$\sigma_{\theta\theta}(a, \theta) = \sigma_0(1 - 2\cos 2\theta) \tag{4.72}$$

Its largest tensile and compressive values are shown in Fig. 4.2, i.e.

$$\sigma_{\theta\theta}(a, 0) = \sigma_{\theta\theta}(a, \pi) = -\sigma_0 \tag{4.73a,b}$$

$$\sigma_{\theta\theta}(a, \pi/2) = \sigma_{\theta\theta}(a, -\pi/2) = 3\sigma_0 \tag{4.73c,d}$$

Thus the *stress concentration factor* is 3.

Such a factor would be increased to 4 if a uniform compression in the y-direction is added to the same plate as shown in Fig. 4.3.

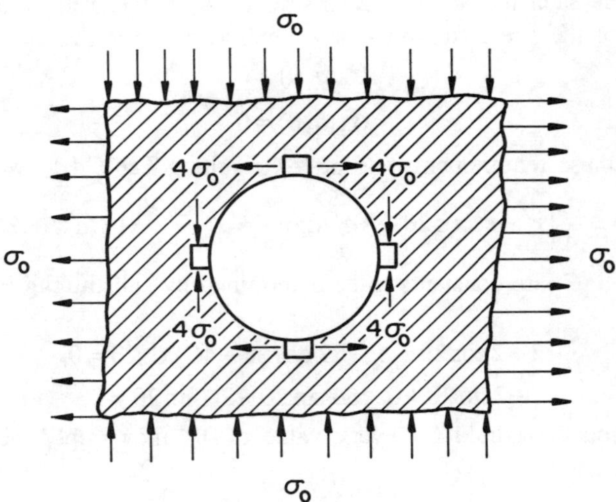

Fig. 4.3 Infinite plane with a hole, under tension in one direction and compression in another orthogonal direction.

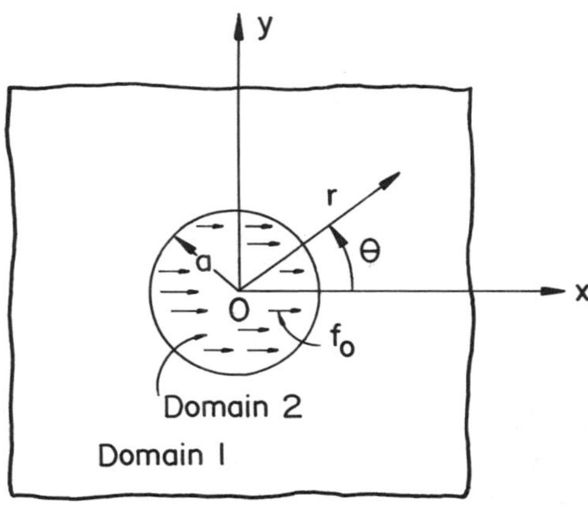

Fig. 4.4 Infinite plane subjected to uniform body force.

The last example to be considered in this section is the problem of an infinite plane subjected to a uniform body force in a circular region as shown in Fig. 4.4. Let f_0 be the uniform intensity of the body force in the x-direction in domain 2, where $0 \leq r < a$. Domain 1 is the region $r > a$. Thus for domain 2 we have

$$X^{(2)} = f_0, \qquad Y^{(2)} = 0.$$

Note that a superscript in the parentheses is used to denote a domain number. A simple check by Eq. 4.17 shows that the body force is conservative, and Eqs. 4.16 give

$$\frac{\partial V^{(2)}}{\partial x} = -f_0, \quad \frac{\partial V^{(2)}}{\partial y} = 0 \tag{4.74a,b}$$

A general solution to Eqs. 4.74 is $V^{(2)} = -f_0 x + C$, where C is an arbitrary constant which can be set to zero without loss of generality, i.e.

$$V^{(2)} = -f_0 x = -f_0 r \cos\theta \tag{4.75}$$

According to Eqs. 4.30 or Table 4.1, the stresses due to $V^{(2)}$ alone are

$$\sigma_{rr} = -f_0 r \cos\theta \tag{4.76a}$$

$$\sigma_{r\theta} = 0 \tag{4.76b}$$

$$\sigma_{\theta\theta} = -f_0 r \cos\theta \tag{4.76c}$$

The unified Hooke's law gives the strains

$$\frac{\partial u_r}{\partial r} = -f_0 \frac{\kappa - 1}{4\mu} r \cos\theta \tag{4.77a}$$

$$r\frac{\partial}{\partial r}\left(\frac{u_\theta}{r}\right)+\frac{1}{r}\frac{\partial u_r}{\partial \theta}=0 \tag{4.77b}$$

$$\frac{1}{r}\frac{\partial u_\theta}{\partial \theta}+\frac{u_r}{r}=-f_0\frac{\kappa-1}{4\mu}r\cos\theta \tag{4.77c}$$

Integrating Eqs. 4.77 yields u_r and u_θ as

$$2\mu u_r=-\frac{1}{4}(\kappa-1)f_0 r^2\cos\theta \tag{4.78a}$$

$$2\mu u_\theta=-\frac{1}{4}(\kappa-1)f_0 r^2\sin\theta \tag{4.78b}$$

which are the same as listed in Table 4.1. While $V^{(1)}$ vanishes, $V^{(2)}$ does not but $\nabla^2 V^{(2)}=0$; hence the governing equation for both domains 1 and 2 is $\nabla^4\phi=0$. The conditions at the interface $(r=a)$ between these two domains are

$$\sigma_{rr}^{(1)}(a,\theta)=\sigma_{rr}^{(2)}(a,\theta) \tag{4.79a}$$

$$\sigma_{r\theta}^{(1)}(a,\theta)=\sigma_{r\theta}^{(2)}(a,\theta) \tag{4.79b}$$

$$u_r^{(1)}(a,\theta)=u_r^{(2)}(a,\theta) \tag{4.79c}$$

$$u_\theta^{(1)}(a,\theta)=u_\theta^{(2)}(a,\theta) \tag{4.79d}$$

The pattern of stress and displacement fields in domain 2 due to f_0 is given in Eqs. 4.76 and Eqs. 4.78. In order to satisfy Eqs. 4.79 we need to investigate the following biharmonic functions listed in Table 4.1

$$r^3\cos\theta,\quad r\theta\sin\theta,\quad r\log r\cos\theta,\quad \cos\theta/r.$$

Since the stresses must be bounded as $r\to\infty$ in domain 1, and as $r\to 0$ in domain 2, we take

$$\phi^{(1)}=f_0a^2\left(2Ar\theta\sin\theta+Br\log r\cos\theta+Ca^2\frac{\cos\theta}{r}\right) \tag{4.80a}$$

$$\phi^{(2)}=f_0Dr^3\cos\theta \tag{4.80b}$$

where A, B, C and D are constants to be determined from the boundary conditions Eqs. 4.79. From the table, we can put the stresses and displacements for the domain 1 as

$$\sigma_{rr}^{(1)}=f_0a^2(2A\cos\theta/r+B\cos\theta/r-2Ca^2\cos\theta/r^3) \tag{4.81a}$$

$$\sigma_{r\theta}^{(1)}=f_0a^2(B\sin\theta/r-2Ca^2\sin\theta/r^3) \tag{4.81b}$$

$$\sigma_{\theta\theta}^{(1)}=f_0a^2(B\cos\theta/r+2Ca^2\cos\theta/r^3) \tag{4.81c}$$

$$2\mu u_r^{(1)} = f_0 a^2\{[A(\kappa-1)+B(\kappa+1)]\theta\sin\theta/2$$
$$+[A(\kappa+1)+B(\kappa-1)]\log r\cos\theta/2-(A+B)\cos\theta/2$$
$$+Ca^2\cos\theta/r^2 - C_1^{(1)}\cos\theta\} \quad (4.81d)$$

$$2\mu u_\theta^{(1)} = f_0 a^2\{[A(\kappa-1)+B(\kappa+1)]\theta\cos\theta/2$$
$$-[A(\kappa+1)+B(\kappa-1)]\log r\sin\theta/2-(A+B)\sin\theta/2$$
$$+Ca^2\sin\theta/r^2 - C_1^{(1)}\sin\theta\} \quad (4.81e)$$

in which $C_1^{(1)}$ corresponds to the appropriate rigid-body displacement. In domain 2, the Airy function is as given in Eq. 4.80b, for which the stresses and displacements can be obtained from Table 4.1. These quantities together with those due to the applied body force f_0 in Eqs. 4.76 and 4.78 are listed below.

$$\sigma_{rr}^{(2)} = f_0(2D-1)r\cos\theta \quad (4.82a)$$

$$\sigma_{r\theta}^{(2)} = 2f_0 Dr\sin\theta \quad (4.82b)$$

$$\sigma_{\theta\theta}^{(2)} = f_0(6D-1)r\cos\theta \quad (4.82c)$$

$$2\mu u_r^{(2)} = f_0\left\{\left[D(\kappa-2)-\frac{1}{4}(\kappa-1)\right]r^2\cos\theta + a^2C_1^{(2)}\cos\theta\right\} \quad (4.82d)$$

$$2\mu u_\theta^{(2)} = f_0\left\{\left[D(\kappa+2)-\frac{1}{4}(\kappa-1)\right]r^2\sin\theta - a^2C_1^{(2)}\sin\theta\right\} \quad (4.82e)$$

Before applying the boundary conditions, a closer look at Eqs. 4.81d and e reveals multivalued displacement terms, $\theta\sin\theta$ and $\theta\cos\theta$, which are not admissible in the present problem. In order to delete such terms, we should set their coefficients to zero, i.e.

$$A(\kappa-1)+B(\kappa+1)=0 \quad (4.83)$$

Substituting the general expressions of stresses and displacements into the boundary conditions Eqs. 4.79, and using Eq. 4.83, we find

$$2A+B-2C=2D-1 \quad (4.84a)$$

$$B-2C=2D \quad (4.84b)$$

$$\frac{1}{2}[A(\kappa+1)+B(\kappa-1)]\log a-\frac{1}{2}(A+B)+C+(C_1^{(1)}-C_1^{(2)})=D(\kappa-2)-\frac{1}{4}(\kappa-1) \quad (4.84c)$$

$$-\frac{1}{2}[A(\kappa+1)+B(\kappa-1)]\log a-\frac{1}{2}(A+B)+C-(C_1^{(1)}-C_1^{(2)})=D(\kappa+2)-\frac{1}{4}(\kappa-1) \quad (4.84d)$$

Equations 4.83 and 4.84 form a set of 5 linear simultaneous equations for determining 5 constants, i.e. A, B, C, D and $C_1^{(1)} - C_1^{(2)}$. The solution is

$$A = -\frac{1}{2}, \qquad B = \frac{\kappa - 1}{2(\kappa + 1)} \tag{4.85a,b}$$

$$C = -\frac{1}{4(\kappa + 1)}, \quad D = \frac{\kappa}{4(\kappa + 1)} \tag{4.85c,d}$$

$$C_1^{(1)} - C_1^{(2)} = \frac{\kappa^2 - \kappa + 1}{2(\kappa + 1)} \tag{4.85e}$$

The term $C_1^{(1)} - C_1^{(2)}$ signifies the indefinite displacement. The displacement has a 'weak' singularity as $\log r$ as r tends to infinity. Such peculiar situation of the indefiniteness and unboundedness of the displacement occurs only in statics of infinite planes.

The solution for a *concentrated force* P applied to an infinite plane, as shown in Fig. 4.5, can be obtained by taking the limit $a \to 0$, and assuming that

$$\lim_{a \to 0} (\pi a^2 f_0) = P.$$

Equation 4.80a gives the limiting potential

$$\phi = \frac{P}{2\pi(\kappa + 1)}[-(\kappa + 1) r\theta \sin\theta + (\kappa - 1) r \log r \cos\theta] \tag{4.86}$$

the stresses and displacements

$$\sigma_{rr} = -\frac{(\kappa + 3)P}{2\pi(\kappa + 1)} \frac{\cos\theta}{r} \tag{4.87a}$$

$$\sigma_{r\theta} = \frac{(\kappa - 1)P}{2\pi(\kappa + 1)} \frac{\sin\theta}{r} \tag{4.87b}$$

$$\sigma_{\theta\theta} = \frac{(\kappa - 1)P}{2\pi(\kappa + 1)} \frac{\cos\theta}{r} \tag{4.87c}$$

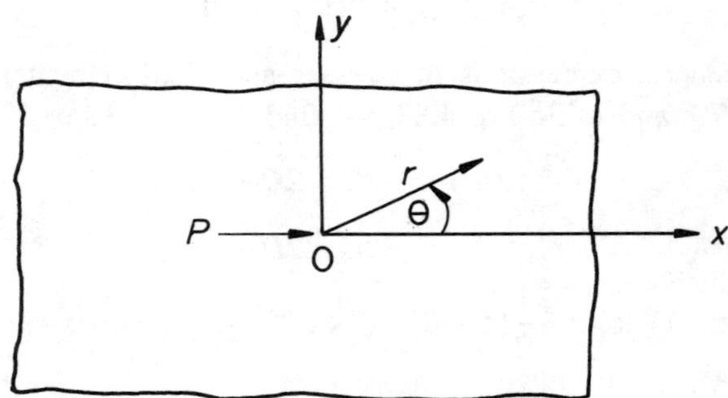

Fig. 4.5 Force at a point inside infinite plane.

$$2\mu u_r = \frac{P}{2\pi(\kappa+1)}(-2\kappa \log r + 1)\cos\theta + C_1 \cos\theta \tag{4.87d}$$

$$2\mu u_\theta = \frac{P}{2\pi(\kappa+1)}(2\kappa \log r + 1)\sin\theta - C_1 \sin\theta \tag{4.87e}$$

Observe that the stress decays as r^{-1} when r increases and is singular at the origin. The displacement has a *weak* singularity of $\log r$ both at the origin and at infinity.

4.6 PARTICULAR SOLUTIONS FOR CONCENTRATED FORCES

The function ϕ given by Eq. 4.86 can be taken as a particular solution to any problem of a concentrated force inside a plane, whether finite or infinite. The first term gives the appropriate singularity at the point where the force is concentrated, while the second term suppresses the multivaluedness of the displacement components due to the first term.

In a problem where the concentrated force is applied at a point on a smooth boundary of the material domain as shown in Fig. 4.6, a particular solution can assume the form

$$\phi = Ar\theta \sin\theta \tag{4.88}$$

The constant A must be determined from the conditions that the force resultant in the x-direction is P and that in the y-direction is zero, i.e.

$$\lim_{r\to 0} r\int_{-\alpha}^{\pi-\alpha} \overline{X}(r,\theta)d\theta + P = 0 \tag{4.89a}$$

$$\lim_{r\to 0} r\int_{-\alpha}^{\pi} \overline{Y}(r,\theta)d\theta = 0 \tag{4.89b}$$

Here $\overline{X}(r,\theta)$ and $\overline{Y}(r,\theta)$ are components of the stress vector on the boundary where r is a constant, i.e.

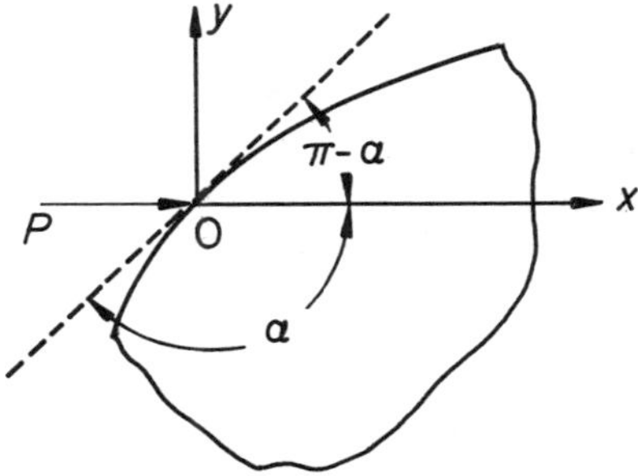

Fig. 4.6 Force at a point on smooth boundary.

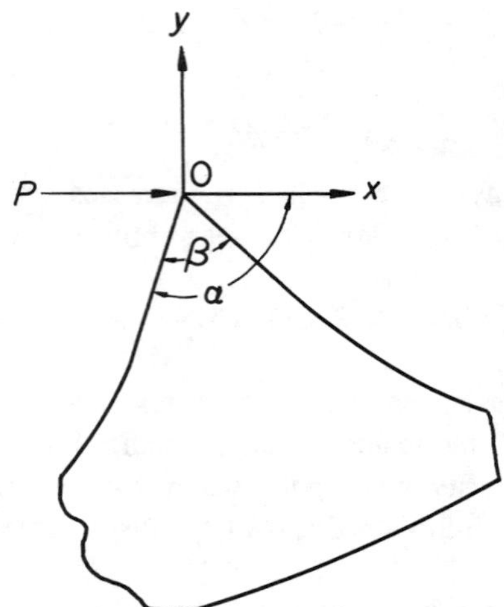

Fig. 4.7 Force at a cornered boundary.

$$\overline{X}(r,\theta) = \sigma_{xx}(r,\theta)\cos\theta + \sigma_{xy}(r,\theta)\sin\theta \tag{4.90a}$$

$$\overline{Y}(r,\theta) = \sigma_{yy}(r,\theta)\sin\theta + \sigma_{xy}(r,\theta)\cos\theta \tag{4.90b}$$

In writing Eqs. 4.89, we have incorporated the fact that ϕ as given by Eq. 4.88 yields $\sigma_{r\theta} = \sigma_{\theta\theta} = 0$ everywhere. The Cartesian stress components according to Table 4.1 are

$$\sigma_{xx} = 2A\cos^3\theta r^{-1}, \quad \sigma_{xy} = 2A\sin\theta\cos^2\theta r^{-1} \tag{4.91a,b}$$

$$\sigma_{yy} = 2A\cos\theta(1-\cos^2\theta)r^{-1} \tag{4.91c}$$

Substituting Eqs. 4.91 into Eq. 4.89a yields $A = -P/\pi$, while Eq. 4.89b is satisfied automatically. Thus a particular solution is

$$\phi = -\frac{P}{\pi} r\theta\sin\theta \tag{4.92}$$

For a concentrated force acting at a cornered boundary, as shown in Fig 4.7, a particular solution can be found to be

$$\phi = -\frac{Pr\theta}{1-\cos 2\beta - 2\beta^2}\{[\sin 2(\beta-\alpha) + \sin 2\alpha - 2\beta]\sin\theta + [\cos 2(\beta-\alpha) - \cos 2\alpha]\cos\theta\} \tag{4.93}$$

This satisfies the conditions near the point of application of the force, i.e.

$$\lim_{r\to 0} r\int_{-\alpha}^{\beta-\alpha} \overline{X}(r,\theta)d\theta + P = 0 \tag{4.94a}$$

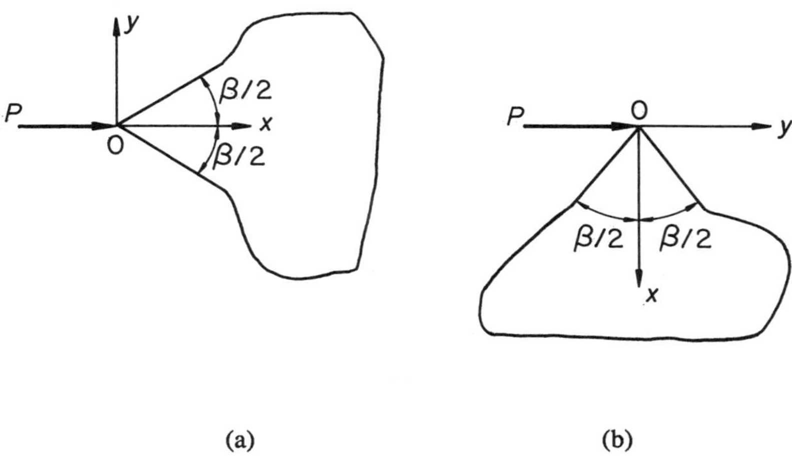

Fig. 4.8 Concentrated force P in special alignments with the cornered boundary.

$$\operatorname*{Lim}_{r\to 0} r\int_{-\alpha}^{\beta-\alpha}\overline{Y}(r,\theta)\mathrm{d}\theta = 0 \tag{4.94b}$$

If $\beta=\pi$, the boundary becomes smooth, the problem is as shown in Fig. 4.6, and Eq. 4.93 is specialized into Eq. 4.92.

If the force P has special alignments as shown in Figs. 4.8a and b, particular solutions are

$$\phi=\frac{2P(\beta-\sin\beta)r\theta\sin\theta}{1-\cos 2\beta-2\beta^2},\quad \text{for Fig. 4.8(a)} \tag{4.95}$$

$$\phi=-\frac{2P(\beta+\sin\beta)r\theta\cos\theta}{1-\cos 2\beta-2\beta^2},\quad \text{for Fig. 4.8(b)} \tag{4.96}$$

Putting $\beta=\pi$ in Eqs. 4.95, we can get a particular solution for the case of a concentrated force normal to a smooth boundary, as shown in Fig. 4.9, namely

$$\phi=-\frac{P}{\pi}r\theta\sin\theta \tag{4.97}$$

On the other hand, Eq. 4.96 can be specialized to give a particular solution for the case of a shear concentrated force on a smooth boundary, as shown in Fig. 4.10, namely

$$\phi=\frac{P}{\pi}r\theta\cos\theta \tag{4.98}$$

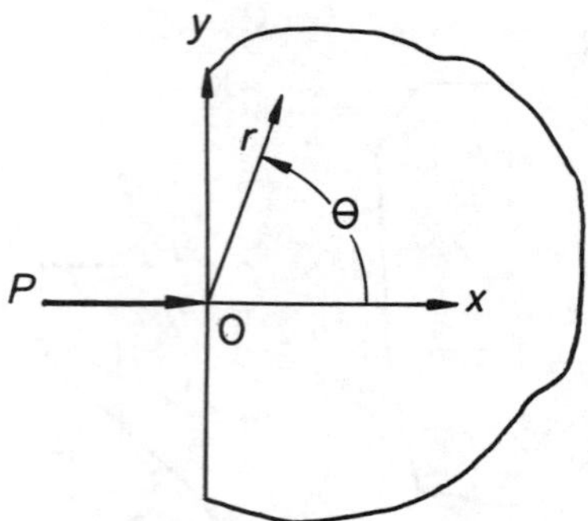

Fig. 4.9 Concentrated force normal to smooth boundary.

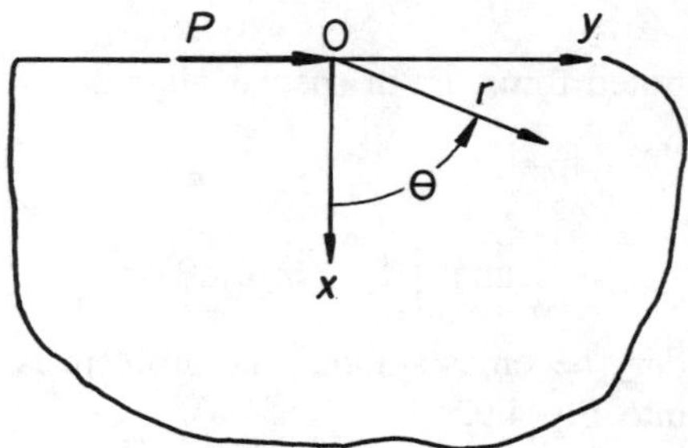

Fig. 4.10 Shear concentrated force on smooth boundary.

4.7 EXERCISE PROBLEMS

With permission of the respective publishers, Problems 4.7.5, 4.7.10 to 4.7.14 were taken from the book by Boresi and Lynn (1974), and Problems 4.7.18, 4.7.19, and 4.7.21 to 4.7.28 from the book by Timoshenko and Goodier (1970).

4.7.1 Show that the governing equation of an axisymmetric problem without any body forces has the form

$$\frac{1}{r}\frac{\mathrm{d}}{\mathrm{d}r}\left\{ r\frac{\mathrm{d}}{\mathrm{d}r}\left[\frac{1}{r}\frac{\mathrm{d}}{\mathrm{d}r}\left(r\frac{\mathrm{d}\phi}{\mathrm{d}r}\right)\right]\right\} = 0.$$

Solve the problem of a thick-walled cylinder whose outer surface is fixed while the inner surface is subjected to a uniform fluid pressure.

Hint: Integrating successively yields

$$\phi = Ar^2 \log r + Br^2 + C \log r + D,$$

$A = 0$, since $r^2 \log r$ gives a multivalued u_θ, and $D = 0$ since it contributes zero stresses. The remaining constants B and C are to be determined from boundary conditions.

4.7.2 A composite thick cylinder is made of two different isotropic elastic materials. The radii of the hole, the interface and the outer circumference are a, b and c respectively. Solve the problem when the pressure inside the cylinder is p.

4.7.3 A steel disk of radius b is welded to a rigid concentric circular core of radius a. A magnetic pole is placed at the center of the core. The magnetic force per unit volume of steel is equal to k times the reciprocal of the square of the distance from the pole. Is the magnetic force conservative? If so, find the solution.
Hint:

$$X_r = -\frac{k}{r^2}, \qquad V = -\frac{k}{r},$$

$$\phi = \frac{2(\kappa - 1)kr}{(\kappa + 1)} + Br^2 + C \log r.$$

Show that due to the first term of ϕ above and V,

$$u_r = -\frac{k(\kappa - 1)}{\mu(\kappa + 1)}.$$

4.7.4 Solve the plane stress problem of a ring under a uniform shearing stress applied at the outer surface of the ring while its inner surface is fixed. Hint: The Airy stress function ϕ is a function of θ only.

4.7.5 For a ring loaded as shown in Fig. 4.11, the Airy stress function has the form

$$\phi = \left(Ar^2 + Br^4 + \frac{C}{r^2} + D \right) \cos 2\theta + Fr^2 + H \log r.$$

Determine the constants A, B, C, D, F and H.

4.7.6 An infinite plate whose middle plane is xy has a circular rigid core of radius a. If the plate is subjected to a uniform tension σ_0 at $x = \pm\infty$, what are the stresses and displacements? Hint:

$$\phi = \frac{\sigma_0}{4} \left(r^2 - r^2 \cos 2\theta + Aa^2 \log r + Ba^4 \frac{\cos 2\theta}{r^2} + Ca^2 \cos 2\theta \right).$$

Set 3 equations for determining 3 unknowns A, B and C.

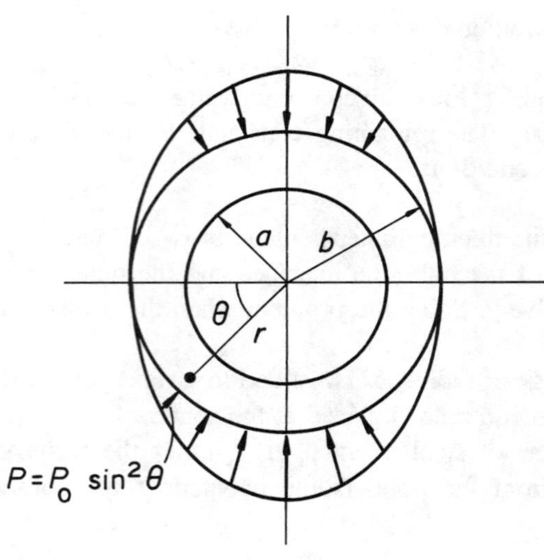

Fig. 4.11

4.7.7 Solve the previous problem, if the core is not rigid, but made of an isotropic material with different properties from the plate. Hint: For domain 1, $a < r < \infty$,

$$\phi^{(1)} = \frac{\sigma_0}{4}\left(r^2 - r^2\cos 2\theta + Aa^2\log r + Ba^4\frac{\cos 2\theta}{r^2} + Ca^2\cos 2\theta\right).$$

For domain 2, $0 \leq r < a$,

$$\phi^{(2)} = \frac{\sigma_0}{4}\left(Dr^2 + Fr^2\cos 2\theta + H\frac{r^4}{a^2}\cos 2\theta\right).$$

Obtain 6 equations for the 6 unknowns A, B, C, D, F and H.

4.7.8 A heavy circular isotropic elastic disk of mass density ρ, radius b and thickness h; has a rigid concentric circular core of radius a. The core supports the disk in such a way that the disk is in the xy-plane and the gravitational force is in the x-direction. Find stresses and displacements. If the support becomes a point, what is the solution? Hint:

$$V = -f_0 r\cos\theta,$$

$$\phi = f_0 a^2\left(Ar\theta\sin\theta + Br\log r\cos\theta + Ca^2\frac{\cos\theta}{r} + D\frac{r^3}{a^2}\cos\theta\right).$$

Obtain 5 equations for the 5 unknowns A, B, C, D and C_1, where C_1 is a rigid-body coefficient in Eqs. 4.32.

4.7.9 A homogeneous isotropic elastic plane consists of two regions, as shown in

Fig. 4.12. The inner region is subjected to a body force in the x-direction equal to f_0 per unit volume. The edge of the plane is fixed. Find the stresses and displacements. Hint: For domain 1, $a < r < b$,

$$\phi^{(1)} = f_0 a^2 \left(A r\theta \sin\theta + B r \log r \cos\theta + C a^2 \frac{\cos\theta}{r} + D \frac{r^3}{a^2} \cos\theta \right).$$

For domain 2, $0 \leq r < a$,

$$V = -f_0 r \cos\theta,$$
$$\phi^{(2)} = f_0 F r^3 \cos\theta.$$

Obtain 7 equations for the 7 unknowns A, B, C, D, F, $C_1^{(1)}$ and $C_1^{(2)}$.

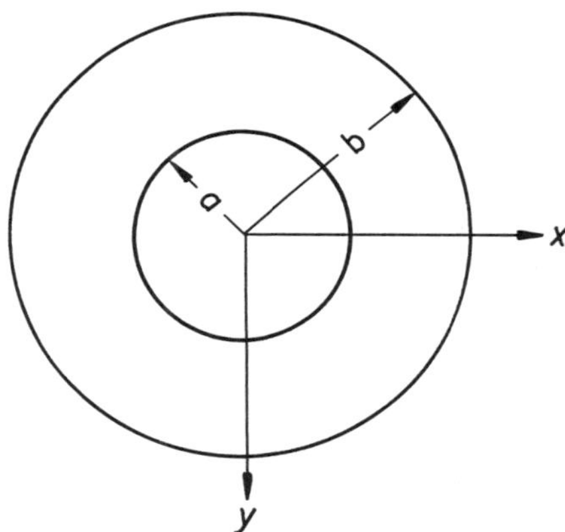

Fig 4.12

4.7.10 Pure bending of curved bars. The Airy stress function

$$\phi = A \log r + B r^2 \log r + C r^2 + D$$

may be used to study the problem of pure bending of curved bars (Fig. 4.13). The constants A, B and C are determined from the boundary conditions

$$\sigma_{rr}(a,\theta) = \sigma_{rr}(b,\theta) = 0,$$

$$\int_a^b \sigma_{\theta\theta} dr = 0, \qquad \int_a^b \sigma_{\theta\theta} r dr = -M,$$

$$\sigma_{r\theta} = 0, \quad \text{on all boundaries.}$$

The constants are given by

$$A = -\frac{4M}{N} a^2 b^2 \log\frac{b}{a},$$

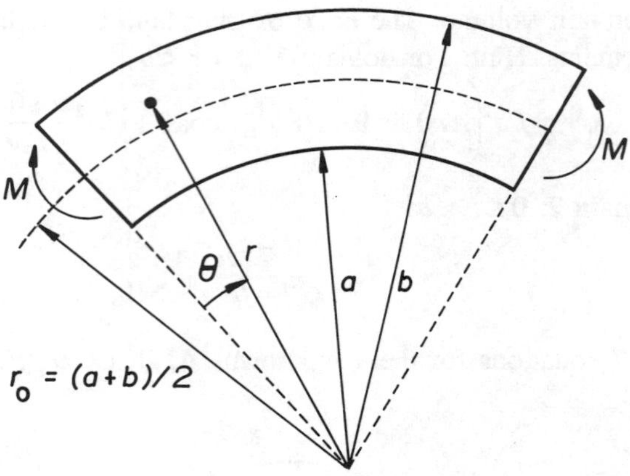

Fig. 4.13

$$B = -\frac{2M}{N}(b^2 - a^2),$$

$$C = \frac{M}{N}[b^2 - a^2 + 2(b^2 \log b - a^2 \log a)],$$

$$N = (b^2 - a^2)^2 - 4a^2 b^2 \left(\log \frac{b}{a}\right)^2.$$

4.7.11 Circular cantilever beam. The Airy stress function

$$\phi(r, \theta) = f(r) \sin \theta$$

may be used to study the problem of the circular cantilever beam subject to end shear (Fig. 4.14), with

$$f(r) = Ar^3 + \frac{B}{r} + Cr + Dr \log r.$$

The stress components are

$$\sigma_{rr} = \left(2Ar - \frac{2B}{r^3} + \frac{D}{r}\right) \sin \theta,$$

$$\sigma_{\theta\theta} = \left(6Ar + \frac{2B}{r^3} + \frac{D}{r}\right) \sin \theta,$$

$$\sigma_{r\theta} = -\left(2Ar - \frac{2B}{r^3} + \frac{D}{r}\right) \cos \theta.$$

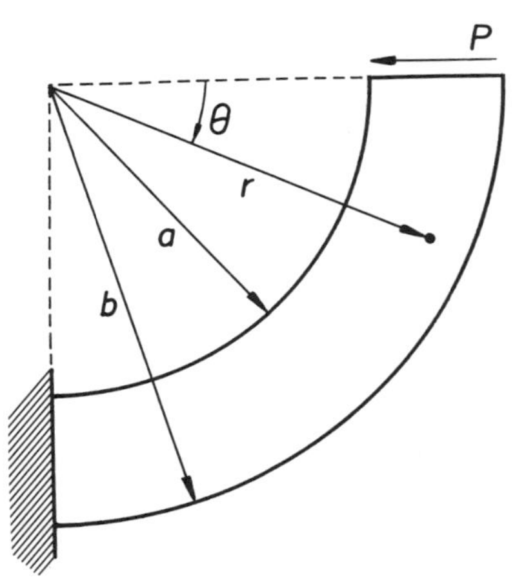

Fig. 4.14

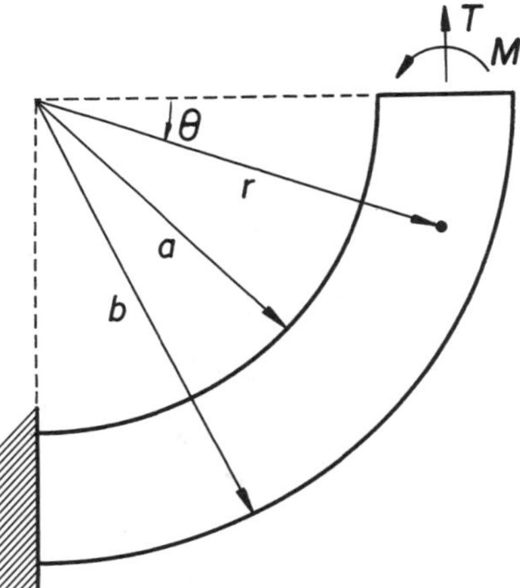

Fig 4.15

With boundary conditions,

$$\sigma_{rr} = \sigma_{r\theta} = 0 \text{ at } r = a,\ b$$

and the end shear condition,

$$\int_a^b \sigma_{r\theta} \mathrm{d}r = P$$

we get

$$A = \frac{P}{2N},\quad B = -\frac{P a^2 b^2}{2N},\quad D = -\frac{P}{N}(a^2 + b^2),$$

$$N = a^2 - b^2 + (a^2 + b^2)\log\frac{b}{a}.$$

It may be shown that the Airy stress function

$$\phi = f(r)\cos\theta$$

yields a solution to the circular cantilever beam subjected to end tension T and end moment M (Fig. 4.15). By appropriate superposition we may obtain solutions of the problems illustrated in Figs. 4.16 and 4.17.

4.7.12 A circular cantilever beam is loaded in pure bending (Fig. 4.18). Determine the displacement of the end (point A). For $r = (a + b)/2$, $\theta = 0$, let the radial and tangential displacement components vanish,

$$u_r = u_\theta = 0 \text{ and } \frac{\partial u_r}{\partial \theta} = 0.$$

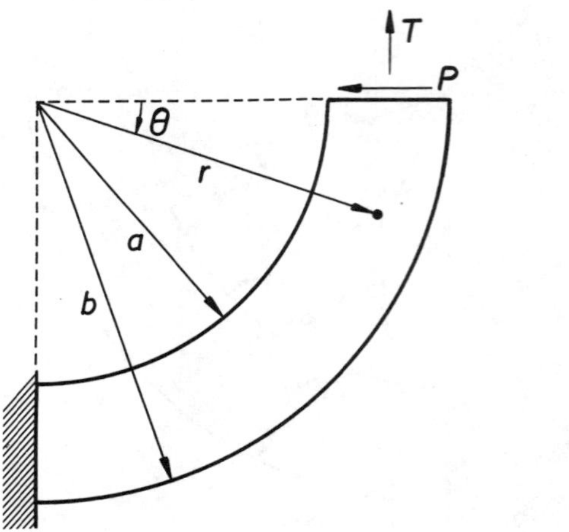

Fig. 4.16

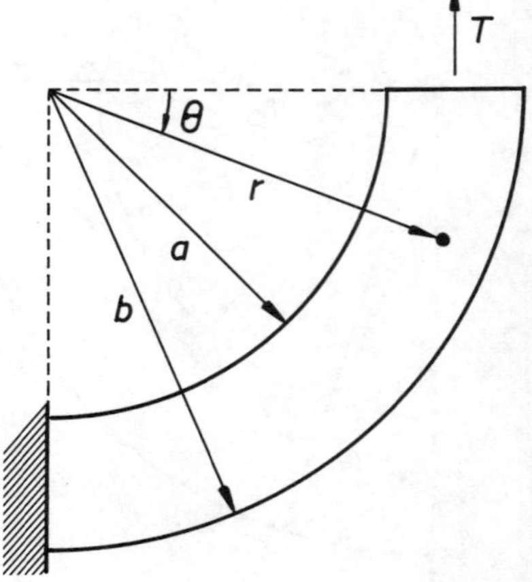

Fig 4.17

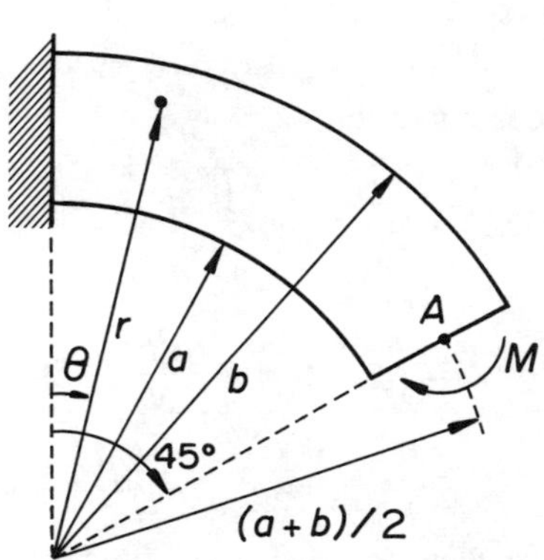

Fig. 4.18

4.7.13 A thick rectangular plate is rolled into a cylindrical shape (Fig. 4.19). Residual stresses due to the rolling process are removed by annealing. After annealing, the end planes are a small angle α apart. The end planes are then brought together by applying a moment M to each plane, and the faces are welded together. Then uniform internal pressure p_i and external pressure p_0 are applied to the lateral surface of the cylinder. Derive expressions for radial and tangential stress components σ_{rr} and $\sigma_{\theta\theta}$. Hint: Before applying pressure, show that the

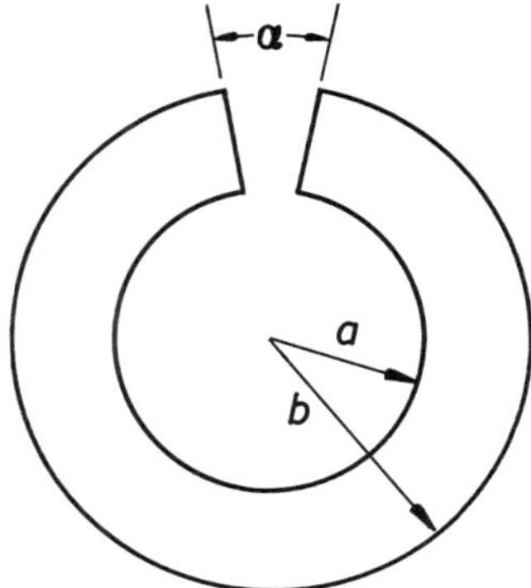

Fig 4.19

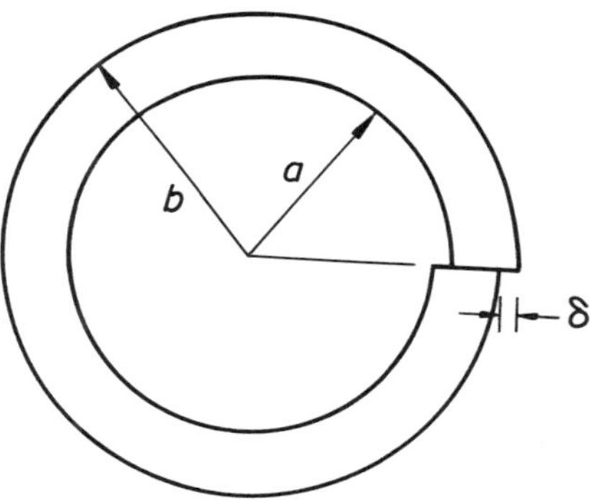

Fig 4.20

solution is the same as Problem 4.7.10, with the moment

$$M = -\frac{\mu \alpha N}{2\pi(b^2 - a^2)(\kappa + 1)}$$

where N is as given in Problem 4.7.10

4.7.14 For a circular ring with a radial slit, δ, as shown in Fig. 4.20, the function $r \log r \sin\theta$ is one of the possible candidates for the Airy stress function ϕ. Find the stresses and displacements. Hint: Try

$$\phi = Ar \log r \sin\theta + Br\theta\cos\theta + Ca^2\frac{\sin\theta}{r} + \frac{Dr^3}{a^2}\sin\theta.$$

Show that

$$A = 2\mu\delta/[\pi(\kappa+1)], \quad B = 0,$$
$$C = Ab^2/[2(a^2+b^2)], \quad D = -Aa^2/[2(a^2+b^2)].$$

Find u_θ, the shear force P, the normal force T and the moment M, at each side of the slit. Note that the two rigid-body displacement constants can be determined by setting u_θ at the slit to zero at two points only, not at every point.

4.7.15 Normal point load on edge of half plane. The problem of the point load P on the half plane boundary (Fig. 4.21), with zero stresses as $r \to \infty$, may be analyzed by means of the stress function

$$\phi(r,\theta) = -\frac{P}{\pi} r\theta \sin\theta.$$

4.7.16 Shear point load on edge of the half plane (Fig. 4.22). Show that

$$\phi = \frac{Q}{\pi} r\theta \cos\theta$$

is the solution.

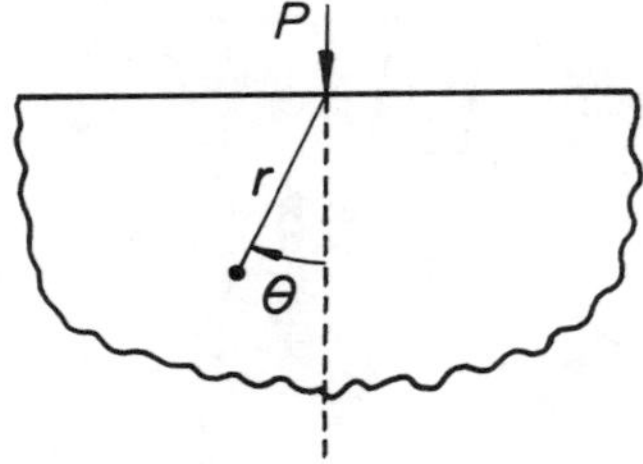

Fig. 4.21

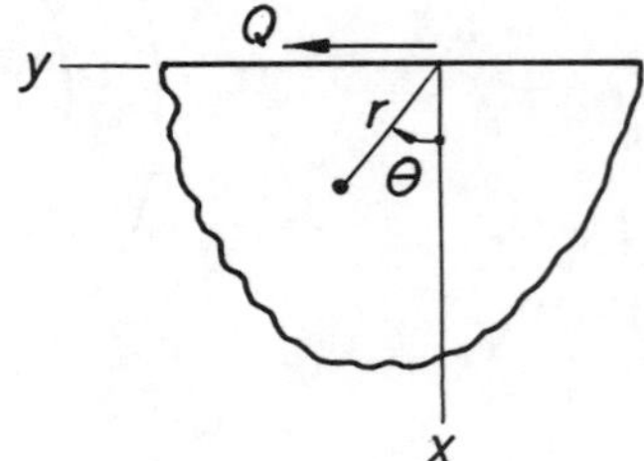

Fig 4.22

4.7.17 Verify Hertz's solution

$$\phi = \frac{P}{\pi}\left(-r_1\theta_1 \sin\theta_1 + r_2\theta_2 \sin\theta_2 + \frac{r^2}{2a}\right),$$

where the notation is shown in Fig. 4.23. Find its σ_{yy} along the x-axis. Answer: $\sigma_{yy}(x,0) = P/(\pi a)$, constant tension.

4.7.18 A circular isotropic elastic plate is subjected to two equal but opposite forces as shown in Fig. 4.24. The solution can be written in the form

$$\phi = -\frac{P}{\pi}\left[r_1\theta_1 \sin\theta_1 - r_2\theta_2 \sin\theta_2 + A\frac{r^2}{a}\right].$$

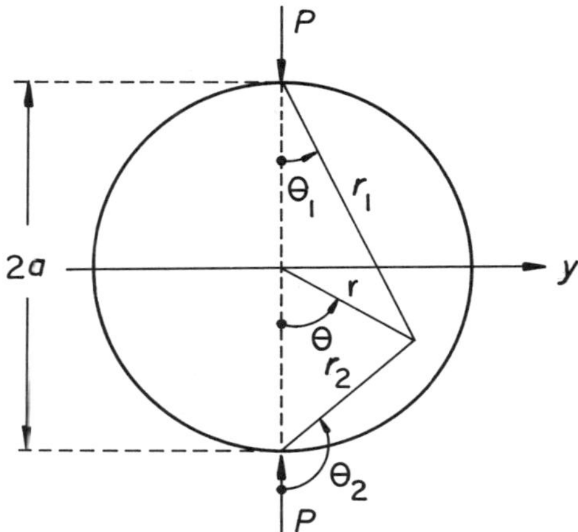

Fig. 4.23

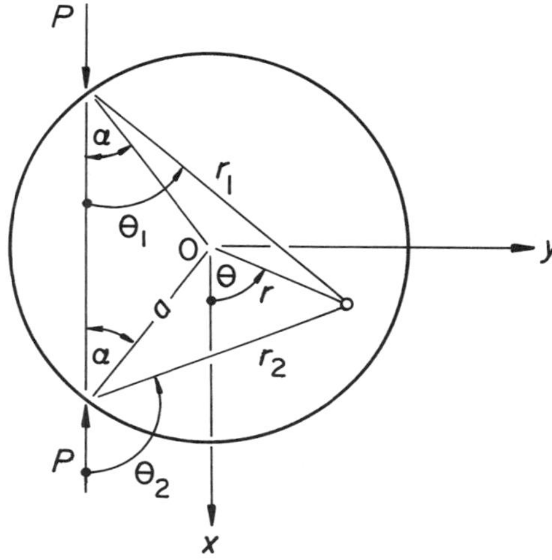

Fig. 4.24

Determine the constant A from the boundary conditions at $r = a$. Find also the hoop stress $\sigma_{\theta\theta}(a,\theta)$. Answer:

$$A = -\frac{\cos\alpha}{2}, \quad \sigma_{\theta\theta}(a,\theta) = \frac{P\sin 2\alpha}{\pi a(\sin\theta + \sin\alpha)}.$$

4.7.19 Verify that the Airy stress function

$$\phi = -\frac{P}{\pi}\left\{ \theta_1 r \sin\theta - \frac{1}{4}(1-\nu) r \log r \cos\theta - \frac{1}{2} r\theta \sin\theta + \frac{d}{4}\log r - \frac{d^2}{32}(3-\nu)\frac{1}{r}\cos\theta \right\}$$

is the solution of the problem of a force P acting in a hole in an infinite plate as shown in Fig. 4.25. Furthermore, show that the circumferential stress around the hole is

$$\frac{P}{\pi d}[2+(3-\nu)\cos\theta]$$

except at the point where the force is applied, and that the function ϕ corresponds to single-valued displacements. Hint: The question whether $\nabla^4\phi = 0$ should be answered only for the term $\theta_1 r \sin\theta$, which is equal to $\theta_1 r_1 \sin\theta_1$. Thus $\nabla^4(\theta_1 r_1 \sin\theta_1) = 0$, where ∇^4 is a biharmonic operator in the $r_1\theta_1$-system. Note that only $\sigma_{r_1 r_1}$ is the non-zero stress due to this stress function term. In order to settle the question of single-valuedness of the displacements, we need to examine the terms

$$\theta_1 r_1 \sin\theta_1 - \frac{1}{4}(1-\nu) r \log r \cos\theta - \frac{1}{2} r\theta \sin\theta,$$

which give possible multivalued displacement components proportional to

$$\left\{ \begin{array}{c} u_r(\theta,\theta_1) \\ ----- \\ u_\theta(\theta,\theta_1) \end{array} \right\} = \frac{1}{2}(\kappa-1)\theta_1 \left\{ \begin{array}{c} \sin\theta \\ --- \\ \cos\theta \end{array} \right\} + \left[-\frac{1}{8}(1-\nu)(\kappa+1) - \frac{1}{4}(\kappa-1) \right] \theta \left\{ \begin{array}{c} \sin\theta \\ --- \\ \cos\theta \end{array} \right\}.$$

Noting that $\nu = (3-\kappa)/(\kappa+1)$ for the present plane stress problem, we can show that

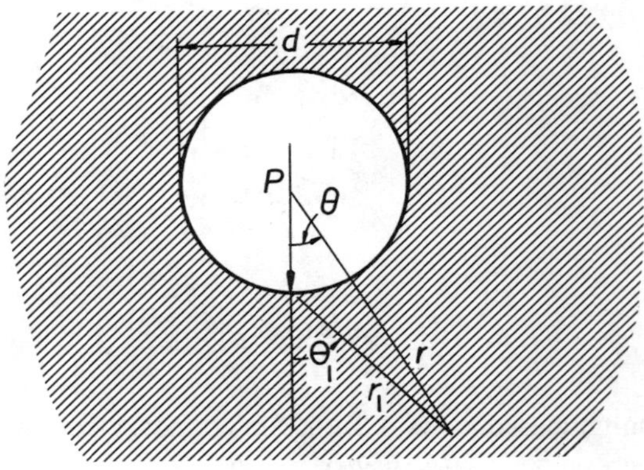

Fig. 4.25

$$u_r(\theta,\theta_1) = u_r(\theta + 2n\pi, \theta_1 + 2n\pi),$$
$$u_\theta(\theta,\theta_1) = u_\theta(\theta + 2n\pi, \theta_1 + 2n\pi),$$

where n is an integer. Thus there is no multivalued displacement.

4.7.20 The Airy stress function

$$\phi = \frac{Q}{\pi}\left[r_1\theta_1\cos\theta_1 + \frac{A(\kappa-1)}{2(\kappa+1)}r\log r\sin\theta - \frac{B}{2}r\theta\cos\theta + C\theta d + Dd^2r^{-1}\sin\theta\right]$$

is the solution to the problem of a force acting tangentially in a hole of an infinite plate as shown in Fig. 4.26. Find the constants A, B, C and D from appropriate conditions. Furthermore, find the circumferential stress around the hole. Hint: For singlevalued displacements,

$$2 - A - B = 0.$$

Answer: $A = B = 1$, $C = 1/4$, $D = \kappa/[8(\kappa+1)]$.

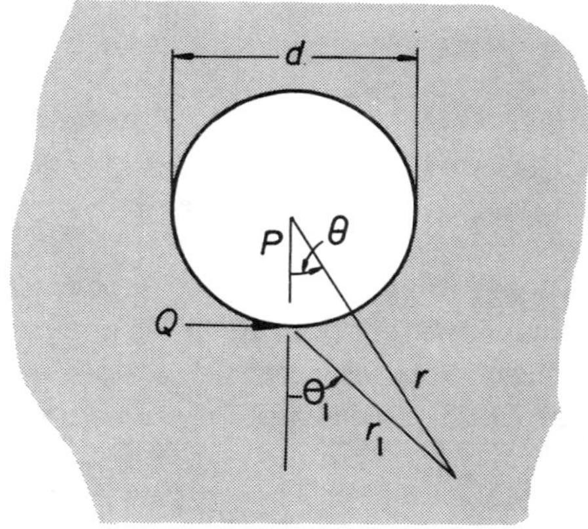

Fig. 4.26

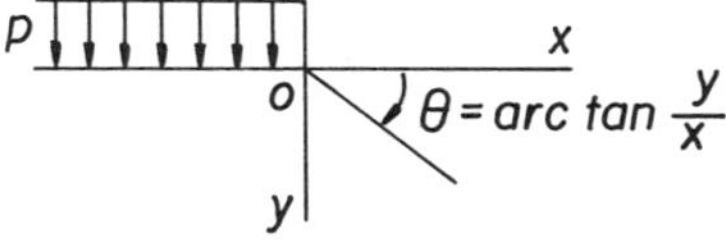

Fig. 4.27

4.7.21 Show that the stress function

$$\phi = -\frac{P}{2\pi}\left[(x^2+y^2)\arctan\frac{y}{x} - xy\right]$$

is the solution to the problem of a half plane $(-\infty < x < \infty, 0 \le y < \infty)$ subjected to a uniform pressure p on the surface $y = 0$, as shown in Fig. 4.27. The load extends indefinitely to the left.

4.7.22 Show that the function

$$\phi = \frac{s}{\pi}\left[\frac{1}{2}y^2\log(x^2+y^2) + xy\arctan\frac{y}{x} - y^2\right]$$

is biharmonic, and is the solution to the problem of a half plane subjected to a uniform shear load on the surface $y = 0$, as shown in Fig. 4.28. The load extends indefinitely to the left. Hint: Note the function defined by Eq. 4.64a is harmonic.

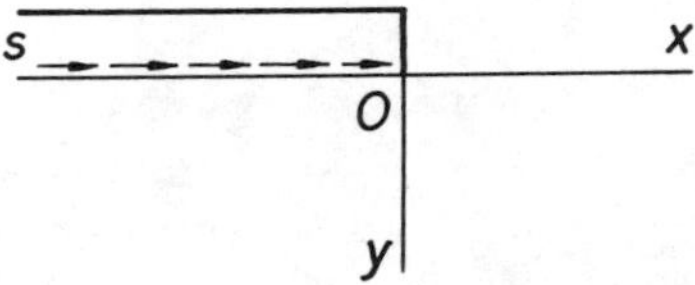

Fig. 4.28

4.7.23 By superposition, using the results of Problem 4.7.21, obtain stress and displacement components for pressure p on a segment $-a < x < a$ on the surface of the half plane. Note the indefiniteness of the displacement, and the 'weak' singularity of the displacement as r tends to infinity. Answer:

$$\sigma_{xy} = -\frac{p}{\pi}\frac{4axy^2}{[(x-a)^2+y^2]\,[(x+a)^2+y^2]}$$

4.7.24 If the pressure loading of Problem 4.7.23 is replaced by a shear load s, find stress and displacement components.

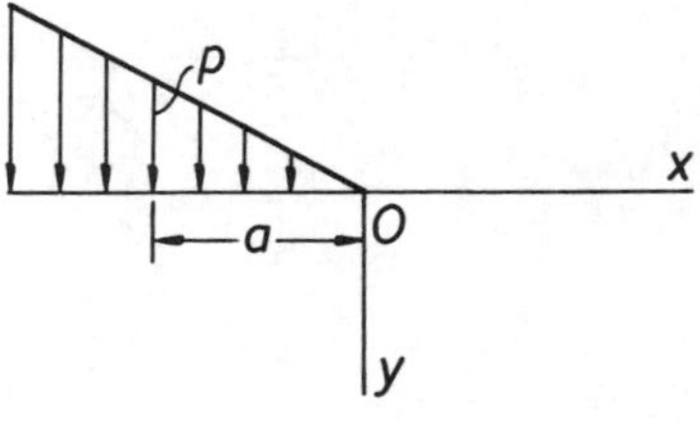

Fig. 4.29

4.7.25 Show that the function

$$\phi = \frac{p}{2\pi a}\left[\left(\frac{1}{3}x^3 + xy^2\right)\arctan\frac{y}{x} + \frac{1}{3}y^3\log(x^2+y^2) - \frac{1}{3}x^2y\right]$$

is biharmonic and is the solution to the problem of a half plane subjected to a linearly increasing pressure p which extends indefinitely to the left, as shown in Fig. 4.29. Hint: Note the functions defined by Eqs. 4.64a and c are harmonic.

4.7.26 Show that if the pressure loading of Problem 4.7.25 is replaced by a shear load s, the appropriate stress function is

$$\phi = \frac{s}{2\pi a}\left[xy^2\log(x^2+y^2) + (x^2y - y^3)\arctan\frac{y}{x} - 3xy^2\right].$$

Hint: Note that the functions defined by Eqs. 4.64b and d are harmonic.

4.7.27 Show how the distribution of load indicated in Fig. 4.30 may be obtained by superposition of loadings of the type shown in Fig. 4.29.

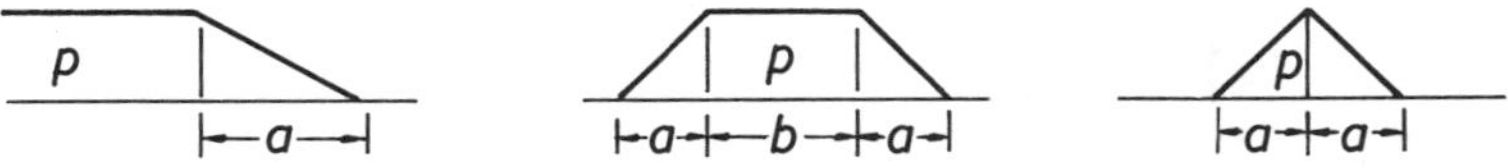

Fig. 4.30

4.7.28 Show that the solution to the half plane subjected to the parabolic loading as shown in Fig. 4.31 is given by the stress function

$$-\frac{p}{\pi}\left\{-\frac{xy^3}{3a^2}\log\frac{r_2^2}{r_1^2} - \left[\frac{a^2}{4} + \frac{1}{2}(x^2+y^2)\left(1 - \frac{x^2}{6a^2} + \frac{y^2}{2a^2}\right)\right]\alpha + \frac{2}{3}ax\beta + \frac{1}{2}ay\left(1 - \frac{x^2}{3a^2} + \frac{y^2}{a^2}\right)\right\}$$

for pressure, and

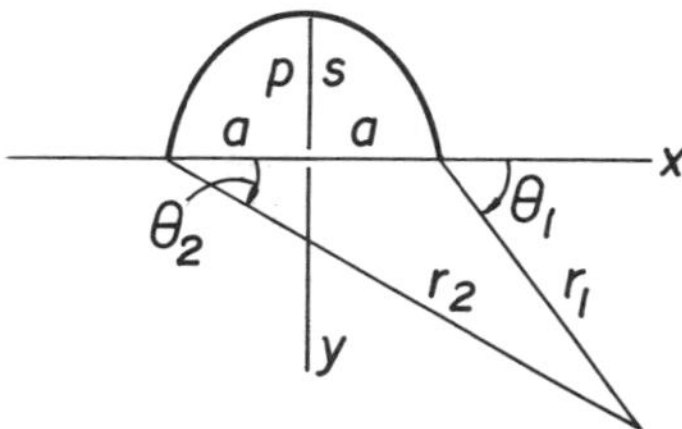

Fig. 4.31

$$\frac{s}{\pi}\left\{\frac{y^2}{6a^2}(3a^2-3x^2+y^2)\log\frac{r_2^2}{r_1^2}+\frac{2}{3}ay\beta+\frac{xy}{3a^2}(x^2-3y^2-3a^2)\alpha+\frac{4xy^2}{3a}\right\}$$

for shear, where

$$r_1^2=(x-a)^2+y^2,\quad r_2^2=(x+a)^2+y^2,$$

$$\alpha=\theta_1-\theta_2=\arctan\frac{2ay}{x^2+y^2-a^2},\quad \beta=\theta_1+\theta_2=\arctan\frac{2xy}{x^2-y^2-a^2}.$$

Table 4.1 Complementary and Particular Solutions for Elastostatics of Isotropic Planes.

Complementary solutions, $\nabla^4\phi = 0$.

ϕ	σ_{rr}	$\sigma_{r\theta}$	$\sigma_{\theta\theta}$	$2\mu u_r$	$2\mu u_\theta$
1 *	0	0	0	0	0
$r\cos\theta$ *	0	0	0	0	0
$r\sin\theta$ *	0	0	0	0	0
r^2	2	0	2	$(\kappa-1)r$	0
$\log r$ *	$1/r^2$	0	$-1/r^2$	$-1/r$	0
θ *	0	$1/r^2$	0	0	$-1/r$
$r^2\log r$	$2\log r+1$	0	$2\log r+3$	$(\kappa-1)r\log r-r$	$(\kappa+1)r\theta$
$r^2\theta$	2θ	-1	2θ	$(\kappa-1)r\theta$	$-(\kappa+1)r\log r$
$\cos\theta/r$ *	$-2\cos\theta/r^3$	$-2\sin\theta/r^3$	$2\cos\theta/r^3$	$\cos\theta/r^2$	$\sin\theta/r^2$
$\sin\theta/r$ *	$-2\sin\theta/r^3$	$2\cos\theta/r^3$	$2\sin\theta/r^3$	$\sin\theta/r^2$	$-\cos\theta/r^2$
$r^2\cos 2\theta$ *	$-2\cos 2\theta$	$2\sin 2\theta$	$2\cos 2\theta$	$-2r\cos 2\theta$	$2r\sin 2\theta$
$r^2\sin 2\theta$ *	$-2\sin 2\theta$	$-2\cos 2\theta$	$2\sin 2\theta$	$-2r\sin 2\theta$	$-2r\cos 2\theta$
$\cos 2\theta/r^2$ *	$-6\cos 2\theta/r^4$	$-6\sin 2\theta/r^4$	$6\cos 2\theta/r^4$	$2\cos 2\theta/r^3$	$2\sin 2\theta/r^3$
$\sin 2\theta/r^2$ *	$-6\sin 2\theta/r^4$	$6\cos 2\theta/r^4$	$6\sin 2\theta/r^4$	$2\sin 2\theta/r^3$	$-2\cos 2\theta/r^3$
$\cos 2\theta$	$-4\cos 2\theta/r^2$	$-2\sin 2\theta/r^2$	0	$(\kappa+1)\cos 2\theta/r$	$-(\kappa-1)\sin 2\theta/r$
$\sin 2\theta$	$-4\sin 2\theta/r^2$	$2\cos 2\theta/r^2$	0	$(\kappa+1)\sin 2\theta/r$	$(\kappa-1)\cos 2\theta/r$
$r^4\cos 2\theta$	0	$6r^2\sin 2\theta$	$12r^2\cos 2\theta$	$-(3-\kappa)r^3\cos 2\theta$	$(3+\kappa)r^3\sin 2\theta$
$r^4\sin 2\theta$	0	$-6r^2\cos 2\theta$	$12r^2\sin 2\theta$	$-(3-\kappa)r^3\sin 2\theta$	$-(3+\kappa)r^3\cos 2\theta$

* Harmonic functions, $\nabla^2\phi = 0$.

Complementary solutions, $\nabla^4\phi = 0$.

ϕ	σ_{rr}	$\sigma_{r\theta}$	$\sigma_{\theta\theta}$	$2\mu u_r$	$2\mu u_\theta$
$r^n \cos n\theta$ *	$-n(n-1)r^{n-2}\cos n\theta$	$n(n-1)r^{n-2}\sin n\theta$	$n(n-1)r^{n-2}\cos n\theta$	$-nr^{n-1}\cos n\theta$	$nr^{n-1}\sin n\theta$
$r^n \sin n\theta$ *	$-n(n-1)r^{n-2}\sin n\theta$	$-n(n-1)r^{n-2}\cos n\theta$	$n(n-1)r^{n-2}\sin n\theta$	$-nr^{n-1}\sin n\theta$	$-nr^{n-1}\cos n\theta$
$r^{n+2}\cos n\theta$	$-(n+1)(n-2)r^n\cos n\theta$	$(n+1)nr^n\sin n\theta$	$(n+2)(n+1)r^n\cos n\theta$	$-(n+1-\kappa)r^{n+1}\cos n\theta$	$(n+1+\kappa)r^{n+1}\sin n\theta$
$r^{n+2}\sin n\theta$	$-(n+1)(n-2)r^n\sin n\theta$	$-(n+1)nr^n\cos n\theta$	$(n+2)(n+1)r^n\sin n\theta$	$-(n+1-\kappa)r^{n+1}\sin n\theta$	$-(n+1+\kappa)r^{n+1}\cos n\theta$
$\frac{\cos n\theta}{r^n}$ *	$-\frac{(n+1)n\cos n\theta}{r^{n+2}}$	$-\frac{(n+1)n\sin n\theta}{r^{n+2}}$	$\frac{(n+1)n\cos n\theta}{r^{n+2}}$	$\frac{n\cos n\theta}{r^{n+1}}$	$\frac{n\sin n\theta}{r^{n+1}}$
$\frac{\sin n\theta}{r^n}$ *	$-\frac{(n+1)n\sin n\theta}{r^{n+2}}$	$-\frac{(n+1)n\cos n\theta}{r^{n+2}}$	$\frac{(n+1)n\sin n\theta}{r^{n+2}}$	$\frac{n\sin n\theta}{r^{n+1}}$	$-\frac{n\cos n\theta}{r^{n+1}}$
$\frac{\cos n\theta}{r^{n-2}}$	$-\frac{(n+2)(n-1)\cos n\theta}{r^n}$	$-\frac{n(n-1)\sin n\theta}{r^n}$	$\frac{(n-1)(n-2)\cos n\theta}{r^n}$	$\frac{(n-1+\kappa)\cos n\theta}{r^{n-1}}$	$\frac{(n-1-\kappa)\sin n\theta}{r^{n-1}}$
$\frac{\sin n\theta}{r^{n-2}}$	$-\frac{(n+2)(n-1)\sin n\theta}{r^n}$	$\frac{n(n-1)\cos n\theta}{r^n}$	$\frac{(n-1)(n-2)\sin n\theta}{r^n}$	$\frac{(n-1+\kappa)\sin n\theta}{r^{n-1}}$	$-\frac{(n-1-\kappa)\cos n\theta}{r^{n-1}}$

* Harmonic functions, $\nabla^2\phi = 0$.

Complementary solutions, $\nabla^4\phi = 0$.

ϕ	σ_{rr}	$\sigma_{r\theta}$	$\sigma_{\theta\theta}$	$2\mu u_r$	$2\mu u_\theta$
$r^3\cos\theta$	$2r\cos\theta$	$2r\sin\theta$	$6r\cos\theta$	$(\kappa-2)r^2\cos\theta$	$(\kappa+2)r^2\sin\theta$
$r^3\sin\theta$	$2r\sin\theta$	$-2r\cos\theta$	$6r\sin\theta$	$(\kappa-2)r^2\sin\theta$	$-(\kappa+2)r^2\cos\theta$
$r\theta\sin\theta$	$2\cos\theta/r$	0	0	$\frac{1}{2}[(\kappa+1)\log r\cos\theta - \cos\theta+(\kappa-1)\theta\sin\theta]$	$\frac{1}{2}[-(\kappa+1)\log r\sin\theta - \sin\theta+(\kappa-1)\theta\cos\theta]$
$r\log r\cos\theta$	$\cos\theta/r$	$\sin\theta/r$	$\cos\theta/r$	$\frac{1}{2}[(\kappa-1)\log r\cos\theta - \cos\theta+(\kappa+1)\theta\sin\theta]$	$\frac{1}{2}[-(\kappa-1)\log r\sin\theta - \sin\theta+(\kappa+1)\theta\cos\theta]$
$r\theta\cos\theta$	$-2\sin\theta/r$	0	0	$\frac{1}{2}[-(\kappa+1)\log r\sin\theta + \sin\theta+(\kappa-1)\theta\cos\theta]$	$\frac{1}{2}[-(\kappa+1)\log r\cos\theta - \cos\theta-(\kappa-1)\theta\sin\theta]$
$r\log r\sin\theta$	$\sin\theta/r$	$-\cos\theta/r$	$\sin\theta/r$	$\frac{1}{2}[(\kappa-1)\log r\sin\theta - \sin\theta-(\kappa+1)\theta\cos\theta]$	$\frac{1}{2}[(\kappa-1)\log r\cos\theta + \cos\theta+(\kappa+1)\theta\sin\theta]$

Complementary solutions, $\nabla^4\phi = 0$.

ϕ	σ_{xx}	σ_{xy}	σ_{yy}	$2\mu u$	$2\mu v$
θ *	$-\frac{2xy}{r^4}$	$-\frac{1}{r^2}+\frac{2x^2}{r^4}$	$\frac{2xy}{r^4}$	$\frac{y}{r^2}$	$-\frac{x}{r^2}$
$\log r$ *	$-\frac{1}{r^2}+\frac{2x^2}{r^4}$	$\frac{2xy}{r^4}$	$\frac{1}{r^2}-\frac{2x^2}{r^4}$	$-\frac{x}{r^2}$	$-\frac{y}{r^2}$
$r\theta\cos\theta$	$-\frac{2x^2y}{r^4}$	$-\frac{2x}{r^2}+\frac{2x^3}{r^4}$	$-\frac{2y}{r^2}+\frac{2x^2y}{r^4}$	$\frac{1}{2}(\kappa-1)\theta+\frac{xy}{r^2}$	$-\frac{1}{2}(\kappa+1)\log r-\frac{x^2}{r^2}$
$r\theta\sin\theta$	$\frac{2x^3}{r^4}$	$\frac{2x^2y}{r^4}$	$\frac{2x}{r^2}-\frac{2x^3}{r^4}$	$\frac{1}{2}(\kappa+1)\log r-\frac{x^2}{r^2}$	$\frac{1}{2}(\kappa-1)\theta-\frac{xy}{r^2}$
$r\log r\cos\theta$	$-\frac{x}{r^2}+\frac{2x^3}{r^4}$	$-\frac{y}{r^2}+\frac{2x^2y}{r^4}$	$\frac{3x}{r^2}-\frac{2x^3}{r^4}$	$\frac{1}{2}(\kappa-1)\log r-\frac{x^2}{r^2}$	$\frac{1}{2}(\kappa+1)\theta-\frac{xy}{r^2}$
$r\log r\sin\theta$	$\frac{y}{r^2}+\frac{2x^2y}{r^4}$	$\frac{x}{r^2}-\frac{2x^3}{r^4}$	$\frac{y}{r^2}-\frac{2x^2y}{r^4}$	$-\frac{1}{2}(\kappa+1)\theta-\frac{xy}{r^2}$	$\frac{1}{2}(\kappa-1)\log r+\frac{x^2}{r^2}$
$\frac{\cos\theta}{r}$ *	$\frac{6x}{r^4}-\frac{8x^3}{r^6}$	$\frac{2y}{r^4}-\frac{8x^2y}{r^6}$	$-\frac{6x}{r^4}+\frac{8x^3}{r^6}$	$-\frac{1}{r^2}+\frac{2x^2}{r^4}$	$\frac{2xy}{r^4}$
$\frac{\sin\theta}{r}$ *	$\frac{2y}{r^4}-\frac{8x^2y}{r^6}$	$-\frac{6x}{r^4}+\frac{8x^3}{r^6}$	$-\frac{2y}{r^4}+\frac{8x^2y}{r^6}$	$\frac{2xy}{r^4}$	$\frac{1}{r^2}-\frac{2x^2}{r^4}$

* Harmonic functions, $\nabla^2\phi = 0$.

Complementary solutions, $\nabla^4\phi = 0$.

ϕ	σ_{xx}	σ_{xy}	σ_{yy}	$2\mu u$	$2\mu v$
$\cos 2\theta$	$\frac{12x^2}{r^4} - \frac{16x^4}{r^6}$	$\frac{8xy}{r^4} - \frac{16x^3y}{r^6}$	$\frac{4}{r^2} - \frac{20x^2}{r^4} + \frac{16x^4}{r^6}$	$-(3-\kappa)\frac{x}{r^2} + \frac{4x^3}{r^4}$	$-(\kappa+1)\frac{y}{r^2} + \frac{4x^2y}{r^4}$
$\sin 2\theta$	$\frac{4xy}{r^4} - \frac{16x^3y}{r^6}$	$\frac{2}{r^2} - \frac{16x^2}{r^4} + \frac{16x^4}{r^6}$	$-\frac{12xy}{r^4} + \frac{16x^3y}{r^6}$	$(\kappa-1)\frac{y}{r^2} + \frac{4x^2y}{r^4}$	$(3+\kappa)\frac{x}{r^2} - \frac{4x^3}{r^4}$

Particular solutions, $\nabla^4\phi = -2\frac{\kappa-1}{\kappa+1}\nabla^2 V$.

V	1	r^{-2}	$\log r$	$r^{-1}\cos\theta$	$r^{-1}\sin\theta$
$-\frac{\kappa+1}{2(\kappa-1)}\phi$	0	$\frac{(\log r)^2}{2}$	0	0	0
σ_{rr}	0	$\left[1 - \frac{2(\kappa-1)\log r}{(\kappa+1)}\right] r^{-2}$	$\log r$	$r^{-1}\cos\theta$	$r^{-1}\sin\theta$
$\sigma_{r\theta}$	0	0	0	0	0
$\sigma_{\theta\theta}$	0	$\left[1 - \frac{2(\kappa-1)(1-\log r)}{(\kappa+1)}\right] r^{-2}$	$\log r$	$r^{-1}\cos\theta$	$r^{-1}\sin\theta$
$2\mu u_r$	0	$\frac{2(\kappa-1)\log r}{(\kappa+1)} r^{-1}$	$\frac{1}{2}(\kappa-1)(\log r - 1)r$	$\frac{\kappa-1}{2}(\log r\cos\theta + \theta\sin\theta)$	$\frac{\kappa-1}{2}(\log r\sin\theta - \theta\cos\theta)$
$2\mu u_\theta$	0	0	$\frac{1}{2}(\kappa-1)r\theta$	$-\frac{\kappa-1}{2}(\log r\sin\theta - \theta\cos\theta)$	$\frac{\kappa-1}{2}(\log r\cos\theta + \theta\sin\theta)$

Particular solutions, $\nabla^4\phi = -2\frac{\kappa-1}{\kappa+1}\nabla^2 V$.

V	r^n $(n \neq 0 \text{ or } -2)$	$\log r \cos 2\theta$	$\log r \sin 2\theta$
$-\frac{\kappa+1}{2(\kappa-1)}\phi$	$\frac{r^{n+2}}{(n+2)^2}$	$\frac{(2\log r-1)}{16} r^2 \log r \cos 2\theta$	$\frac{(2\log r-1)}{16} r^2 \log r \sin 2\theta$
σ_{rr}	$\left[1-\frac{2(\kappa-1)}{(\kappa+1)(n+2)}\right] r^n$	$\left\{\frac{(\kappa-1)}{8(\kappa+1)}[1+2(2\log r-3)\log r] + \log r\right\} \cos 2\theta$	$\left\{\frac{(\kappa-1)}{8(\kappa+1)}[1+2(2\log r-3)\log r] + \log r\right\} \sin 2\theta$
$\sigma_{\theta\theta}$	$\left[1-\frac{2(\kappa-1)(n+1)}{(\kappa+1)(n+2)}\right] r^n$	$\left\{-\frac{(\kappa-1)}{8(\kappa+1)}[2\log r(2\log r+5)+1] + \log r\right\} \cos 2\theta$	$\left\{-\frac{(\kappa-1)}{8(\kappa+1)}[2\log r(2\log r+5)+1] + \log r\right\} \sin 2\theta$
$\sigma_{r\theta}$	0	$\frac{(\kappa-1)[1-(3+2\log r)\log r] \sin 2\theta}{4(\kappa+1)}$	$\frac{(\kappa-1)[(3+2\log r)\log r-1] \cos 2\theta}{4(\kappa+1)}$
$2\mu u_r$	$\frac{2(\kappa-1)r^{n+1}}{(\kappa+1)(n+2)}$	$\frac{(\kappa-1)[2(2\log r+1)\log r-1] r \cos 2\theta}{8(\kappa+1)}$	$\frac{(\kappa-1)[2(2\log r+1)\log r-1] r \sin 2\theta}{8(\kappa+1)}$
$2\mu u_\theta$	0	$\frac{(\kappa-1)(1-2\log r) r \log r \sin 2\theta}{4(\kappa+1)}$	$-\frac{(\kappa-1)(1-2\log r) r \log r \cos 2\theta}{4(\kappa+1)}$

Particular solutions, $\nabla^4\phi = -2\frac{\kappa-1}{\kappa+1}\nabla^2 V$.

$\frac{V}{(\cos m\theta, \sin m\theta)}$	r^m $(m \neq 0 \text{ or } -1)$	r^{m-2} $(m \neq 0 \text{ or } 1)$	$\log r$ $(m \neq 0 \text{ or } \pm 2)$	r^n $(n \neq m, n \neq \pm m-2, m \neq 0)$
$\frac{-(\kappa+1)\phi}{2(\kappa-1)(\cos m\theta, \sin m\theta)}$	0	$\frac{r^m \log r}{2m}$	$-\frac{[4+(m^2-4)\log r]r^2}{(m^2-4)^2}$	$\frac{r^{n+2}}{(n+2)^2-m^2}$
$\frac{\sigma_{rr}}{(\cos m\theta, \sin m\theta)}$	r^m	$r^{m-2}\left\{1+\frac{(1-\kappa)}{m(\kappa+1)}[1 - m(m-1)\log r]\right\}$	$\frac{2(\kappa-1)}{(\kappa+1)(m^2-4)^2}[4-3m^2 -(m^2-4)(m^2-2)\log r]+\log r$	$\left\{1-\frac{2(\kappa-1)(n+2-m^2)}{(\kappa+1)[(n+2)^2-m^2]}\right\}r^n$
$\frac{\sigma_{r\theta}}{(\sin m\theta, -\cos m\theta)}$	0	$\frac{(1-\kappa)[1+(m-1)\log r]r^{m-2}}{\kappa+1}$	$\frac{2(\kappa-1)m[m^2+(m^2-4)\log r]}{(\kappa+1)(m^2-4)^2}$	$-\frac{2(\kappa-1)m(n+1)r^n)}{(\kappa+1)[(n+2)^2-m^2]}$
$\frac{\sigma_{\theta\theta}}{(\cos m\theta, \sin m\theta)}$	r^m	$r^{m-2}\left\{1-\frac{\kappa-1}{m(\kappa+1)}[-1 +m(m-1)\log r+2m]\right\}$	$\frac{2(\kappa-1)}{(\kappa+1)(m^2-4)^2}[-4+3m^2 +2(m^2-4)\log r]+\log r$	$\left\{1-\frac{2(\kappa-1)(n+2)(n+1)}{(\kappa+1)[(n+2)^2-m^2]}\right\}r^n$
$\frac{2\mu u_r}{(\cos m\theta, \sin m\theta)}$	$\frac{(\kappa-1)r^{m+1}}{2(m+1)}$	$\frac{(\kappa-1)(1+m\log r)r^{m-1}}{m(\kappa+1)}$	$\frac{2(\kappa-1)}{(\kappa+1)(m^2-4)^2}r[-m^2-4 -2(m^2-4)\log r]$	$\frac{2(\kappa-1)(n+2)r^{n+1}}{(\kappa+1)[(n+2)^2-m^2]}$
$\frac{2\mu u_\theta}{(\sin m\theta, -\cos m\theta)}$	$\frac{(\kappa-1)r^{m+1}}{2(m+1)}$	$-\frac{(\kappa-1)r^{m-1}\log r}{\kappa+1}$	$\frac{2(\kappa-1)m[4+(m^2-4)\log r]r}{(\kappa+1)(m^2-4)^2}$	$-\frac{2(\kappa-1)mr^{n+1}}{(\kappa+1)[(n+2)^2-m^2]}$

CHAPTER V

BENDING OF ELASTIC THIN PLATES

5.1 BASIC ASSUMPTIONS

The preceding chapters have been concerned, in large part, with so-called theory of elasticity solution. It was assumed that there were three displacements u, v, w, and each was, in general, a function of x, y, z. The displacements and corresponding strains and stresses had to satisfy the stress-strain relations and the equilibrium equations which, when put all together, gave Navier equations of motion.

This chapter, on the other hand, describes an approximate engineering theory. For a thin plate in bending, shown in Fig. 5.1, which has a z-dimension (thickness) much smaller than its other dimensions; we assume that a normal to the undeformed middle plane of the plate remains normal after deformation (Fig. 5.2), and the *deflection* w (displacement in the z-direction) is constant across the plate thickness. Thus

$$u(x,y,z) = u^0(x,y) - z\frac{\partial w(x,y)}{\partial x} \tag{5.1a}$$

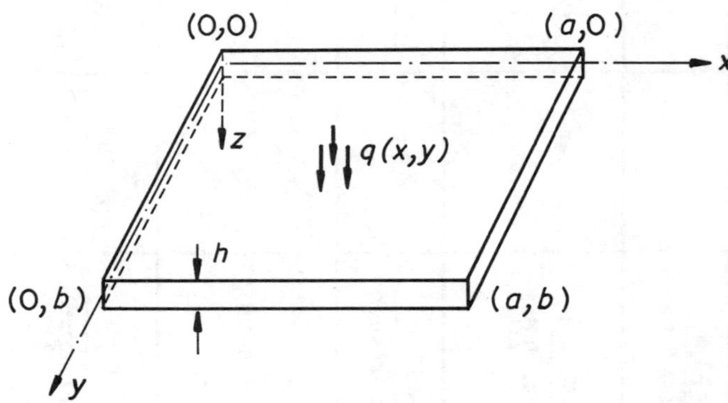

Fig. 5.1 Rectangular plate with span lengths equal to a and b in x and y directions respectively.

$$v(x,y,z) = v^0(x,y) - z\frac{\partial w(x,y)}{\partial y} \tag{5.1b}$$

Here the origin of the Cartesian system is located on the plate middle plane, and the superscript zero stands for the quantities there. Using Eqs. 5.1, we can show that shear distortions γ_{xz} and γ_{yz} are neglected.

The only non-zero strains are

$$\begin{aligned} \varepsilon_{xx} &= \varepsilon_{xx}^0 - z\frac{\partial^2 w}{\partial x^2} \\ \varepsilon_{yy} &= \varepsilon_{yy}^0 - z\frac{\partial^2 w}{\partial y^2} \\ \gamma_{xy} &= \gamma_{xy}^0 - 2z\frac{\partial^2 w}{\partial x \partial y} \end{aligned} \tag{5.2}$$

Note that ε_{zz} also vanishes and all compatibility conditions are satisfied exactly. Moreover, it is normal to further assume that the middle plane remains unstrained due to bending, i.e. $\varepsilon_{xx}^0 = \varepsilon_{yy}^0 = \gamma_{xy}^0 = 0$. This leads to

$$\begin{aligned} \varepsilon_{xx} &= -z\kappa_{xx} \\ \varepsilon_{yy} &= -z\kappa_{yy} \\ \gamma_{xy} &= -2z\kappa_{xy} \end{aligned} \tag{5.3}$$

where κ stands for plate *curvatures,* i.e.

$$\kappa_{xx} = \frac{\partial^2 w}{\partial x^2} \tag{5.4a}$$

$$\kappa_{yy} = \frac{\partial^2 w}{\partial y^2} \tag{5.4b}$$

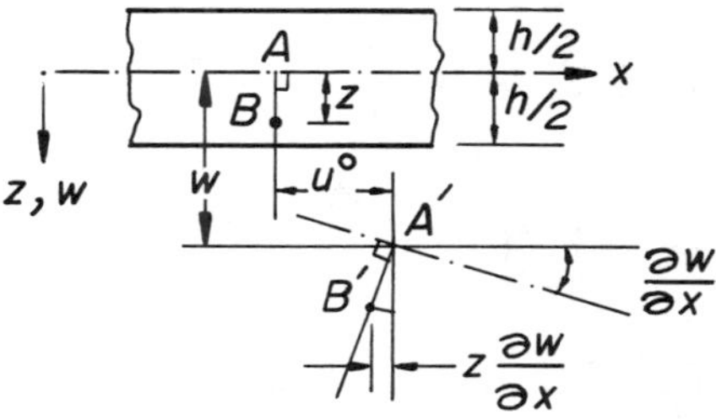

Fig. 5.2 Original and displaced configurations of an xz-section together with a normal (AB) to middle plane of the plate.

$$\kappa_{xy} = \frac{\partial^2 w}{\partial x \partial y} \tag{5.4c}$$

5.2 EQUILIBRIUM, BOUNDARY CONDITIONS, AND STRESS RESULTANTS

The equilibrium and well-posed boundary conditions of a rectangular plate as shown in Fig. 5.1 can be derived from the following virtual work principle,

$$W_i = W_e \tag{5.5}$$

where W_i and W_e are the internal and external virtual works, respectively, and can be expressed by

$$W_i = \int_A \int_{-h/2}^{h/2} (\sigma_{xx}\delta\varepsilon_{xx} + \sigma_{yy}\delta\varepsilon_{yy} + \sigma_{xy}\delta\gamma_{xy})\mathrm{d}z\,\mathrm{d}A \tag{5.6}$$

$$W_e = \int_A q\,\delta w\,\mathrm{d}A + \left[\int_0^b \int_{-h/2}^{h/2} (\bar{\sigma}_{xx}\delta u + \bar{\sigma}_{xy}\delta v + \bar{\sigma}_{xz}\delta w)\mathrm{d}z\,\mathrm{d}y\right]_{x=0}^{x=a}$$

$$+\left[\int_0^a \int_{-h/2}^{h/2} (\bar{\sigma}_{yy}\delta v + \bar{\sigma}_{yx}\delta u + \bar{\sigma}_{yz}\delta w)\mathrm{d}z\,\mathrm{d}x\right]_{y=0}^{y=b} \tag{5.7}$$

where δ is the variational symbol signifying a virtual quantity, $q(x,y)$ is the applied transverse load, a bar $(^{-})$ denotes a boundary traction but with a stress sign convention, and $[f(x)]_{x=a_1}^{x=a_2} \equiv f(a_2) - f(a_1)$. Performing integration with respect to z in Eqs. 5.6 and 5.7 yields

$$W_i = -\int_A (M_x\delta\kappa_{xx} + M_y\delta\kappa_{yy} + 2M_{xy}\delta\kappa_{xy})\mathrm{d}A \tag{5.8}$$

$$W_e = \int_A q\,\delta w\,\mathrm{d}A - \left[\int_0^b \left\{\bar{M}_x\delta\left(\frac{\partial w}{\partial x}\right) + \bar{M}_{xy}\delta\left(\frac{\partial w}{\partial y}\right) - \bar{Q}_x\delta w\right\}\mathrm{d}y\right]_{x=0}^{x=a}$$

$$-\left[\int_0^a \left\{\bar{M}_y\delta\left(\frac{\partial w}{\partial y}\right) + \bar{M}_{xy}\delta\left(\frac{\partial w}{\partial x}\right) - \bar{Q}_y\delta w\right\}\mathrm{d}x\right]_{y=0}^{y=b}$$

$$= \int_A q\,\delta w\,\mathrm{d}A - \oint_C \left[\bar{M}_x\delta\left(\frac{\partial w}{\partial x}\right) + \bar{M}_{xy}\delta\left(\frac{\partial w}{\partial y}\right) - \bar{Q}_x\delta w\right]\mathrm{d}y$$

$$+\oint_C \left[\bar{M}_y\delta\left(\frac{\partial w}{\partial y}\right) + \bar{M}_{xy}\delta\left(\frac{\partial w}{\partial x}\right) - \bar{Q}_y\delta w\right]\mathrm{d}x \tag{5.9}$$

where $\oint_C$ denotes the integral along the *simple closed curve* C enclosing the plate middle plane, and *stress resultants* are introduced for the first time, i.e.

$$(M_x, M_y, M_{xy}) = \int_{-h/2}^{h/2} z(\sigma_{xx}, \sigma_{yy}, \sigma_{xy})\mathrm{d}z \tag{5.10a,b,c}$$

$$(Q_x, Q_y) = \int_{-h/2}^{h/2} (\sigma_{xz}, \sigma_{yz})\mathrm{d}z \tag{5.10d,e}$$

Note the following identities;

$$M_x\kappa_{xx} + M_{xy}\kappa_{xy} = \left[\frac{\partial}{\partial x}\left(M_x\frac{\partial w}{\partial x}\right) + \frac{\partial}{\partial y}\left(M_{xy}\frac{\partial w}{\partial x}\right)\right] - \left[\frac{\partial M_x}{\partial x}\frac{\partial w}{\partial x} + \frac{\partial M_{xy}}{\partial y}\frac{\partial w}{\partial x}\right] \tag{5.11}$$

and

$$\frac{\partial M_x}{\partial x}\frac{\partial w}{\partial x} + \frac{\partial M_{xy}}{\partial x}\frac{\partial w}{\partial y} = \left[\frac{\partial}{\partial x}\left(\frac{\partial M_x}{\partial x}w\right) + \frac{\partial}{\partial y}\left(\frac{\partial M_{xy}}{\partial x}w\right)\right] - \left[\frac{\partial^2 M_x}{\partial x^2}w + \frac{\partial^2 M_{xy}}{\partial x \partial y}w\right] \tag{5.12}$$

In view of Eq. 5.11 and the Green's theorem (Eq. 1.89), we have

$$\int_A (M_x\delta\kappa_{xx} + M_{xy}\delta\kappa_{xy})\mathrm{d}A = -\oint_C\left[M_{xy}\delta\left(\frac{\partial w}{\partial x}\right)\mathrm{d}x - M_x\delta\left(\frac{\partial w}{\partial x}\right)\mathrm{d}y\right]$$
$$-\int_A\left[\frac{\partial M_x}{\partial x}\delta\left(\frac{\partial w}{\partial x}\right) + \frac{\partial M_{xy}}{\partial y}\delta\left(\frac{\partial w}{\partial x}\right)\right]\mathrm{d}A \tag{5.13}$$

Similarly, we can get

$$\int_A (M_y\delta\kappa_{yy} + M_{xy}\delta\kappa_{xy})\mathrm{d}A = \oint_C\left[M_{xy}\delta\left(\frac{\partial w}{\partial y}\right)\mathrm{d}y - M_y\delta\left(\frac{\partial w}{\partial y}\right)\mathrm{d}x\right]$$
$$-\int_A\left[\frac{\partial M_y}{\partial y}\delta\left(\frac{\partial w}{\partial y}\right) + \frac{\partial M_{xy}}{\partial x}\delta\left(\frac{\partial w}{\partial y}\right)\right]\mathrm{d}A \tag{5.14}$$

Due to Eq. 5.12, applying the Green's theorem one more time, we have

$$\int_A\left[\frac{\partial M_x}{\partial x}\delta\left(\frac{\partial w}{\partial x}\right) + \frac{\partial M_{xy}}{\partial x}\delta\left(\frac{\partial w}{\partial y}\right)\right]\mathrm{d}A = -\oint_C\left(\frac{\partial M_{xy}}{\partial x}\delta w\mathrm{d}x - \frac{\partial M_x}{\partial x}\delta w\mathrm{d}y\right)$$
$$-\int_A\left(\frac{\partial^2 M_x}{\partial x^2}\delta w + \frac{\partial^2 M_{xy}}{\partial x \partial y}\delta w\right)\mathrm{d}A \tag{5.15}$$

Similarly, we can get

$$\int_A\left[\frac{\partial M_y}{\partial y}\delta\left(\frac{\partial w}{\partial y}\right) + \frac{\partial M_{xy}}{\partial y}\delta\left(\frac{\partial w}{\partial x}\right)\right]\mathrm{d}A = \oint_C\left(\frac{\partial M_{xy}}{\partial y}\delta w\mathrm{d}y - \frac{\partial M_y}{\partial y}\delta w\mathrm{d}x\right)$$
$$-\int_A\left(\frac{\partial^2 M_y}{\partial y^2}\delta w + \frac{\partial^2 M_{xy}}{\partial x \partial y}\delta w\right)\mathrm{d}A \tag{5.16}$$

In addition, due to integration by parts, we can show that

$$\oint_C M_{xy}\delta\left(\frac{\partial w}{\partial x}\right)dx = -\oint_C \frac{\partial M_{xy}}{\partial x}\delta w dx + M_{xy}(a,0)\delta w(a,0) - M_{xy}(0,0)\delta w(0,0)$$

$$+ M_{xy}(0,b)\delta w(0,b) - M_{xy}(a,b)\delta w(a,b) \tag{5.17}$$

Similarly,

$$\oint_C M_{xy}\delta\left(\frac{\partial w}{\partial y}\right)dy = -\oint_C \frac{\partial M_{xy}}{\partial y}\delta w dy + M_{xy}(a,b)\delta w(a,b) - M_{xy}(a,0)\delta w(a,0)$$

$$+ M_{xy}(0,0)\delta w(0,0) - M_{xy}(0,b)\delta w(0,b) \tag{5.18}$$

Due to Eqs. 5.13 to 5.18, Eq. 5.8 becomes

$$W_i = -\int_A \left(\frac{\partial^2 M_x}{\partial x^2} + 2\frac{\partial^2 M_{xy}}{\partial x \partial y} + \frac{\partial^2 M_y}{\partial y^2}\right)\delta w dA$$

$$-\oint_C M_x \delta\left(\frac{\partial w}{\partial x}\right)dy + \oint_C M_y \delta\left(\frac{\partial w}{\partial y}\right)dx$$

$$+\oint_C \left(Q_x + \frac{\partial M_{xy}}{\partial y}\right)\delta w dy - \oint_C \left(Q_y + \frac{\partial M_{xy}}{\partial x}\right)\delta w dx$$

$$-2M_{xy}(a,b)\delta w(a,b) + 2M_{xy}(0,b)\delta w(0,b)$$

$$+2M_{xy}(a,0)\delta w(a,0) - 2M_{xy}(0,0)\delta w(0,0) \tag{5.19}$$

where Q_x and Q_y are defined as

$$Q_x = \frac{\partial M_x}{\partial x} + \frac{\partial M_{xy}}{\partial y} \tag{5.20a}$$

$$Q_y = \frac{\partial M_y}{\partial y} + \frac{\partial M_{xy}}{\partial x} \tag{5.20b}$$

It will be shown later that Q_x and Q_y are *shear stress resultants* in the same manner as Eqs. 5.10d and e. In view of Eqs. 5.17 and 5.18, Eq. 5.9 becomes

$$W_e = \int_A \int q\delta w dA - \oint_C \overline{M}_x \delta\left(\frac{\partial w}{\partial x}\right)dy + \oint_C \overline{M}_y \delta\left(\frac{\partial w}{\partial y}\right)dx + \oint_C \left(\overline{Q}_x + \frac{\partial \overline{M}_{xy}}{\partial y}\right)\delta w dy$$

$$-\oint_C \left(\overline{Q}_y + \frac{\partial \overline{M}_{xy}}{\partial x}\right)\delta w dx - 2\overline{M}_{xy}(a,b)\delta w(a,b) + 2\overline{M}_{xy}(0,b)\delta w(0,b)$$

$$+2\overline{M}_{xy}(a,0)\delta w(a,0) - 2\overline{M}_{xy}(0,0)\delta w(0,0) \tag{5.21}$$

Substituting Eqs. 5.19 and 5.21 into Eq. 5.5 leads to

$$\frac{\partial^2 M_x}{\partial x^2}+2\frac{\partial^2 M_{xy}}{\partial x \partial y}+\frac{\partial^2 M_y}{\partial y^2}+q=0; \quad \text{in } A \tag{5.22}$$

$$M_x=\overline{M}_x \quad \text{or} \quad \frac{\partial w}{\partial x}=\text{constant}; \quad \text{on boundaries } x=0 \text{ and } a \tag{5.23a}$$

$$M_y=\overline{M}_y \quad \text{or} \quad \frac{\partial w}{\partial y}=\text{constant}; \quad \text{on boundaries } y=0 \text{ and } b \tag{5.23b}$$

$$V_x=\overline{V}_x \quad \text{or} \quad w=\text{constant}; \quad \text{on boundaries } x=0 \text{ and } a \tag{5.24a}$$

$$V_y=\overline{V}_y \quad \text{or} \quad w=\text{constant}; \quad \text{on boundaries } y=0 \text{ and } b \tag{5.24b}$$

$$R=\overline{R} \quad \text{or} \quad w=\text{constant}; \quad \text{at four corners} \tag{5.25}$$

Equation 5.22 is the *equilibrium equation,* and Eqs. 5.23 to 5.25 are *well-posed boundary conditions*. In these equations,

$$V_x=Q_x+\frac{\partial M_{xy}}{\partial y} \tag{5.26a}$$

$$V_y=Q_y+\frac{\partial M_{yx}}{\partial x} \tag{5.26b}$$

$$R=-M_{xy}-M_{yx}=-2M_{xy}; \quad \text{for corners } (0,0) \text{ and } (a,b) \tag{5.27a}$$

$$=M_{xy}+M_{yx}=2M_{xy}; \quad \text{for corners } (a,0) \text{ and } (0,b) \tag{5.27b}$$

It will be shown later that M_x and M_y are *bending moments,* M_{xy} *twisting moment,* V_x and V_y *Kirchhoff* or *supplemented shear forces* (G. Kirchhoff, 1850), and R a *corner force.*

5.3 PHYSICAL MEANING OF STRESS RESULTANTS

According to Eqs. 5.10, we can illustrate the stress resultants involved on a small element of the plate middle plane as in Fig. 5.3. For the equilibrium of forces in z-direction, it can be shown that

$$\frac{\partial Q_x}{\partial x}+\frac{\partial Q_y}{\partial y}+q=0 \tag{5.28}$$

while the equilibrium of moments about x and y axes gives Eqs. 5.20b and 5.20a, respectively; which upon substituting into Eq. 5.28 give the same equilibrium equation as Eq. 5.22. For the *supplemented shear forces* on plate boundaries, consider the boundary $x=a$, as an example. The shear force Q_x and the twisting moment M_{xy} are, according to Eqs. 5.10, as shown on the Fig. 5.4(a), while Fig. 5.4(b) depicts an equivalent system, i.e. a twisting moment is taken as a couple of equal and opposite

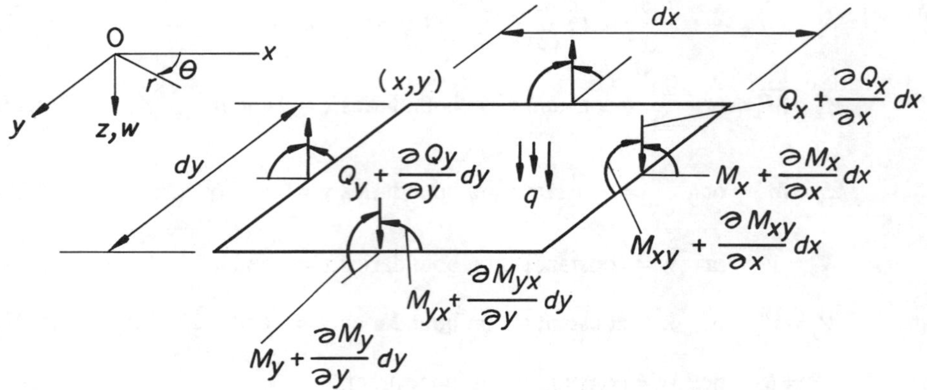

Fig. 5.3 Small plate element subjected to stress resultants.

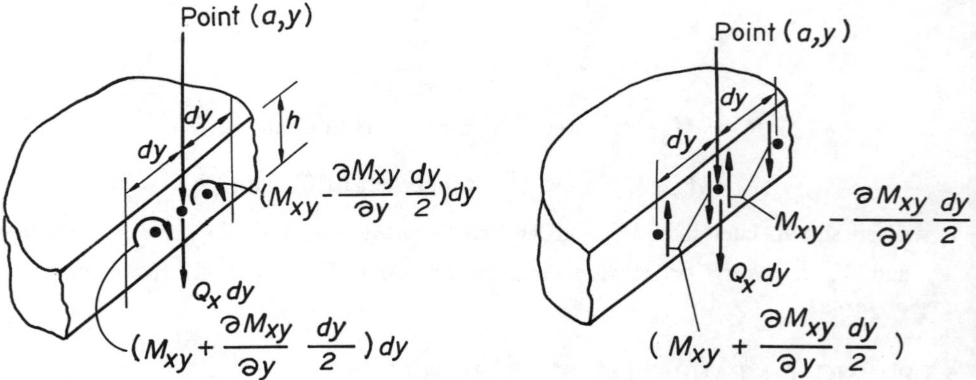

(a) Plate edge subjected to transverse shear force Q_x and twisting moment M_{xy}.

(b) Equivalent couple of shear forces for M_{xy}.

Fig. 5.4 Supplemented or Kirchhoff shear force on a plate edge perpendicular to x-axis.

forces. Thus the equivalent force at point (a, y) per unit length of the plate edge is

$$\left[Q_x \mathrm{d}y + \left(M_{xy} + \frac{\partial M_{xy}}{\partial y}\frac{\mathrm{d}y}{2}\right) - \left(M_{xy} - \frac{\partial M_{xy}}{\partial y}\frac{\mathrm{d}y}{2}\right)\right]\frac{1}{\mathrm{d}y} = Q_x + \frac{\partial M_{xy}}{\partial y},$$

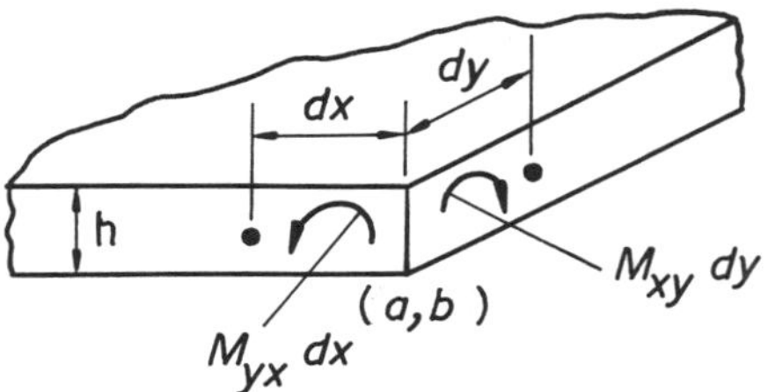

(a) Twisting moments near a corner (a,b).

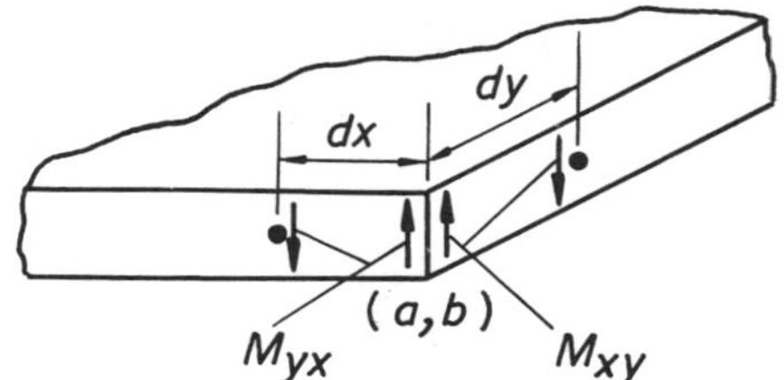

(b) Equivalent couple of shear forces for each twisting moment near the corner.

Fig. 5.5 Corner force at corner (a,b).

which is defined as V_x in Eq. 5.26a. For the *corner force,* consider the corner (a,b) as an example. The total twisting moments $M_{xy}dy$ and $M_{yx}dx$ act on edges near the corner as shown in Fig. 5.5(a), while Fig. 5.5(b) depicts an equivalent system, i.e. a twisting moment is taken as a couple of equal and opposite forces. The *free* force at the corner is thus

$$-M_{xy} - M_{yx},$$

which is defined as R in Eq. 5.27a.

For a non-right angle $(\neq \pi/2)$ corner as shown in Fig. 5.6, the corner force can be shown to be

$$R = -M_{x'y'} + M_{x''y''} \tag{5.29}$$

where x', y', z and x'', y'', z belong to right-hand systems with x' and x'' normal to two different edges forming the corner.

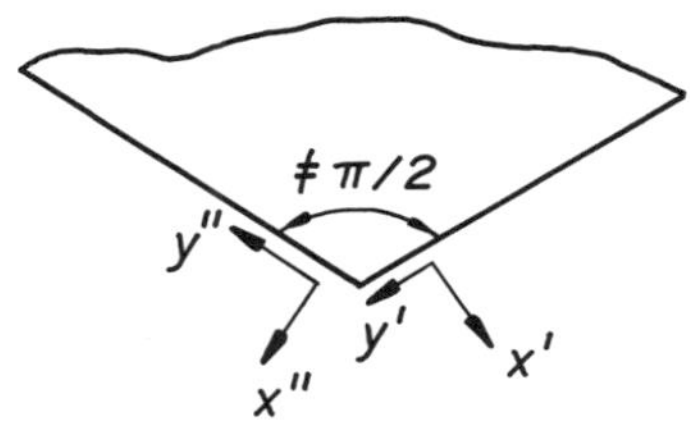

Fig. 5.6 Formulation of corner force at non-right angle.

5.4 GOVERNING CONDITIONS FOR ISOTROPIC PLATES

For isotropic elastic plane stress problems, the stress-strain relations, Eqs. 4.7, are

$$\sigma_{xx} = \frac{E}{(1-\nu^2)}(\varepsilon_{xx} + \nu\varepsilon_{yy}) \tag{5.30a}$$

$$\sigma_{yy} = \frac{E}{(1-\nu^2)}(\varepsilon_{yy} + \nu\varepsilon_{xx}) \tag{5.30b}$$

$$\sigma_{xy} = \frac{E}{2(1+\nu)}\gamma_{xy} \tag{5.30c}$$

Substituting these equations into Eqs. 5.10a to c leads to

$$M_x = -D(\kappa_{xx} + \nu\kappa_{yy}) \tag{5.31a}$$

$$M_y = -D(\kappa_{yy} + \nu\kappa_{xx}) \tag{5.31b}$$

$$M_{xy} = M_{yx} = -D(1-\nu)\kappa_{xy} \tag{5.31c}$$

where D is the *flexural rigidity in bending*, i.e.

$$D = \frac{Eh^3}{12(1-\nu^2)} \tag{5.32}$$

and κ's are curvatures as defined in Eqs. 5.4. Thus

$$M_x = -D\left(\frac{\partial^2 w}{\partial x^2} + \nu\frac{\partial^2 w}{\partial y^2}\right)$$

$$M_y = -D\left(\frac{\partial^2 w}{\partial y^2} + \nu\frac{\partial^2 w}{\partial x^2}\right) \tag{5.33}$$

$$M_{xy} = M_{yx} = -D(1-\nu)\frac{\partial^2 w}{\partial x \partial y}$$

For a plate with uniform properties; the equilibrium (Eq. 5.22), shear stress resultants (Eqs. 5.20 and 5.26) and the corner force (Eq. 5.27), can be put in terms of w as

$$\nabla^4 w = \frac{q}{D} \tag{5.34}$$

$$Q_x = -D\frac{\partial}{\partial x}(\nabla^2 w) \tag{5.35a}$$

$$Q_y = -D\frac{\partial}{\partial y}(\nabla^2 w) \tag{5.35b}$$

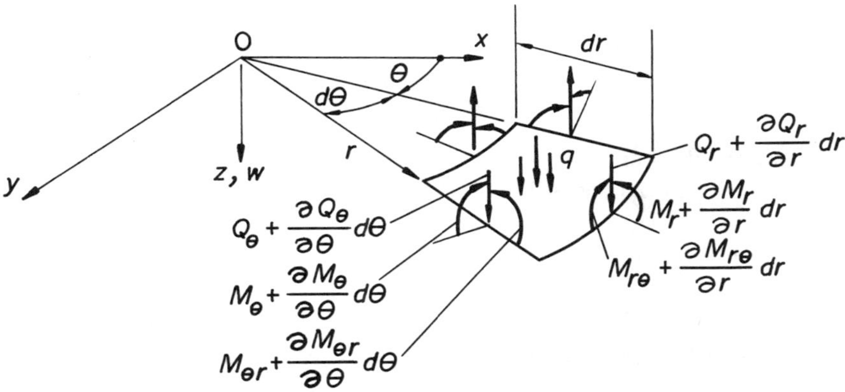

Fig. 5.7 Stress resultants in polar coordinates acting on small plate element.

$$V_x = -D\frac{\partial}{\partial x}\left[\frac{\partial^2 w}{\partial x^2} + (2-\nu)\frac{\partial^2 w}{\partial y^2}\right] \tag{5.36a}$$

$$V_y = -D\frac{\partial}{\partial y}\left[\frac{\partial^2 w}{\partial y^2} + (2-\nu)\frac{\partial^2 w}{\partial x^2}\right] \tag{5.36b}$$

$$R = \pm 2D(1-\nu)\frac{\partial^2 w}{\partial x \partial y} \tag{5.37}$$

In term of polar coordinates, the equilibrium equation is Eq. 5.34, while the stress resultants acting on the plate element as shown in Fig. 5.7 are

$$\begin{aligned} M_r &= -D\left[(1-\nu)\frac{\partial^2 w}{\partial r^2} + \nu\nabla^2 w\right] \\ M_\theta &= -D\left[\nabla^2 w - (1-\nu)\frac{\partial^2 w}{\partial r^2}\right] \\ M_{r\theta} &= M_{\theta r} = -D(1-\nu)\frac{\partial}{\partial r}\left(\frac{1}{r}\frac{\partial w}{\partial \theta}\right) \end{aligned} \tag{5.38}$$

$$Q_r = -D\frac{\partial}{\partial r}(\nabla^2 w) \tag{5.39a}$$

$$Q_\theta = -D\frac{1}{r}\frac{\partial}{\partial \theta}(\nabla^2 w) \tag{5.39b}$$

and the supplemented or Kirchhoff shear forces are

$$V_r = Q_r + \frac{1}{r}\frac{\partial M_{r\theta}}{\partial \theta} \tag{5.40a}$$

$$V_\theta = Q_\theta + \frac{\partial M_{\theta r}}{\partial r} \tag{5.40b}$$

In these equations the Laplace operator ∇^2 and the biharmonic operator ∇^4 are given, respectively, by Eqs. 4.26 and 4.21.

5.5 SOLUTIONS FOR RECTANGULAR PLATES

Consider an infinitely long plate of width a, i.e. occupying $-\infty < y < \infty$, $0 \le x \le a$, with a load q which depends only on x. The field equation, Eq. 5.34, is

$$\frac{d^4 w}{dx^4} = \frac{q(x)}{D} \tag{5.41}$$

The solution of w can be written as the sum of a *complementary solution* w_c and a *particular solution* w_p, i.e.

$$w = w_c + w_p \tag{5.42}$$

Thus w_c must satisfy the homogeneous differential equation

$$\frac{d^4 w_c}{dx^4} = 0 \tag{5.43}$$

the solution of which is

$$w_c = C_1 + C_2 x + C_3 x^2 + C_4 x^3 \tag{5.44}$$

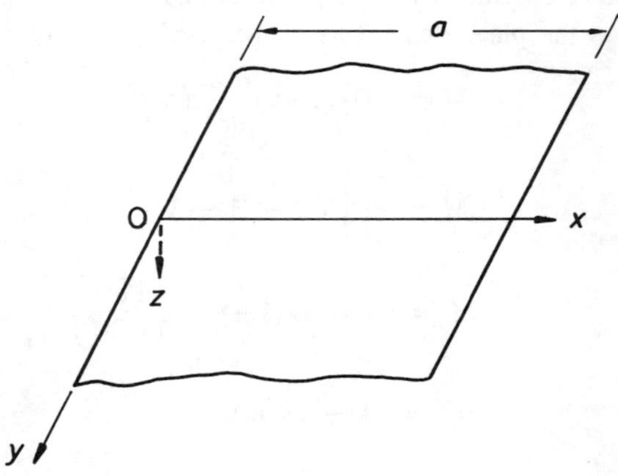

Fig. 5.8 An infinitely long plate.

in which C_1 to C_4 are arbitrary constants to be determined by means of four boundary conditions along the plate edges, two at $x=0$ and two at $x=a$. The particular solution is any solution of the original field equation, i.e.

$$\frac{\mathrm{d}^4 w_p}{\mathrm{d}x^4}=\frac{q(x)}{D} \tag{5.45}$$

The present problem is commonly known as *cylindrical bending*. As an example, consider the case where $q(x)=q_0$, one edge of the plate is built-in, while the other is free. The particular solution w_p can be chosen as

$$w_p=\frac{q_0 x^4}{24D}$$

and the pertinent boundary conditions are

$$w(0)=0, \quad \left.\frac{\mathrm{d}w}{\mathrm{d}x}\right|_{x=0}=0,$$

$$M_x(a)=0, \quad V_x(a)=0.$$

These give the four equations

$$C_1=0, \quad C_2=0,$$

$$2C_3+6aC_4=-\frac{q_0}{2D}a^2, \quad C_4=-\frac{q_0}{6D}a.$$

Next, consider the problem of a rectangular plate supported at three corners and loaded by a force P at the remaining corner, as shown in Fig. 5.9. The field equation is

$$\nabla^4 w=0 \tag{5.46}$$

and boundary conditions are

$$M_x(0,y)=0, \quad M_x(a,y)=0 \tag{5.47a,b}$$

$$V_x(0,y)=0, \quad V_x(a,y)=0 \tag{5.47c,d}$$

$$M_y(x,0)=0, \quad M_y(x,b)=0 \tag{5.47e,f}$$

$$V_y(x,0)=0, \quad V_y(x,b)=0 \tag{5.47g,h}$$

$$w(0,0)=w(a,0)=w(0,b)=0 \tag{5.47i,j,k}$$

$$-2M_{xy}(a,b)=P \tag{5.48}$$

The solution is found to be

$$w=Axy \tag{5.49}$$

This satifies Eqs. 5.46 and 5.47. The arbitrary constant A can be determined from

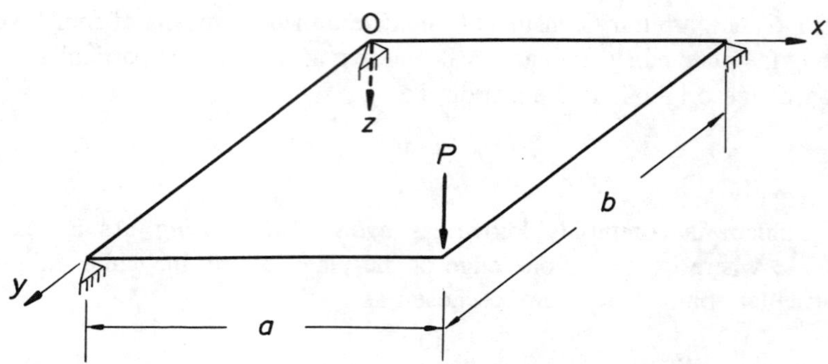

Fig. 5.9 Anticlastic bending.

Eq. 5.48, to give

$$w = \frac{Pxy}{2D(1-\nu)} \tag{5.50}$$

Note that all stress resultants vanish everywhere and M_{xy} is constant; the problem has been named *pure torsion bending* or *anticlastic bending*.

Consider the problem of a rectangular plate simply supported along all four edges, as shown in Fig. 5.10. The field equation is

$$\nabla^4 w = \frac{q(x,y)}{D} \tag{5.51}$$

and boundary conditions are

$$w(0,y) = 0, \quad w(a,y) = 0 \tag{5.52a,b}$$

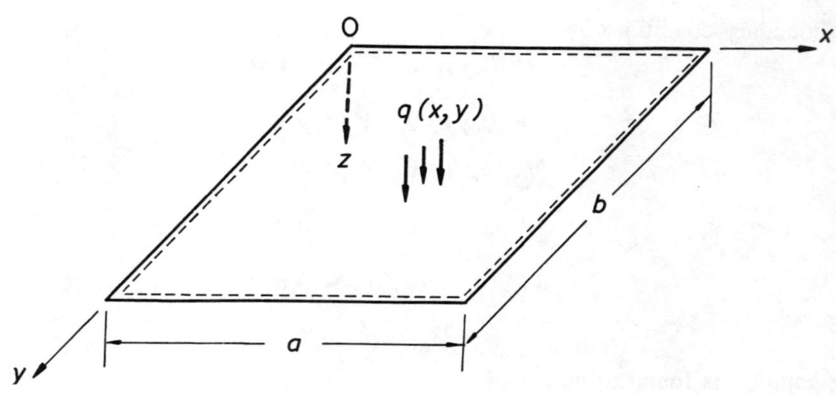

Fig. 5.10 Rectangular plate simply supported along all edges.

$$M_x(0,y)=0, \quad M_x(a,y)=0 \tag{5.52c,d}$$

$$w(x,0)=0, \quad w(x,b)=0 \tag{5.52e,f}$$

$$M_y(x,0)=0, \quad M_y(x,b)=0 \tag{5.52g,h}$$

In order to satisfy Eqs. 5.52 we use the series solution

$$w(x,y)=\sum_{m=1}^{\infty}\sum_{n=1}^{\infty} w_{mn}\sin\frac{m\pi x}{a}\sin\frac{n\pi y}{b} \tag{5.53}$$

The loading function can be expressed as a similar double Fourier series as

$$q(x,y)=\sum_{m=1}^{\infty}\sum_{n=1}^{\infty} q_{mn}\sin\frac{m\pi x}{a}\sin\frac{n\pi y}{b} \tag{5.54}$$

where

$$q_{mn}=\frac{4}{ab}\int_0^a\int_0^b q(x,y)\sin\frac{m\pi x}{a}\sin\frac{n\pi y}{b}\,dy\,dx \tag{5.55}$$

Substituting Eqs. 5.53 and 5.54 into Eq. 5.51 yields, for each pair of harmonics m and n,

$$w_{mn}=\frac{q_{mn}}{D\pi^4(m^2/a^2+n^2/b^2)^2} \tag{5.56}$$

and the solution is complete. The solution in the form of Eq. 5.53 is commonly known as *Navier solution* (C. L. M. H. Navier, 1823 and 1827). It is noteworthy that corner forces exist in this case.

The technique of Fourier series can be applied to any rectangular plate simply supported on two opposite edges, as shown in Fig. 5.11. In order to satisfy the simply

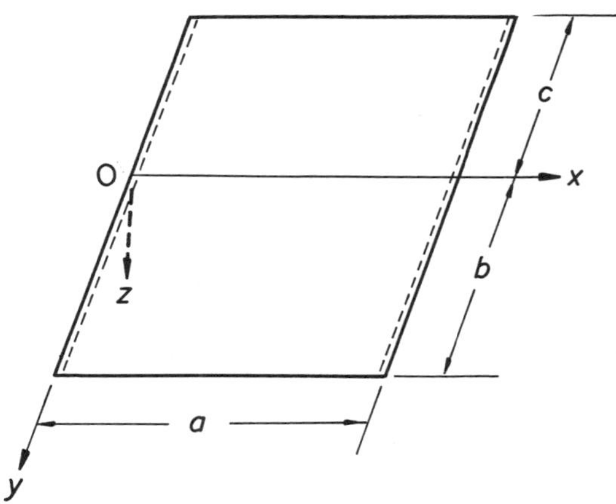

Fig. 5.11 Rectangular plate simply supported along two opposite edges.

supported conditions along the edges where $x=0$ and $x=a$ we take the single series solution

$$w(x,y)=\sum_{m=1}^{\infty} w_m(y)\sin\frac{m\pi x}{a} \tag{5.57}$$

The loading function can be expanded in a similar Fourier series

$$q(x,y)=\sum_{m=1}^{\infty} q_m(y)\sin\frac{m\pi x}{a} \tag{5.58}$$

where

$$q_m(y)=\frac{2}{a}\int_0^a q(x,y)\sin\frac{m\pi x}{a}\mathrm{d}x \tag{5.59}$$

Substituting Eqs. 5.57 and 5.58 into the field equation, Eq. 5.51, leads to, for each harmonic m, an ordinary differential equation of the form

$$\frac{\mathrm{d}^4 w_m}{\mathrm{d}y^4}-2\frac{m^2\pi^2}{a^2}\frac{\mathrm{d}^2 w_m}{\mathrm{d}y^2}+\frac{m^4\pi^4}{a^4}w_m=\frac{q_m}{D} \tag{5.60}$$

the solution of which can be taken as a sum of complementary and particular solutions

$$w_m(y)=w_{mc}(y)+w_{mp}(y) \tag{5.61}$$

where w_{mc} satisfying Eq. 5.60 with $q_m=0$ takes the form

$$w_{mc}(y)=A_m\cosh\frac{m\pi y}{a}+B_m\frac{m\pi y}{a}\sinh\frac{m\pi y}{a}+C_m\sinh\frac{m\pi y}{a}+D_m\frac{m\pi y}{a}\cosh\frac{m\pi y}{a} \tag{5.62a}$$

or

$$w_{mc}(y)=A_m\mathrm{e}^{-m\pi y/a}+B_m\frac{m\pi y}{a}\mathrm{e}^{-m\pi y/a}+C_m\mathrm{e}^{m\pi y/a}+D_m\frac{m\pi y}{a}\mathrm{e}^{m\pi y/a} \tag{5.62b}$$

Here A_m to D_m are arbitrary constants to be determined by means of four boundary conditions along the remaining two edges where $y=b$ and $y=-c$. Equation 5.62a is convenient in the consideration of the symmetry with respect to the x-axis, while Eq. 5.62b is more appropriate in the discussion of the boundedness of the solution when the y-dimension of the plate extends to infinity. The solution form in the present case is known as *Levy-Nadai solution* (M. Levy, 1899 and A. Nadai, 1925).

Solutions for several useful boundary value problems concerning rectangular plates are compiled by Timoshenko and Woinowsky-Krieger (1959).

5.6 CLOSED FORM SOLUTIONS FOR CIRCULAR PLATES

Consider axisymmetrical bending of a circular plate subjected to a uniform load of an intensity q_0 in the domain $0\le r<b<a$ and simply supported along its edge where $r=a$. We divide the plate into two domains by a circle $r=b$, then write the field equations

$$\nabla^4 w_1 = \frac{q_0}{D} \qquad (0 < r < b) \tag{5.63a}$$

and

$$\nabla^4 w_2 = 0 \qquad (b < r < a) \tag{5.63b}$$

Here subscripts 1 and 2 are introduced to denote domain 1 where $0 < r < b$, and domain 2 where $b < r < a$, respectively. The biharmonic operator ∇^4 for the present case which is axisymmetrical assumes the simple form

$$\nabla^4 = \frac{1}{r}\frac{\mathrm{d}}{\mathrm{d}r}\left\{r\frac{\mathrm{d}}{\mathrm{d}r}\left[\frac{1}{r}\frac{\mathrm{d}}{\mathrm{d}r}\left(r\frac{\mathrm{d}}{\mathrm{d}r}\right)\right]\right\} \tag{5.64}$$

For Eq. 5.63a, the solution is

$$w_1 = \frac{q_0}{64D}(r^4 + A_1 + B_1 r^2 + C_1 \log r + D_1 r^2 \log r) \tag{5.65}$$

where A_1 to D_1 are arbitrary constants to be determined from boundary conditions. To be bounded everywhere inside the domain where it belongs, Eq. 5.65 must not contain the terms $\log r$ and $r^2 \log r$ since they are singular and/or give singular stress resultants at $r = 0$. Thus $C_1 = D_1 = 0$ and

$$w_1 = \frac{q_0}{64D}(r^4 + A_1 + B_1 r^2), \quad (0 \le r < b) \tag{5.66}$$

At this stage, we can say that a particular solution for any case that has a uniform load intensity q_0 is

$$w_p = \frac{q_0 r^4}{64D} \tag{5.67}$$

For domain 2 which is governed by Eq. 5.63b, there is only the complementary solution, i.e.

$$w_2 = \frac{q_0}{64D}(A_2 + B_2 r^2 + C_2 \log r + D_2 r^2 \log r), \quad (b < r < a) \tag{5.68}$$

The remaining task is to determine the constants A_1, B_1, A_2, B_2, C_2 and D_2 from the boundary conditions at $r = a$ and the continuity conditions at the interface $r = b$ between the two domains. These conditions are

$$w_1|_{r=b} = w_2|_{r=b}, \qquad \left.\frac{\mathrm{d}w_1}{\mathrm{d}r}\right|_{r=b} = \left.\frac{\mathrm{d}w_2}{\mathrm{d}r}\right|_{r=b} \tag{5.69a,b}$$

$$M_{r1}|_{r=b} = M_{r2}|_{r=b}, \quad V_{r1}|_{r=b} = V_{r2}|_{r=b} \tag{5.69c,d}$$

$$w_2|_{r=a} = 0, \qquad M_{r2}|_{r=a} = 0 \tag{5.69e,f}$$

Note that V_r and Q_r are identical in axisymmetrical cases. Table 5.1, at the end of the present chapter, shows that Eqs. 5.69 may be written

$$b^4+A_1+b^2B_1=A_2+b^2B_2+\log b\,C_2+b^2\log b\,D_2 \tag{5.70a}$$

$$4b^3+2bB_1=2bB_2+b^{-1}C_2+b(1+2\log b)D_2 \tag{5.70b}$$

$$-4(3+\nu)b^2-2(1+\nu)B_1=-2(1+\nu)B_2+(1-\nu)b^{-2}C_2$$

$$-[2(1+\nu)\log b+3+\nu]D_2 \tag{5.70c}$$

$$-32b=-4b^{-1}D_2 \tag{5.70d}$$

$$A_2+a^2B_2+\log a\,C_2+a^2\log a\,D_2=0 \tag{5.70e}$$

$$-2(1+\nu)B_2+(1-\nu)a^{-2}C_2-[2(1+\nu)\log a+3+\nu]D_2=0 \tag{5.70f}$$

Equations 5.70 form a set of six linear simultaneous equations for determining the six constants A_1 to D_2.

Now consider the axisymmetrical bending of a plate subjected to a concentrated force P at the origin of the reference coordinates, as shown in Fig. 5.12. The pertinent field equation is

$$\nabla^4 w=0, \quad (r>0).$$

The general solution is the complementary solution, i.e.

$$w=\frac{1}{D}(A_1+A_2r^2+A_3\log r+A_4r^2\log r).$$

Table 5.1 contains slopes and stress resultants for variaus common deflection functions. From that table, we note that the term $\log r$ is *inadmissible* since it is singular at $r=0$, thus the constant A_3 must be set to zero. The term $r^2\log r$, though it gives singular moments and a singular shear stress resultant, should be retained since it is appropriate for a concentrated load. Consider a small neighborhood of the point 0, as shown in Fig. 5.13. The equilibrium of the neighborhood demands

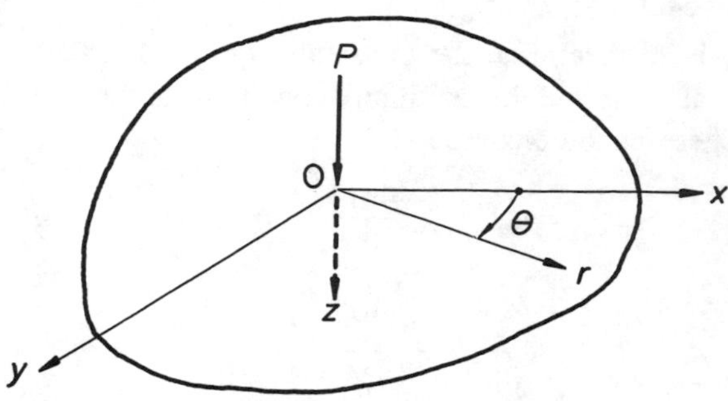

Fig. 5.12 Plate subjected to a concentrated force.

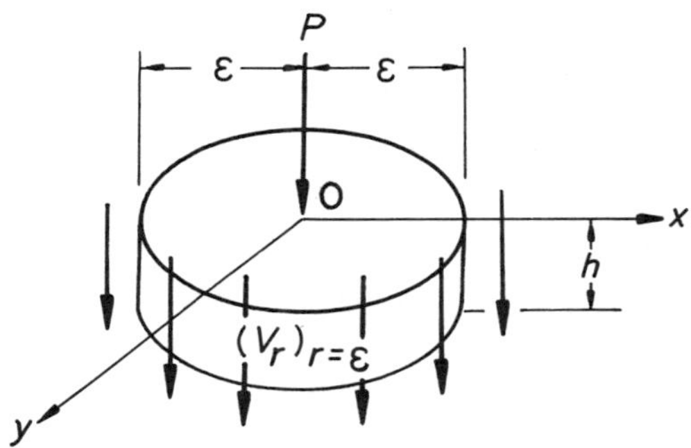

Fig. 5.13 Equilibrium near application point of the concentrated load.

$$P + \lim_{\varepsilon \to 0} \int_0^{2\pi} (rV_r)_{r=\varepsilon} \mathrm{d}\theta = 0 \tag{5.71}$$

With the help of Table 5.1, we can obtain

$$P + A_1(0) + A_2(0) + 2\pi A_4 \varepsilon \left(-\frac{4}{\varepsilon}\right) = 0,$$

or

$$A_4 = \frac{P}{8\pi}.$$

We can also verify that any other *admissible* displacement functions (axisymmetric as well as non-axisymmetric) contribute nothing to the second term in Eq. 5.71. Thus without loss of generality, we can conclude that the term

$$\frac{P}{8\pi D} r^2 \log r \tag{5.72}$$

is a particular solution for axisymmetric as well as non-axisymmetric problems relating to a plate subjected to a concentrated force P at the origin.

We may use the particular solution Eq. 5.72 in *Michell's problem,* the problem of a circular plate clamped along its edge and subjected to a concentrated load P as shown in Fig. 5.14. Michell (1902) used the method of inversion to derive the closed form solution

$$w = \frac{P}{8\pi D} \left\{ r_1^2 \log r_1 + \left[\frac{1}{2} (\beta^2 r_2^2 - r_1^2) - r_1^2 \log \beta \right] - r_1^2 \log r_2 \right\} \tag{5.73}$$

The first term is a particular solution Eq. 5.72, and the remaining terms are admissible complementary solution. Even $r_1^2 \log r_2$ is indeed admissible inside the plate, since $r_2 > 0$ there. There remains the task of checking the boundary conditions

$$w|_{r=a} = 0, \quad (\partial w / \partial r)|_{r=a} = 0 \tag{5.74a,b}$$

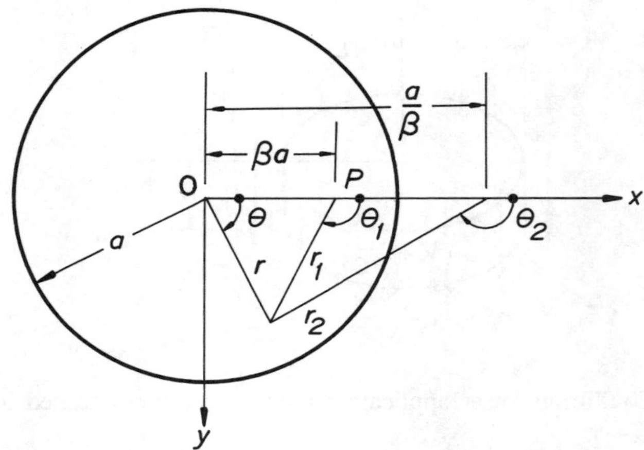

Fig. 5.14 Michell's solution.

We start with the observation that Eq. 5.73 expresses w as a function of r_1 and r_2, i.e. $w(r_1, r_2)$. Accordingly, its derivatives are

$$\frac{\partial w}{\partial r} = \frac{\partial w}{\partial r_1}\frac{\partial r_1}{\partial r} + \frac{\partial w}{\partial r_2}\frac{\partial r_2}{\partial r} \tag{5.75a}$$

$$\frac{\partial w}{\partial r_1} = \frac{P}{4\pi D}\left(r_1 \log\frac{r_1}{\beta r_2}\right) \tag{5.75b}$$

$$\frac{\partial w}{\partial r_2} = \frac{P}{8\pi D}\left(\frac{\beta^2 r_2^2 - r_1^2}{r_2}\right) \tag{5.75c}$$

We note that

$$r_1^2 = r^2 + \beta^2 a^2 - 2\beta a r \cos\theta \tag{5.76a}$$

$$r_2^2 = r^2 + \frac{a^2}{\beta^2} - 2\frac{a}{\beta} r \cos\theta \tag{5.76b}$$

$$r \sin\theta = r_1 \sin\theta_1 = r_2 \sin\theta_2 \tag{5.76c,d}$$

$$r_1 \cos\theta_1 = r\cos\theta - \beta a \tag{5.76e}$$

$$r_2 \cos\theta_2 = r\cos\theta - \frac{a}{\beta} \tag{5.76f}$$

$$r_1\cos\theta_1 - r_2\cos\theta_2 = \frac{a}{\beta} - \beta a \tag{5.76g}$$

and in particular for $r = a$,

$$r_1 = \beta r_2 \tag{5.77}$$

Equations 5.75 and 5.77 show that w in Eq. 5.73 satisfies Eqs. 5.74

Next, consider the problem of a clamped plate subjected to a concentrated moment M_1 as shown in Fig. 5.15(a). We may consider the moment M_1 to be produced by two equal and opposite forces at two neighboring points as shown in Fig. 5.15(b), where

$$\lim_{\varepsilon \to 0}(P\varepsilon) = M_1 \tag{5.78}$$

The solution for the force P acting at the point 0_1 is given by Eq. 5.73, i.e.

$$w^{+}(r,\theta;\, \beta a) = \frac{P}{8\pi D} w_0(r_1, r_2, \beta) \tag{5.79}$$

where

$$w_0(r_1, r_2, \beta) = r_1^2 \log\frac{r_1}{\beta r_2} + \frac{1}{2}(\beta^2 r_2^2 - r_1^2) \tag{5.80}$$

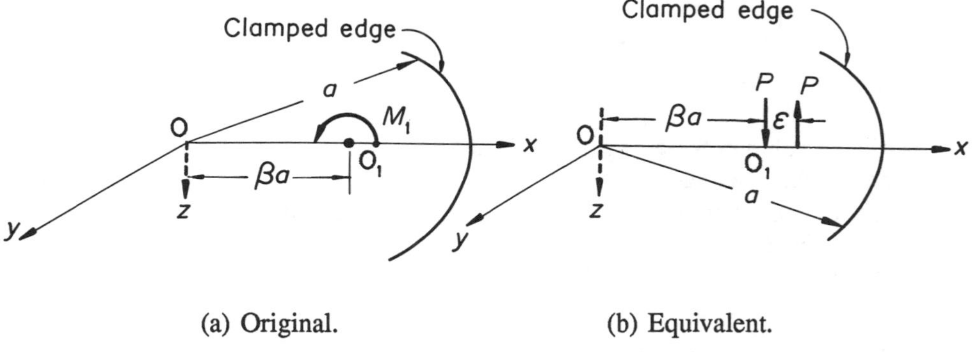

(a) Original. (b) Equivalent.

Fig. 5.15 Clamped circular plate subjected to a concentrated moment M_1.

Differentiating Eqs. 5.76a and b, and using Eqs. 5.76e and f, we find

$$\frac{\partial r_1}{\partial \beta} = -a\cos\theta_1 \tag{5.81a}$$

$$\frac{\partial r_2}{\partial \beta} = a\beta^{-2}\cos\theta_2 \tag{5.81b}$$

The solution for the other force $-P$ at the point very close to 0_1 in Fig. 5.15(b) is

$$w^{-}(r,\theta;\, \beta a + \varepsilon) = -w^{+}(r,\theta;\, \beta a) - \frac{\partial w^{+}}{\partial \beta}\frac{\varepsilon}{a} \tag{5.82}$$

Adding the last equation to Eq. 5.79, and using Eq. 5.78, we find the required solution for the applied moment M_1 to be

$$w = -\frac{M_1}{8\pi D a}\frac{\partial w_0(r_1, r_2, \beta)}{\partial \beta}$$

$$= -\frac{M_1}{8\pi D a}\left(\frac{\partial w_0}{\partial r_1}\frac{\partial r_1}{\partial \beta}+\frac{\partial w_0}{\partial r_2}\frac{\partial r_2}{\partial \beta}+\frac{\partial w_0}{\partial \beta}\right) \tag{5.83}$$

Because of Eqs. 5.80 and 5.81, Eq. 5.83 is

$$w = \frac{M_1}{8\pi D}\left[2r_1\log\left(\frac{r_1}{\beta r_2}\right)\cos\theta_1 + (r_1^2-\beta^2 r_2^2)\left(\frac{\cos\theta_2}{\beta^2 r_2}+\frac{1}{\beta a}\right)\right] \tag{5.84a}$$

or, due to the identity,

$$\frac{\cos\theta_2}{\beta^2 r_2}+\frac{1}{\beta a} = -\frac{r\cos\theta}{\beta^2 r_2^2}+\frac{r^2}{\beta a r_2^2},$$

derived from Eqs. 5.76b and 5.76f,

$$w = \frac{M_1}{8\pi D}\left[2r_1\log\left(\frac{r_1}{\beta r_2}\right)\cos\theta_1 + (r_1^2-\beta^2 r_2^2)\left(\frac{r^2}{\beta a r_2^2}-\frac{r\cos\theta}{\beta^2 r_2^2}\right)\right] \tag{5.84b}$$

Being the only singular term in the solution above, the term

$$\frac{M_1}{4\pi D} r_1 \log r_1 \cos\theta_1 \tag{5.85}$$

is a particular solution for any plate subjected to a concentrated moment M_1 about the y-axis and at the origin 0_1.

For a circular plate axisymmetrically supported along its edge and subjected to a concentrated force P at its center, the solution takes the form

$$w(r) = \frac{P}{8\pi D}(r^2\log r + A_1 + A_2 r^2) \tag{5.86}$$

in which axisymmetry and boundedness of the solution at $r = 0$ have been taken into account. The arbitrary constants A_1 and A_2 can be determined from the boundary conditions Eqs. 5.74 for clamped plates, or

$$w|_{r=a} = 0 \tag{5.87a}$$

$$M_r|_{r=a} = 0 \tag{5.87b}$$

for simply supported plates. Use Table 5.1.

When the same plate is subjected, at its center, to a concentrated moment M_1 about the y-axis, its solution takes the form

$$w(r,\theta) = \frac{M_1}{4\pi D}(r\log r + A_1 r + A_2 r^3)\cos\theta \tag{5.88}$$

Here we have used Table 5.1 to select the complementary solution and have accounted for the symmetry and the boundedness of the solution at $r = 0$. The boundary conditions,

Eqs. 5.74 or 5.87, determine the arbitrary constants A_1 and A_2.

The eventual results are the following: For a circular plate simply supported along its edge and subjected, at its center, to a concentrated force P,

$$w = \frac{P}{8\pi D}\left[r^2 \log\frac{r}{a} + \frac{(3+\nu)(a^2-r^2)}{2(1+\nu)}\right] \tag{5.89}$$

while that subjected to a concentrated moment M_1 about the y-axis,

$$w = \frac{M_1}{4\pi D}\left[r \log\frac{r}{a}\cos\theta + \frac{(1+\nu)(a^2-r^2)r\cos\theta}{2(3+\nu)a^2}\right] \tag{5.90}$$

For a circular plate clamped along its edge and subjected, at its center, to a concentrated force P,

$$w = \frac{P}{8\pi D}\left[r^2 \log\frac{r}{a} + \frac{1}{2}(a^2-r^2)\right] \tag{5.91}$$

while that subjected to a concentrated moment M_1 about the y-axis,

$$w = \frac{M_1}{4\pi D}\left[r \log\frac{r}{a}\cos\theta + \frac{(a^2-r^2)}{2a^2}r\cos\theta\right] \tag{5.92}$$

Equations 5.91 and 5.92 can be obtained by specializing Eqs. 5.73 and 5.84, respectively, and using

$$\beta \to 0, \quad r_1 \to r, \quad \theta_1 \to \theta,$$

$$\theta_2 \to \pi, \quad r_2 \to \infty, \quad \beta r_2 \to a.$$

Note that the limiting case of Eq. 5.84a involves an indeterminate quantity $(\infty - \infty)$, but Eq. 5.84b is well behaved.

If the concentrated force P and concentrated moment M_1 are applied at an eccentric point and the plate edge is not clamped, closed form solutions (like Eqs. 5.73 and 5.84) seem not possible. For such cases we use the particular solution given by Eq. 5.72 or 5.85 with a complementary solution in Fourier series form, as described in the following section.

5.7 SERIES SOLUTION FOR CIRCULAR PLATES

The deflection and load functions are expanded in Fourier cosine series, i.e.

$$w(r,\theta) = \sum_{m=0}^{\infty} w_m(r)\cos m\theta \tag{5.93a}$$

$$q(r,\theta) = \sum_{m=0}^{\infty} q_m(r)\cos m\theta \tag{5.93b}$$

where

$$q_0(r) = \frac{1}{\pi}\int_0^{\pi} q(r,\theta)d\theta \tag{5.94a}$$

$$q_m(r) = \frac{2}{\pi}\int_0^{\pi} q(r,\theta)\cos m\theta \mathrm{d}\theta, \quad (m > 0) \tag{5.94b}$$

Substituting Eqs. 5.93 and 5.94 into the field equation, Eq. 5.34, yields

$$\nabla_m^4 w_m(r) = \frac{q_m(r)}{D} \tag{5.95}$$

where

$$\nabla_m^2 = \frac{1}{r}\frac{\mathrm{d}}{\mathrm{d}r}\left(r\frac{\mathrm{d}}{\mathrm{d}r}\right) - \frac{m^2}{r^2} \tag{5.96}$$

The total solution to Eq. 5.95 can be taken as

$$w_m = w_{pm} + w_{cm} \tag{5.97}$$

where w_{pm} is a particular solution satisfying Eq. 5.95, and w_{cm} is the complementary solution satisfying

$$\nabla_m^4 w_{cm}(r) = 0 \tag{5.98}$$

For each m, w_{cm} can be obtained as

$$w_{c0}(r) = A_0 + B_0 r^2 + C_0 \log r + D_0 r^2 \log r \tag{5.99a}$$

$$w_{c1}(r) = A_1 r + B_1 r^3 + C_1 r^{-1} + D_1 r \log r \tag{5.99b}$$

$$w_{cm}(r) = A_m r^m + B_m r^{-m} + C_m r^{m+2} + D_m r^{-m+2}, \quad (m > 1) \tag{5.99c}$$

where the constants A_m, B_m, C_m and D_m are to be determined so that the complete solution satisfies the boundary conditions. We may obtain a similar sine series solution. Table 5.1 (at the end of the present chapter) contains $\partial w/\partial r$ and the stress resultants for all these complementary and particular solutions.

As an example, consider an annular plate with two concentric boundary circles of radii a and b $(a > b)$. The load on the plate is uniform with an intensity q_0. The plate is supported by k equally spaced columns along its outer edge. Each column is $2\alpha a$ wide, where 2α is the angle subtended by each column at the center of the boundary circles. It is assumed that the column reaction is only in the form of a force uniformly distributed over each column width. Choose the reference coordinate system to have its origin at the center of the boundary circles and the x-axis passing through the center of a column. The total load on the plate is $\pi(a^2 - b^2)q_0$, thus the reaction intensity on each column is $\pi(a^2 - b^2)q_0/(2k\alpha a)$. This reaction can be written as a periodic but discontinuous function of θ as follows.

$$f(\theta) = \frac{\pi(a^2 - b^2)q_0}{2k\alpha a}, \quad \left(\frac{2\pi j}{k} - \alpha < \theta < \frac{2\pi j}{k} + \alpha, \quad j = 0, 1, 2, \cdots, k\right) \tag{5.100a}$$

$$= 0, \qquad \text{(elsewhere)} \tag{5.100b}$$

The period is equal to $2\pi/k$. The function $f(\theta)$ is expanded in a Fourier cosine series with such a period, namely

$$f(\theta)=\frac{a_0}{2}+\sum_{n=1}^{\infty}a_n\cos nk\theta,$$

where

$$a_n=\frac{2k}{\pi}\int_0^{\pi/k}f(\theta)\cos nk\theta d\theta,$$

or, due to Eqs. 5.100,

$$a_n=\frac{(a^2-b^2)q_0}{\alpha a}\int_0^{\alpha}\cos nk\theta d\theta$$

$$=\frac{(a^2-b^2)q_0\sin nk\alpha}{\alpha ank}.$$

Replacing nk by m in the series above, we get

$$f(\theta)=\frac{a_0}{2}+\sum_{m=k,2k,3k,\cdots}^{\infty}a_m\cos m\theta \tag{5.101}$$

where

$$a_m=\frac{(a^2-b^2)q_0\sin m\alpha}{am\alpha} \tag{5.102a}$$

$$a_0=(a^2-b^2)q_0/a \tag{5.102b}$$

The solution can then be written in a similar series as

$$w(r,\theta)=\frac{q_0}{64D}\left(r^4+\sum_{m=0,k,2k,\cdots}^{\infty}w_{cm}\cos m\theta\right) \tag{5.103}$$

in which the particular solution given in Eq. 5.67 has been adopted and $w_{cm}(r)$ is as defined by Eqs. 5.99. Boundary conditions for the present case are

$$V_r(a,\theta)=-f(\theta),\quad M_r(a,\theta)=0 \tag{5.104a,b}$$

$$V_r(b,\theta)=0,\quad M_r(b,\theta)=0 \tag{5.104c,d}$$

Substituting Eqs. 5.101 and 5.103 into Eqs. 5.104, using Eqs. 5.38 to 5.40 (or Table 5.1) and 5.102, and equating the coefficients for each m in the resulting series, we arrive at the simultaneous equations; for $m=0$,

$$-32a-4a^{-1}D_0=-32(a^2-b^2)/a \tag{5.105a}$$

$$-4(3+\nu)a^2-2(1+\nu)B_0+(1-\nu)a^{-2}C_0-[2(1+\nu)\log a+3+\nu]D_0=0 \tag{5.105b}$$

$$-32b-4b^{-1}D_0=0 \tag{5.105c}$$

$$-4(3+\nu)b^2-2(1+\nu)B_0+(1-\nu)b^{-2}C_0-[2(1+\nu)\log b+3+\nu]D_0=0 \tag{5.105d}$$

and for $m \geq 2$,

$$(1-\nu)m^2(m-1)a^{m-3}A_m-(1-\nu)m^2(m+1)a^{-m-3}B_m+(m+1)m[(1-\nu)m$$

$$-4]a^{m-1}C_m+(1-m)m[4+(1-\nu)m]a^{-m-1}D_m=-64(a^2-b^2)\frac{\sin m\alpha}{am\alpha} \quad (5.106a)$$

$$-(1-\nu)m(m-1)a^{m-2}A_m-(1-\nu)m(m+1)a^{-m-2}B_m-(m+1)[4\nu$$

$$+(1-\nu)(m+2)]a^mC_m+(m-1)[(1-\nu)(2-m)+4\nu]a^{-m}D_m=0 \quad (5.106b)$$

$$(1-\nu)m^2(m-1)b^{m-3}A_m-(1-\nu)m^2(m+1)b^{-m-3}B_m+(m+1)m[(1-\nu)m$$

$$-4]b^{m-1}C_m+(1-m)m[4+(1-\nu)m]b^{-m-1}D_m=0 \quad (5.106c)$$

$$-(1-\nu)m(m-1)b^{m-2}A_m-(1-\nu)m(m+1)b^{-m-2}B_m-(m+1)[4\nu$$

$$+(1-\nu)(m+2)]b^mC_m+(m-1)[(1-\nu)(2-m)+4\nu]b^{-m}D_m=0 \quad (5.106d)$$

Noting that Eqs. 5.105a and c are identical, and that A_0 does not appear in any conditions above, we need to impose another condition to determine this constant, which corresponds to a rigid-body deflection of the plate. We choose the condition of zero deflection at the center of a column, e.g. $w(a,0)=0$ which leads to

$$a^4+A_0+B_0a^2+C_0\log a+D_0a^2\log a$$

$$+\sum_{m=k,2k,\cdots}^{\infty}(A_ma^m+B_ma^{-m}+C_ma^{m+2}+D_ma^{-m+2})=0 \quad (5.107)$$

Then D_0 can be determined from Eq. 5.105a or c, B_0 and C_0 from Eqs. 5.105b and d, A_m to D_m ($m \geq 2$) from Eqs. 5.106a to d, and A_0 from Eq. 5.107. If there is a single supporting column, the constants for $m=1$ vanish, i.e.

$$A_1=B_1=C_1=D_1=0.$$

but this case is not valid since the structure is unstable.

Several problems of an annular plate supported by two concentric rings of equally spaced columns were analyzed by Agabein, Lee and Dundurs (1967). Later, problems of annular plates supported by columns on the outer boundary and fixed along the inner boundary were solved by Kosherick, Dundurs and Lee (1968).

5.8 POLYGONAL PLATES SUPPORTED AT CORNERS

This section deals with problems of polygonal slabs or plates supported directly on columns at the corners. Such problems are frequently encountered in practice. The solutions to be presented are approximate, but in closed forms.

As an example, consider the paper by Lee and Ballesteros (1960), which is a study on a rectangular plate subjected to a uniform load of an intensity q_0 and supported at its four corners, as shown in Fig. 5.16. For an approximate solution, incorporating the relevant symmetry, we take a polynomial

$$w(x,y)=C_1+C_2x^2+C_3y^2+C_4x^4+C_5x^2y^2+C_6y^4 \tag{5.108}$$

Substituting Eq. 5.108 into the field equation, Eq. 5.34, for $q(x,y)=q_0$, yields

$$24C_4+8C_5+24C_6=\frac{q_0}{D} \tag{5.109}$$

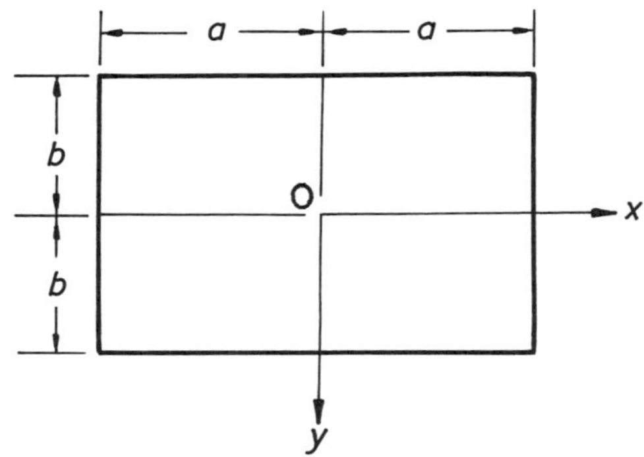

Fig. 5.16 Rectangular plate supported at corners.

The boundary condition that there be no shear stress resultants along the free edges is expressed by

$$V_x(a,y)=0, \quad V_y(x,b)=0 \tag{5.110a,b}$$

Substituting Eq. 5.108 into Eqs. 5.110, and using Eqs. 5.36, we find

$$6C_4+(2-\nu)C_5=0 \tag{5.111a}$$

$$(2-\nu)C_5+6C_6=0 \tag{5.111b}$$

Solving Eqs. 5.109 and 5.111 for C_4, C_5 and C_6 and substituting results in Eq. 5.108 yields

$$w(x,y)=C_1+C_2x^2+C_3y^2+\frac{q_0}{48(1-\nu)D}[(2-\nu)(x^4+y^4)-6x^2y^2] \tag{5.112}$$

The twisting moments M_{xy} and M_{yx} given by Eq. 5.33c are

$$M_{xy}=M_{yx}=\frac{q_0xy}{2} \tag{5.113}$$

and the force at each plate corner given by Eqs. 5.27 is

$$R=-qab \tag{5.114}$$

Note that, irrespective of C_1, C_2 and C_3, Eq. 5.112 satisfies the field equation, the condition of zero shear stress resultants at the free edges, and the condition of corner forces. The remaining unknowns C_1, C_2 and C_3 can be determined from the condition of zero deflection at a corner, i.e.

$$w(a,b)=0 \tag{5.115}$$

and two conditions stating that the mean moments on the edges are zero,

$$\int_0^b M_x(a,y)\mathrm{d}y=0 \tag{5.116a}$$

$$\int_0^a M_y(x,b)\mathrm{d}x=0 \tag{5.116b}$$

Equations 5.116 are approximations to the exact boundary conditions at the free edges, i.e. $M_x(a,y)=0$ and $M_y(x,b)=0$. Substitute Eq. 5.112 into Eqs. 5.115 and 5.116, and use Eqs. 5.33a and 5.33b. We find

$$C_1=\frac{q_0a^2b^2}{48(1-\nu^2)D}\left[(10+\nu-\nu^2)\left(\frac{a^2}{b^2}+\frac{b^2}{a^2}\right)-2(7\nu-1)\right] \tag{5.117a}$$

$$C_2=\frac{q_0ab}{24(1-\nu^2)D}\left[(1+5\nu)\frac{b}{a}-(6+\nu-\nu^2)\frac{a}{b}\right] \tag{5.117b}$$

$$C_3=\frac{q_0ab}{24(1-\nu^2)D}\left[(1+5\nu)\frac{a}{b}-(6+\nu-\nu^2)\frac{b}{a}\right] \tag{5.117c}$$

Numerical results obtained from by this approximate solution agree closely with those obtained numerically by Nadai (1922) and Marcus (1932), and by experiments on prestressed concrete slabs by Scordelis, Pister and Lin (1956).

A similar approach was adopted by Vijakkhana, Karasudhi and Lee (1973) for equilateral triangular plates subjected to uniform loads and supported at the plate corners.

Suriyamongkol (1985) studied the bending of regular polygonal plates of general form, with 3, 4, 5, $\cdots$ sides, subjected to uniformly distributed loads and central concentrated loads, and having simply supported edges, or free edges with corner supports. The solution is a sum of a particular solution and a complementary solution. The particular solution is given by Eq. 5.67 for a uniform load and by Eq. 5.72 for a concentrated force. The complementary solution is taken as a series. The *Gram-Schmidt orthogonalization process* (Wylie, 1975) is applied to solve the series equations. Exact solutions are obtained for simply supported edges. Approximate moment conditions (Eqs. 5.116) have to be adopted for free edges. For practical purposes, the series may be truncated at the third term. Results agree closely with others found by different reliable methods.

5.9 PLATES ON ELASTIC FOUNDATION

When a transversely loaded plate rests on a yielding foundation, the reaction of the foundation opposes the plate deflection. If the foundation is elastic, the resistance is proportional to the deflection, i.e. γw, where γ is the modulus of the foundation and expressed in force per unit of plate area per unit of deflection. The only change brought about by the presence of the elastic foundation is that the equilibrium equation for forces in the z-direction, Eq. 5.28, becomes

$$\frac{\partial Q_x}{\partial x}+\frac{\partial Q_y}{\partial y}-\gamma w+q=0 \tag{5.118}$$

Consequently, the field equation in terms of the deflection for a uniform isotropic plate, Eq. 5.34, becomes

$$\nabla^4 w+\xi^4 w=\frac{q}{D} \tag{5.119}$$

where

$$\xi=+(\gamma/D)^{1/4} \tag{5.120}$$

For cylindrical bending of very long rectangular plates, Eq. 5.41 becomes

$$\frac{\mathrm{d}^4 w}{\mathrm{d}x^4}+\xi^4 w=\frac{q(x)}{D} \tag{5.121}$$

and the complementary solution, in place of Eq. 5.44, is

$$w_c=C_1\cos\xi x+C_2\sin\xi x+C_3\xi x\cos\xi x+C_4\xi x\sin\xi x \tag{5.122}$$

The plate deflection function for anticlastic or pure torsion bending (Fig. 5.9) is given by Eq. 5.50. Substituting that equation into Eq. 5.119 yields

$$q(x,y)=\frac{\gamma Pxy}{2D(1-\nu)} \tag{5.123}$$

Thus the anticlastic or pure torsion bending for plates on elastic foundation is possible if the distributed load is related to the applied corner force by the equation above.

When a rectangular plate is simply supported along all four edges as in Fig. 5.10, the Navier solution scheme given by Eqs. 5.52 to 5.55 may be applied. Substituting these equations into Eqs. 5.119 yields

$$w_{mn}=\frac{q_{mn}}{D\pi^4(m^2/a^2+n^2/b^2)^2+\gamma} \tag{5.124}$$

When a rectangular plate is simply supported along two opposite edges as in Fig. 5.11, the Levy-Nadai solution scheme as given by Eqs. 5.57 to 5.59 is also applicable. Substituting those equations into Eq. 5.119 yields, in place of Eq. 5.60, the following ordinary differential equation

$$\frac{\mathrm{d}^4 w_m}{\mathrm{d}y^4}-\frac{2m^2\pi^2}{a^2}\frac{\mathrm{d}^2 w_m}{\mathrm{d}y^2}+\left(\frac{m^4\pi^4}{a^4}+\xi^4\right)w_m=\frac{q_m}{D} \tag{5.125}$$

The complementary solution for this equation is

$$w_m(y) = A_m \cosh \beta_m y \cos \alpha_m y + B_m \sinh \beta_m y \sin \alpha_m y$$

$$+C_m \sinh \beta_m y \cos \alpha_m y + D_m \cosh \beta_m y \sin \alpha_m y \tag{5.126a}$$

or

$$w_m(y) = (A_m \cos \alpha_m y + B_m \sin \alpha_m y) e^{-\beta_m y}$$

$$+(C_m \cos \alpha_m y + D_m \sin \alpha_m y) e^{\beta_m y} \tag{5.126b}$$

where

$$2\beta_m^2 = [(m\pi/a)^4 + \xi^4]^{1/2} + (m\pi/a)^2 \tag{5.127a}$$

$$2\alpha_m^2 = [(m\pi/a)^4 + \xi^4]^{1/2} - (m\pi/a)^2 \tag{5.127b}$$

In Eqs. 5.126, A_m to D_m are arbitrary constants to be determined by means of four boundary conditions along the remaining two edges where $y = b$ and $y = -c$. Equation 5.126a is convenient in the consideration of the symmetry with respect to the x-axis, while Eq. 5.126b is more apropriate in the discussion of the boundedness of the solution when the y-dimension of the plate extends to infinity.

For circular plates, the solution may assume the form of Fourier series as in Eqs. 5.93, but now $w_m(r)$ is governed by the ordinary differential equation

$$(\nabla_m^4 + \xi^4) w_m(r) = \frac{q_m(r)}{D} \tag{5.128}$$

The complete solution to the equation is given by Eq. 5.97, where w_{pm} and w_{cm} are particular and complementary solutions for Eq. 5.128. For each m, w_{cm} can be expressed in terms of Kelvin functions (Section 1.10); thus

$$w_{cm}(r) = A_m ber_m \xi r + B_m bei_m \xi r + C_m ker_m \xi r + D_m kei_m \xi r \tag{5.129}$$

Note that the following deflection function

$$-\frac{P}{2\pi D \xi^2} kei_0 \xi r \tag{5.130}$$

is a particular solution for the problem of a plate on an elastic foundation subjected to a concentrated force P at the origin, as in Fig. 5.12, since it has the characteristic singularity of Eq. 5.72 as r tends to zero. The same process of reasoning can be used in the search for a particular solution to the problem of a concentrated moment. Specifically for $m = 1$, Eq. 5.129 can be rewritten as

$$w_{c1}(r) = A_1 ber_1 \xi r + B_1 bei_1 \xi r + C_1 (ker_1 \xi r + kei_1 \xi r) + D_1 (ker_1 \xi r - kei_1 \xi r) \tag{5.131}$$

The term within the last pair of parentheses in the equation above has the correct characteristic singularity (Eq. 5.85) as r tends to zero. Thus a particular solution for the deflection function of the problem of a moment M_1 about the y-axis and concentrated at the origin of the reference axes is

$$\frac{\sqrt{2}M_1}{4\pi D\xi}(ker_1\xi r - kei_1\xi r) \tag{5.132}$$

Computing Kelvin functions is complicated and time consuming, especially for high order functions. Karasudhi, Jou and Tansirikongkol (1976) avoided this type of analytical functions by employing, for each harmonic m, central finite differences with respect to r directly in the field equations and all pertinent boundary conditions.

5.10 EXERCISE PROBLEMS

5.10.1 The infinitely long plate shown in Fig. 5.8 has properties in domain 1 where $0 \le y < \infty$ different from those in domain 2 where $-\infty < y \le 0$. Find the solution if the plate is simply supported along both long edges $x = 0$ and a, and the load is a point force P concentrated at $x = d$ and $y = 0$, and $0 < d < a$. Hint: Adopt the Levy-Nadai solution (Eq. 5.57) for both domains. The (complementary) solutions are modifications of Eq. 5.62b taking into consideration the boundedness of the solutions as $y \to \infty$ and $y \to -\infty$ for domains 1 and 2, respectively. The four pertinent boundary conditions at $y = 0$, where the two domains join, are

$$w_1(x,0) = w_2(x,0), \quad \left.\frac{\partial w_1(x,y)}{\partial y}\right|_{y=0} = \left.\frac{\partial w_2(x,y)}{\partial y}\right|_{y=0},$$

$$M_{y1}(x,0) = M_{y2}(x,0), \quad V_{y2}(x,0) - V_{y1}(x,0) = f(x),$$

where $f(x)$ is the Dirac-delta distribution (Section 1.9) of the force P along the x-axis, i.e.

$$f(x) = P\delta(x-d).$$

Note that if the two domains have the same properties, we can take advantage of symmetry and find the solution for domain 1 only, and use only two conditions at $y = 0$, i.e.

$$\left.\frac{\partial w_1(x,y)}{\partial y}\right|_{y=0} = 0, \quad V_{y1}(x,0) = -\frac{f(x)}{2}.$$

5.10.2 If the plate in Problem 5.10.1 is subjected to a concentrated moment M_1 about an axis parallel to the y-axis, the solution can be obtained as in the previous problem by changing $f(x)$ into

$$f(x) = \lim_{\varepsilon \to 0}\left\{\frac{M_1}{\varepsilon}[\delta(x-d) - \delta(x-d-\varepsilon)]\right\} = M_1\delta^{(1)}(x-d).$$

5.10.3 If the concentrated moment M_1 in the preceding problem is about the x-axis instead, the pertinent boundary conditions are

$$w_1(x,0)=w_2(x,0), \quad \left.\frac{\partial w_1(x,y)}{\partial y}\right|_{y=0}=\left.\frac{\partial w_2(x,y)}{\partial y}\right|_{y=0},$$

$$M_{y1}(x,0)-M_{y2}(x,0)=m(x), \quad V_{y1}(x,0)=V_{y2}(x,0),$$

where $m(x)$ is the Dirac-delta distribution of M_1 along the x-axis, i.e.

$$m(x)=M_1\delta(x-d).$$

5.10.4 If the y-spans of the plate in Problem 5.10.1 are finite as in Fig. 5.11, we can use solutions given by Eq. 5.62a or b. In addition to the four boundary conditions at $y=0$, two more boundary conditions must be prescribed along the edge $y=b$ and two more on $y=-c$. If both domains are identical, take advantage of symmetry and find the solution of domain 1 only.

5.10.5 Solve Problem 5.10.4 if the plate is subjected to concentrated moments.

5.10.6 A rectangular plate simply supported along all edges as shown in Fig. 5.10, is subjected to a concentrated force P at $x=c$, $y=d$. Find the solution and note the existence of corner forces. Hint: Adopt the Navier solution (Eq. 5.53) and take the loading function as

$$q(x,y)=P\delta(x-c)\delta(y-d).$$

5.10.7 If the plate in Problem 5.10.6 is subjected to concentrated moments instead, find solutions. Hint: For the case of a concentrated moment M_1 about an axis parallel to the y-axis, express the load function as

$$q(x,y)=M_1\delta^{(1)}(x-c)\delta(y-d).$$

5.10.8 For the plate as shown in Fig. 5.14, find Michell's solution for a concentrated moment M_1 about the x-axis. Hint:

$$w(r,\theta)=-\frac{M_1}{8\pi D\,\beta a}\frac{\partial w_0}{\partial\theta},$$

where w_0 is as defined in Eq. 5.80.

5.10.9 Find a series solution for the circular plate shown in Fig. 5.14. Hint: Consider two domains; domain 1 where $0\le r\le\beta a$ and domain 2 where $\beta a\le r\le a$. Adopt the cosine series solution (Eqs. 5.93) for both domains, but the boundedness of the solution at $r=0$ must be incorporated in domain 1. The four pertinent conditions at the interface $r=\beta a$, between the domains, are

$$w_1(\beta a,\theta)=w_2(\beta a,\theta),\quad \left.\frac{\partial w_1(r,\theta)}{\partial r}\right|_{r=\beta a}=\left.\frac{\partial w_2(r,\theta)}{\partial r}\right|_{r=\beta a},$$

$$M_{r1}(\beta a,\theta)=M_{r2}(\beta a,\theta),\quad V_{r1}(\beta a,\theta)-V_{r2}(\beta a,\theta)=f(\theta),$$

Here $f(\theta)$ is the Dirac-delta distribution of the concentrated force P along the circumference of radius βa, i.e.

$$f(\theta)=\frac{P}{\beta a}\delta(\theta).$$

In addition, two more boundary conditions must be prescribed at the plate edge, i.e. Eqs. 5.74 for the clamped edge or Eqs. 5.87 for the simply supported edge.

5.10.10 If the plate in the previous problem is subjected to a concentrated moment M_1 about the x-axis instead, find the solution. Hint: Adopt the sine series solution, and write the load distribution in the form

$$f(\theta)=-\frac{M_1}{\beta^2 a^2}\delta^{(1)}(\theta),$$

then proceed in the same manner as in the previous problem.

5.10.11 If the concentrated moment M_1 is about an axis parallel to the y-axis, what is the solution? Hint: Adopt a cosine series solution. The conditions at the interface between the two domains are

$$w_1(\beta a,\theta)=w_2(\beta a,\theta),\quad \left.\frac{\partial w_1(r,\theta)}{\partial r}\right|_{r=\beta a}=\left.\frac{\partial w_2(r,\theta)}{\partial r}\right|_{r=\beta a},$$

$$M_{r1}(\beta a,\theta)-M_{r2}(\beta a,\theta)=m(\theta),\quad V_{r1}(\beta a,\theta)=V_{r2}(\beta a,\theta),$$

where

$$m(\theta)=\frac{M_1}{\beta a}\delta(\theta).$$

5.10.12 A circular plate is simply supported along its edge and subjected to a linearly varying load

$$q(r,\theta)=pr\cos\theta/a,$$

where p is a constant and a is the plate radius. Find the solution. Hint: Table 5.1 gives a particular solution as

$$w_p(r,\theta)=pr^5\cos\theta/(192Da).$$

Find a complementary solution of the form

$$w_c(r,\theta)=w_1(r)\cos\theta,$$

where $w_1(r)$ is as defined in Eq. 5.99b, but with C_1 and D_1 zero to make the solution bounded at $r=0$. The remaining two arbitrary constants can be determined from the boundary conditions at $r=a$, i.e.

$$w(a,\theta)=0, \quad M_r(a,\theta)=0.$$

Answer:

$$w(r,\theta)=\frac{pr(a^2-r^2)[a^2(7+\nu)-(3+\nu)r^2]}{192(3+\nu)Da}\cos\theta.$$

5.10.13 Solve the preceding problem, if the plate outer edge is free, and the plate region $0\le r\le b$ $(b<a)$ is a rigid core and fixed in space.

5.10.14 Suppose that the plate in Problem 5.10.13 has its outer edge simply supported and the rigid core is rotated by a moment M_1 about the y-axis through a small angle ϕ. Find the relationship between M_1 and ϕ. Hint: The (complementary) solution takes the same form as in Problem 5.10.12, but $w_1(r)$ is given by Eq. 5.99b. Boundary conditions along the circumference of the core, $r=b$, are

$$w(b,\theta)=-b\phi\cos\theta, \quad \left.\frac{\partial w(r,\theta)}{\partial r}\right|_{r=b}=-\phi\cos\theta.$$

The relationship between M_1 and ϕ can be obtained by considering the moment equilibrium of the core, i.e.

$$M_1=b\int_0^{2\pi}[bV_r(b,\theta)-M_r(b,\theta)]\cos\theta d\theta.$$

5.10.15 Solve the preceding problem when the outer edge is fixed. Then specialize both problems for the cases of a concentrated moment M_1 at the center of the plate, to prove Eqs. 5.90 and 5.92.

5.10.16 An annular plate is segmented along two radii in such a way that the straight edges of the plate subtend an angle α at the plate center. Formulate the solution when the plate is subjected to the load $q(r,\theta)$ and the straight edges of the plate are simply supported. Hint: Set the coordinate system so that one straight edge is at $\theta=0$ and the other at $\theta=\alpha$, and one circular edge is at $r=a$ and the other at $r=b$. Adopt a 'polar' Levy-Nadai solution, i.e.

$$w(r,\theta)=\sum_{m=1}^{\infty}w_m(r)\sin\frac{m\pi\theta}{\alpha},$$

which satisfies the boundary conditions along the straight edges exactly. Expand $q(r,\theta)$ in a similar sine series. Determine arbitrary constants by means of two boundary conditions along each circular edge.

5.10.17 Find the solution for the same annular plate of Section 5.7 when the inner edge is fixed instead. Hint: The column reaction distribution in Eq. 5.100a should be modified to read

$$f(\theta)=\frac{\eta\pi(a^2-b^2)q_0}{2k\alpha a},$$

where η is the fraction of the total load to be carried by the columns. Arbitrary constants involved are to be determined from the boundary conditions;

$$w(b,\theta)=0,\quad \left.\frac{\partial w(r,\theta)}{\partial r}\right|_{r=b}=0,$$
$$M_r(a,\theta)=0,\quad V_r(a,\theta)=-f(\theta).$$

The last unknown constant η can be evaluated from $w(a,0)=0$. Reference: Kosherick, Dundurs and Lee (1968).

5.10.18 Solve the same annular plate of Section 5.7 when the supports are two concentric rings of equally spaced columns, situated in between and concentrically with the boundary circles. The number of columns in each ring is k. Each pair of opposite columns in different rings subtends the same angle 2α at the center of the boundary circles.

5.10.19 Find the solution to the problem of an annular plate on elastic foundation and supporting a circular ring of k equally spaced columns. The ring and the boundary circles are concentric. The radius of the ring is c. Hint: Set the reference coordinate system as in the example of Section 5.7. The column force P on the plate can be put as a distribution in the form

$$f(\theta)=P/(2k\alpha c),\quad \left(\frac{2\pi j}{k}-\alpha<\theta<\frac{2\pi j}{k}+\alpha,\quad j=0,1,2,\cdots,k\right)$$
$$=0,\quad (\text{ elsewhere }).$$

The solution can assume the following series

$$w(r,\theta)=\sum_{m=0,k,2k,\cdots}^{\infty} w_{cm}(r)\cos m\theta,$$

where $w_{cm}(r)$ is as given in Eq. 5.129. Boundary conditions are free conditions along both circular edges, and four more conditions along the column ring. The latter are

$$w_1(c,\theta)=w_2(c,\theta),\quad \left.\frac{\partial w_1(r,\theta)}{\partial r}\right|_{r=c}=\left.\frac{\partial w_2(r,\theta)}{\partial r}\right|_{r=c},$$

$$M_{r1}(c,\theta)=M_{r2}(c,\theta),\quad V_{r1}(c,\theta)-V_{r2}(c,\theta)=f(\theta),$$

in which subscripts 1 and 2 denote respective domains; domain 1 where $b\le r\le c$, and domain 2 where $c\le r\le a$.

5.10.20 Study the bending by a uniformly distributed load of a very long rectangular flat slab supported on three parallel rows of equidistant columns as shown in Fig. 5.17. It is assumed that the columns offer no restraint against rotation of the slab and that the reaction of each column is a line load uniformly distributed over a small interval in the same direction as the longitudinal column lines. Reference: Karasudhi, Alam and Lee (1973).

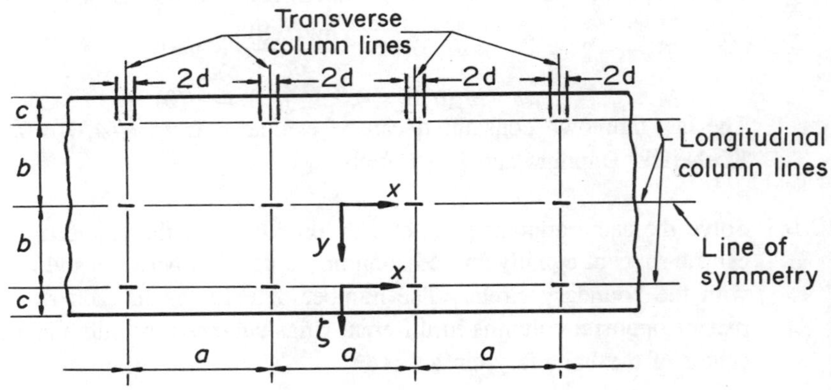

Fig. 5.17 Long flat slab on three rows of columns.

5.10.21 Formulate a rational analysis for a rectangular plate with eccentric stiffeners as shown in Fig. 5.18. Then approximate the governing equations to obtain a single fourth order differential equation similar to that for bending of an ordinary plate, Eq. 5.34. Reference: Hovichitr, Karasudhi, Nishino and Lee (1977).

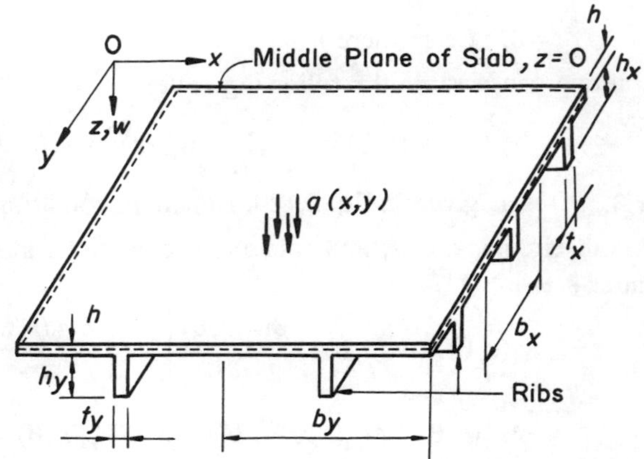

Fig. 5.18 Plate with eccentric stiffening ribs.

5.10.22 Since Reissner (1948, 1949) formulated a small deflection theory for bending of isotropic sandwich plates, there has been a rapid development in the theory and its application to the design of structural elements. A detailed literature survey up to 1956 was made by Habip. Kao (1965, 1969, 1970, 1973) treated problems of circular sandwich plates. Folie (1970) solved the bending of orthotropic sandwich plates by means of a numerical integration scheme. Discussions on bending and buckling of sandwich systems were made by Plantema (1966). Karasudhi, Ng and Lee (1977) studied axisymmetric bending of isotropic sandwich plates on elastic foundations as illustrated in Fig. 5.19. The main assumptions in the analysis of bending of an elastic sandwich plate are that in the core there are only transverse shear stress components uniformly distributed across the core thickness, and in the facings there are only in-plane stress components uniformly distributed across respective thicknesses. Solve Problem 5.10.19 for isotropic sandwich plates.

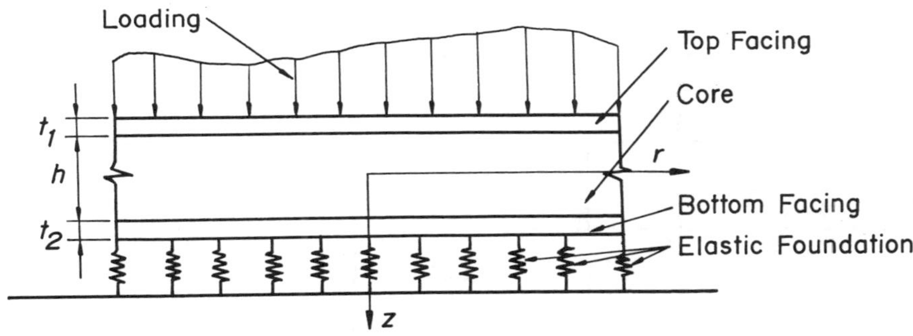

Fig. 5.19 Sandwich plate on elastic foundation.

5.10.23 Find an approximate solution to the problem of uniformly loaded rectangular orthotropic plates supported at the corners. Reference: Lee, Karasudhi, Zakeria and Chan (1971).

5.10.24 Solve Problem 5.10.16 if the plate is an orthotropic sandwich plate. Reference: Karasudhi, Jou and Tansirikongkol (1976).

Table 5.1 Complementary and Particular Solutions for Elastostatic Bending of Thin Isotropic Plates.

Complementary solutions, $\nabla^4 w = 0$.

w	dw/dr	M_r/D	M_θ/D	$M_{r\theta}/D$	Q_r/D	Q_θ/D
1	0	0	0	0	0	0
r^2	$2r$	$-2(1+\nu)$	$-2(1+\nu)$	0	0	0
$\log r$	r^{-1}	$(1-\nu)r^{-2}$	$-(1-\nu)r^{-2}$	0	0	0
$r^2 \log r$	$r(1+2\log r)$	$-2(1+\nu)\log r$ $-3-\nu$	$-2(1+\nu)\log r$ $-1-3\nu$	0	$-4r^{-1}$	0

$\dfrac{w}{(\cos\theta, \sin\theta)}$	r	$1/r$	r^3	$r\log r$
$\dfrac{\partial w/\partial r}{(\cos\theta, \sin\theta)}$	1	$-\dfrac{1}{r^2}$	$3r^2$	$\log r + 1$
$\dfrac{M_r/D}{(\cos\theta, \sin\theta)}$	0	$-\dfrac{2(1-\nu)}{r^3}$	$-2(3+\nu)r$	$-\dfrac{(1+\nu)}{r}$
$\dfrac{M_\theta/D}{(\cos\theta, \sin\theta)}$	0	$\dfrac{2(1-\nu)}{r^3}$	$-2(1+3\nu)r$	$-\dfrac{(1+\nu)}{r}$
$\dfrac{M_{r\theta}/D}{(\sin\theta, -\cos\theta)}$	0	$-\dfrac{2(1-\nu)}{r^3}$	$2(1-\nu)r$	$\dfrac{(1-\nu)}{r}$
$\dfrac{Q_r/D}{(\cos\theta, \sin\theta)}$	0	0	-8	$\dfrac{2}{r^2}$
$\dfrac{Q_\theta/D}{(\sin\theta, -\cos\theta)}$	0	0	8	$\dfrac{2}{r^2}$
$\dfrac{V_r/D}{(\cos\theta, \sin\theta)}$	0	$-\dfrac{2(1-\nu)}{r^4}$	$-2(3+\nu)$	$\dfrac{(3-\nu)}{r^2}$
$\dfrac{V_\theta/D}{(\sin\theta, -\cos\theta)}$	0	$\dfrac{6(1-\nu)}{r^4}$	$2(5-\nu)$	$\dfrac{(1+\nu)}{r^2}$

Complementary solutions, $\nabla^4 w = 0$, $(m \neq -1, 0, 1)$.

$\dfrac{w}{(\cos m\theta, \sin m\theta)}$	r^m	r^{-m}	r^{m+2}	r^{-m+2}
$\dfrac{\partial w/\partial r}{(\cos m\theta, \sin m\theta)}$	mr^{m-1}	$-mr^{-m-1}$	$(m+2)r^{m+1}$	$(-m+2)r^{-m+1}$
$\dfrac{M_r/D}{(\cos m\theta, \sin m\theta)}$	$-(1-\nu)m(m-1)r^{m-2}$	$-(1-\nu)m(m+1)r^{-m-2}$	$-(m+1)[4\nu + (1-\nu)(m+2)]r^m$	$(m-1)[4\nu + (1-\nu)(2-m)]r^{-m}$
$\dfrac{M_\theta/D}{(\cos m\theta, \sin m\theta)}$	$(1-\nu)m(m-1)r^{m-2}$	$(1-\nu)m(m+1)r^{-m-2}$	$(m+1)[-4 + (1-\nu)(m+2)]r^m$	$(m-1)[4 - (1-\nu)(2-m)]r^{-m}$
$\dfrac{M_{r\theta}/D}{(\sin m\theta, -\cos m\theta)}$	$(1-\nu)m(m-1)r^{m-2}$	$-(1-\nu)m(m+1)r^{-m-2}$	$(1-\nu)m(m+1)r^m$	$(1-\nu)m(1-m)r^{-m}$
$\dfrac{Q_r/D}{(\cos m\theta, \sin m\theta)}$	0	0	$-4(m+1)mr^{m-1}$	$4(1-m)mr^{-m-1}$
$\dfrac{Q_\theta/D}{(\sin m\theta, -\cos m\theta)}$	0	0	$4(m+1)mr^{m-1}$	$4(1-m)mr^{-m-1}$
$\dfrac{V_r/D}{(\cos m\theta, \sin m\theta)}$	$\dfrac{(1-\nu)m^2(m-1)}{r^{3-m}}$	$-\dfrac{(1-\nu)m^2(m+1)}{r^{m+3}}$	$\dfrac{(m+1)m[(1-\nu)m-4]}{r^{1-m}}$	$\dfrac{(1-m)m[4+(1-\nu)m]}{r^{m+1}}$
$\dfrac{V_\theta/D}{(\sin m\theta, -\cos m\theta)}$	$\dfrac{(1-\nu)m(m-1)(m-2)}{r^{3-m}}$	$\dfrac{(1-\nu)m(m+1)(m+2)}{r^{m+3}}$	$\dfrac{(m+1)m[(1-\nu)m+4]}{r^{1-m}}$	$\dfrac{(1-m)m[4-(1-\nu)m]}{r^{m+1}}$

Particular solutions, $\nabla^4 w = q/D \neq 0$.

q/D	w	dw/dr	M_r/D	M_θ/D	$M_{r\theta}/D$	Q_r/D	Q_θ/D
1	$\frac{r^4}{64}$	$\frac{r^3}{16}$	$-\frac{(3+\nu)r^2}{16}$	$-\frac{(1+3\nu)r^2}{16}$	0	$-\frac{r}{2}$	0
r	$\frac{r^5}{225}$	$\frac{r^4}{45}$	$-\frac{(4+\nu)r^3}{45}$	$-\frac{(1+4\nu)r^3}{45}$	0	$-\frac{r^2}{3}$	0

$\frac{q/D}{(\cos m\theta, \sin m\theta)}$	1 $(m \neq 0, \pm 2, \pm 4)$	r $(m \neq 0, \pm 3, \pm 5)$
$\frac{w}{(\cos m\theta, \sin m\theta)}$	$\frac{r^4}{64-20m^2+m^4}$	$\frac{r^5}{225-34m^2+m^4}$
$\frac{\partial w/\partial r}{(\cos m\theta, \sin m\theta)}$	$\frac{4r^3}{64-20m^2+m^4}$	$\frac{5r^4}{225-34m^2+m^4}$
$\frac{M_r/D}{(\cos m\theta, \sin m\theta)}$	$\frac{(\nu m^2-4\nu-12)r^2}{64-20m^2+m^4}$	$\frac{(\nu m^2-5\nu-20)r^3}{225-34m^2+m^4}$
$\frac{M_\theta/D}{(\cos m\theta, \sin m\theta)}$	$\frac{(m^2-12\nu-4)r^2}{64-20m^2+m^4}$	$\frac{(m^2-20\nu-5)r^3}{225-34m^2+m^4}$
$\frac{M_{r\theta}/D}{(\sin m\theta, -\cos m\theta)}$	$\frac{3m(1-\nu)r^2}{64-20m^2+m^4}$	$\frac{4m(1-\nu)r^3}{225-34m^2+m^4}$
$\frac{Q_r/D}{(\cos m\theta, \sin m\theta)}$	$\frac{2r}{m^2-4}$	$\frac{3r^2}{m^2-9}$
$\frac{Q_\theta/D}{(\sin m\theta, -\cos m\theta)}$	$\frac{-mr}{m^2-4}$	$\frac{-mr^2}{m^2-9}$
$\frac{V_r/D}{(\cos m\theta, \sin m\theta)}$	$\frac{[(5-3\nu)m^2-32]r}{64-20m^2+m^4}$	$\frac{[(7-4\nu)m^2-75]r^2}{225-34m^2+m^4}$
$\frac{V_\theta/D}{(\sin m\theta, -\cos m\theta)}$	$\frac{m(22-6\nu-m^2)r}{64-20m^2+m^4}$	$\frac{m(37-12\nu-m^2)r^2}{225-34m^2+m^4}$

Particular solutions, $\nabla^4 w = q/D \neq 0$.

$\dfrac{q/D}{(\cos 2\theta, \sin 2\theta)}$	1
$\dfrac{w}{(\cos 2\theta, \sin 2\theta)}$	$\dfrac{r^4 \log r}{48}$
$\dfrac{\partial w/\partial r}{(\cos 2\theta, \sin 2\theta)}$	$\dfrac{(4\log r + 1)r^3}{48}$
$\dfrac{M_r/D}{(\cos 2\theta, \sin 2\theta)}$	$-\dfrac{(7+\nu+12\log r)r^2}{48}$
$\dfrac{M_\theta/D}{(\cos 2\theta, \sin 2\theta)}$	$-\dfrac{(1+7\nu+12\nu\log r)r^2}{48}$
$\dfrac{M_{r\theta}/D}{(\sin 2\theta, -\cos 2\theta)}$	$\dfrac{(1-\nu)(3\log r + 1)r^2}{24}$
$\dfrac{Q_r/D}{(\cos 2\theta, \sin 2\theta)}$	$-\dfrac{(7+6\log r)r}{12}$
$\dfrac{Q_\theta/D}{(\sin 2\theta, -\cos 2\theta)}$	$\dfrac{(2+3\log r)r}{6}$
$\dfrac{12V_r/D}{(\cos 2\theta, \sin 2\theta)}$	$-[3(1+\nu)\log r$ $+6+\nu]r$
$\dfrac{24V_\theta/D}{(\sin 2\theta, -\cos 2\theta)}$	$[6(3-\nu)\log r$ $+13-5\nu]r$

$\dfrac{q/D}{(\cos 4\theta, \sin 4\theta)}$	1
$\dfrac{w}{(\cos 4\theta, \sin 4\theta)}$	$-\dfrac{r^4 \log r}{96}$
$\dfrac{\partial w/\partial r}{(\cos 4\theta, \sin 4\theta)}$	$-\dfrac{(4\log r + 1)r^3}{96}$
$\dfrac{96M_r/D}{(\cos 4\theta, \sin 4\theta)}$	$[12(1-\nu)\log r$ $+7+\nu]r^2$
$\dfrac{96M_\theta/D}{(\cos 4\theta, \sin 4\theta)}$	$[12(\nu-1)\log r$ $+1+7\nu]r^2$
$\dfrac{M_{r\theta}/D}{(\sin 4\theta, -\cos 4\theta)}$	$-\dfrac{(1-\nu)(3\log r + 1)r^2}{24}$
$\dfrac{Q_r/D}{(\cos 4\theta, \sin 4\theta)}$	$\dfrac{r}{6}$
$\dfrac{Q_\theta/D}{(\sin 4\theta, -\cos 4\theta)}$	$-\dfrac{r}{3}$
$\dfrac{V_r/D}{(\cos 4\theta, \sin 4\theta)}$	$\dfrac{[\nu-3(1-\nu)\log r]r}{6}$
$\dfrac{24V_\theta/D}{(\sin 4\theta, -\cos 4\theta)}$	$[6(\nu-1)\log r$ $+5\nu-13]r$

Particular solutions, $\nabla^4 w = q/D \neq 0$.

$\dfrac{q/D}{(\cos 3\theta, \sin 3\theta)}$	r
$\dfrac{w}{(\cos 3\theta, \sin 3\theta)}$	$\dfrac{r^5 \log r}{96}$
$\dfrac{\partial w/\partial r}{(\cos 3\theta, \sin 3\theta)}$	$\dfrac{(5\log r + 1)r^4}{96}$
$\dfrac{96M_r/D}{(\cos 3\theta, \sin 3\theta)}$	$-[4(5-\nu)\log r + 9 + \nu]r^3$
$\dfrac{96M_\theta/D}{(\cos 3\theta, \sin 3\theta)}$	$[4(1-5\nu)\log r - 1 - 9\nu]r^3$
$\dfrac{M_{r\theta}/D}{(\sin 3\theta, -\cos 3\theta)}$	$\dfrac{(1-\nu)(4\log r + 1)r^3}{32}$
$\dfrac{Q_r/D}{(\cos 3\theta, \sin 3\theta)}$	$-\dfrac{(23 + 24\log r)r^2}{48}$
$\dfrac{Q_\theta/D}{(\sin 3\theta, -\cos 3\theta)}$	$\dfrac{(5 + 8\log r)r^2}{16}$
$\dfrac{96V_r/D}{(\cos 3\theta, \sin 3\theta)}$	$-[12(1+3\nu)\log r + 9\nu + 37]r^2$
$\dfrac{32V_\theta/D}{(\sin 3\theta, -\cos 3\theta)}$	$[4(7-3\nu)\log r + 17 - 7\nu]r^2$

$\dfrac{q/D}{(\cos 5\theta, \sin 5\theta)}$	r
$\dfrac{w}{(\cos 5\theta, \sin 5\theta)}$	$-\dfrac{r^5 \log r}{160}$
$\dfrac{\partial w/\partial r}{(\cos 5\theta, \sin 5\theta)}$	$-\dfrac{(5\log r + 1)r^4}{160}$
$\dfrac{160M_r/D}{(\cos 5\theta, \sin 5\theta)}$	$[20(1-\nu)\log r + 9 + \nu]r^3$
$\dfrac{160M_\theta/D}{(\cos 5\theta, \sin 5\theta)}$	$[20(\nu-1)\log r + 1 + 9\nu]r^3$
$\dfrac{M_{r\theta}/D}{(\sin 5\theta, -\cos 5\theta)}$	$-\dfrac{(1-\nu)(4\log r + 1)r^3}{32}$
$\dfrac{Q_r/D}{(\cos 5\theta, \sin 5\theta)}$	$\dfrac{3r^2}{16}$
$\dfrac{Q_\theta/D}{(\sin 5\theta, -\cos 5\theta)}$	$-\dfrac{5r^2}{16}$
$\dfrac{32V_r/D}{(\cos 5\theta, \sin 5\theta)}$	$[20(\nu-1)\log r + 1 + 5\nu]r^2$
$\dfrac{32V_\theta/D}{(\sin 5\theta, -\cos 5\theta)}$	$[12(\nu-1)\log r + 7\nu - 17]r^2$

Particular solutions, $\nabla^4 w = q/D \neq 0$.

$\frac{q}{D}$	$r^{-3}\begin{Bmatrix}\cos\theta \\ --- \\ \sin\theta\end{Bmatrix}$	$\log r\begin{Bmatrix}\cos 2\theta \\ ---- \\ \sin 2\theta\end{Bmatrix}$	$\log r\begin{Bmatrix}\cos 4\theta \\ ---- \\ \sin 4\theta\end{Bmatrix}$
w	$-\frac{r(\log r)^2}{8}\begin{Bmatrix}\cos\theta \\ --- \\ \sin\theta\end{Bmatrix}$	$\frac{(6\log r - 12)r^4\log r}{576}\begin{Bmatrix}\cos 2\theta \\ ---- \\ \sin 2\theta\end{Bmatrix}$	$-\frac{(12\log r + 5)r^4\log r}{2304}\begin{Bmatrix}\cos 4\theta \\ ---- \\ \sin 4\theta\end{Bmatrix}$

$\frac{q}{D}$	r^n $(n \neq -2 \text{ or } -4)$	$\frac{1}{r^2}$	$\frac{1}{r^4}$
w	$\frac{r^{n+4}}{(n+2)^2(n+4)^2}$	$\frac{r^2(\log r)^2}{8}$	$\frac{(\log r)^2}{8}$

$\frac{q/D}{(\cos m\theta, \sin m\theta)}$	$\frac{w}{(\cos m\theta, \sin m\theta)}$
r^n $(n \neq \pm m - 2 \text{ or } -1)$	$\frac{r^{n+4}}{[(n+2)^2 - m^2]\,[(n+4)^2 - m^2]}$
r^{m-2} $(m \neq 0 \text{ or } -1)$	$\frac{r^{m+2}\log r}{8m(m+1)}$
r^{m-4} $(m \neq 0 \text{ or } 1)$	$-\frac{r^m \log r}{8m(m-1)}$
$\log r$ $(m \neq \pm 2 \text{ or } \pm 4)$	$\frac{[12(m^2-8) + (m^2-4)(m^2-16)\log r]r^4}{(m^2-4)^2(m^2-16)^2}$

CHAPTER VI

ELASTOSTATICS WITH DISPLACEMENTS AS UNKNOWNS

6.1 FIELD EQUATIONS FOR PLANE PROBLEMS

The Cartesian field equations for the elastostatics of an isotropic solid in plane strain may be obtained as a special case of Eq. 3.66, namely

$$\mu\left(\nabla^2 u_i + \frac{1}{1-2\nu}\varepsilon_{kk,i}\right) + X_i = 0, \qquad (i,k=1,2) \tag{6.1}$$

where

$$\nabla^2 = \frac{\partial^2}{\partial x_1^2} + \frac{\partial^2}{\partial x_2^2} \tag{6.2a}$$

$$\varepsilon_{kk} = \frac{\partial u_1}{\partial x_1} + \frac{\partial u_2}{\partial x_2} \tag{6.2b}$$

Due to Eq. 4.8b, we have

$$1-2\nu = \frac{1}{2}(\kappa - 1) \tag{6.3}$$

Substituting Eq. 6.3 into Eq. 6.1 yields field equations for plane problems as

$$\mu\left(\nabla^2 u_i + \frac{2}{\kappa-1}\varepsilon_{kk,i}\right) + X_i = 0 \tag{6.4}$$

The equation above holds for plane stress as well as plane strain problems depending on which definition of κ in Eqs. 4.8 is used.

Similarly, we can obtain field equations in terms of cylindrical displacement components by specializing from Eqs. 3.67; thus

$$\mu\left[\frac{2}{\kappa-1}\frac{\partial}{\partial r}(\varepsilon_{kk}) + \nabla^2 u_r - \frac{u_r}{r^2} - \frac{2}{r^2}\frac{\partial u_\theta}{\partial \theta}\right] + X_r = 0 \tag{6.5a}$$

$$\mu\left[\frac{2}{\kappa-1}\frac{1}{r}\frac{\partial}{\partial \theta}(\varepsilon_{kk}) + \nabla^2 u_\theta + \frac{2}{r^2}\frac{\partial u_r}{\partial \theta} - \frac{u_\theta}{r^2}\right] + X_\theta = 0 \tag{6.5b}$$

where

$$\nabla^2 = \frac{1}{r}\frac{\partial}{\partial r}\left(r\frac{\partial}{\partial r}\right) + \frac{1}{r^2}\frac{\partial^2}{\partial \theta^2} \tag{6.6a}$$

$$\varepsilon_{kk} = \frac{1}{r}\left[\frac{\partial}{\partial r}(ru_r) + \frac{\partial u_\theta}{\partial \theta}\right] \tag{6.6b}$$

For axisymmetric torsion-free deformation, u_θ must vanish, and there is no dependence on θ. Thus Eqs. 6.5 reduce to

$$\mu\frac{\kappa+1}{\kappa-1}\left\{\frac{\mathrm{d}}{\mathrm{d}r}\left[\frac{1}{r}\frac{\mathrm{d}}{\mathrm{d}r}(ru_r)\right]\right\} + X_r = 0 \tag{6.7}$$

the general solution of which can be obtained readily by successive integration with respect to r. In fact, its complementary solution is a linear combination of r and r^{-1}. Problems 6.9.1 and 6.9.2 should serve as good examples of this approach.

For torsion about the z-axis, u_r must vanish, and all functions are independent of θ. The non-trivial field equation is

$$\mu\frac{\mathrm{d}}{\mathrm{d}r}\left[\frac{1}{r}\frac{\mathrm{d}}{\mathrm{d}r}(ru_\theta)\right] + X_\theta = 0 \tag{6.8}$$

which is similar to Eq. 6.7. Problem 6.9.3 should serve as a good example of this approach.

A solid medium is said to be in an *antiplane* state (of shear) with respect to the xy-plane if $u = v = 0$, $w = w(x,y)$. There is just one field equation

$$\mu\nabla^2 w + Z = 0 \tag{6.9}$$

which has any plane harmonic function as a complementary solution, i.e. $\nabla^2 w = 0$.

6.2. SOLUTION SCHEME FOR LARGE PLANES

In this section we shall consider the elastostatics of an infinitely extended plane. This may be an *infinite plane* ($-\infty < x,\ y < \infty$), a *half plane* ($-\infty < x < \infty$, $0 \le y < \infty$), or a layer infinitely extended in the x-direction ($-\infty < x < \infty$) but finite and constant in the y-direction.

The field equations for *in-plane* (plane stress or plane strain) problems, Eq. 6.4, in the absence of the body forces, i.e. with

$$X_1 = X_2 = 0 \tag{6.10}$$

can be expressed in explicit symbols as

$$\nabla^2 u + \frac{2}{\kappa-1}\frac{\partial}{\partial x}(\varepsilon_{xx} + \varepsilon_{yy}) = 0 \tag{6.11a}$$

$$\nabla^2 v + \frac{2}{\kappa-1}\frac{\partial}{\partial y}(\varepsilon_{xx} + \varepsilon_{yy}) = 0 \tag{6.11b}$$

We seek solutions to these equations in the form of *Fourier integrals* (Sneddon, 1951)

$$u(x,y)=\frac{1}{\pi}\int_{-\infty}^{\infty}\tilde{u}(\xi,y)e^{-i\xi x}d\xi \tag{6.12a}$$

$$v(x,y)=\frac{1}{\pi}\int_{-\infty}^{\infty}\tilde{v}(\xi,y)e^{-i\xi x}d\xi \tag{6.12b}$$

in which a tilde ($\tilde{}$) denotes a *Fourier transform*. Note that the *Riemann-Lebesgue lemma* (Whittaker and Watson, 1965) ensures the boundedness of the solution as $|x|\to\infty$. Substituting Eqs. 6.12 into Eqs. 6.11, we find

$$-\xi^2\tilde{u}+\frac{d^2\tilde{u}}{dy^2}+\frac{2}{\kappa-1}\left(-\xi^2\tilde{u}-i\xi\frac{d\tilde{v}}{dy}\right)=0 \tag{6.13a}$$

$$\xi\tilde{v}+\frac{d^2\tilde{v}}{dy^2}+\frac{2}{\kappa-1}\left(-i\xi\frac{d\tilde{u}}{dy}+\frac{d^2\tilde{v}}{dy^2}\right)=0 \tag{6.13b}$$

Seeking $\tilde{u}(\xi,y)$ and $\tilde{v}(\xi,y)$ in the form

$$\tilde{u}\sim e^{\alpha y} \text{ and } \tilde{v}\sim e^{\alpha y},$$

we find

$$(\alpha^2-\xi^2)^2=0,$$

or

$$\alpha=-\xi,\ -\xi,\ \xi,\ \xi.$$

Because of the double roots above, the functions $\tilde{u}(\xi,y)$ and $\tilde{v}(\xi,y)$ must be linear combinations of the functions

$$e^{-\xi y},\ ye^{-\xi y},\ e^{\xi y} \text{ and } ye^{\xi y}.$$

Equations 6.13 show that if

$$\tilde{u}(\xi,y)=(A+By)e^{-\xi y}+(C+Dy)e^{\xi y} \tag{6.14a}$$

then

$$\tilde{v}(\xi,y)=i\{-[A+B(\kappa\xi^{-1}+y)]e^{-\xi y}+[C+D(-\kappa\xi^{-1}+y)]e^{\xi y}\} \tag{6.14b}$$

The constants A, B, C and D must be determined from four boundary conditions at the two boundaries, say at $y=y_1$ and $y=y_1+h$, if h is the thickness of the layer.

When the loading is symmetrical with respect to the y-axis, we seek the solution in the form

$$u(x,y)=\frac{2}{\pi}\int_0^{\infty}\tilde{u}(\xi,y)\sin\xi x d\xi \tag{6.15a}$$

$$v(x,y)=\frac{2}{\pi}\int_0^{\infty}\tilde{v}(\xi,y)\cos\xi x d\xi \tag{6.15b}$$

where

$$\bar{u}(\xi,y)=\int_0^\infty u(x,y)\sin\xi x\,\mathrm{d}x \tag{6.16a}$$

$$=(A+By)\mathrm{e}^{-\xi y}+(C+Dy)\mathrm{e}^{\xi y} \tag{6.16b}$$

$$\bar{v}(\xi,y)=\int_0^\infty v(x,y)\cos\xi x\,\mathrm{d}x \tag{6.17a}$$

$$=[A+B(\kappa\xi^{-1}+y)]\mathrm{e}^{-\xi y}-[C+D(-\kappa\xi^{-1}+y)]\mathrm{e}^{\xi y} \tag{6.17b}$$

The unified Hooke's law, Eqs. 4.6, gives the stress components

$$\sigma_{xx}(x,y)=\frac{2}{\pi}\int_0^\infty \tilde{\sigma}_x(\xi,y)\cos\xi x\,\mathrm{d}\xi \tag{6.18a}$$

$$\sigma_{xy}(x,y)=\frac{2}{\pi}\int_0^\infty \tilde{\tau}(\xi,y)\sin\xi x\,\mathrm{d}\xi \tag{6.18b}$$

$$\sigma_{yy}(x,y)=\frac{2}{\pi}\int_0^\infty \tilde{\sigma}(\xi,y)\cos\xi x\,\mathrm{d}\xi \tag{6.18c}$$

where

$$\tilde{\sigma}_x(\xi,y)=\frac{\mu}{\kappa-1}\left[(\kappa+1)\xi\bar{u}+(3-\kappa)\frac{\mathrm{d}\bar{v}}{\mathrm{d}y}\right] \tag{6.19a}$$

$$\tilde{\tau}(\xi,y)=\mu\left(\frac{\mathrm{d}\bar{u}}{\mathrm{d}y}-\xi\bar{v}\right) \tag{6.19b}$$

$$\tilde{\sigma}(\xi,y)=\frac{\mu}{\kappa-1}\left[(\kappa+1)\frac{\mathrm{d}\bar{v}}{\mathrm{d}y}+(3-\kappa)\xi\bar{u}\right] \tag{6.19c}$$

Consider a plane being composed of N parallel homogeneous layers, with differing elastic moduli. The solution for layer n will involve four arbitrary constants; A_n, B_n, C_n and D_n. Altogether, there will be $4N$ arbitrary constants to be determined from boundary conditions at the two surfaces of the whole plane and at the $N-1$ interfaces between layers. Considering the interface between layer $n-1(y_{n-2}\le y\le y_{n-1})$ and layer $n(y_{n-1}\le y\le y_n)$. If there are no applied tractions at the interface, u, v, σ_{yy} and σ_{xy} must be continuous, i.e.

$$\bar{u}_n(\xi,y)=\bar{u}_{n-1}(\xi,y) \tag{6.20a}$$

$$\bar{v}_n(\xi,y)=\bar{v}_{n-1}(\xi,y) \tag{6.20b}$$

$$\tilde{\sigma}_n(\xi,y)=\tilde{\sigma}_{n-1}(\xi,y) \tag{6.20c}$$

$$\tilde{\tau}_n(\xi,y)=\tilde{\tau}_{n-1}(\xi,y) \tag{6.20d}$$

Here the subscripts $n-1$ and n refer to the respective layers, and y is identical to y_{n-1}. If tractions are applied to the interface between layers $n-1$ and n, which may or may not have different elastic moduli, then the requisite conditions are

$$\tilde{u}_n(\xi, y) = \tilde{u}_{n-1}(\xi, y) \tag{6.21a}$$

$$\tilde{v}_n(\xi, y) = \tilde{v}_{n-1}(\xi, y) \tag{6.21b}$$

$$\tilde{\sigma}_n(\xi, y) + \tilde{\overline{Y}}(\xi, y) = \tilde{\sigma}_{n-1}(\xi, y) \tag{6.21c}$$

$$\tilde{\tau}_n(\xi, y) + \tilde{\overline{X}}(\xi, y) = \tilde{\tau}_{n-1}(\xi, y) \tag{6.21d}$$

Here $\tilde{\overline{X}}$ is the Fourier sine transform of the applied traction in the x-direction, i.e.

$$\tilde{\overline{X}}(\xi, y) = \int_0^\infty \overline{X}(x, y) \sin \xi x dx \tag{6.22a}$$

and, similarly,

$$\tilde{\overline{Y}}(\xi, y) = \int_0^\infty \overline{Y}(x, y) \cos \xi x dx \tag{6.22b}$$

At the top surface of the first layer ($n = 1, y_0 \leq y \leq y_1$) which is also the top surface of the whole plane, the boundary conditions will involve

$$\tilde{u}_1(\xi, y_0) \text{ or } \tilde{\tau}_1(\xi, y_0) \tag{6.23a}$$

$$\tilde{v}_1(\xi, y_0) \text{ or } \tilde{\sigma}_1(\xi, y_0) \tag{6.23b}$$

However, if this particular layer is an overlying half plane, the arbitrary constants associated with exponentials of negative y must be set to zero to assure the boundedness of the solution as $y \to y_0 = -\infty$, i.e. Eqs. 6.23 must be replaced by

$$A_1 = B_1 = 0 \tag{6.24a,b}$$

At the lower surface of the last layer ($n = N, y_{N-1} \leq y \leq y_N$) which is also the bottom of the whole plane, the boundary conditions will involve

$$\tilde{u}_N(\xi, y_N) \text{ or } \tilde{\tau}_N(\xi, y_N) \tag{6.25a}$$

$$\tilde{v}_N(\xi, y_N) \text{ or } \tilde{\sigma}_N(\xi, y_N) \tag{6.25b}$$

However, if this particular layer is an underlying half plane, the arbitrary constants associated with exponentials of positive y must be set to zero to assure the boundedness of the solutions as $y \to y_N = \infty$, i.e. Eqs. 6.25 must be replaced by

$$C_N = D_N = 0 \tag{6.26a,b}$$

If the loading is antisymmetrical with respect to the y-axis, the displacement and the stress functions should take the form

$$u(x, y) = \frac{2}{\pi} \int_0^\infty \tilde{u}(\xi, y) \cos \xi x d\xi \tag{6.27a}$$

$$v(x,y)=\frac{2}{\pi}\int_0^\infty \bar{v}(\xi,y)\sin\xi x\,\mathrm{d}\xi \tag{6.27b}$$

$$\sigma_{xx}(x,y)=\frac{2}{\pi}\int_0^\infty \tilde{\sigma}_x(\xi,y)\sin\xi x\,\mathrm{d}\xi \tag{6.27c}$$

$$\sigma_{xy}(x,y)=\frac{2}{\pi}\int_0^\infty \tilde{\tau}(\xi,y)\cos\xi x\,\mathrm{d}\xi \tag{6.27d}$$

$$\sigma_{yy}(x,y)=\frac{2}{\pi}\int_0^\infty \tilde{\sigma}(\xi,y)\sin\xi x\,\mathrm{d}\xi \tag{6.27e}$$

where

$$\bar{u}(\xi,y)=(A+By)\mathrm{e}^{-\xi y}+(C+Dy)\mathrm{e}^{\xi y} \tag{6.28a}$$

$$\bar{v}(\xi,y)=-[A+B(\kappa\xi^{-1}+y)]\mathrm{e}^{-\xi y}+[C+D(-\kappa\xi^{-1}+y)]\mathrm{e}^{\xi y} \tag{6.28b}$$

$$\tilde{\sigma}_x(\xi,y)=\frac{\mu}{\kappa-1}\left[-(\kappa+1)\xi\bar{u}+(3-\kappa)\frac{\mathrm{d}\bar{v}}{\mathrm{d}y}\right] \tag{6.28c}$$

$$\tilde{\tau}(\xi,y)=\mu\left(\frac{\mathrm{d}\bar{u}}{\mathrm{d}y}+\xi\bar{v}\right) \tag{6.28d}$$

$$\tilde{\sigma}(\xi,y)=\frac{\mu}{\kappa-1}\left[(\kappa+1)\frac{\mathrm{d}\bar{v}}{\mathrm{d}y}-(3-\kappa)\xi\bar{u}\right] \tag{6.28e}$$

In addition, the Fourier transforms of the applied tractions are

$$\tilde{\bar{X}}(\xi,y)=\int_0^\infty \bar{X}(x,y)\cos\xi x\,\mathrm{d}x \tag{6.29a}$$

$$\tilde{\bar{Y}}(\xi,y)=\int_0^\infty \bar{Y}(x,y)\sin\xi x\,\mathrm{d}x \tag{6.29b}$$

The relevant boundary conditions for this case are the same as in the previous case, i.e. Eqs. 6.20 or 6.21, 6.23 or 6.24, and 6.25 or 6.26.

For antiplane problems where loading is symmetrical with respect to the y-axis, the solution to the field equation (Eq. 6.9) for $Z=0$ takes the form

$$w(x,y)=\frac{2}{\pi}\int_0^\infty \bar{w}(\xi,y)\cos\xi x\,\mathrm{d}\xi \tag{6.30a}$$

and the only stresses are

$$\sigma_{yz}(x,y)=\frac{2}{\pi}\int_0^\infty \tilde{\tau}(\xi,y)\cos\xi x\,\mathrm{d}\xi \tag{6.30b}$$

$$\sigma_{xz}(x,y)=\frac{2}{\pi}\int_0^\infty \tilde{\tau}_x(\xi,y)\sin\xi x\,\mathrm{d}\xi \tag{6.30c}$$

where

$$\bar{w}(\xi, y) = Ae^{-\xi y} + Be^{\xi y} \tag{6.31a}$$

$$\bar{\tau}(\xi, y) = \mu \frac{d\bar{w}}{dy} \tag{6.31b}$$

$$\bar{\tau}_x(\xi, y) = -\mu\xi\bar{w} \tag{6.31c}$$

At the interface between the layers $n-1$ and n, boundary conditions in the absence of any applied traction are

$$\bar{w}_n(\xi, y) = \bar{w}_{n-1}(\xi, y) \tag{6.32a}$$

$$\bar{\tau}_n(\xi, y) = \bar{\tau}_{n-1}(\xi, y) \tag{6.32b}$$

If a traction in the z-direction is applied at this interface, Eqs. 6.32 have to be replaced by

$$\bar{w}_n(\xi, y) = \bar{w}_{n-1}(\xi, y) \tag{6.33a}$$

$$\bar{\tau}_n(\xi, y) + \bar{\bar{Z}}(\xi, y) = \bar{\tau}_{n-1}(\xi, y) \tag{6.33b}$$

where

$$\bar{\bar{Z}}(\xi, y) = \int_0^\infty \bar{Z}(x, y) \cos \xi x d\xi \tag{6.34}$$

in which $\bar{Z}(x, y)$ is the applied traction. At the top surface of the first layer ($n = 1$), boundary conditions are in

$$\bar{w}_1(\xi, y_0) \text{ or } \bar{\tau}_1(\xi, y_0) \tag{6.35}$$

However, if this particular layer is an overlying half plane, the equation above must be replaced by

$$A_1 = 0 \tag{6.36}$$

At the lower surface of the last layer ($n = N$), boundary conditions are in

$$\bar{w}_N(\xi, y_N) \text{ or } \bar{\tau}_N(\xi, y_N) \tag{6.37}$$

However, if this particular layer is an underlying half plane, the equation above must be replaced by

$$B_N = 0 \tag{6.38}$$

For antiplane problems in which the loading is antisymmetrical with respect to the y-axis, the displacement and stress functions take the form

$$w(x, y) = \frac{2}{\pi} \int_0^\infty \bar{w}(\xi, y) \sin \xi x d\xi \tag{6.39a}$$

$$\sigma_{yz}(x, y) = \frac{2}{\pi} \int_0^\infty \bar{\tau}(\xi, y) \sin \xi x d\xi \tag{6.39b}$$

$$\sigma_{xz}(x,y)=\frac{2}{\pi}\int_0^\infty \tilde{\tau}_x(\xi,y)\cos\xi x\mathrm{d}\xi \tag{6.39c}$$

where

$$\tilde{w}(\xi,y)=A\mathrm{e}^{-\xi y}+B\mathrm{e}^{\xi y} \tag{6.40a}$$

$$\tilde{\tau}(\xi,y)=\mu\frac{\mathrm{d}\tilde{w}}{\mathrm{d}y} \tag{6.40b}$$

$$\tilde{\tau}_x(\xi,y)=\mu\xi\tilde{w} \tag{6.40c}$$

In addition, the Fourier transform of the applied traction is

$$\tilde{\bar{Z}}(\xi,y)=\int_0^\infty \bar{Z}(x,y)\sin\xi x\mathrm{d}x \tag{6.41}$$

The relevant boundary conditions for this case are the same as before, i.e. Eqs. 6.32 or 6.33, 6.35 or 6.36, and 6.37 or 6.38.

Technically, the solution scheme described so far is complete for any multilayered planes. However, the exact closed form solution is only possible for a homogeneous half plane subjected to simple loading types. For more complicated cases, approximate solutions may be approached in similar manners as in the three dimensional cases which are discussed in Sections 6.5, 6.6 and 6.7.

6.3 SOLUTION SCHEME FOR LARGE SPACES

This section follows the work of Muki (1960) to develop a method applicable to the elastostatics of an isotropic solid domain which is infinitely extended radially ($0\le r<\infty$). The domain may be an *infinite space* ($-\infty<z<\infty$), a half space ($0\le z<\infty$), or an infinitely large plate with a finite constant thickness ($z_0\le z\le z_N$). In cylindrical coordinates (r,θ,z), the equations of equilibrium, in the absence of the body force, are given by Eqs. 3.67 with

$$X_r=X_\theta=X_z=0 \tag{6.42}$$

It may be verified by direct substitution that these statical field equations are satisfied by the displacements

$$u_r=-\frac{\partial^2\Phi}{\partial r\partial z}+\frac{2}{r}\frac{\partial\Psi}{\partial\theta} \tag{6.43a}$$

$$u_\theta=-\frac{\partial^2\Phi}{r\partial\theta\partial z}-2\frac{\partial\Psi}{\partial r} \tag{6.43b}$$

$$u_z=2(1-\nu)\nabla^2\Phi-\frac{\partial^2\Phi}{\partial z^2} \tag{6.43c}$$

provided that the displacement potentials Φ and Ψ satisfy

$$\nabla^4\Phi = 0, \quad \nabla^2\Psi = 0 \tag{6.44a,b}$$

In these equations, ν is the Poisson's ratio, ∇^2 the Laplace operator (Eq. 3.69b), and ∇^4 the biharmonic operator, i.e. $\nabla^4 = \nabla^2\nabla^2$. For the solution to be singlevalued with respect to θ, $\Phi(r,\theta,z)$ and $\Psi(r,\theta,z)$ can be written in Fourier series of θ with 2π as the period, i.e.

$$\Phi(r,\theta,z) = \sum_{m=0}^{\infty} [\Phi_m(r,z)\cos m\theta - \overline{\Phi}_m(r,z)\sin m\theta] \tag{6.45a}$$

$$\Psi(r,\theta,z) = \sum_{m=0}^{\infty} [\Psi_m(r,z)\sin m\theta + \overline{\Psi}_m(r,z)\cos m\theta] \tag{6.45b}$$

We will assume that $\overline{\Phi}_m = \overline{\Psi}_m = 0$, and develop solutions corresponding to Φ_m and Ψ_m. If needed, the solution corresponding to $\overline{\Phi}_m$ and $\overline{\Psi}_m$ may be obtained by making the replacements

Φ_m by $\overline{\Phi}_m$, $\quad\quad$ Ψ_m by $\overline{\Psi}_m$,

$\cos m\theta$ by $-\sin m\theta$, $\quad$ and $\sin m\theta$ by $\cos m\theta$.

Substituting Eqs. 6.45 into Eqs. 6.44, we see that Φ_m and Ψ_m satisfy

$$\nabla_m^4\Phi_m(r,z) = 0 \tag{6.46a}$$

$$\nabla_m^2\Psi_m(r,z) = 0 \tag{6.46b}$$

where

$$\nabla_m^2 = \frac{1}{r}\frac{\partial}{\partial r}\left(r\frac{\partial}{\partial r}\right) - \frac{m^2}{r^2} + \frac{\partial^2}{\partial z^2} \tag{6.47}$$

Consider the case that a homogeneous material layer is unbounded in the radial dimension but bounded by two infinite planes normal to the z-axis, say $z_{n-1} \le z \le z_n$. If the solution to Eqs. 6.46 is to be nonsingular for the whole range of r ($0 \le r < \infty$), then it must take the form

$$\Phi_m(r,z) = \int_0^{\infty} \xi\tilde{\Phi}_m(\xi,z)J_m(\xi r)d\xi \tag{6.48a}$$

$$\Psi_m(r,z) = \int_0^{\infty} \xi\tilde{\Psi}_m(\xi,z)J_m(\xi r)d\xi \tag{6.48b}$$

where $J_m(\ )$ is a Bessel function of the first kind (see Section 1.10), and the upper tilde ($\tilde{}$) denotes a *Hankel transform* (Sneddon, 1951), i.e.

$$\tilde{\Phi}_m(\xi,z) = \int_0^{\infty} r\Phi_m(r,z)J_m(\xi r)dr \tag{6.49a}$$

$$\tilde{\Psi}_m(\xi,z) = \int_0^{\infty} r\Psi_m(r,z)J_m(\xi r)dr \tag{6.49b}$$

Substituting Eqs. 6.48 into Eq. 6.46, and using the Bessel equation (Eq. 1.111), we find that $\tilde{\Phi}_m$ and $\tilde{\Psi}_m$ must satisfy the ordinary differential equations

$$\left(\frac{d^2}{dz^2}-\xi^2\right)^2\tilde{\Phi}_m=0 \tag{6.50a}$$

$$\left(\frac{d^2}{dz^2}-\xi^2\right)\tilde{\Psi}_m=0 \tag{6.50b}$$

The general solutions of these equations are

$$\tilde{\Phi}_m(\xi,z)=(A_m+B_m z)e^{\xi z}+(C_m+D_m z)e^{-\xi z} \tag{6.51a}$$

$$\tilde{\Psi}_m(\xi,z)=E_m e^{\xi z}+F_m e^{-\xi z} \tag{6.51b}$$

Here A_m to F_m are arbitrary constants to be determined from the Hankel transform of the given boundary conditions on the two bounding infinite planes $z=z_{n-1}$ and $z=z_n$. The displacement and stress components can be put in terms of $\tilde{\Phi}_m$ and $\tilde{\Psi}_m$ as follows:

$$u_r=\frac{1}{2}\sum_{m=0}^{\infty}[U_{m+1}(r,z)-V_{m-1}(r,z)]\cos m\theta \tag{6.52a}$$

$$u_\theta=\frac{1}{2}\sum_{m=0}^{\infty}[U_{m+1}(r,z)+V_{m-1}(r,z)]\sin m\theta \tag{6.52b}$$

$$u_z=\sum_{m=0}^{\infty}\left\{\int_0^\infty\left[(1-2\nu)\frac{d^2\tilde{\Phi}_m}{dz^2}-2(1-\nu)\xi^2\tilde{\Phi}_m\right]\xi J_m(\xi r)d\xi\right\}\cos m\theta \tag{6.52c}$$

$$\frac{\sigma_{rr}}{2\mu}=\sum_{m=0}^{\infty}\left\{\int_0^\infty\left[\nu\frac{d^3\tilde{\Phi}_m}{dz^3}+(1-\nu)\xi^2\frac{d\tilde{\Phi}_m}{dz}\right]\xi J_m(\xi r)d\xi\right.$$
$$\left.-\frac{(m+1)}{2r}U_{m+1}-\frac{(m-1)}{2r}V_{m-1}\right\}\cos m\theta \tag{6.53a}$$

$$\frac{\sigma_{\theta\theta}}{2\mu}=\sum_{m=0}^{\infty}\left[\nu\int_0^\infty\left(\frac{d^3\tilde{\Phi}_m}{dz^3}-\xi^2\frac{d\tilde{\Phi}_m}{dz}\right)\xi J_m(\xi r)d\xi\right.$$
$$\left.+\frac{(m+1)}{2r}U_{m+1}+\frac{(m-1)}{2r}V_{m-1}\right]\cos m\theta \tag{6.53b}$$

$$\frac{\sigma_{zz}}{2\mu}=\sum_{m=0}^{\infty}\left\{\int_0^\infty\left[(1-\nu)\frac{d^3\tilde{\Phi}_m}{dz^3}-(2-\nu)\xi^2\frac{d\tilde{\Phi}_m}{dz}\right]\xi J_m(\xi r)d\xi\right\}\cos m\theta \tag{6.53c}$$

$$\frac{\sigma_{\theta z}}{2\mu}=\frac{1}{2}\sum_{m=0}^{\infty}\left\{\int_0^\infty\left[\nu\frac{d^2\tilde{\Phi}_m}{dz^2}+(1-\nu)\xi^2\tilde{\Phi}_m+\frac{d\tilde{\Psi}_m}{dz}\right]\xi^2 J_{m+1}(\xi r)d\xi\right.$$
$$\left.+\int_0^\infty\left[\nu\frac{d^2\tilde{\Phi}_m}{dz^2}+(1-\nu)\xi^2\tilde{\Phi}_m-\frac{d\tilde{\Psi}_m}{dz}\right]\xi^2 J_{m-1}(\xi r)d\xi\right\}\sin m\theta \tag{6.53d}$$

$$\frac{\sigma_{zr}}{2\mu}=\frac{1}{2}\sum_{m=0}^{\infty}\left\{\int_0^{\infty}\left[\nu\frac{d^2\tilde{\Phi}_m}{dz^2}+(1-\nu)\xi^2\tilde{\Phi}_m+\frac{d\tilde{\Psi}_m}{dz}\right]\xi^2 J_{m+1}(\xi r)d\xi\right.$$

$$\left.-\int_0^{\infty}\left[\nu\frac{d^2\tilde{\Phi}_m}{dz^2}+(1-\nu)\xi^2\tilde{\Phi}_m-\frac{d\tilde{\Psi}_m}{dz}\right]\xi^2 J_{m-1}(\xi r)d\xi\right\}\cos m\theta \quad (6.53e)$$

$$\frac{\sigma_{r\theta}}{2\mu}=\sum_{m=0}^{\infty}\left[\int_0^{\infty}\tilde{\Psi}_m\xi^3 J_m(\xi r)d\xi-\frac{(m+1)}{2r}U_{m+1}+\frac{(m-1)}{2r}V_{m-1}\right]\sin m\theta \quad (6.53f)$$

where

$$U_{m+1}(r,z)=\int_0^{\infty}\left(\frac{d\tilde{\Phi}_m}{dz}+2\tilde{\Psi}_m\right)\xi^2 J_{m+1}(\xi r)d\xi, \quad (6.54a)$$

$$V_{m-1}(r,z)=\int_0^{\infty}\left(\frac{d\tilde{\Phi}_m}{dz}-2\tilde{\Psi}_m\right)\xi^2 J_{m-1}(\xi r)d\xi. \quad (6.54b)$$

At a point on each of the boundary planes (where $z=z_{n-1}$ or z_n), there are three boundary conditions, i.e. involving the prescription of either

$$u_r \text{ or } \sigma_{zr},$$

$$u_\theta \text{ or } \sigma_{z\theta},$$

$$\text{and } u_z \text{ or } \sigma_{zz}$$

Before applying these boundary conditions, we find it convenient to rewrite Eqs. 6.52a, 6.52b, 6.53d and 6.53e so that the mth equation involves only $J_m(\xi r)$; thus

$$\frac{u_{rm}}{\cos m\theta}=\int_0^{\infty}\left[-\frac{d\tilde{\Phi}_m}{dz}\frac{dJ_m(\xi r)}{d(\xi r)}+2m\tilde{\Psi}_m\frac{J_m(\xi r)}{\xi r}\right]\xi^2 d\xi \quad (6.55a)$$

$$\frac{u_{\theta m}}{\sin m\theta}=\int_0^{\infty}\left[m\frac{d\tilde{\Phi}_m}{dz}\frac{J_m(\xi r)}{\xi r}-2\tilde{\Psi}_m\frac{dJ_m(\xi r)}{d(\xi r)}\right]\xi^2 d\xi \quad (6.55b)$$

$$\frac{\sigma_{\theta zm}}{\sin m\theta}=\int_0^{\infty}\left\{m\left[\nu\frac{d^2\tilde{\Phi}_m}{dz^2}+(1-\nu)\xi^2\tilde{\Phi}_m\right]\frac{J_m(\xi r)}{\xi r}-\frac{d\tilde{\Psi}_m}{dz}\frac{dJ_m(\xi r)}{d(\xi r)}\right\}\xi^2 d\xi \quad (6.56a)$$

$$\frac{\sigma_{zrm}}{\cos m\theta}=\int_0^{\infty}\left\{-\left[\nu\frac{d^2\tilde{\Phi}_m}{dz^2}+(1-\nu)\xi^2\tilde{\Phi}_m\right]\frac{dJ_m(\xi r)}{d(\xi r)}+m\frac{d\tilde{\Psi}_m}{dz}\frac{J_m(\xi r)}{\xi r}\right\}\xi^2 d\xi \quad (6.56b)$$

For axisymmetric torsion-free deformation, u_θ must vanish, all functions must be independent of θ, so that $\Phi=\Phi_0$, $\Psi=0$.

For torsion about the z-axis, u_r and u_z must vanish, all functions must be independent of θ, and the only non-trivial equilibrium equation is

$$\nabla_1^2 u_\theta(r,z)=0 \quad (6.57)$$

in which ∇_1^2 is defined by Eq. 6.47 with $m=1$. Thus $u_\theta(r,z)$ assumes the solution in the form

$$u_\theta(r,z)=\int_0^\infty \xi(A\mathrm{e}^{\xi z}+C\mathrm{e}^{-\xi z})J_1(\xi r)\mathrm{d}\xi \tag{6.58}$$

where A and C are arbitrary constants to be determined from boundary conditions. The non-vanishing stress components are

$$\sigma_{\theta z}=\mu\int_0^\infty \xi^2(A\mathrm{e}^{\xi z}-C\mathrm{e}^{-\xi z})J_1(\xi r)\mathrm{d}\xi \tag{6.59a}$$

$$\sigma_{r\theta}=-\mu\int_0^\infty \xi^2(A\mathrm{e}^{\xi z}+C\mathrm{e}^{-\xi z})J_2(\xi r)\mathrm{d}\xi \tag{6.59b}$$

6.4 HOMOGENEOUS HALF PLANES AND HALF SPACES

The first example to be discussed in this section is the problem of a homogeneous isotropic elastic half plane $-\infty<x<\infty$, $0\le y<\infty$ subjected to a normal indentation by a rigid plate of finite width $2a$. The contact area between the rigid plate and the half plane is assumed to be smooth. The Fourier transforms of displacement components given in Eqs. 6.16 and 6.17 are

$$\bar{u}(\xi,y)=(A+By)\mathrm{e}^{-\xi y} \tag{6.60a}$$

$$\bar{v}(\xi,y)=[A+B(\kappa\xi^{-1}+y)]\mathrm{e}^{-\xi y} \tag{6.60b}$$

and the boundary conditions are

$$v(x,0)=v_0, \qquad (|x|\le a) \tag{6.61a}$$

$$\sigma_{yy}(x,0)=0, \qquad ((|x|>a) \tag{6.61b}$$

$$\sigma_{xy}(x,0)=0, \qquad (0\le|x|<\infty) \tag{6.61c}$$

where v_0 is the prescribed uniform indentation. The last equation above gives $\tilde{\tau}(\xi,0)=0$ or, because of Eqs. 6.60 and 6.19b,

$$2\xi A+(\kappa-1)B=0.$$

On the other hand, Eq. 6.19c gives

$$2\xi A+(\kappa+1)B=\frac{\tilde{\sigma}(\xi,0)}{\mu}.$$

Solving the last two equations above yields

$$A=-\frac{(\kappa-1)\tilde{\sigma}(\xi,0)}{4\xi\mu} \tag{6.62a}$$

$$B=\frac{\tilde{\sigma}(\xi,0)}{2\mu} \tag{6.62b}$$

Using these results in the remaining boundary conditions, Eqs. 6.61a and b, we get respectively

$$\int_0^\infty \tilde{\sigma}(\xi,0)\xi^{-1}\cos\xi x\,d\xi = \frac{2\pi\mu v_0}{\kappa+1}, \qquad (0 \le x \le a) \tag{6.63a}$$

$$\int_0^\infty \tilde{\sigma}(\xi,0)\cos\xi x\,d\xi = 0, \qquad (a < x < \infty) \tag{6.63b}$$

Equations 6.63 are called *dual integral equations*; they may be solved by schemes described by Sneddon (1951 and 1966). Note that Eq. 6.63a states that $v(x,0) = \text{constant}$, for $|x| \le a$. However, since any static half plane problem has a rigid body displacement solution, the actual value of the constant cannot be found. So we replace this equation by

$$\frac{dv(x,0)}{dx} = 0$$

in the same region, i.e.

$$\int_0^\infty \tilde{\sigma}(\xi,0)\sin\xi x\,d\xi = 0, \qquad (0 \le x \le a) \tag{6.64}$$

The general solution of Eqs. 6.63b and 6.64 is

$$\tilde{\sigma}(\xi,0) = C_0 J_0(\xi a) \tag{6.65}$$

where J_0 is the Bessel function of the first kind and of zero order, and C_0 is a constant to be determined in terms of the total applied force P_0. Substituting Eq. 6.65 into Eq. 6.18c for $y = 0$ yields the contact normal stress

$$\sigma_{yy}(x,0) = \frac{2C_0}{\pi(a^2 - x^2)^{1/2}}, \qquad (0 \le x < a) \tag{6.66a}$$

$$= 0, \qquad (a < x < \infty) \tag{6.66b}$$

Note that the contact normal stress is singular at the edges of the rigid plate. The total applied force P_0 is

$$P_0 = -2\int_0^a \sigma_{yy}(x,0)dx,$$

or

$$C_0 = -\frac{P_0}{2} \tag{6.67}$$

The results are identical to those obtained by Sadowsky (1928).

The second example to be discussed is an antiplane problem of a homogeneous isotropic elastic half plane. The solid region is denoted by $-\infty < x < \infty$, $0 \le y < \infty$. A traction is applied on the half plane surface such that

$$\frac{dw(x,0)}{dx} = k, \qquad (|x| \le a) \tag{6.68a}$$

$$\sigma_{yz}(x,0) = 0, \qquad (|x| > a) \tag{6.68b}$$

Since the loading is antisymmetric with respect to the y-axis, the solution is given by Eqs. 6.39 to 6.41. Since the solution must be bounded as y tends to infinity, B in those equations must be set to zero, thus

$$\tilde{w}(\xi, y) = A e^{-\xi y},$$

$$\tilde{\tau}(\xi, 0) = -\mu \xi A.$$

Using these results in boundary conditions, Eqs. 6.68, we get

$$\int_0^\infty \tilde{\tau}(\xi, 0) \cos \xi x \, d\xi = -\frac{\pi k \mu}{2}, \qquad (0 \le x < a) \tag{6.69a}$$

$$\int_0^\infty \tilde{\tau}(\xi, 0) \sin \xi x \, d\xi = 0, \qquad (a < x < \infty) \tag{6.69b}$$

These dual integral equations have the solution

$$\tilde{\tau}(\xi, 0) = -\frac{\pi k a \mu J_1(\xi a)}{2} \tag{6.70}$$

which when substituted into Eq. 6.39b for $y = 0$ yields the contact shear stress

$$\sigma_{yz}(x, 0) = \frac{-\mu k x}{(a^2 - x^2)^{1/2}}, \qquad (0 \le x < a) \tag{6.71a}$$

$$= 0, \qquad (a < x < \infty) \tag{6.71b}$$

Note that the contact shear stress is singular at $x = a$.

The third and last example to be discussed in this section is a three dimensional problem of a homogeneous isotropic elastic half space. The solid domain is denoted by $0 \le r < \infty$, $0 \le z < \infty$. A normal indentation is made on the surface of the half space by means of a smooth rigid circular plate of radius a. Thus boundary conditions are

$$u_z(r, \theta, 0) = w_0, \qquad (0 \le r \le a) \tag{6.72a}$$

$$\sigma_{zz}(r, \theta, 0) = 0, \qquad (a < r < \infty) \tag{6.72b}$$

$$\sigma_{zr}(r, \theta, 0) = 0, \qquad (0 \le r < \infty) \tag{6.72c}$$

where w_0 is the prescribed uniform indentation. This is an axisymmetric torsion-free problem, thus u_θ must vanish, and all functions are independent of θ. Its solution can be obtained from Eqs. 6.52 to 6.56 by setting $m = 0$ and $\tilde{\Psi}_m = 0$. Since the solution must be bounded as z tends to infinity, Eq. 6.51a simplifies to

$$\tilde{\Phi}(\xi, z) = (C + Dz) e^{-\xi z} \tag{6.73}$$

The last boundary condition, Eq. 6.72c, gives

$$\xi C - 2\nu D = 0 \tag{6.74a}$$

On the other hand, Eq. 6.53c gives

$$\xi^3 C + (1-2\nu)\xi^2 D = \frac{\tilde{\sigma}_{zz}(\xi,0)}{2\mu} \tag{6.74b}$$

where $\tilde{\sigma}_{zz}(\xi,0)$ is the Hankel transform of $\sigma_{zz}(r,\theta,0)$. Solving the last two equations above yields

$$C = \frac{\nu\tilde{\sigma}_{zz}(\xi,0)}{\mu\xi^3} \tag{6.75a}$$

$$D = \frac{\tilde{\sigma}_{zz}(\xi,0)}{2\mu\xi^2} \tag{6.75b}$$

Using these results in the remaining boundary conditions, Eqs. 6.72a and b, we get respectively

$$\int_0^\infty \tilde{\sigma}_{zz}(\xi,0)J_0(\xi r)\mathrm{d}\xi = -\frac{\mu w_0}{1-\nu}, \qquad (0 \le r \le a) \tag{6.76a}$$

$$\int_0^\infty \tilde{\sigma}_{zz}(\xi,0)\xi J_0(\xi r)\mathrm{d}\xi = 0, \qquad (r > a) \tag{6.76b}$$

The solution of these dual integral equations is

$$\tilde{\sigma}_{zz}(\xi,0) = -\frac{2\mu w_0 \sin\xi a}{\pi(1-\nu)\xi} \tag{6.77}$$

which when substituted into Eqs. 6.53c for $z=0$ yields the normal contact stress

$$\sigma_{zz}(r,\theta,0) = -\frac{2\mu w_0}{\pi(1-\nu)(a^2-r^2)^{1/2}}, \qquad (0 \le r < a) \tag{6.78a}$$

$$= 0, \qquad (r > a) \tag{6.78b}$$

Again note that the normal contact stress is singular at the edge of the rigid plate. The total applied force P_0 to produce the indentation w_0 is

$$P_0 = -\int_0^a 2\pi r\sigma_{zz}(r,\theta,0)\mathrm{d}r$$

$$= \frac{4\mu w_0}{(1-\nu)}\int_0^a \frac{r\mathrm{d}r}{(a^2-r^2)},$$

or

$$P_0 = \frac{4\mu a w_0}{(1-\nu)} \tag{6.79}$$

The results are identical to those obtained by Sneddon (1951). Note that for the half space it is possible to obtain the 'stiffness equation', Eq. 6.79, linking P_0 and w_0. It was impossible to find such a result for the half plane because it had an arbitrary rigid-body displacement.

6.5 CONCENTRATED FORCE INSIDE A HALF SPACE

Mindlin (1936) was the first to study the problem of a force at a point in the interior of a homogeneous elastic half space. The solution was obtained by the method of synthesis and superposition. It requires ingeneous guesses of proper potentials. Mindlin (1953) reconsidered the same problem and solved it by means of Papkovich potentials. The same potentials were applied by Rongved (1955) to find the solution for the force in the interior of two joined semi-infinite solids. In fact, these solutions can be obtained by using the formulation described in Section 6.3. The present section follows the work by Chan, Karasudhi and Lee (1974) who solved the problem of a force at a point in the interior of a layered half space, as shown in Fig. 6.1. Chan *et al* considered four cases, relating to a concentrated force applied either vertically or horizontally in the interior of the upper layer or in the underlying half space. Only the case of a horizontal force in the upper layer will be presented here to illustrate the use of the formulation described in Section 6.3.

The force P acting in the x-direction at a depth $z = z'$ below the free surface produces a traction on the horizontal plane at that depth given by

$$q(r) = \frac{P\delta(r-\varepsilon)}{2\pi r}, \quad (\varepsilon \to 0) \tag{6.80}$$

where $\delta(r)$ is the Dirac-delta function described in Section 1.9, and can now be represented by a Hankel transform

$$q(r) = \int_0^\infty \xi \bar{q}(\xi) J_0(\xi r) \mathrm{d}\xi \tag{6.81a}$$

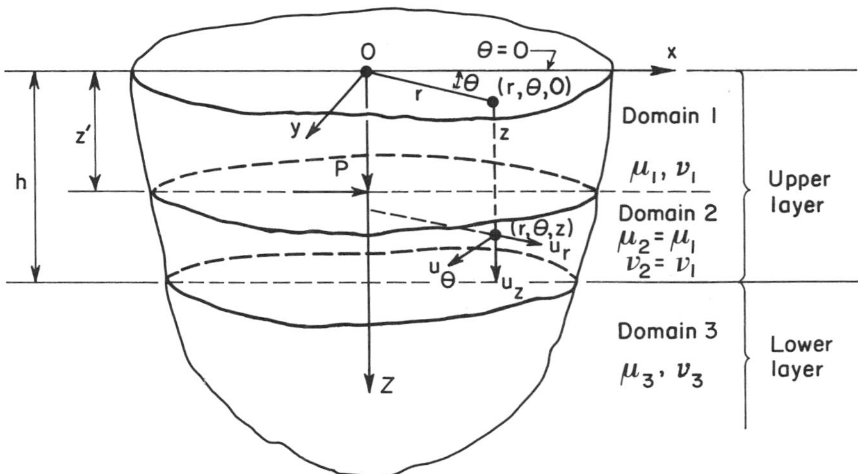

Fig. 6.1 A concentrated force in a layered half space.

or, due to Eq. 1.112c,

$$q(r)=\int_0^\infty \xi\bar{q}(\xi)\left[\frac{J_1(\xi r)}{\xi r}+\frac{\mathrm{d}J_1(\xi r)}{\mathrm{d}(\xi r)}\right]\mathrm{d}\xi \tag{6.81b}$$

where

$$\bar{q}(\xi)=\int_0^\infty rq(r)J_0(\xi r)\mathrm{d}r=\frac{P}{2\pi} \tag{6.82}$$

Imagine a horizontal plane passing through the point of application of the force, and dividing the material system into three domains; $0<z<z'$ for domain 1, $z'<z<h$ for domain 2, and $h<z<\infty$ for domain 3. Domains 1 and 2 have the same elastic properties. For the boundedness of the solution as z tends to infinity, coefficients associated with $e^{\xi z}$ in Eqs. 6.51 must be set to zero for the domain 3, i.e. $A_{m3}=B_{m3}=E_{m3}=0$. The free boundary conditions at the plane $z=0$ are

$$\sigma_{zz1}(r,\theta,0)=0,\quad \sigma_{zr1}(r,\theta,0)=0,\quad \sigma_{z\theta1}(r,\theta,0)=0 \tag{6.83a,b,c}$$

The continuity conditions at the plane $z=z'$ are

$$u_{r1}(r,\theta,z')=u_{r2}(r,\theta,z') \tag{6.83d}$$

$$u_{\theta1}(r,\theta,z')=u_{\theta2}(r,\theta,z') \tag{6.83e}$$

$$u_{z1}(r,\theta,z')=u_{z2}(r,\theta,z') \tag{6.83f}$$

$$\sigma_{zz1}(r,\theta,z')=\sigma_{zz2}(r,\theta,z') \tag{6.83g}$$

$$\sigma_{z\theta1}(r,\theta,z')-\sigma_{z\theta2}(r,\theta,z')=-q(r)\sin\theta \tag{6.83h}$$

$$\sigma_{zr1}(r,\theta,z')-\sigma_{zr2}(r,\theta,z')=q(r)\cos\theta \tag{6.83i}$$

The interface continuity conditions at the plane $z=h$ are

$$u_{r2}(r,\theta,h)=u_{r3}(r,\theta,h) \tag{6.83j}$$

$$u_{\theta2}(r,\theta,h)=u_{\theta3}(r,\theta,h) \tag{6.83k}$$

$$u_{z2}(r,\theta,h)=u_{z3}(r,\theta,h) \tag{6.83l}$$

$$\sigma_{zz2}(r,\theta,h)=\sigma_{zz3}(r,\theta,h) \tag{6.83m}$$

$$\sigma_{z\theta2}(r,\theta,h)=\sigma_{z\theta3}(r,\theta,h) \tag{6.83n}$$

$$\sigma_{zr2}(r,\theta,h)=\sigma_{zr3}(r,\theta,h) \tag{6.83o}$$

The only nonhomogeneous conditions, Eqs. 6.83h and 6.83i, indicate that the solution involves only the harmonic term $m=1$. For simplicity, the subscript m will be omitted. Substituting Eqs. 6.52 to 6.56 into Eqs. 6.83 yields a set of fifteen equations for the fifteen coefficients, A_1 to F_1, A_2 to F_2 and C_3, D_3 and F_3. Introducing the dimensionless terms

$$\alpha = \xi h, \quad \zeta = z/h, \quad \rho = r/h, \quad \beta = z'/h \tag{6.84a,b,c,d}$$

we obtain the displacement components in the dimensionless forms

$$\frac{\mu_1 h}{P\cos\theta} u_{rn}(\rho,\theta,\zeta) = \frac{1}{2}[U_n(\rho,\zeta) + V_n(\rho,\zeta)] \tag{6.85a}$$

$$\frac{\mu_1 h}{P\sin\theta} u_{\theta n}(\rho,\theta,\zeta) = \frac{1}{2}[U_n(\rho,\zeta) - V_n(\rho,\zeta)] \tag{6.85b}$$

$$\frac{\mu_1 h}{P\cos\theta} u_{zn}(\rho,\theta,\zeta) = \int_0^\infty \left[w_n^*(\alpha,\zeta) + \frac{\overline{w}_n(\alpha,\zeta)}{D(\alpha)} \right] J_1(\rho\alpha)\mathrm{d}\alpha \tag{6.85c}$$

where the subscript n denotes the domain number 1, 2 or 3, and

$$U_n(\rho,\zeta) = \int_0^\infty \left\{ [U_n^*(\alpha,\zeta) + V_n^*(\alpha,\zeta)] + \left[\frac{\overline{U}_n(\alpha,\zeta)}{D(\alpha)} + \frac{\overline{V}_n(\alpha,\zeta)}{H(\alpha)} \right] \right\} J_2(\rho\alpha)\mathrm{d}\alpha \tag{6.86a}$$

$$V_n(\rho,\zeta) = \int_0^\infty \left\{ [-U_n^*(\alpha,\zeta) + V_n^*(\alpha,\zeta)] + \left[\frac{-\overline{U}_n(\alpha,\zeta)}{D(\alpha)} + \frac{\overline{V}_n(\alpha,\zeta)}{H(\alpha)} \right] \right\} J_0(\rho\alpha)\mathrm{d}\alpha \tag{6.86b}$$

$$D(\alpha) = 1 - (a + b + 4b\alpha^2)\mathrm{e}^{-2\alpha} + ab\,\mathrm{e}^{-4\alpha} \tag{6.87a}$$

$$H(\alpha) = 1 - \frac{1-\mu_0}{1+\mu_0}\mathrm{e}^{-2\alpha} \tag{6.87b}$$

$$a = \frac{(3-4\nu_3) - \mu_0(3-4\nu_1)}{3-4\nu_3+\mu_0} \tag{6.88a}$$

$$b = \frac{(1-\mu_0)}{1+\mu_0(3-4\nu_1)} \tag{6.88b}$$

$$\mu_0 = \frac{\mu_3}{\mu_1} \tag{6.88c}$$

For $n = 1$ and 2,

$$U_n^*(\alpha,\zeta) = \frac{1}{16\pi\gamma_0}[-\gamma_1 + (\beta-\zeta)\alpha]\mathrm{e}^{-|\beta-\zeta|\alpha} \tag{6.89a}$$

$$V_n^*(\alpha,\zeta) = \frac{1}{4\pi}\mathrm{e}^{-|\beta-\zeta|\alpha} \tag{6.89b}$$

$$w_n^*(\alpha,\zeta) = \frac{-1}{16\pi\gamma_0}(\beta-\zeta)\alpha\mathrm{e}^{-|\beta-\zeta|\alpha} \tag{6.89c}$$

$$
\begin{aligned}
\overline{U}_n(\alpha,\zeta) = \frac{1}{16\pi\gamma_0}\Bigg(& [-(4\gamma_0\gamma_2+1)+\gamma_1(\beta+\zeta)\alpha-2\beta\zeta\alpha^2]\,e^{-(\beta+\zeta)\alpha} \\
& +\left[\frac{1}{2}(a+b\gamma_1^2)+b\gamma_1(2-\beta-\zeta)\alpha+2b(1-\beta)(1-\zeta)\alpha^2\right]e^{-(2+\beta+\zeta)\alpha} \\
& +ab[\gamma_1+(\beta-\zeta)\alpha]\,e^{-(4-\beta+\zeta)\alpha}+\Bigg\{-\frac{\gamma_1}{2}(a+b)+[a\zeta+b(8\gamma_0\gamma_1 \\
& +2-\beta)]\alpha-2b\gamma_1(1-\beta+\zeta)\alpha^2+4b(1-\beta)\zeta\alpha^3\Bigg\}\ e^{-(2-\beta+\zeta)\alpha} \\
& +ab[\gamma_1-(\beta-\zeta)\alpha]\,e^{-(4+\beta-\zeta)\alpha}+\Bigg\{-\frac{\gamma_1}{2}(a+b)+[a\beta+2b(4\gamma_0\gamma_2 \\
& +1)-b\zeta]\alpha-2b\gamma_1(1+\beta-\zeta)\alpha^2+4b\beta(1-\zeta)\alpha^3\Bigg\}\ e^{-(2+\beta-\zeta)\alpha} \\
& +ab[(4\gamma_0\gamma_2+1)+\gamma_1(\beta+\zeta)\alpha+2\beta\zeta\alpha^2]\,e^{-(4-\beta-\zeta)\alpha}+\Bigg[-\frac{1}{2}(a+b\gamma_1^2) \\
& +b\gamma_1(2-\beta-\zeta)\alpha-2b(1-\beta)(1-\zeta)\alpha^2\Bigg]e^{-(2-\beta-\zeta)\alpha}\Bigg)
\end{aligned}
\tag{6.90a}
$$

$$
\overline{V}_n(\alpha,\zeta) = \frac{1}{4\pi}\left\{e^{-(\beta+\zeta)\alpha}+\frac{1-\mu_0}{1+\mu_0}[e^{-(2-\beta+\zeta)\alpha}+e^{-(2+\beta-\zeta)\alpha}+e^{-(2-\beta-\zeta)\alpha}]\right\}
\tag{6.90b}
$$

$$
\begin{aligned}
\overline{w}_n(\alpha,\zeta) = \frac{1}{16\pi\gamma_0}\Bigg(& [4\gamma_0\gamma_2-\gamma_1(\beta-\zeta)\alpha-2\beta\zeta\alpha^2]\,e^{-(\beta+\zeta)\alpha} \\
& +\left[\frac{1}{2}(a-b\gamma_1^2)+b\gamma_1(\beta-\zeta)\alpha+2b(1-\beta)(1-\zeta)\alpha^2\right]e^{-(2+\beta+\zeta)\alpha} \\
& +ab(\beta-\zeta)\alpha e^{-(4-\beta+\zeta)\alpha}+\Bigg\{\frac{\gamma_2}{2}(a-b)+[-b(\beta+8\gamma_0\gamma_2)+a\zeta]\alpha \\
& +2b\gamma_1(1-\beta-\zeta)\alpha^2+4b(1-\beta)\zeta\alpha^3\Bigg\}\ e^{-(2-\beta+\zeta)\alpha} \\
& +ab(\beta-\zeta)\alpha e^{-(4+\beta-\zeta)\alpha}+\Bigg[\frac{\gamma_1}{2}(a-b)+(b\zeta-a\beta+8\gamma_0\gamma_2)\alpha \\
& +2b\gamma_1(1-\beta-\zeta)\alpha^2-4b\beta(1-\zeta)\alpha^3\Bigg]\ e^{-(2+\beta-\zeta)\alpha}
\end{aligned}
$$

$$+ab[4\gamma_0\gamma_2+\gamma_1(\beta-\zeta)\alpha-2\beta\zeta\alpha^2]e^{-(4-\beta-\zeta)\alpha}$$

$$+\left[\frac{1}{2}(a-b\gamma_1^2)-b\gamma_1(\beta-\zeta)\alpha+2b(1-\beta)(1-\zeta)\alpha^2\right]e^{-(2-\beta-\zeta)\alpha}\Bigg) \qquad (6.90c)$$

and for $n=3$,

$$U_n^*(\alpha,\zeta)=V_n^*(\alpha,\zeta)=w_n^*(\alpha,\zeta)=0 \qquad (6.91a,b,c)$$

$$\overline{U}_n(\alpha,\zeta)=\frac{1}{8\pi}\Bigg(\{-(c+d\gamma_1\gamma_3)+2[\gamma_1(c-d+d\zeta)+d\beta\gamma_3]\alpha$$

$$-4\beta(c-d+d\zeta)\alpha^2\}e^{-(\zeta+\beta)\alpha}+\{(ac-bd\gamma_1\gamma_3)$$

$$+2bd[\gamma_1(1-\zeta)+\gamma_3(1-\beta)]\alpha+4bd(1-\beta)(1-\zeta)\alpha^2\}e^{-(2+\zeta+\beta)\alpha}$$

$$+\{(-c\gamma_1+d\gamma_3)+2[c(1-\beta)-d(1-\zeta)]\alpha\}e^{-(\zeta-\beta)\alpha}$$

$$+\{(ac\gamma_1-bd\gamma_3)+2[ac\beta+bd(\gamma_1\gamma_3+1-\zeta)]\alpha$$

$$+4bd[\gamma_1(1-\zeta)+\gamma_3\beta]\alpha^2+8bd\beta(1-\zeta)\alpha^3\}e^{-(2+\zeta-\beta)\alpha}\Bigg) \qquad (6.92a)$$

$$\overline{V}_n(\alpha,\zeta)=\frac{\mu_0}{2\pi(1+\mu_0)}[e^{-(\zeta+\beta)\alpha}+e^{-(\zeta-\beta)\alpha}] \qquad (6.92b)$$

$$\overline{w}_n(\alpha,\zeta)=\frac{1}{8\pi}\Bigg(\{-c+d\gamma_1\gamma_3+2[\gamma_1(c-d+d\zeta)-d\beta\gamma_3]\alpha$$

$$-4\beta(c-d+d\zeta)\alpha^2\}e^{-(\zeta+\beta)\alpha}+\{(ac-bd\gamma_1\gamma_3)$$

$$+2bd[\gamma_1(1-\zeta)-\gamma_3(1-\beta)]\alpha+4bd(1-\beta)(1-\zeta)\alpha^2\}e^{-(2+\zeta+\beta)\alpha}$$

$$+\{(-c\gamma_1+d\gamma_3)+2[c(1-\beta)-d(1-\zeta)]\alpha\}e^{-(\zeta-\beta)\alpha}$$

$$+\{(ac\gamma_1-bd\gamma_3)+2[ac\beta-bd(\gamma_1\gamma_3-1+\zeta)]\alpha$$

$$+4bd[\gamma_1(1-\zeta)-\gamma_3\beta]\alpha^2+8bd\beta(1-\zeta)\alpha^3\}e^{-(2+\zeta-\beta)\alpha}\Bigg) \qquad (6.92c)$$

In Eqs. 6.89 to 6.92,

$$\gamma_0=1-\nu_1,\quad \gamma_1=3-4\nu_1,\quad \gamma_2=1-2\nu_1,\quad \gamma_3=3-4\nu_3 \qquad (6.93a,b,c,d)$$

$$c=\frac{1}{1+\gamma_1\mu_0},\quad d=\frac{1}{\gamma_3+\mu_0} \qquad (6.93e,f)$$

Note that the solution functions take the same form for $n=1$ and 2. Thus the solution is given in the form of a two-domain problem. The integrals given by $U_n^*(\alpha,\zeta)$,

$V_n^*(\alpha,\zeta)$ and $w_n^*(\alpha,\zeta)$ correspond to the singularity of Mindlin's solution (1936), while the integrals given by $\overline{U}_n(\alpha,\zeta)$, $\overline{V}_n(\alpha,\zeta)$ and $\overline{w}_n(\alpha,\zeta)$ are nonsingular. For a homogeneous half space, $D(\alpha)$ and $H(\alpha)$ (Eqs. 6.85c, 8.86 and 8.87) are equal to unity, and the solution is Mindlin's solution (Problem 6.9.5 of the present chapter). To avoid laborious numerical integration in Eqs. 6.85c and 6.86, the reciprocals of $D(\alpha)$ and $H(\alpha)$ for the layered half space will be approximated in such a way that the integrals assume the standard forms.

Since $D(\alpha)$ is positive for all values of α from zero to infinity and asymptotic to unity as α approaches infinity, it is reasonable to approximate $1/D(\alpha)$ by a series of exponential functions of the form

$$\frac{1}{D(\alpha)} \cong 1 + R(\alpha) \tag{6.94}$$

where

$$R(\alpha) = \sum_{j=1}^{s} k_j e^{-p_j \alpha} \tag{6.95}$$

$$k_1 = \frac{a+b-ab}{1-a-b+ab} - \sum_{j=2}^{s} k_j \tag{6.96}$$

s is a positive integer, k_j are constants, and p_j are real numbers greater than zero. The approximate function proposed in Eq. 6.94 is smooth and continuous and asymptotically approaches exact values as $\alpha \to \infty$ and at $\alpha = 0$. For a given set of s and p_j we may find the constants k_j for $j = 2$ to s by defining

$$\varepsilon(\alpha) = 1 - D(\alpha)[1 + R(\alpha)] \tag{6.97}$$

and choosing the k_j so that

$$\int_0^\infty [\varepsilon(\alpha)]^2 d\alpha = \text{minimum} \tag{6.98}$$

Following the same procedure used in approximating $1/D(\alpha)$, $1/H(\alpha)$ is approximated by

$$\frac{1}{H(\alpha)} \cong 1 + R'(\alpha) \tag{6.99}$$

where

$$R'(\alpha) = \sum_{j=1}^{s'} k_j' e^{-p_j' \alpha} \tag{6.100}$$

$$k_1' = \frac{1-\mu_0}{2\mu_0} - \sum_{j=2}^{s'} k_j' \tag{6.101}$$

Since the approximations of $1/D(\alpha)$ and $1/H(\alpha)$ yield exact values at $\alpha = 0$, the integration of shear traction in the x-direction on a horizontal plane is equal to zero when the plane is above the point of application of the force P, and equal to P when the plane is below. The approximate solutions for each layer given by Eqs. 6.85 can be expressed in the general form

$$I = I^* + I^c \tag{6.102}$$

where

$$I^* = \sum_n c_n \int_0^\infty \alpha^n e^{-|\beta-\zeta|} J_m(\rho\alpha) d\alpha \tag{6.103a}$$

$$I^c = \sum_l \sum_n c_{ln} \int_0^\infty \alpha^n e^{-f_l \alpha} J_m(\rho\alpha) d\alpha \tag{6.103b}$$

where the indices l and n are either zero or positive integers, the coefficients c_n and c_{ln} are point functions and f_l is a positive point function. Thus the solution involves infinite integrals of a standard form

$$G(m,n,p) = \int_0^\infty \alpha^n e^{-p\alpha} J_m(\rho\alpha) d\alpha, \qquad (p \geq 0) \tag{6.104}$$

which can be evaluated in closed forms. The integral I^* has a singularity at the point of application of the concentrated force, corresponding to Mindlin's solution; the integral I^c is non-singular. Thus the proposed solution is in a closed form. In the limiting case of a homogeneous half space, $\mu_0 = 1$, $\nu_1 = \nu_3$ and $k_j = k_j' = 0$ for every j, and the integral I^c simplifies into the non-singular part of the Mindlin's solution.

Chan *et al* (1974) solved other plane strain and plane stress problems in which the applied force is a line force uniformly distributed on the line parallel to the y-axis and through the point $x = 0$, $y = 0$, $z = c$. The solution can be obtained from that for a concentrated force P, by performing an appropriate integration with respect to y from $y = -\infty$ to $y = \infty$.

6.6 LOAD TRANSFER PROBLEMS

Apirathvorakij and Karasudhi (1980), Niumpradit and Karasudhi (1981), Karasudhi, Rajapakse and Hwang (1984) and Karasudhi, Rajapakse and Liyanage (1984) have solved several problems of long cylindrical elastic bars partially embedded in elastic half spaces. The problem of torsion of a long bar embedded in a layered half space, which has been solved by Karasudhi, Rajapakse and Hwang (1984), will be presented here to illustrate a solution technique. Following the scheme proposed by Muki and Sternberg (1970) for the treatment of an axial load transfer problem, the system of the present study depicted in Fig. 6.2 is decomposed into an extended half space and a fictitious bar as shown in Figs. 6.3(a) and (b). Based on an appropriate compatibility condition between the two latter systems the solution is derived from a *Fredholm integral equation of the second kind* (Sneddon, 1966). The torque transfer and the torque-twist relationship can be obtained numerically for various slenderness ratios of the bar and ratios of shear moduli.

It should be recalled that the nontrivial displacement and stress components in a torsion problem such as this are given by Eqs. 6.58 and 6.59. For reasons of convenience, it is appropriate to non-dimensionalize all functions and variables with length dimensions by a, which is the radius of the embedded bar.

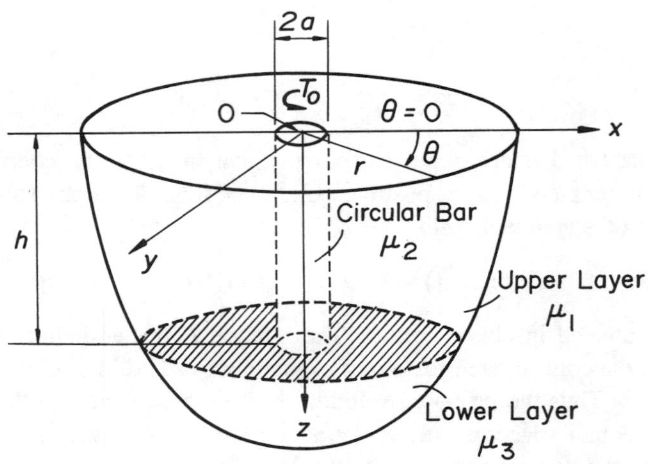

Fig. 6.2 Geometry of bar and embedding medium in torsion problem.

Before solving the main problem, one may wonder what is the *factor of stress singularity* at the base perimeter of the bar. It has been found by Luco (1976), in the problem of torsion of a rigid cylindrical bar partially embedded in a layered elastic half space, as shown in Fig. 6.2, that the shear stresses at the base perimeter of the bar are in the following forms

$$\sigma_{\theta z}(r,h) \sim r(1-r^2)^{k-0.5} \tag{6.105a}$$

$$\sigma_{r\theta}(r,h) \sim \left[1-\left(\frac{z}{h}\right)^2\right]^{k-0.5} \tag{6.105b}$$

where k is a stress singularity factor in the range $0 \le k < 0.5$ and h is the nondimensionalized length of the bar. For the current problem of an elastic bar, the factor k can be taken as a function of β_1 and β_2 where

$$\beta_1 = \frac{\mu_1}{\mu_2}, \quad \beta_2 = \frac{\mu_1}{\mu_3} \tag{6.106}$$

in which μ_1 is the shear modulus of the upper layer, μ_2 of the bar, and μ_3 of the underlying half space. The method developed by Williams (1952) can be applied to estimate $k(\beta_1, \beta_2)$ by considering an analogous antiplane system, as shown in Fig. 6.4, consisting of two *quarter planes* with shear moduli μ_1 and μ_2 and a half plane with a

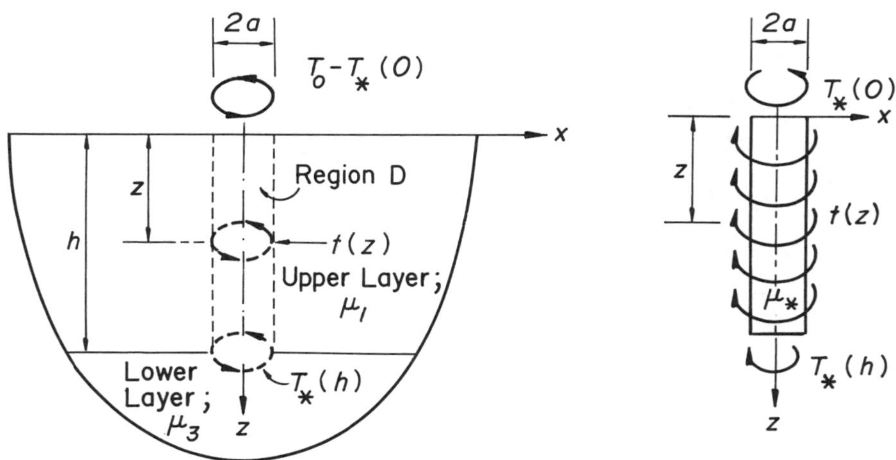

(a) Extended half space B. (b) Fictitious bar B_*.

Fig. 6.3 Composition of the torsion problem.

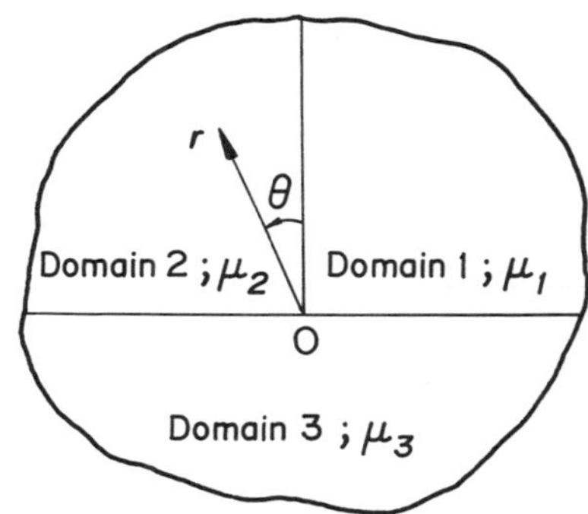

Fig. 6.4 Geometry of analogous system for estimating stress singularity factor.

shear modulus μ_3. The equilibrium equation is given by Eq. 6.9 with zero body force, i.e. $\nabla^2 w = 0$. The solution for each constituent plane for small r can be put in the form

$$w_i(r,\theta) = r^{\gamma}[A_i \sin(\gamma\theta) + B_i \cos(\gamma\theta)] \tag{6.107}$$

where

$$\gamma = k + 0.5 \tag{6.108}$$

i denotes a domain 1, 2 or 3 in the material system (Fig. 6.4), and A_i and B_i are arbitrary constants. It should be noted that while displacements $w_i(r,\theta)$ are bounded as r tends to zero, stresses $\sigma_{\theta zi}(r,\theta)$ and $\sigma_{zri}(r,\theta)$ are singular. Continuity conditions appropriate for the present antiplane case are

$$w_1(r,0) = w_2(r,0), \qquad \sigma_{\theta z1}(r,0) = \sigma_{\theta z2}(r,0) \tag{6.109a,b}$$

$$w_2(r,\pi/2) = w_3(r,\pi/2), \qquad \sigma_{\theta z2}(r,\pi/2) = \sigma_{\theta z3}(r,\pi/2) \tag{6.109c,d}$$

$$w_3(r,3\pi/2) = w_1(r,3\pi/2), \quad \sigma_{\theta z3}(r,3\pi/2) = \sigma_{\theta z1}(r,3\pi/2) \tag{6.109e,f}$$

Substituting Eq. 6.107 into Eqs. 6.109 leads to a set of six homogeneous equations, i.e. an eigenvalue problem in which $k(\beta_1,\beta_2)$ is the smallest positive real eigenvalue and less than 0.5. In general, a trial-and-error process has to be adopted. For some values of β_1 and β_2, the stress singularity does not even exist. For the case of an infinitely rigid bar ($\beta_1 = 0$), it can be found that

$$k(0,\beta_2) = \pi^{-1}\arcsin[\beta_2/(1+\beta_2)], \qquad (0 \le \beta_2 < 1) \tag{6.110a}$$

$$= \frac{1}{6}, \qquad (\beta_2 \ge 1) \tag{6.110b}$$

For a homogeneous half space ($\beta_2 = 1$) and an elastic bar, the stress singularity factor is found to be the same as that of an infinitely rigid bar embedded in a layered half space with $\beta_2 \ge 1$, i.e.

$$k(\beta_1, 1) = \frac{1}{6} \tag{6.111}$$

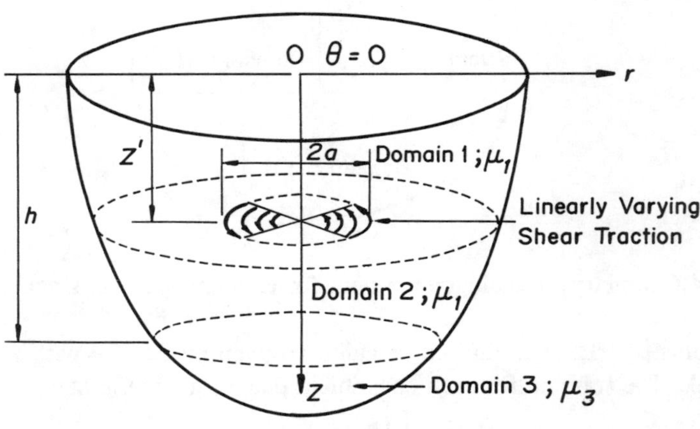

Fig. 6.5 Unit torque in the interior of layered half space.

Now, it is necessary to consider the case of a shear traction equivalent to a unit torque acting over a circular area of radius a in the interior of the half space at $z = z'$ as shown in Fig. 6.5. This auxiliary solution can be found by considering the half space to be divided into three domains, namely, domain 1 ($0 \le z \le z'$), domain 2 ($z' \le z \le h$) and domain 3 ($h \le z < \infty$) as shown. The solution for each domain is given by Eq. 6.58, with two constants of integration A_n and C_n. The subscript n is used to denote a domain number. The constant A_3 must vanish in order to guarantee the boundedness of the solution as z tends to infinity. Thus there are five constants; A_1, C_1, A_2, C_2 and C_3. The boundary and continuity conditions are:

$$\sigma_{\theta z1}(r,0) = 0 \tag{6.112a}$$

$$u_{\theta 1}(r,z') = u_{\theta 2}(r,z') \tag{6.112b}$$

$$u_{\theta 2}(r,h) = u_{\theta 3}(r,h) \tag{6.112c}$$

$$\sigma_{\theta z2}(r,h) = \sigma_{\theta z3}(r,h) \tag{6.112d}$$

$$\sigma_{z\theta 1}(r,z') - \sigma_{z\theta 2}(r,z') = -\frac{2r}{\pi}, \quad (r<1) \tag{6.112e}$$

$$= 0, \qquad (r>1) \tag{6.112f}$$

or

$$\sigma_{z\theta 1}(r,z') - \sigma_{z\theta 2}(r,z') = -\frac{2}{\pi}\int_0^\infty J_1(\xi r)J_2(\xi)\mathrm{d}\xi, \quad (0 \le r < \infty) \tag{6.112g}$$

Equations 6.112a to d and g can be used to determine the five constants of integration A_1, C_1, A_2, C_2 and C_3. However, to incorporate the notion of the stress singularity of the stress field in the form of Eqs. 6.105 in the main problem, the applied shear traction in the fundamental solution for $z' = h$ should be, in place of the linear Eqs. 6.112e and f, in the form

$$\sigma_{\theta z1}(r,h) - \sigma_{\theta z3}(r,h) = -\frac{2}{\pi}\alpha_1(k) r (1-r^2)^{k-0.5}, \quad (r<1) \tag{6.113a}$$

$$= 0, \quad (r>1) \tag{6.113b}$$

or

$$\sigma_{\theta z1}(r,h) - \sigma_{\theta z3}(r,h) = -\frac{2}{\pi}\alpha_2(k)\int_0^\infty \xi^{0.5-k} J_1(\xi r) J_{k+1.5}(\xi)\mathrm{d}\xi, \quad (0 \le r < \infty) \tag{6.113c}$$

where

$$\alpha_1(k) = \frac{1}{2}(k+0.5)(k+1.5) \tag{6.114a}$$

$$\alpha_2(k) = 2^{k+0.5}\alpha_1(k)\Gamma(k+0.5) \tag{6.114b}$$

in which Γ is a gamma function. In addition, other boundary conditions for the fundamental solution in case $z' = h$ are

$$\sigma_{\theta z1}(r,0) = 0 \tag{6.115a}$$

$$u_{\theta 1}(r,h) = u_{\theta 3}(r,h) \tag{6.115b}$$

Equations 6.113c and 6.115a and b can be used to determine the three constants of integration A_1, C_1 and C_3.

The problem of axially symmetric torsion of an elastic cylindrical bar partially embedded in a layered elastic half space is depicted in Fig. 6.2. In addition to those previously defined, the following symbols are introduced; T_0 the torque applied at the top end of the bar which is flush with the surface of the half space; and x, y and z Cartesian components of position vector $\mathbf{x}$; thus x_3 is identical to z. The top layer ($r > 1$, $0 < z < h$) and the underlying half space ($z > h$) are perfectly bonded on the contact surface ($r > 1$, $z = h$). The cylinder is bonded to the half space in the area ($r < 1$, $z = h$) and along the surface of the cylinder ($r = 1$, $0 \leq z \leq h$). Following the approximate scheme similarly used by Muki and Sternberg (1970) in the elastostatic axial load transfer, the system in Fig. 6.2 is decomposed into two systems; an extended half space B as shown in Fig. 6.3(a) and a fictitious bar B_* as shown in Fig. 6.3(b) with a shear moduli μ_* equal to the difference between the shear moduli of the real bar μ_2 and the half space μ_1, i.e.

$$\mu_* = \mu_2 - \mu_1 \geq 0 \tag{6.116}$$

The extended half space is subjected to a distributed torque $t(z)$ which is exerted by B_* on B at $x_3 = z$ in a region D in place of the bar. In addition, B is also subjected to end torques $T_0 - T_*(0)$ and $T_*(h)$ applied at the terminal cross sections as shown in Fig. 6.3(a). The bond torque $t(z)$ and the torque $T_0 - T_*(0)$ are assumed to be distributed linearly in the form of Eqs. 6.112e over their respective cross sections $A_z(0 < z < h)$ and A_0, while the torque $T_*(h)$ in the singular form of Eq. 6.113a over the cross section A_h. Conversely, the bond torque and end torques are exerted by the extended half space B on the fictitious bar B_*, which may be treated as a one-dimensional elastic continuum, for which the equation of equilibrium is

$$t(z) = -\frac{dT_*(z)}{dz} \tag{6.117}$$

The torque-twist relation of the fictitious bar is

$$T_*(z) = \frac{\pi\mu_*}{2}\frac{d\phi_*(z)}{dz} \tag{6.118}$$

where $T_*(z)$ and $\phi_*(z)$ are fictitious torque and angle of twist of the bar respectively.

The governing integral equation can be derived using the following compatibility condition between B_* and the edge of the region D:

$$\frac{d\phi_*(z)}{dz} = \frac{du_\theta(1^-, z)}{dz}, \quad (0 \le z \le h) \tag{6.119}$$

Here $u_\theta(r,z)$ is the displacement in the θ-direction at a point $\mathbf{x}(r,\theta,z)$ in domain B, and 1^- denotes a value infinitesimally less than unity. Muki and Sternberg (1970) originally demanded compatibility between B_* and the corresponding average over a cross section of region D. Karasudhi, Rajapakse and Hwang (1984) found that this was not sufficiently accurate. The function $u_\theta(r,z)$ can be expressed in the form

$$u_\theta(r,z) = [T_0 - T_*(0)]\, u_{\theta T}(r,z;0) + T_*(h) u_{\theta T}(r,z;h) + \int_0^h u_{\theta T}(r,z;z')t(z')dz' \tag{6.120}$$

where $u_{\theta T}(r,z;z')$ is the auxiliary solution, i.e. the displacement in the θ-direction at a point $\mathbf{x}(r,\theta,z)$ due to a unit torque applied at a depth z' (Fig. 6.5). Substitution of Eqs. 6.117, 6.118 and 6.120 in Eq. 6.119 results in,

$$\frac{2T_*(z)}{\pi\mu_*} = [T_0 - T_*(0)]\, \phi^0_{Tz}(z,0) + T_*(h)\phi^0_{Tz}(z,h) - \int_0^h \phi^0_{Tz}(z,z')\frac{dT(z')}{dz'}dz' \tag{6.121}$$

where $\phi^0_{Tz}(z,z')$ is an influence function defined as

$$\phi^0_{Tz}(z,z') = \frac{\partial u_{\theta T}(1^-, z; z')}{\partial z} \tag{6.122}$$

It should be noted that this influence function is smooth and continuous everywhere except at $z = z'$, where the magnitude of the discontinuity according to Eq. 6.112e is equal to $2/\pi\mu_1$ for $0 < z < h$. Incorporation of the stress singularity at the end of the bar, $\phi^0_{Tz}(z,z')$ when $z = z' = h$ requires appropriate handling in the numerical solution scheme.

Integrating the integrals in Eq. 6.121 by parts, while taking proper account of the discontinuity, we find

$$\frac{2T_*(z)}{\pi}\left(\frac{1}{\mu_*} + \frac{1}{\mu_1}\right) - \int_0^h T_*(z')\frac{\partial\phi^0_{Tz}(z,z')}{\partial z'}dz' = T_0\phi^0_{Tz}(z,0) \tag{6.123}$$

Equation 6.123 is a Fredholm integral equation of the second kind governing the distribution of $T_*(z)$ along the fictitious bar.

The real bar torque $T(z)$ can be obtained by combining the fictitious torque $T_*(z)$ with the corresponding area integral of the shear stress $\sigma_{\theta z}$ in the region D of B, i.e.

$$T(z) = T_*(z) + \int_{A_z} r\sigma_{\theta z}(r,z)dA, \quad (0 \le z \le h) \tag{6.124}$$

The angle of twist of the real bar taken as equal to that of the fictitious bar, thus

$$\phi_*(z) = u_\theta(1,z), \quad (0 \le z \le h) \tag{6.125}$$

Using Eq. 6.120 and performing appropriate integrations in Eqs. 6.124 and 6.125, we obtain

$$T(z) = T_0\tau^0_{\theta z}(z,0) + \int_0^h T_*(z')\frac{\partial\tau^0_{\theta z}(z,z')}{\partial z'}dz' \tag{6.126}$$

$$\phi_*(z) = T_0\phi_T^0(z,0) + \int_0^h T_*(z')\frac{\partial\phi_T^0(z,z')}{\partial z'}dz' \tag{6.127}$$

where $\tau_{\theta z}^0(z,z')$ and $\phi_T^0(z,z')$ are influence functions defined as

$$\tau_{\theta z}^0(z,z') = \int_{A_z} r\sigma_{\theta zT}(r,z;z')dA \tag{6.128}$$

$$\phi_T^0(z,z') = u_{\theta T}(1,z;z') \tag{6.129}$$

Here $\sigma_{\theta zT}(r,z;z')$ is a shear stress in the auxiliary solution depicted in Fig. 6.5.

The Fredholm integral equation for $T_*(z)$ has an integrable logarithmic singularity at $z = z'$, and is solved numerically. We use a scheme which is similar to that employed by Hopkins and Hamming (1957) and Lee and Rogers (1963) in the solution of Volterra integral equations. Divide the interval of integration into n equal partitions and denote the mesh points by z_j; $j = 1, 2, \ldots, n+1$, with $z_1 = 0$ and $z_{n+1} = h$. Equation 6.123 becomes

$$\frac{2T_*(z_i)}{\pi}\left(\frac{1}{\mu_*}+\frac{1}{\mu_1}\right) - \sum_{j=1}^{n}\frac{1}{2}[T_*(z_j)+T_*(z_{j+1})]\,[\phi_{Tz}^0(z_i,z_{j+1})-\phi_{Tz}^0(z_i,z_j)] = T_0\phi_{Tz}^0(z_i,0) \tag{6.130}$$

which is a set of $n+1$ simultaneous equations, the solution of which gives the fictitious torque $T_*(z)$ at all mesh points.

In order to find the stress singularity at the base perimeter of the bar, we note that $\phi_{Tz}^0(h,h)$ contains a singular term of the form

$$\int_0^\infty \xi^{0.5-k}J_1(\xi r)J_{k+1.5}(\xi)d\xi \sim r(1-r^2)^{k-0.5}, \quad (r = 1^-) \tag{6.131}$$

Thus there exists in Eq. 6.130 an indeterminate term

$$T_*(h)\phi_{Tz}^0(h,h).$$

As z tends to h, we speculate that $T_*(h)$ approaches zero as

$$T_*(z) = \left[1-\left(\frac{z}{h}\right)^2\right]^{0.5+k}, \quad \left(\frac{z}{h}\to 1^-\right) \tag{6.132}$$

In the numerical solution to Eq. 6.130, the numerical value of 1^- in Eq. 6.131 is increased towards unity gradually until $T_*(h)$ is sufficiently close to zero and the results become stable. Moreover, the physical interpretation that the shear stress along the mantle of the cylindrical bar is proportional to the bond torque, i.e.

$$\sigma_{r\theta}(1,z) \sim t(z),$$

together with Eqs. 6.117 and 6.132 can help us to confirm the singular form of $\sigma_{r\theta}(1,z)$ as Eq. 6.105b.

Numerical results by the present approach were compared with those by Luco (1976), who had been more rigorous but restrictive to the case of rigid bars ($\beta_1 = 0$). The agreement between the two approaches was good and became better as the bar length increased. The significance of the singular stress field condition at the base perimeter of the bar was found to be small, especially when the main concern was on the relationship between the applied torque T_0 and the angle of twist $\phi_*(0)$. This type of situation is more pronounced in the problem of a flexible bar.

6.7 INFINITE ELEMENTS FOR MULTILAYERED HALF SPACES

Analytical methods such as that described in the preceding section become impractical if not impossible when extended to problems of multilayered half spaces. In order to bypass such difficulties, some research workers employ *finite element methods*, i.e. truncating the semi-infinite domain into a large but finite field which is then discretized into *finite elements*. However, such methods are computationally inefficient. As an example, Muki and Dong (1980) found that a finite element field of at least 750 degrees of freedom is necessary to obtain accurate results for a layered half space subjected to a uniform normal pressure on the surface of the half space.

The present section follows Rajapakse and Karasudhi (1985) who have shown that the computational efficiency can be achieved by properly using *infinite elements* in conjunction with finite elements. The whole material domain is separated into a *near field* and a *far field* as shown in Fig. 6.6. The near field, consisting of the partially embedded bar and a finite region of the half space around it, is discretized into conventional finite elements. The far field covering the rest of the half space is discretized

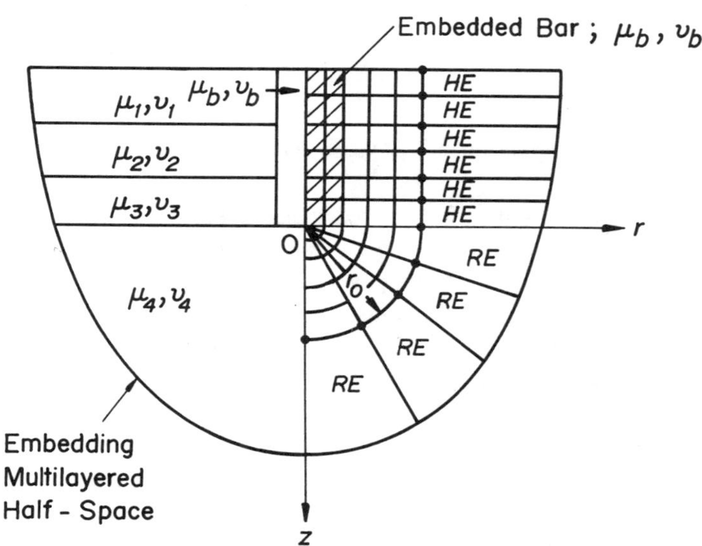

Fig. 6.6 Discretization of multilayered half space.

into infinite elements. Near the surface of the half space are horizontal infinite elements (HE), while the rest of the far field are occupied by radiating infinite elements (RE). Every infinite element has point nodes only on the interface between the near field and the far field. An obvious major advantage of this approach is that nonlinearity effects can be easily incorporated in the near field finite elements while all infinite elements always remain linearly elastic. We now describe the development of each type of infinite element.

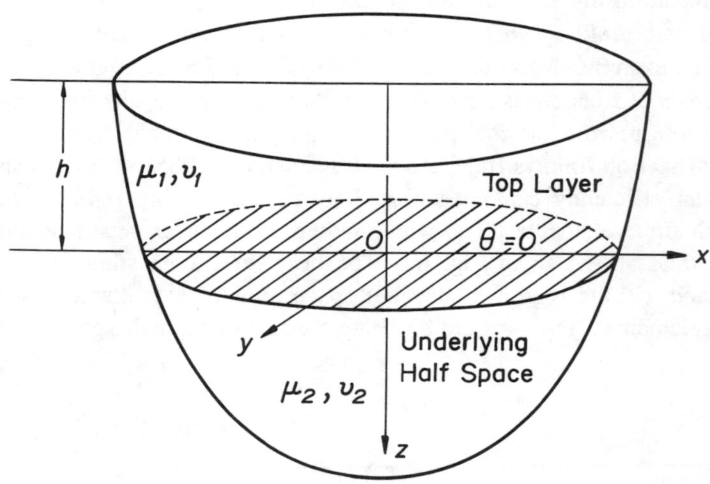

Fig. 6.7 Layered half space.

Rajapakse and Karasudhi (1985) based their development of infinite elements on the analytical solution of a layered half space, shown in Fig. 6.7, subjected to four types of loading on its surface. The loading types are shown in Fig. 6.8. The orders of magnitude of the displacement components at a large distance from the loads are shown in Table 6.1; in terms of r for the top layer, and in terms of R (spherical radial coordinate) for the underlying half space. Experience with the tedious algebra involved with layered half spaces tells us that an analytical derivation of far field displacement functions for a multilayered half space like Fig. 6.6 is rather impractical. However, we should expect that the far field displacement functions for horizontal infinite elements should be qualitatively the same as those for the layer in the table, and those for radiating infinite elements the same as those for the underlying half space in the table. It is important to note that the origin of the coordinate R should be at the intersection between the axis of symmetry and the last layer-half space interface.

Table 6.1

Types of loading	Top layer or half space	Displacements		
		u_r	u_θ	u_z
Torsion	Layer	–	$r^{-2}+0(r^{-3})$	–
(Fig. 6.8a)	Half space	–	$R^{-2}+0(R^{-3})$	–
Vertical	Layer	$r^{-1}+0(r^{-2})$	–	$r^{-1}+0(r^{-2})$
(Fig. 6.8b)	Half space	$R^{-1}+0(R^{-2})$	–	$R^{-1}+0(R^{-2})$
Horizontal	Layer	$\cos\theta[r^{-1}+0(r^{-2})]$	$\sin\theta[r^{-1}+0(r^{-2})]$	$\cos\theta[r^{-1}+0(r^{-2})]$
(Fig. 6.8c)	Half space	$\cos\theta[R^{-1}+0(R^{-2})]$	$\sin\theta[R^{-1}+0(R^{-2})]$	$\cos\theta[R^{-1}+0(R^{-2})]$
Moment	Layer	$\cos\theta[r^{-2}+0(r^{-3})]$	$\sin\theta[r^{-2}+0(r^{-3})]$	$\cos\theta[r^{-2}+0(r^{-3})]$
(Fig. 6.8d)	Half space	$\cos\theta[R^{-2}+0(R^{-3})]$	$\sin\theta[R^{-2}+0(R^{-3})]$	$\cos\theta[R^{-2}+0(R^{-3})]$

Note: 0(   ) = order of truncation error.

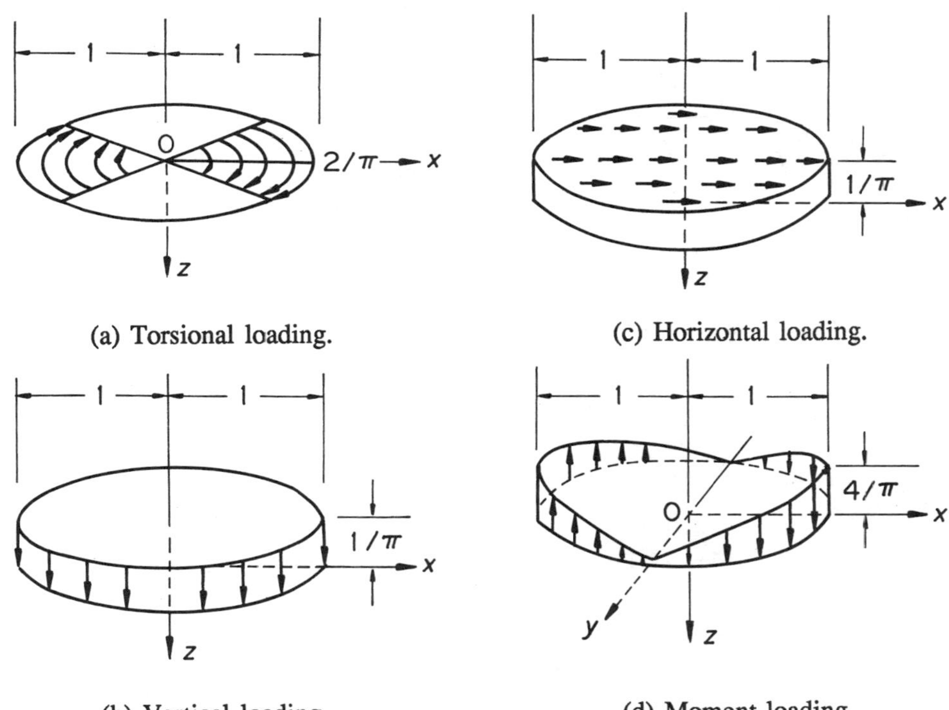

(a) Torsional loading.

(c) Horizontal loading.

(b) Vertical loading.

(d) Moment loading.

Fig. 6.8 Types of loading.

Any infinite element in the rz-plane can be contracted into a finite element in a new $\xi\eta$ plane by employing a transformation which is singular at $\xi=+1$. The coordinate mapping for a horizontal infinite element is

$$r=\sum_j 2(1-\xi)^{-1}L_j(\eta)r_j \tag{6.133a}$$

$$z=\sum_j L_j(\eta)z_j \tag{6.133b}$$

and for a radiating infinite element

$$r=\sum_j 2(1-\xi)^{-1}L_j(\eta)r_j \tag{6.134a}$$

$$z=\sum_j 2(1-\xi)^{-1}L_j(\eta)z_j \tag{6.134b}$$

where $L_j(\eta)$ is the *Lagrange polynomial* for node j. In short, a finite element is obtained by a *singular contraction* of infinite elements as depicted in Fig. 6.9.

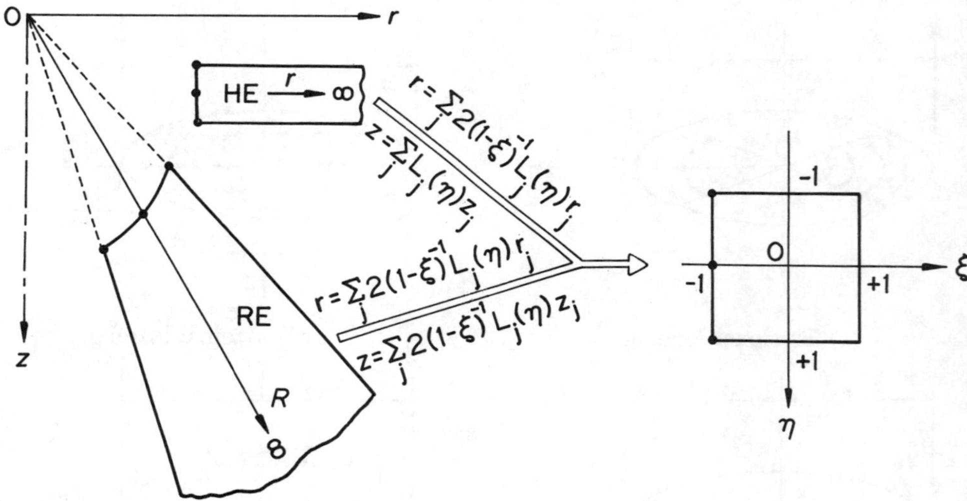

Fig. 6.9 Finite elements by singular contraction.

The relationship between displacement components inside an infinite element and their nodal values has the usual form

$$u_r(\xi,\eta)=\sum_j N_j^r(\xi,\eta)u_r^j \tag{6.135a}$$

$$u_\theta(\xi,\eta)=\sum_j N_j^\theta(\xi,\eta)u_\theta^j \tag{6.135b}$$

$$u_z(\xi,\eta) = \sum_j N_j^z(\xi,\eta)u_z^j \tag{6.135c}$$

in which a superscript or a subscript j denotes the node j, and N_j^r, N_j^θ and N_j^z are *displacement interpolation functions.* Being consistent with the analytical far field displacements, these displacement interpolation functions can be expanded, for horizontal loading as,

$$N_j^r(\xi,\eta) = N_j^\theta(\xi,\eta) = N_j^z(\xi,\eta) = \frac{1}{2}(1-\xi)L_j(\eta) \tag{6.136}$$

and for moment loading as,

$$N_j^r(\xi,\eta) = N_j^\theta(\xi,\eta) = N_j^z(\xi,\eta) = \frac{1}{4}(1-\xi)^2L_j(\eta) \tag{6.137}$$

For vertical loading, $N_j^\theta(\xi,\eta)$ does not exist and $N_j^r(\xi,\eta)$ and $N_j^z(\xi,\eta)$ take the form given by Eq. 6.136. For torsional loading, only $N_j^\theta(\xi,\eta)$ exists and is given by Eq. 6.137. All elements developed are compatible in the sense that there is continuity between the relevant first derivatives of N_j within the elements. The stiffnesses of these elements involve double integrals (with respect to ξ and η), which can be accurately evaluated by standard *Gauss quadrature formula.* Another major advantage of this algorithm is that we do not need any special integration scheme and can use any standard finite element subroutines, since eventually we are dealing with finite elements (*not* infinite elements).

6.8 SATURATED LARGE SPACES

The Navier equations and Darcy's law for quasi-statics of porous isotropic elastic solids completely saturated with fluids have been derived in Section 3.6.7 for Cartesian system; see Eqs 3.123 and 3.124. For cylindrical coordinates, the Navier equations are given by Eqs. 3.67 with the body force components there being replaced by the excess pore pressure gradients. Replace

$$X_r \text{ by } \frac{\partial p_f}{\partial r} \tag{6.138a}$$

$$X_\theta \text{ by } \frac{1}{r}\frac{\partial p_f}{\partial \theta} \tag{6.138b}$$

$$X_z \text{ by } \frac{\partial p_f}{\partial z} \tag{6.138c}$$

Darcy's law, Eq. 3.124 remains invariant. To uncouple these four field equations. Schiffman and Fungaroli (1965) introduced three displacement potentials, so that

$$u_r = -\frac{\partial^2 \Phi}{\partial r \partial z} + \frac{2}{r}\frac{\partial \Psi}{\partial \theta} + z\frac{\partial \chi}{\partial r} \tag{6.139a}$$

$$u_\theta = -\frac{\partial^2\Phi}{r\partial\theta\partial z} - 2\frac{\partial\Psi}{\partial r} + \frac{z}{r}\frac{\partial\chi}{\partial\theta} \tag{6.139b}$$

$$u_z = -\frac{\partial^2\Phi}{\partial z^2} + z\frac{\partial\chi}{\partial z} - \chi \tag{6.139c}$$

$$p_f = (\lambda + 2\mu)\frac{\partial}{\partial z}(\nabla^2\Phi) - 2\mu\frac{\partial\chi}{\partial z} \tag{6.139d}$$

It can be verified that the four field equations can be transformed into the following three uncoupled equations

$$c\nabla^4\Phi = \nabla^2\left(\frac{\partial\Phi}{\partial t}\right) \tag{6.140a}$$

$$\nabla^2\Psi = 0 \tag{6.140b}$$

$$\nabla^2\chi = 0 \tag{6.140c}$$

If the material domain is infinitely extended radially ($0 \le r < \infty$), the solution scheme for solids described in Section 6.3 can be adapted for the present purpose to solve Eqs. 6.140. However, it is now necessary to use Laplace transforms with respect to the time t. The solution scheme starts by expanding Φ, Ψ and χ in Fourier series, thus

$$\Phi(r,\theta,z,t) = \sum_{m=0}^{\infty} \Phi_m(r,z,t)\cos m\theta \tag{6.141a}$$

$$\Psi(r,\theta,z,t) = \sum_{m=0}^{\infty} \Psi_m(r,z,t)\sin m\theta \tag{6.141b}$$

$$\chi(r,\theta,z,t) = \sum_{m=0}^{\infty} \chi_m(r,z,t)\cos m\theta \tag{6.141c}$$

Eventually, the solution to Eqs. 6.140 can be written in the form

$$\tilde{\Phi}_m(\xi,z,p) = A_m e^{-\xi z} + B_m e^{\xi z} + C_m e^{-\gamma z} + D_m e^{\gamma z} \tag{6.142a}$$

$$\tilde{\Psi}_m(\xi,z,p) = E_m e^{-\xi z} + F_m e^{\xi z} \tag{6.142b}$$

$$\tilde{\chi}_m(\xi,z,p) = G_m e^{-\xi z} + H_m e^{\xi z} \tag{6.142c}$$

where a super tilde ($\tilde{}$) denotes a Hankel-Laplace transform, ξ is the Hankel transform parameter versus r, p the Laplace transform parameter versus t, A_m to H_m are functions of ξ and p, and

$$\gamma = \left(\xi^2 + \frac{p}{c}\right)^{1/2} \tag{6.143}$$

Functions A_m to H_m can be determined from well-posed boundary conditions similar to Eqs. 3.127 already derived for the Cartesian coordinate system. The boundary conditions are prescribed on each boundary where z is constant and involve the prescription of either

$$\bar{u}_r \text{ or } \tilde{\sigma}_{zr} \tag{6.144a}$$

$$\bar{u}_\theta \text{ or } \tilde{\sigma}_{z\theta} \tag{6.144b}$$

$$\bar{u}_z \text{ or } \tilde{\sigma}_{zz} \tag{6.144c}$$

$$\text{and } \bar{p}_f \text{ or } k\frac{\partial \bar{p}_f}{\partial z} \tag{6.144d}$$

Apirathvorakij and Karasudhi (1980), and Niumpradit and Karasudhi (1981) applied this theory to load transfer problems in a homogeneous soil domain, and found that the inverse Laplace transform formula given by Schapery, Eq. 1.164, is accurate enough. Later, Karasudhi, Karunasena and Puswewala (1987) proposed an infinite element algorithm similar to that described in Section 6.7 for ideal solids. Most recently, Karasudhi and Prechaverakul (1987) presented an analytical prediction of land subsidence due to water pumping from deep wells. A development of an infinite element algorithm for such subsidence in multilayered soil media was carried out by Karasudhi and Alvappillai (1988).

6.9 EXERCISE PROBLEMS

6.9.1 Use Eq. 6.7 to solve Problems 4.7.1 to 4.7.3.

6.9.2 A long mine tunnel of radius a is cut in deep rock. Before the tunnel is cut the rock is subjected to a uniform pressure p. Considering the rock to be an infinite, homogeneous, isotropic elastic medium, determine the inward radial displacement at the tunnel surface due to the excavation.

6.9.3 Use Eq. 6.8 to solve Problem 4.7.4.

6.9.4 An infinite composite plate made of two elastic materials of distinct properties is supported by a smooth rigid horizontal base. The thickness of the top and bottom elastic layers are h_1 and h_2 respectively. The top surface of the plate is loaded by a uniform normal pressure q and the top elastic layer is subjected to a uniform body force f_0 in the downward vertical direction. Find displacements and stresses.
Hint: With reference to the axis z pointing downwards, only u_z is nonzero in this problem. The only Navier equation for the top layer can be shown to be

$$(\lambda_1+2\mu_1)\frac{d^2u_{z1}}{dz^2}+f_0=0.$$

6.9.5 Using the formulation of Section 6.3, find Mindlin's solution which is the solution to the problem of a force at a point in the interior of a homogeneous isotropic elastic half space.

Answer: With reference to the coordinate system as shown in Fig. 6.10, the pertinent displacement components are; for the case of the force P in the z-direction,

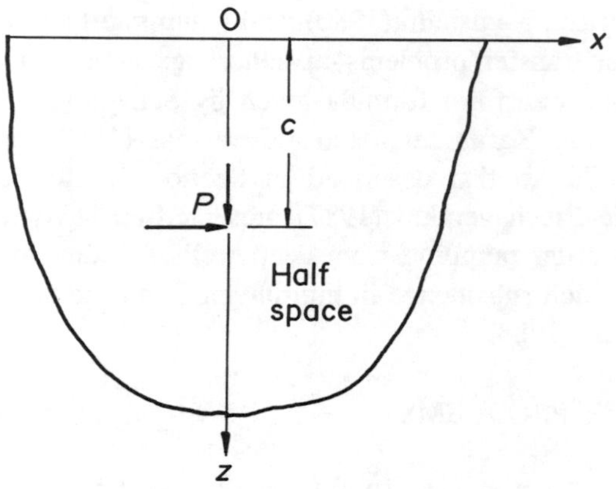

Fig. 6.10 Mindlin's solution.

$$u_r=\frac{Pr}{16\pi\mu(1-\nu)}\left[\frac{z-c}{R_1^3}+\frac{(3-4\nu)(z-c)}{R_2^3}-\frac{4(1-\nu)(1-2\nu)}{R_2(R_2+z+c)}+\frac{6cz(z+c)}{R_2^5}\right],$$

$$u_z=\frac{P}{16\pi\mu(1-\nu)}\left[\frac{3-4\nu}{R_1}+\frac{8(1-\nu)^2-(3-4\nu)}{R_2}+\frac{(z-c)^2}{R_1^3}\right.$$

$$\left.+\frac{(3-4\nu)(z+c)^2-2cz}{R_2^3}+\frac{6cz(z+c)^2}{R_2^5}\right],$$

and for the force P in the x-direction,

$$u=\frac{P}{16\pi\mu(1-\nu)}\left[\frac{3-4\nu}{R_1}+\frac{1}{R_2}+\frac{x^2}{R_1^3}+\frac{(3-4\nu)x^2}{R_2^3}+\frac{2cz}{R_2^3}\left(1-\frac{3x^2}{R_2^2}\right)\right.$$

$$\left.+\frac{4(1-\nu)(1-2\nu)}{R_2+z+c}\left(1-\frac{x^2}{R_2(R_2+z+c)}\right)\right],$$

$$v = \frac{Pxy}{16\pi\mu(1-\nu)}\left[\frac{1}{R_1^3} + \frac{3-4\nu}{R_2^3} - \frac{6cz}{R_2^5} - \frac{4(1-\nu)(1-2\nu)}{R_2(R_2+z+c)^2}\right],$$

$$w = \frac{Px}{16\pi\mu(1-\nu)}\left[\frac{z-c}{R_1^3} + \frac{(3-4\nu)(z-c)}{R_2^3} - \frac{6cz(z+c)}{R_2^5} + \frac{4(1-\nu)(1-2\nu)}{R_2(R_2+z+c)}\right],$$

where

$$R_1 = [r^2 + (z-c)^2]^{1/2}, \quad R_2 = [r^2 + (z+c)^2]^{1/2}.$$

6.9.6 A homogeneous isotropic elastic half plane $-\infty < x < \infty$, $0 \le y < \infty$, is subjected to the following tractions on its surface $y = 0$ in the region $-a < x < a$: (a) a uniformly distributed normal force, (b) a uniformly distributed shear force, and (c) a normal force distributed as a linear homogeneous function of x. Find the stresses.

6.9.7 With the same geometry as in the preceding problem, the displacement in the x-direction is prescribed uniformly in the region $-a < x < a$, while the normal traction is assumed to vanish everywhere on the half plane surface. Find the distribution of the contract shear stress in terms of the applied shear force P_0.
Answer:

$$\sigma_{xy}(x,0) = -\frac{P_0}{\pi(a^2-x^2)^{1/2}}, \qquad (|x| < a)$$

$$= 0, \qquad (|x| > a).$$

6.9.8 With the same geometry as in the preceding problem, a uniform rocking (rotation about the z axis) is prescribed in the region $-a < x < a$, while shear traction is assumed to vanish everywhere on the half plane surface. Find the distribution of the contact normal stress in terms of the prescribed angle of rocking α.
Answer:

$$\sigma_{yy}(x,0) = \frac{-\mu\alpha x}{(1-\nu)(a^2-x^2)^{1/2}}, \qquad (|x| < a)$$

$$= 0, \qquad (|x| > a).$$

6.9.9 A homogeneous isotropic elastic half space $0 \le r < \infty$, $0 \le z < \infty$, is subjected to the following tractions on its surface where $z = 0$ in the region $0 \le r < a$: (a) a uniformly distributed normal force, (b) a uniformly distributed shear force in the x-direction, and (c) a normal force distributed as a linear homogeneous function of x. Find the displacements.

6.9.10 With the same geometry as in the preceding problem, the displacement in the x-direction is prescribed uniformly in the region $0 \le r < a$, while the normal traction and the shear traction in the y-direction are assumed to vanish everywhere on the half space surface. Moreover, the shear traction in the x-direction vanishes outside the loaded area. For zero Poisson's ratio, find the distribution of the contact shear stress in terms of the applied shear force P_0, and the relationship between P_0 and the prescribed displacement u_0.
Answer:

$$\sigma_{zx}(r,\theta,0) = -\frac{P_0}{2\pi a(r^2-a^2)^{1/2}}, \qquad (0 \le r < a)$$

$$= 0, \qquad (r > a),$$

$$P_0 = 8\pi\mu a u_0.$$

6.9.11 With the same geometry as in the preceding problem, a uniform angle of twist (rotation about the z axis) is prescribed in the region $0 \le r < a$. Find the distribution of the contract shear stress in terms of the prescribed angle of twist α, and the relationship between the applied torque T and α.
Answer: This is a pure torsion problem.

$$\sigma_{z\theta}(r,0) = \frac{-4\mu\alpha r}{\pi(a^2-r^2)^{1/2}}, \qquad (0 \le r < a)$$

$$= 0, \qquad (r > a),$$

$$T = \frac{16\mu\alpha a^3}{3}.$$

CHAPTER VII

LINEAR VISCOELASTICITY

7.1 LINEAR ELASTICITY AND NEWTONIAN VISCOSITY

Measurements of load-displacement relationships show that the theory of linear elasticity holds for most metals under small strains. However, the simple fact that the vibrations of metal systems decay even *in vacuo* indicates that there is some energy dissipation. Hooke's law is insufficient, and we need to consider others. This need has become more evident since the development of polymeric materials. These materials possess a capacity to store some and dissipate the rest of the energy input. While the theory of elasticity accounts for materials which are capable of storing energy without any dissipation, the theory of *Newtonian viscous fluids* governs those which dissipate the whole energy input. Deformed elastic bodies return to a natural or undeformed state upon removal of applied loads. Viscous fluids, however, possess no capacity at all for deformational recovery. In an elastic solid, stress is related directly to deformation; in a viscous fluid, stress depends (except for its hydrostatic component) upon the rate of deformation. Hookean elastic solids and Newtonian viscous fluids represent the ends of a spectrum. Between these extremes there are materials which incorporate a blend of both linear elastic and (Newtonian) viscous characteristics. Appropriately, such materials are named *linearly viscoelastic*.

Linear elastic solids retain linearity between load and deformation, and the linear relationship is independent of time. Any current state of deformation of a linear elastic solid can be determined completely without the knowledge of the entire loading history. In other words, a linear elastic solid remembers only the undeformed state of its body. Other solids do not behave in such a simple way. Their current state of deformation cannot be determined completely unless the entire loading history is known. Linearly viscoelastic materials remember the past, but their memory belongs to a relatively simple class. For such materials, linearity between load and deformation is maintained, but the linear relationship depends on time t. The principle of superposition still holds. Thus if $\varepsilon_1(t)$ and $\varepsilon_2(t)$ are deformations due to loads $\sigma_1(t)$ and $\sigma_2(t)$ respectively, then the deformation due to the load $A_1\sigma_1(t)+A_2\sigma_2(t)$ is

$$\varepsilon(t)=A_1\varepsilon_1(t)+A_2\varepsilon_2(t) \tag{7.1}$$

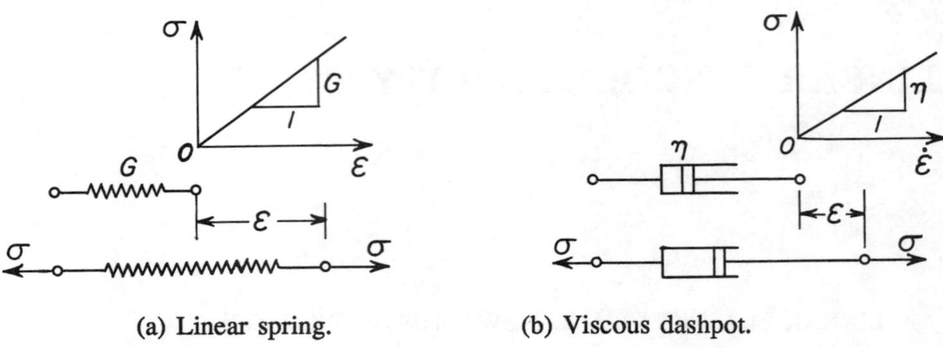

(a) Linear spring. (b) Viscous dashpot.

Fig. 7.1 Basic elements in mechanical models.

where A_1 and A_2 are constants.

Constitutive laws for linear viscoelasticity could be introduced in a simple way through the use of *one-dimensional mechanical models* made of massless linear springs and viscous dashpots. The springs characterize linear elasticity, while the dashpots incorporate Newtonian viscosity. See Fig. 7.1. In a spring with a spring constant G, the stress σ is related to its strain ε by

$$\sigma = G\varepsilon \tag{7.2}$$

The analogous equation for the dashpot is

$$\sigma = \eta\dot{\varepsilon} \tag{7.3}$$

where η is a viscosity constant and $\dot{\varepsilon} = \mathrm{d}\varepsilon/\mathrm{d}t$. By direct integration, we can rewrite Eq. 7.3 in the form

$$\varepsilon(t) = \frac{1}{\eta}\int_{0^-}^{t} \sigma(s)\mathrm{d}s \tag{7.4}$$

in which the integral lower limit 0^- denotes quiescent initial conditions, Eq. 1.123.

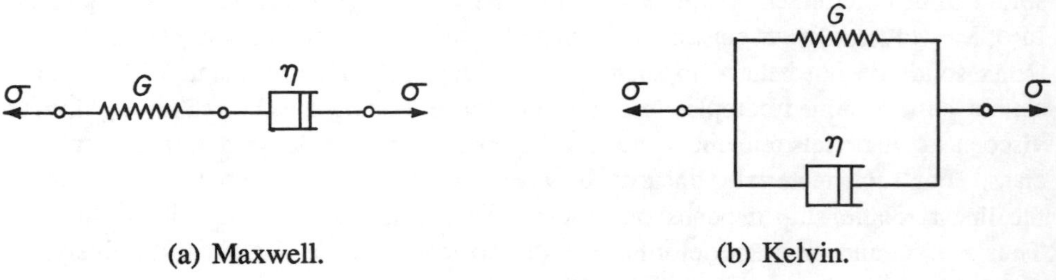

(a) Maxwell. (b) Kelvin.

Fig. 7.2 Two-parameter models.

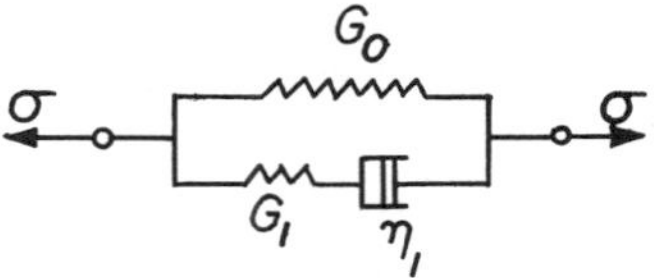

Fig. 7.3 Standard linear solid model.

The *Maxwell model* of viscoelasticity is the combination of a spring and dashpot in series as shown in Fig. 7.2(a). The *Kelvin* or *Voigt model* is the parallel assembly shown in Fig. 7.2(b). For many materials, even these two-parameter models have been found to be inadequate. A three-parameter model constructed from one spring and one Maxwell unit in parallel and known as the *standard linear solid model* is shown in Fig. 7.3. The *generalized Kelvin model* consists of one Maxwell unit and a sequence of Kelvin units in series as depicted by Fig. 7.4. On the other hand, an assembly of one Kelvin unit and a sequence of Maxwell units in parallel as shown in Fig. 7.5 is called a *generalized Maxwell model.*

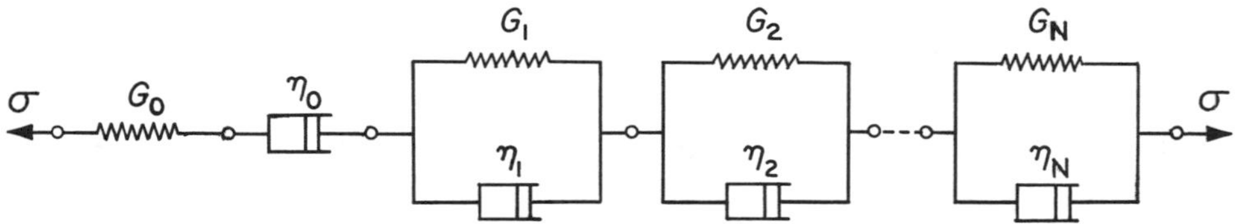

Fig. 7.4 Generalized Kelvin model.

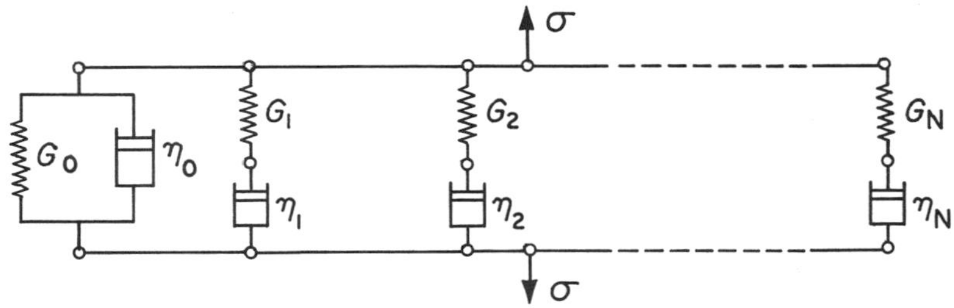

Fig. 7.5 Generalized Maxwell model.

7.2 CREEP AND RELAXATION

If the stress σ is kept constant in any mechanical models presented in the previous section, the strain ε tends to increase with time, i.e. there is *creep*. If the strain ε is kept constant, the stress σ tends to decrease with time, i.e. there is *relaxation.* The *creep compliance* $J(t)$ is defined to be the strain $\varepsilon(t)$ when the stress $\sigma(t)$ is equal to the Heaviside step function $H(t)$. The *relaxation modulus* $\mu(t)$ is defined to be the stress $\sigma(t)$ when the strain $\varepsilon(t)$ is equal to the Heaviside step function $H(t)$. For a given material, $J(t)$ and $\mu(t)$ are related to each other.

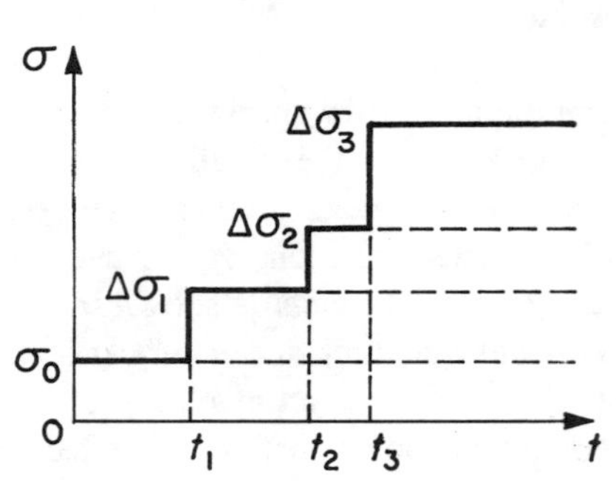

(a) Stepped stress history.

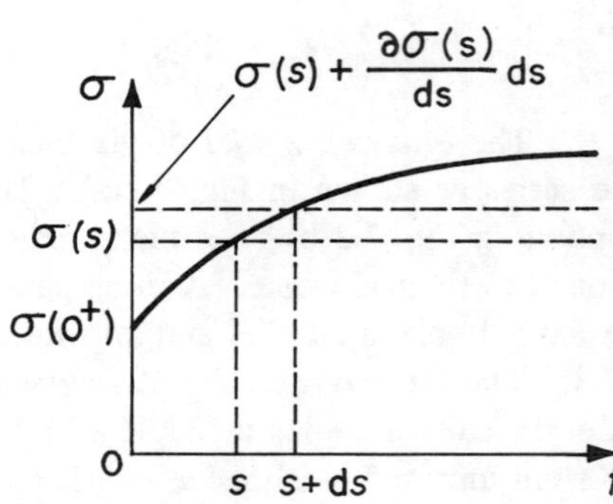

(b) Arbitrary stress history.

Fig. 7.6 Derivation of convolution integral.

Because the superposition principle holds, the strain due to the stepped stress in Fig. 7.6(a) is

$$\varepsilon(t) = \sigma_0 J(t) + \Delta\sigma_1 J(t-t_1) + \Delta\sigma_2 J(t-t_2) + \Delta\sigma_3 J(t-t_3) \tag{7.5}$$

The arbitrary stress history $\sigma(t)$ of Fig. 7.6(b) may be analyzed as an infinity of step loadings, each of magnitude $d\sigma$, so that the strain $\varepsilon(t)$ is

$$\varepsilon(t) = \sigma(0^+)J(t) + \int_{0^+}^{t} J(t-s)\mathrm{d}\sigma(s) \tag{7.6}$$

where $\mathrm{d}\sigma(s) = [\mathrm{d}\sigma(s)/\mathrm{d}s]\,\mathrm{d}s$, and $\sigma(0^+)$ is the value of $\sigma(t)$ at t slightly greater than zero, i.e. the initial value of $\sigma(t)$. Note that all functions involved are zero for $t<0$ (Eq. 1.123). Due to the relationship between the Heaviside step function $H(t)$ and the Dirac-delta function $\delta(t)$, Eq. 1.106, Eq. 7.6 can be rewritten as a single integral with the lower limit infinitesimally less than zero, i.e. 0^-, so that

$$\varepsilon(t) = \int_{0^-}^{t} J(t-s)\mathrm{d}\sigma(s) \tag{7.7}$$

in which $\sigma(s)$ is considered as

$$\sigma(s) = \sigma_c(s)H(s),$$

where $\sigma_c(s)$ is a continuous function of time. In other words $\sigma_c(t)$ is not only equal to $\sigma(t)$ for $t>0$ but also continuous at $t=0$. Figure 7.7 should help to clarify this concept further. The same equation (Eq. 7.7) also holds for other discontinuities of $\sigma(t)$. Suppose there is a discontinuity $\Delta\sigma_0$ at time t_0. The stress function in Eq. 7.7 should be considered as

$$\sigma(s) = \sigma_c(s)H(s) + \Delta\sigma_0 H(s - t_0) \tag{7.8}$$

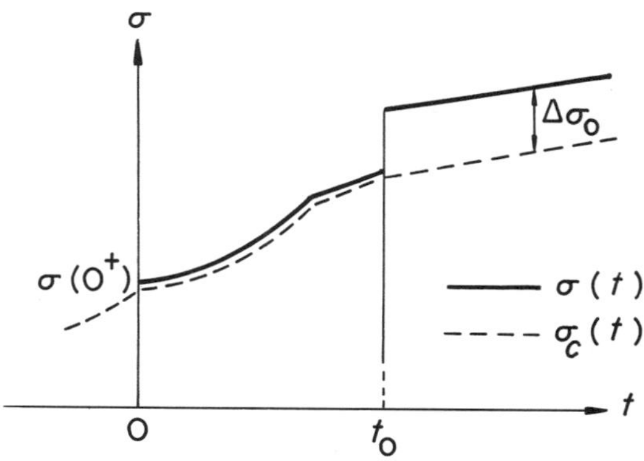

Fig. 7.7 Continuous $\sigma_c(t)$ in comparison with discontinuous $\sigma(t)$.

where $\sigma_c(s)$ is as shown in Fig. 7.7, i.e. it is continuous everywhere and parallel to the $\sigma(s)$ curve ($d\sigma_c/ds = d\sigma/ds$ for $0 < s \leq t_0$ and $t_0 \leq s \leq t$). Substituting Eq. 7.8 into Eq. 7.7 we find

$$\varepsilon(t) = \sigma(0^+)J(t) + \Delta\sigma_0 J(t - t_0) + \int_{0^+}^{t} J(t - s)d\sigma_c(s) \tag{7.9}$$

which can be shown to be identical to that obtained directly by the superposition principle. Readers with a mathematical background will recognize Eq. 7.7 as a *Stieltjes integral*. Moreover, it can be shown that Eq. 7.7 still holds even when the stress function is singular but integrable like the function $\delta(t - t_0)$. In other words, the superposition principle and Eq. 7.7 should give the same result,

$$\varepsilon(t) = \frac{dJ(t - t_0)}{d(t - t_0)}, \qquad [\text{ for } \sigma(t) = \delta(t - t_0)] \tag{7.10}$$

The integral in Eq. 7.7 is often referred to as a *convolution integral* or *hereditary integral* since the 'effect' at any time is seen to depend upon the entire 'cause' history. The commutative law for convolution integrals, Eq. 1.138a, permits the interchange of J and σ in this integral, i.e.

$$\varepsilon(t) = J * d\sigma = \sigma * dJ \tag{7.11}$$

in which the following notation is introduced

$$v_1 * dv_2 = \int_{0^-}^{t} v_1(t - s)dv_2(s) \tag{7.12}$$

Equation 7.11 is also known as the *creep law*. Its Laplace transform is

$$\tilde{\varepsilon}(p) = p\tilde{J}(p)\tilde{\sigma}(p) \tag{7.13}$$

When $\varepsilon(t) = H(t)$, $\sigma(t)$ is (by definition) equal to the relaxation modulus $\mu(t)$. Equation 7.13 gives the relationship between the creep compliance and the relaxation modulus in the form

$$p^2\tilde{J}(p)\tilde{\mu}(p) = 1 \tag{7.14a}$$

or, due to Eqs. 1.124a, 1.132 and 1.135,

$$J{*}\mathrm{d}\mu = \mu{*}\mathrm{d}J = H(t) \tag{7.14b}$$

or, due to Eqs. 1.124i and 1.135,

$$J{*}\mu = \mu{*}J = tH(t) \tag{7.14c}$$

Following similar arguments we can express the *relaxation law* in the form

$$\sigma(t) = \mu{*}\mathrm{d}\varepsilon = \varepsilon{*}\mathrm{d}\mu \tag{7.15a,b}$$

the Laplace transform of which is

$$\tilde{\sigma}(p) = p\tilde{\mu}(p)\tilde{\varepsilon}(p) \tag{7.15c}$$

This yields the same relationship between relaxation modulus and creep compliance given by Eqs. 7.14.

7.3 COMPLIANCE AND MODULUS OF MECHANICAL MODELS

For the *linear spring model* shown in Fig. 7.1(a), the stress-strain relationship is given by Eq. 7.2, the Laplace transform of which is

$$\tilde{\sigma}(p) = G\tilde{\varepsilon}(p) \tag{7.16a}$$

Comparing this equation with Eqs. 7.13 and 7.15c we find

$$\tilde{J}(p) = \frac{p^{-1}}{G} \tag{7.16b}$$

$$\tilde{\mu}(p) = Gp^{-1} \tag{7.16c}$$

Thus the creep compliance and the relaxation modulus for this type of element are

$$J(t) = H(t)/G \tag{7.16d}$$

$$\mu(t) = GH(t) \tag{7.16e}$$

For the *viscous dashpot model* shown in Fig. 7.1(b), the stress-strain rate equation is Eq. 7.3 and we find

$$\tilde{\sigma}(p) = \eta p\tilde{\varepsilon}(p) \tag{7.17a}$$

$$\tilde{J}(p) = \frac{p^{-2}}{\eta} \tag{7.17b}$$

$$\tilde{\mu}(p) = \eta \tag{7.17c}$$

$$J(t)=\frac{t}{\eta}H(t) \tag{7.17d}$$

$$\mu(t)=\eta\delta(t) \tag{7.17e}$$

For the *Maxwell model* shown in Fig. 7.2(a), the Laplace transform of the stress-strain relationship in the spring takes the same form as Eq. 7.16a, i.e.

$$\tilde{\sigma}_G(p)=G\tilde{\varepsilon}_G(p) \tag{7.18a}$$

while that in the dashpot takes the same form as Eq. 7.17a, i.e.

$$\tilde{\sigma}_\eta(p)=\eta p\tilde{\varepsilon}_\eta(p) \tag{7.18b}$$

The stress $\sigma(t)$ must be the same in the spring as in the dashpot, i.e.

$$\tilde{\sigma}(p)=\tilde{\sigma}_G(p)=\tilde{\sigma}_\eta(p) \tag{7.18c}$$

The strain $\varepsilon(t)$ must be the sum of those in the spring and the dashpot, i.e.

$$\tilde{\varepsilon}(p)=\tilde{\varepsilon}_G(p)+\tilde{\varepsilon}_\eta(p) \tag{7.18d}$$

Substituting Eqs. 7.18a and b into Eq. 7.18d, and using Eq. 7.18c, we find the Laplace transform of the stress-strain relationship for a Maxwell model to be

$$\tilde{\varepsilon}(p)=\frac{1}{G}[1+(p\tau)^{-1}]\tilde{\sigma}(p) \tag{7.18e}$$

where

$$\tau=\eta/G \tag{7.18f}$$

Comparing Eq. 7.18e with Eqs. 7.13 and 7.15c, we find

$$\tilde{J}(p)=\frac{1}{G}[p^{-1}+p^{-2}\tau^{-1}] \tag{7.18g}$$

$$\tilde{\mu}(p)=G(p+\tau^{-1})^{-1} \tag{7.18h}$$

Thus the creep compliance and the relaxation modulus of a Maxwell model are

$$J(t)=\frac{1}{G}(1+t\tau^{-1})H(t) \tag{7.18i}$$

$$\mu(t)=Ge^{-t/\tau}H(t) \tag{7.18j}$$

For the *Kelvin* or *Voigt model* shown in Fig. 7.2(b), the Laplace transforms of the stress-strain relationship in the spring and the dashpot are the same as Eqs. 7.18a and b, respectively. The strain $\varepsilon(t)$ must be the same in the spring as in the dashpot, i.e.

$$\tilde{\varepsilon}(p)=\tilde{\varepsilon}_G(p)=\tilde{\varepsilon}_\eta(p) \tag{7.19a}$$

The stress $\sigma(t)$ must be the sum of those in the spring and the dashpot, i.e.

$$\tilde{\sigma}(p)=\tilde{\sigma}_G(p)+\tilde{\sigma}_\eta(p) \tag{7.19b}$$

Substituting Eqs. 7.18a and b into Eq. 7.19b, and using Eq. 7.19a, we find the Laplace transform of the stress-strain relationship for a Kelvin model to be

$$\tilde{\sigma}(p)=G(1+p\tau)\tilde{\varepsilon}(p) \tag{7.19c}$$

Proceeding as before, we find

$$J(p)=\frac{1}{G}[p^{-1}-(p+\tau^{-1})^{-1}] \tag{7.19d}$$

$$\tilde{\mu}(p)=G(p^{-1}+\tau) \tag{7.19e}$$

$$J(t)=\frac{1}{G}(1-e^{-t/\tau})H(t) \tag{7.19f}$$

$$\mu(t)=G[H(t)+\tau\delta(t)] \tag{7.19g}$$

For the *standard linear solid model* shown in Fig. 7.3, we find

$$\tilde{\sigma}_0(p)=G_0\tilde{\varepsilon}_0(p) \tag{7.20a}$$

$$\tilde{\varepsilon}_1(p)=\frac{1}{G_1}[1+(p\tau_1)^{-1}]\tilde{\sigma}_1(p) \tag{7.20b}$$

where $\tau_1=\eta_1/G_1$. But

$$\tilde{\varepsilon}(p)=\tilde{\varepsilon}_0(p)=\tilde{\varepsilon}_1(p) \tag{7.20c}$$

$$\tilde{\sigma}(p)=\tilde{\sigma}_0(p)+\tilde{\sigma}_1(p) \tag{7.20d}$$

Substituting Eqs. 7.20a and b into Eq. 7.20d, and using Eq. 7.20c, we find

$$\tilde{\sigma}(p)=\left[G_0+G_1\frac{\tau_1 p}{1+\tau_1 p}\right]\tilde{\varepsilon}(p) \tag{7.20e}$$

Thus we have

$$\tilde{\mu}(p)=\frac{G_0+(G_0+G_1)\tau_1 p}{p(1+\tau_1 p)} \tag{7.20f}$$

$$\bar{J}(p)=\frac{1}{p^2\tilde{\mu}(p)} \tag{7.20g}$$

$$\mu(t)=\left(G_0+G_1 e^{-t/\tau_1}\right)H(t) \tag{7.20h}$$

$$J(t)=\frac{1}{G_0}\left[1-\left(1-\frac{G_0}{G_0+G_1}\right)\exp\left(-\frac{G_0 t/\tau_1}{G_0+G_1}\right)\right]H(t) \tag{7.20i}$$

For a *generalized Kelvin model*, the governing equations are

$$\tilde{\varepsilon}_0(p)=\frac{1}{G_0}[1+(p\tau_0)^{-1}]\tilde{\sigma}_0(p) \tag{7.21a}$$

$$\tilde{\sigma}_i(p)=G_i(1+p\tau_i)\tilde{\varepsilon}_i(p), \qquad (i=1 \text{ to } N) \tag{7.21b}$$

$$\tilde{\sigma}(p)=\tilde{\sigma}_0(p)=\tilde{\sigma}_i(p), \qquad (i=1 \text{ to } N) \tag{7.21c}$$

$$\tilde{\varepsilon}(p) = \tilde{\varepsilon}_0(p) + \sum_{i=1}^{N} \tilde{\varepsilon}_i(p) \tag{7.21d}$$

$$\tilde{\varepsilon}(p) = \left[\frac{1}{G_0}\left(1 + \frac{1}{\tau_0 p}\right) + \sum_{i=1}^{N} \frac{(1+p\tau_i)^{-1}}{G_i} \right] \tilde{\sigma}(p) \tag{7.21e}$$

$$\tilde{J}(p) = \left[\frac{1}{G_0}\left(1 + \frac{1}{\tau_0 p}\right) + \sum_{i=1}^{N} \frac{(1+p\tau_i)^{-1}}{G_i} \right] p^{-1} \tag{7.21f}$$

$$\tilde{\mu}(p) = \frac{1}{p^2 \tilde{J}(p)} \tag{7.21g}$$

$$J(t) = \left[\frac{1}{G_0} + \frac{t}{\eta_0} + \sum_{i=1}^{N} \frac{(1 - e^{-t/\tau_i})}{G_i} \right] H(t) \tag{7.21h}$$

To obtain an analytical expression for $\mu(t)$, $\tilde{\mu}(p)$ must first be written as the quotient of two polynomial functions of p. Then the inverse Laplace transform can be obtained by means of the *Heaviside expansion formula* (Spiegel, 1965). Otherwise, one may solve Eq. 7.14b or c numerically, or adopt one of the approximate methods for inverse Laplace transforms described in Section 1.12.

For a *generalized Maxwell model*, the governing equations are

$$\tilde{\sigma}_0(p) = G_0(1 + p\tau_0)\tilde{\varepsilon}_0(p) \tag{7.22a}$$

$$\tilde{\varepsilon}_i(p) = \frac{1}{G_i}[1 + (p\tau_i)^{-1}]\tilde{\sigma}_i(p), \qquad (i = 1 \text{ to } N) \tag{7.22b}$$

$$\tilde{\varepsilon}(p) = \tilde{\varepsilon}_0(p) = \tilde{\varepsilon}_i(p), \qquad (i = 1 \text{ to } N) \tag{7.22c}$$

$$\tilde{\sigma}(p) = \tilde{\sigma}_0(p) + \sum_{i=1}^{N} \tilde{\sigma}_i(p) \tag{7.22d}$$

$$\tilde{\sigma}(p) = \left[G_0(1 + \tau_0 p) + \sum_{i=1}^{N} \frac{G_i p \tau_i}{1 + p\tau_i} \right] \tilde{\varepsilon}(p) \tag{7.22e}$$

$$\tilde{\mu}(p) = \left[G_0(1 + \tau_0 p) + \sum_{i=1}^{N} \frac{G_i p \tau_i}{1 + p\tau_i} \right] p^{-1} \tag{7.22f}$$

$$\tilde{J}(p) = \frac{1}{p^2 \tilde{\mu}(p)} \tag{7.22g}$$

$$\mu(t) = (G_0 + \sum_{i=1}^{N} G_i e^{-t/\tau_i}) H(t) + \eta_0 \delta(t) \tag{7.22h}$$

In this case $J(t)$ is not simple, but may still be obtained using the methods described above.

Note that the presence of a dashpot in parallel with other units in a mechanical model introduces a singular (but integrable) function, i.e. $\delta(t)$ into the relaxation modulus. For this reason, many researchers have used a generalized Maxwell model without such a 'free' dashpot. The presence of a dashpot in series with other units in a mechanical model introduces a term which tends to infinity with t into the creep compliance. Note also that the attachment of $H(t)$ to every nonsingular term makes such expression valid even for $t<0$, and underlines the *axiom of nonretroactivity*, i.e. the effect at the *present time* is due to causes in the *past* and *not* in the *future*.

If the generalized Kelvin model is considered as typical for the behavior in creep, then a general form of the creep compliance is

$$J(t)=\left[J_g+\frac{t}{\eta_0}+\psi(t)\right]H(t) \tag{7.23a}$$

in which J_g is known as the *glassy creep compliance*, and $\psi(t)$ the *creep function*. J_g is a constant equal to the initial value of $J(t)$, i.e.

$$J_g=J(0^+) \tag{7.23b}$$

and $\psi(t)$ is a monotonically increasing function with a zero initial value and a constant asymptote as $t\rightarrow\infty$, as shown in Fig. 7.8. The stress-strain relationship for this case is given more simply by the creep law (Eq. 7.11) than by the relaxation law (Eqs. 7.15).

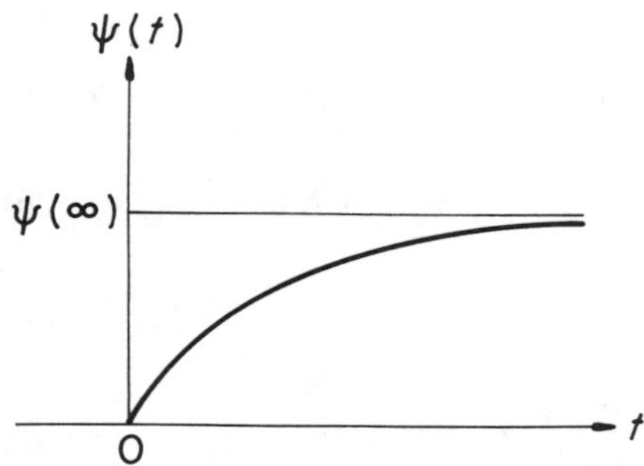

Fig. 7.8 Creep function.

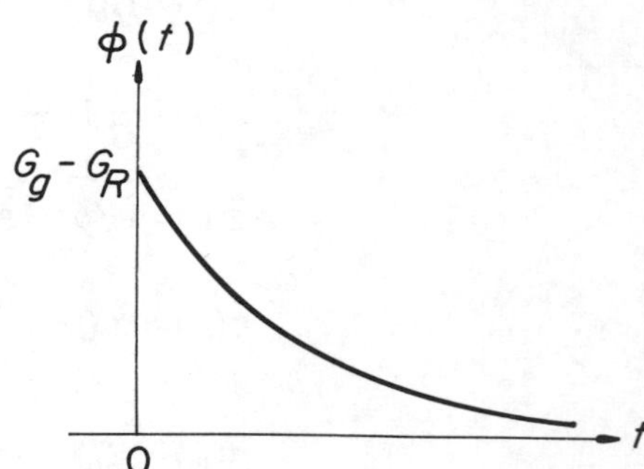

Fig. 7.9 Relaxation function.

On the other hand, if the generalized Maxwell model is considered as typical for the behavior in relaxation, then a general form of the relaxation modulus is

$$\mu(t)=[G_R+\phi(t)]H(t)+\eta_0\delta(t) \tag{7.24a}$$

where G_R is termed the *rubbery modulus*, and $\phi(t)$ the *relaxation function*. As depicted in Fig. 7.9, $\phi(t)$ monotonically decreases from an initial value and tends to zero as $t\rightarrow\infty$. The initial value of $\phi(t)$ is

$$\phi(0^+) = G_g - G_R \tag{7.24b}$$

where G_g is the *glassy modulus.* G_g and G_R are the initial and final values of $\mu(t)$, i.e.

$$G_g = \mu(0^+) \tag{7.24c}$$

$$G_R = \mu(\infty) = G_0 \tag{7.24d}$$

The stress-strain relationship for this case is given more simply by the relaxation law (Eqs. 7.15) than by the creep law (Eq. 7.11). Substituting Eq. 7.24a into Eq. 7.15a we find

$$\sigma(t) = \eta_0 \int_{0^-}^{t} \delta(t-s)\mathrm{d}\varepsilon(s) + \int_{0^-}^{t} [G_R + \phi(t-s)]\,\mathrm{d}\varepsilon(s).$$

Recalling properties of the Dirac-delta function and the argument used to put Eq. 7.6 as identical to Eq. 7.7, we can rewrite the equation above showing more explicitly the discontinuity of $\varepsilon(t)$ at $t = 0$ as

$$\sigma(t) = \eta_0 \varepsilon^{(1)}(t) + \varepsilon(0^+)\mu(t) + \int_{0^+}^{t} \mu(t-s)\mathrm{d}\varepsilon(s), \quad (t > 0) \tag{7.24e}$$

where $\varepsilon^{(1)}(t) = \mathrm{d}\varepsilon(t)/\mathrm{d}t$.

A special case of Eq. 7.23a is Eq. 7.19f of the Kelvin model for which we may write

$$J(\tau) = (1 - \mathrm{e}^{-1})J(\infty) \tag{7.25}$$

τ is called the *retardation time.* Analogous to this is the *relaxation time* τ which is the time for $\mu(t)$ in the Maxwell model to reduce to e^{-1} of its original value (see Eq. 7.18j which is a special case of Eq. 7.22h), i.e.

$$\mu(t+\tau) = \mathrm{e}^{-1}\mu(t) \tag{7.26}$$

7.4 DIFFERENTIAL EQUATIONS FOR STRESS-STRAIN RELATIONSHIP

It should be possible to relate $\sigma(t)$ to $\varepsilon(t)$ by means of a differential equation, rather an integral one. If initial conditions (at $t = 0^+$) are prescribed properly, the solution to that differential equation must agree with the creep law given by Eq. 7.11 as well as with the relaxation law, Eqs. 7.15. A general form of such differential equation would be

$$P(D)\sigma(t) = Q(D)\varepsilon(t) \tag{7.27a}$$

where $P(D)$ and $Q(D)$ are differential operators, i.e. they are polynomial functions of D which denotes differentiation with respect to time t,

$$D = \frac{\mathrm{d}}{\mathrm{d}t} \tag{7.27b}$$

The Laplace transform of Eq. 7.27a is

$$P(p)\tilde{\sigma}(p) = Q(p)\tilde{\varepsilon}(p) \tag{7.28a}$$

Comparing Eq. 7.28a with the creep law (Eq. 7.11) and the relaxation law (Eqs. 7.15), we find

$$P(p) = pQ(p)\bar{J}(p) \tag{7.28b}$$

and

$$Q(p) = pP(p)\tilde{\mu}(p) \tag{7.28c}$$

Naturally, the two equations above are consistent with the relationship between $\bar{J}(p)$ and $\tilde{\mu}(p)$ as given in Eq. 7.14a.

The first four theorems to follow are related to the generalized Maxwell model.

Theorem 7.1. The relaxation function of the generalized Maxwell model satisfies the following differential equation

$$P(D)\mu(t) = a_0 G_R \tag{7.29a}$$

where

$$P(D) = \sum_{i=0}^{N} a_i D^i \tag{7.29b}$$

Here $a_i (i = 0 \text{ to } N)$ are constants and N is the number of the Maxwell units in the model.

Proof of Theorem 7.1. Using Eqs. 7.22h and 7.29b we can put $P(D)G(t)$ for $t > 0$ in the form

$$P(D)\mu(t) = a_0 G_0 + \sum_{i=1}^{N} G_i e^{-t/\tau_i} \left[\sum_{k=0}^{N} (-1)^k a_k \alpha_i^k \right] \tag{7.29c}$$

where G_0 is identical to G_R (Eq. 7.24d), and

$$\alpha_i = \tau_i^{-1} = \frac{G_i}{\eta_i} \tag{7.29d}$$

Equations 7.29a and c are identical provided a_k $(k = 0 \text{ to } N)$ satisfy the equation

$$\sum_{k=0}^{N} (-1)^k a_k \alpha_i^k = 0 \tag{7.29e}$$

Note that α_i are the roots of the following polynomial equation

$$\prod_{i=1}^{N} (\alpha - \alpha_i) = 0 \tag{7.29f}$$

where

$$\prod_{i=1}^{N} (\alpha - \alpha_i) = (\alpha - \alpha_1)(\alpha - \alpha_2).....(\alpha - \alpha_N) \tag{7.29g}$$

It is possible to set the equation

$$\sum_{k=0}^{N} (-1)^k a_k \alpha^k = \prod_{i=1}^{N} (\alpha - \alpha_i) \tag{7.29h}$$

then determine each a_k by equating the coefficients of α^k. Values of a_k obtained in such a way will satisfy Eq. 7.29e. Thus the theorem is proved. Lastly, it is interesting to note that

$$P(D)\mu(t) = a_0 G_R = \sum_{n=0}^{N} a_n \mu^{(n)}(0^+) \tag{7.29i}$$

where a superscript in parentheses denotes a time derivative, e.g. $(\ )^{(n)} = \mathrm{d}^n(\ )/\mathrm{d}t^n$.

Theorem 7.2. If $\sigma(t)$ and $\varepsilon(t)$ are continuous and N times continuously differentiable for $t > 0$, then the differential equation for the relationship between $\sigma(t)$ and $\varepsilon(t)$ is

$$P(D)\sigma(t) = Q(D)\varepsilon(t), \quad (t > 0) \tag{7.30a}$$

where $P(D)$ is as defined in Eq. 7.29b, and $Q(D)$ is another differential operator one order higher than $P(D)$,

$$Q(D) = \sum_{i=0}^{N+1} b_i D^i \tag{7.30b}$$

The relations among the coefficients of these operators are

$$b_0 = \sum_{n=0}^{N} a_n \mu^{(n)}(0^+) \tag{7.30c}$$

$$b_r = \eta_0 a_{r-1} + \sum_{n=r}^{N} a_n \mu^{(n-r)}(0^+), \quad (r = 1 \text{ to } N) \tag{7.30d}$$

$$b_{N+1} = \eta_0 a_N \tag{7.30e}$$

where η_0 is the constant of the 'free' dashpot (Fig. 7.5). Moreover, it should be proved that the initial conditions are

$$\sum_{r=k}^{N} a_r \sigma^{(r-k)}(0^+) + \eta_0 a_{k-1} \varepsilon(0^+) = \sum_{r=k}^{N+1} b_r \varepsilon^{(r-k)}(0^+), \quad (k = 1 \text{ to } N) \tag{7.30f}$$

Before proving this theorem, we state the *lemma*

$$\sum_{r=k}^{N} B_{r-k} \sum_{n=r}^{N} A_n C_{n-r} = \sum_{r=k}^{N} A_r \sum_{n=0}^{r-k} B_n C_{r-k-n}, \quad (k = 1 \text{ to } N) \tag{7.30g}$$

which may be proved by induction.

Proof of Theorem 7.2. The relaxation law is given by Eq. 7.15a or 7.24e, the relaxation modulus by Eq. 7.22h or 7.24a, and coefficients a_r by Theorem 7.1. Using the formula for a derivative under an integral sign, Eq. 1.137, we find the time derivative of Eq. 7.24e to be

$$\sigma^{(1)}(t) = \eta_0 \varepsilon^{(2)}(t) + \varepsilon(0^+)\mu^{(1)}(t) + \mu(0^+)\varepsilon^{(1)}(t) + \int_{0^+}^{t} \mu^{(1)}(t-s)\mathrm{d}\varepsilon(s), \quad (t > 0).$$

Using mathematical induction we can show that a time derivative of higher order can be written

$$\sigma^{(k)}(t) = \eta_o \varepsilon^{(k+1)}(t) + \varepsilon(0^+)\mu^{(k)}(t) + \sum_{n=1}^{k} \mu^{(k-n)}(0^+)\varepsilon^{(n)}(t)$$

$$+ \int_{0^+}^{t} \mu^{(k)}(t-s)\mathrm{d}\varepsilon(s), \qquad (t > 0, \ \ k = 1 \text{ to } N) \tag{7.30h}$$

Due to the equation above, we can write for $t>0$,

$$P(D)\sigma(t)=\eta_0 P(D)\varepsilon^{(1)}(t)+\varepsilon(0^+)P(D)\mu(t)+\int_{0^+}^{t}P(D)\mu(t-s)d\varepsilon(s)$$

$$+\sum_{k=1}^{N}a_k\sum_{n=1}^{k}\mu^{(k-n)}(0^+)\varepsilon^{(n)}(t),$$

or, due to Eqs. 7.29i and 7.30g,

$$P(D)\sigma(t)=\eta_0 P(D)\varepsilon^{(1)}(t)+\varepsilon(0^+)\sum_{n=0}^{N}a_n\mu^{(n)}(0^+)$$

$$+\sum_{n=0}^{N}a_n\mu^{(n)}(0^+)\int_{0^+}^{t}d\varepsilon(s)+\sum_{r=1}^{N}\varepsilon^{(r)}(t)\sum_{n=r}^{N}a_n\mu^{(n-r)}(0^+)$$

$$=\eta_0 P(D)\varepsilon^{(1)}(t)+\varepsilon(t)\sum_{n=0}^{N}a_n\mu^{(n)}(0^+)+\sum_{r=1}^{N}\varepsilon^{(r)}(t)\sum_{n=r}^{N}a_n\mu^{(n-r)}(0^+),$$

which is the same as Eq. 7.30a provided that Eqs. 7.30c, d and e hold. To prove initial conditions, Eq. 7.30f, we may start by writing Eq. 7.30h for $t=0^+$, i.e.

$$\sigma^{(k)}(0^+)=\eta_0\varepsilon^{(k+1)}(0^+)+\sum_{n=0}^{k}\mu^{(k-n)}(0^+)\varepsilon^{(n)}(0^+),\quad (k=1\text{ to }N).$$

Replacing k by $r-k$, multiplying the result by a_r, then taking a summation from $r=k$ to N, we have

$$\sum_{r=k}^{N}a_r\sigma^{(r-k)}(0^+)=\eta_0\sum_{r=k}^{N}a_r\varepsilon^{(r-k+1)}(0^+)+\sum_{r=k}^{N}a_r\sum_{n=0}^{r-k}\mu^{(r-k-n)}(0^+)\varepsilon^{(n)}(0^+),\quad (k=1\text{ to }N).$$

Making use of the lemma, Eq. 7.30g, we can put this equation in the form

$$\sum_{r=k}^{N}a_r\sigma^{(r-k)}(0^+)=\eta_0\sum_{r=k}^{N}a_r\varepsilon^{(r-k+1)}(0^+)+\sum_{r=k}^{N}\varepsilon^{(r-k)}(0^+)\sum_{n=r}^{N}a_n\mu^{(n-r)}(0^+),$$

$$=\eta_0\sum_{r=k+1}^{N+1}a_{r-1}\varepsilon^{(r-k)}(0^+)+\sum_{r=k}^{N}\varepsilon^{(r-k)}(0^+)\sum_{n=r}^{N}a_n\mu^{(n-r)}(0^+)$$

$$=\eta_0\sum_{r=k}^{N}a_{r-1}\varepsilon^{(r-k)}(0^+)+\eta_0 a_N\varepsilon^{(N-k+1)}(0^+)-\eta_0 a_{k-1}\varepsilon(0^+)$$

$$+\sum_{r=k}^{N}\varepsilon^{(r-k)}(0^+)\sum_{n=r}^{N}a_n\mu^{(n-r)}(0^+),\qquad (k=1\text{ to }N)$$

$$=\sum_{r=k}^{N}\left[\eta_0 a_{r-1}+\sum_{n=r}^{N}a_n\mu^{(n-r)}(0^+)\right]\varepsilon^{(r-k)}(0^+)$$

$$+\eta_0 a_N\varepsilon^{(N-k+1)}(0^+)-\eta_0 a_{k-1}\varepsilon(0^+),\qquad (k=1\text{ to }N),$$

which is the same as Eq. 7.30f due to Eqs. 7.30d and e. Thus Theorem 7.2 is completely proved.

Theorem 7.3. If the relationship between $\sigma(t)$ and $\varepsilon(t)$ as given by Eq. 7.30a holds (for $t > 0$), and the initial conditions are as in Eq. 7.30f, then the following relationship holds for $t \geq 0$

$$P(D)\sigma(t) = Q(D)\varepsilon(t) \tag{7.31a}$$

where

$$\sigma(t) = \sigma(t)H(t) + \eta_0\varepsilon(0^+)\delta(t) \tag{7.31b}$$

$$\varepsilon(t) = \varepsilon(t)H(t) \tag{7.31c}$$

and

$$\eta_0 = \frac{b_{N+1}}{a_N} \tag{7.31d}$$

Proof of Theorem 7.3. Due to Eq. 1.110d, we can write

$$\frac{d^k}{dt^k}[\sigma(t)H(t)] = \sigma^{(k)}(t)H(t) + \sum_{n=0}^{k-1} \sigma^{(k-1-n)}(0^+)\delta^{(n)}(t), \quad (k \geq 1) \tag{7.31e}$$

Consequently,

$$P(D)[\sigma(t)H(t)] = a_0\sigma(t)H(t) + \sum_{k=1}^{N} a_k\left[\sigma^{(k)}(t)H(t) + \sum_{r=0}^{k-1} \sigma^{(k-1-r)}(0^+)\delta^{(r)}(t)\right],$$

or, due to Eq. 7.29b and the lemma, Eq. 7.30g,

$$P(D)[\sigma(t)H(t)] = [P(D)\sigma(t)]H(t) + \sum_{k=1}^{N} \delta^{(k-1)}(t) \sum_{r=k}^{N} a_r\sigma^{(r-k)}(0^+) \tag{7.31f}$$

Similarly, we can write

$$\begin{aligned} Q(D)\varepsilon(t) &= Q(D)[\varepsilon(t)H(t)] \\ &= [Q(D)\varepsilon(t)]H(t) + \sum_{k=1}^{N+1} \delta^{(k-1)}(t) \sum_{r=k}^{N+1} b_r\varepsilon^{(r-k)}(0^+) \\ &= [Q(D)\varepsilon(t)]H(t) + \sum_{k=1}^{N} \delta^{(k-1)}(t) \sum_{r=k}^{N+1} b_r\varepsilon^{(r-k)}(0^+) + \delta^{(N)}(t)b_{N+1}\varepsilon(0^+) \end{aligned} \tag{7.31g}$$

Due to Eqs. 7.31b, we have

$$P(D)\sigma(t) = P(D)[\sigma(t)H(t)] + \eta_0\varepsilon(0^+)\sum_{k=1}^{N+1} a_{k-1}\delta^{(k-1)}(t) \tag{7.31h}$$

Due to Eqs. 7.30a, 7.30f, 7.31d, 7.31f, 7.31g and 7.31h, Eq. 7.31a is proved. The Laplace transform of Eq. 7.31a is given by Eq. 7.28a, i.e.

$$P(p)\tilde{\sigma}(p) = Q(p)\tilde{\varepsilon}(p) \tag{7.31i}$$

where

$$P(p) = \sum_{r=0}^{N} a_r p^r \tag{7.31j}$$

$$Q(p)=\sum_{r=0}^{N+1} b_r p^r \tag{7.31k}$$

Theorem 7.4. If the relationship between $\sigma(t)$ and $\varepsilon(t)$ as given by Eq. 7.31a holds, then the same relationship can be put in a convolution integral form as Eqs. 7.15, the relaxation modulus of which is the same as that of the generalized Maxwell model, i.e.

$$\mu(t)=\mu(t)H(t)+\eta_0\delta(t) \tag{7.32a}$$

where

$$\eta_0=\frac{b_{N+1}}{a_N} \tag{7.32b}$$

The function $\mu(t)$ can be determined from the differential equation

$$P(D)\mu(t)=b_0,\quad (t>0) \tag{7.32c}$$

and initial conditions

$$\mu(0^+)=\frac{1}{a_N}[b_N-\eta_0 a_{N-1}] \tag{7.32d}$$

$$\mu^{(n)}(0^+)=\frac{1}{a_N}\left[b_{N-n}-\sum_{k=0}^{n-1} a_{N-n+k}\mu^{(n)}(0^+)-\eta_0 a_{N-n-1}\right],\quad (n=1 \text{ to } N-1) \tag{7.32e}$$

Proof of Theorem 7.4. Just as in Eq. 7.31f, $P(D)\mu(t)$ for any time t can be put in the form

$$P(D)[\mu(t)H(t)]=[P(D)\mu(t)]H(t)+\sum_{k=1}^{N}\delta^{(k-1)}(t)\sum_{r=k}^{N}a_r\mu^{(r-k)}(0^+)$$

$$=[P(D)\mu(t)]H(t)+\sum_{k=0}^{N-1}\delta^{(k)}(t)\sum_{r=k+1}^{N}a_r\mu^{(r-k-1)}(0^+),$$

or, due to Eqs. 7.32a and 7.32c,

$$P(D)\mu(t)=b_0H(t)+\sum_{k=0}^{N-1}\delta^{(k)}(t)\left[\sum_{r=k+1}^{N}a_r\mu^{(r-k-1)}(0^+)+\eta_0 a_k\right]+\eta_0 a_N\delta^{(N)}(t) \tag{7.32f}$$

The Laplace transform of the equation above is

$$P(p)\tilde{\mu}(p)=\frac{b_0}{p}+\sum_{k=0}^{N-1}p^k\left[\sum_{r=k+1}^{N}a_r\mu^{(r-k-1)}(0^+)+\eta_0 a_k\right]+\eta_0 a_N p^N \tag{7.32g}$$

To establish Eq. 7.15a or b is the same as to establish its Laplace transform, i.e. Eq. 7.15c. Comparing Eq. 7.31i with Eq. 7.15c, we can see that the equation to be proved is

$$pP(p)\tilde{\mu}(p)=Q(p) \tag{7.32h}$$

or, due to Eqs. 7.31j, 7.31k, 7.32b and 7.32g,

$$b_{k+1}=\sum_{r=k+1}^{N}[a_r\mu^{(r-k-1)}(0^+)+\eta_0 a_k],\quad (k=0 \text{ to } N-1),$$

or

$$b_k = \sum_{r=k}^{N} [a_r \mu^{(r-k)}(0^+) + \eta_0 a_{k-1}, \quad (k = 1 \text{ to } N) \tag{7.32i}$$

which is identical to Eq. 7.30d in Theorem 7.2 as should be expected. For $k = N$, Eq. 7.32i becomes identical to Eq. 7.32d; thus the remaining task is to show that Eq. 7.32i for $k = 1$ to $N - 1$ is the same as Eq. 7.32e. With the introduction of a new symbol $n = N - k$, Eq. 7.32i becomes

$$b_{N-n} = \sum_{r=N-n}^{N-1} [a_r \mu^{(r+n-N)}(0^+) + \eta_0 a_{N-n-1}] + a_N \mu^{(n)}(0^+),$$

or

$$\mu^{(n)}(0^+) = \frac{1}{a_N}\left[b_{N-n} - \sum_{r=N-n}^{N-1} a_r \mu^{(r+n-N)}(0^+) - \eta_0 a_{N-n-1}\right], \quad (n = 1 \text{ to } N-1),$$

or, putting $k = r + n - N$,

$$\mu^{(n)}(0^+) = \frac{1}{a_N}\left[b_{N-n} - \sum_{k=0}^{n-1} a_{N-n+k} \mu^{(k)}(0^+) - \eta_0 a_{N-n-1}\right], \quad (n = 1 \text{ to } N-1),$$

which is identical to Eq. 7.32e. Thus Theorem 7.4 is proved completely.

Being analogous to the preceding four theorems, the following four theorems are related to the generalized Kelvin model which is described by the creep law more easily than by the relaxation law.

Theorem 7.5. The creep compliance of the generalized Kelvin model satisfies the differential equation

$$Q^{(1)}(D)J(t) = \frac{b_0}{\eta_0}, \quad (t > 0) \tag{7.33a}$$

where

$$Q^{(1)}(D) = \frac{\mathrm{d}}{\mathrm{d}t} Q(D) \tag{7.33b}$$

$$Q(D) = \sum_{i=0}^{N} b_i D^i \tag{7.33c}$$

The b_i ($i = 0$ to N) are constants and N is the number of the Kelvin units in the model. In addition η_0 is the constant of the 'free' dashpot (Fig. 7.4).

Proof of Theorem 7.5. In view of Eqs. 7.21h and 7.33c, we can write for $t > 0$,

$$Q(D)J(t) = b_0 \sum_{i=0}^{N} \frac{1}{G_i} + \frac{b_1}{\eta_0} + \frac{b_0}{\eta_0} t - \sum_{i=1}^{N} \frac{e^{-t/\tau_i}}{G_i}\left[\sum_{k=0}^{N} (-1)^k b_k \alpha^k\right] \tag{7.33d}$$

where

$$\alpha_i = \tau_i^{-1} = G_i/\eta_i \tag{7.33e}$$

Consequently,

$$Q^{(1)}(D)J(t)=\frac{b_0}{\eta_0}+\sum_{i=1}^{N}\frac{e^{-t/\tau_i}}{\eta_i}\left[\sum_{k=0}^{N}(-1)^k b_k \alpha_i^k\right] \tag{7.33f}$$

Equations 7.33a and 7.33f are identical provided b_k ($k=0$ to N) satisfy the equation

$$\sum_{k=0}^{N}(-1)^k b_k \alpha_i^k=0 \tag{7.33g}$$

Such values of b_k are possible and can be obtained by the technique used in the proof of Theorem 7.1. Thus Theorem 7.5 is proved. It is interesting to note that

$$Q(D)J(t)=b_0\sum_{i=0}^{N}\frac{1}{G_i}+\frac{b_1}{\eta_0}+\frac{b_0}{\eta_0}t, \quad (t>0) \tag{7.33h}$$

and

$$Q(D)J(0^+)=\sum_{n=0}^{N}b_n J^{(n)}(0^+)=b_0\sum_{i=0}^{N}\frac{1}{G_i}+\frac{b_1}{\eta_0} \tag{7.33i}$$

$$Q^{(1)}(D)J(t)=Q^{(1)}(D)J(0^+)=\sum_{n=0}^{N}b_n J^{(n+1)}(0^+)=\frac{b_0}{\eta_0} \tag{7.33j}$$

Theorem 7.6. If $\sigma(t)$ and $\varepsilon(t)$ are continuous and N times continuously differentiable for $t>0$, then the differential equation for the relationship between $\sigma(t)$ and $\varepsilon(t)$ is

$$P(D)\sigma(t)=Q^{(1)}(D)\varepsilon(t), \quad (t>0) \tag{7.34a}$$

where $Q(D)$ is as defined in Eq. 7.33c and $P(D)$ is another differential operator one order higher than $Q(D)$,

$$P(D)=\sum_{i=0}^{N+1}a_i D^i=\sum_{i=0}^{N+1}a_i\frac{d^i}{dt^i} \tag{7.34b}$$

The relations among the coefficients of these operators are

$$a_0=\frac{b_0}{\eta_0} \tag{7.34c}$$

$$a_{r+1}=\sum_{n=r}^{N}b_N J^{(n-r)}(0^+), \quad (r=0 \text{ to } N) \tag{7.34d}$$

The initial conditions can be shown to be

$$\sum_{r=k}^{N}b_r\varepsilon^{(r-k)}(0^+)=\sum_{r=k}^{N}a_{r+1}\sigma^{(r-k)}(0^+), \quad (k=0 \text{ to } N) \tag{7.34e}$$

In this case, the creep law Eq. 7.11 holds, the creep compliance is given by Eq. 7.21h or Eq. 7.23a, coefficients b_r are given by Theorem 7.5, and the proof follows the lines adopted for Theorem 7.2.

Theorem 7.7. If the relationship between $\sigma(t)$ and $\varepsilon(t)$ given in Eq. 7.34a holds (for $t>0$), and the initial conditions are as in Eq. 7.34e, then the same relationship holds for $t \geq 0$, i.e.

$$P(D)\sigma(t) = Q^{(1)}(D)\varepsilon(t) \tag{7.35a}$$

where

$$\sigma(t) = \sigma(t)H(t) \tag{7.35b}$$

$$\varepsilon(t) = \varepsilon(t)H(t) \tag{7.35c}$$

The proof follows that for Theorem 7.3. The Laplace transform of Eq. 7.35a is

$$P(p)\tilde{\sigma}(p) = pQ(p)\tilde{\varepsilon}(p) \tag{7.35d}$$

where

$$P(p) = \sum_{r=0}^{N+1} a_r p^r \tag{7.35e}$$

$$Q(p) = \sum_{r=0}^{N} b_r p^r \tag{7.35f}$$

Theorem 7.8. If the relationship between $\sigma(t)$ and $\varepsilon(t)$ given in Eq. 7.35a holds, then the same relationship can be put in a convolution integral form Eq. 7.11. The creep compliance is the same as that of the generalized Kelvin model, i.e. Eq. 7.21h or Eq. 7.23a for which

$$\eta_0 = \frac{b_0}{a_0} \tag{7.36a}$$

The function $J(t)$ can be determined from the differential equation

$$Q^{(1)}(D)J(t) = b_0/\eta_0, \quad (t > 0) \tag{7.36b}$$

and initial conditions

$$J(0^+) = \frac{a_{N+1}}{b_N} \tag{7.36c}$$

$$J^{(n)}(0^+) = \frac{1}{b_N}\left[a_{N-n+1} - \sum_{k=0}^{n-1} b_{N-n+k} J^{(k)}(0^+)\right], \quad (n = 1 \text{ to } N) \tag{7.36d}$$

The proof follows that for Theorem 7.4.

7.5 STEADY STATE HARMONIC OSCILLATION

If the stress $\sigma(t)$ is specified as a *steady state harmonic function* of time in the form

$$\sigma(t) = \sigma_0 e^{i\omega t} \tag{7.37a}$$

then the starting transient strain 'dies out' with time and only a steady state harmonic strain is left, i.e.

$$\varepsilon(t) = \varepsilon_0 e^{i(\omega t - \delta)} \tag{7.37b}$$

Without loss of generality, σ_0, ε_0, ω and δ can be taken as positive real constants. More specifically, σ_0 and ε_0 are the *amplitudes* of stress and strain respectively, ω the *angular frequency*, and δ the *phase angle*. The stress-strain relationship may be put in the form

$$\sigma(t) = \mu^*(\omega)\varepsilon(t) \tag{7.38a}$$

or

$$\varepsilon(t) = J^*(\omega)\sigma(t) \tag{7.38b}$$

The quantities $\mu^*(\omega)$ and $J^*(\omega)$ are known as the *complex modulus* and the *complex compliance*, respectively; they are complex functions of the frequency ω. From these equations we deduce

$$\mu^*(\omega) = \frac{\sigma_0}{\varepsilon_0} e^{i\delta} \tag{7.39a}$$

$$J^*(\omega) = \frac{1}{\mu^*(\omega)} \tag{7.39b}$$

$$\tan\delta = \frac{\mu_2(\omega)}{\mu_1(\omega)} \tag{7.39c}$$

$$\frac{\sigma_0}{\varepsilon_0} = +\sqrt{\mu_1^2(\omega) + \mu_2^2(\omega)} = |\mu^*(\omega)| \tag{7.39d}$$

where $\mu_1(\omega)$ and $\mu_2(\omega)$ are the real and imaginary parts of $\mu^*(\omega)$, respectively, i.e.

$$\mu^*(\omega) = \mu_1(\omega) + i\mu_2(\omega) \tag{7.40}$$

Moreover, substituting Eqs. 7.37 into the relaxation law, Eq. 7.15b, yields for large t,

$$\sigma_0 e^{i\omega t} = \varepsilon_0 e^{i(\omega t - \delta)} \int_{0^-}^{t} e^{-i\omega s} d\mu(s), \quad (t \to \infty),$$

or

$$\frac{\sigma_0}{\varepsilon_0} e^{i\delta} = \int_{0^-}^{\infty} e^{-i\omega s} d\mu(s),$$

or, due to Eq. 7.39a,

$$\mu^*(\omega) = \int_{0^-}^{\infty} e^{-i\omega s} d\mu(s) \tag{7.41}$$

Alternatively, substituting Eqs. 7.37 into Eq. 7.15a yields

$$\sigma_0 e^{i\omega t} = \varepsilon_0 e^{-i\delta} i\omega \int_{0^-}^{t} \mu(t-s) e^{i\omega s} ds, \quad (t \to \infty),$$

or, due to the commutative law for convolution integrals (Eq. 1.138a),

$$\sigma_0 e^{i\omega t} = \varepsilon_0 e^{-i\delta} i\omega \int_{0^-}^{t} e^{i\omega(t-s)} \mu(s) ds, \quad (t \to \infty).$$

Alternatively

$$\frac{\sigma_0 e^{i\delta}}{\varepsilon_0} = i\omega \int_{0^-}^{\infty} e^{-i\omega s} \mu(s) ds,$$

or, due to Eq. 7.39a,

$$\mu^*(\omega) = i\omega\mu_*(\omega) \tag{7.42}$$

where $\mu_*(\omega)$ is the Fourier transform of $\mu(t)$, i.e.

$$\mu_*(\omega) = \int_{0^-}^{\infty} e^{-i\omega s} \mu(s) ds \tag{7.43}$$

If the relationship between the Fourier and Laplace transforms as given by Eq. 1.143 holds, i.e.

$$\mu_*(\omega) = \tilde{\mu}(i\omega) \tag{7.44}$$

we can obtain the complex modulus $\mu^*(\omega)$ from the Laplace transform of the relaxation modulus as follows

$$\mu^*(\omega) = i\omega\tilde{\mu}(i\omega) \tag{7.45}$$

The inverse of Eq. 7.43 is

$$\mu(t) = \frac{2}{\pi} \int_{0^-}^{\infty} \mu_*^c(\omega) \cos\omega t \, d\omega \tag{7.46a}$$

or

$$\mu(t) = \frac{2}{\pi} \int_{0^-}^{\infty} \mu_*^s(\omega) \sin\omega t \, d\omega \tag{7.46b}$$

where $\mu_*^c(\omega)$ and $\mu_*^s(\omega)$ are the Fourier cosine transform and Fourier sine transform, respectively, of $\mu(t)$, i.e.

$$\mu_*^c(\omega) = \int_{0^-}^{\infty} \mu(s) \cos\omega s \, ds = \Re[\mu_*(\omega)] \tag{7.47a}$$

$$\mu_*^s(\omega) = \int_{0^-}^{\infty} \mu(s) \sin\omega s \, ds = -\Im[\mu_*(\omega)] \tag{7.47b}$$

and $\Re$ and $\Im$ denote real part and imaginary part, respectively. Later, Eqs. 7.46 will be rewritten for the generalized Maxwell model in a more convenient form to obtain $\mu(t)$ from $\mu^*(\omega)$.

Similarly, we can obtain the following equations by means of the creep law:

$$J^*(\omega) = \int_{0^-}^{\infty} e^{-i\omega s} dJ(s) \tag{7.48a}$$

$$J^*(\omega) = i\omega J_*(\omega) \tag{7.48b}$$

where

$$J_*(\omega) = \int_{0^-}^{\infty} e^{-i\omega s} J(s) ds \tag{7.48c}$$

If

$$J_*(\omega) = \bar{J}(i\omega) \tag{7.48d}$$

then

$$J^*(\omega) = i\omega \bar{J}(i\omega) \tag{7.48e}$$

The inverse of Eq. 7.48c is

$$J(t) = \frac{2}{\pi} \int_{0^-}^{\infty} J_*^c(\omega) \cos\omega t \, d\omega \tag{7.49a}$$

or

$$J(t) = \frac{2}{\pi} \int_{0^-}^{\infty} J_*^s(\omega) \sin\omega t \, d\omega \tag{7.49b}$$

where

$$J_*^c(\omega) = \int_{0^-}^{\infty} J(s) \cos\omega s \, ds = \Re[J_*(\omega)] \tag{7.49c}$$

$$J_*^s(\omega) = \int_{0^-}^{\infty} J(s) \sin\omega s \, ds = -\Im[J_*(\omega)] \tag{7.49d}$$

Later, Eqs. 7.49a and b will be rewritten for the generalized Kelvin model in a more convenient form to obtain $J(t)$ from $J^*(\omega)$.

For the generalized Maxwell model with relaxation function given by Eq. 7.24a, Eq. 7.41 becomes

$$\mu^*(\omega) = \int_{0^-}^{\infty} e^{-i\omega s} [G_R + \phi(s)] \, \delta(s) ds + \int_{0^+}^{\infty} e^{-i\omega s} d\phi + \eta_0 \int_{0^-}^{\infty} e^{-i\omega s} \delta^{(1)}(s) ds,$$

or

$$\mu^*(\omega) = G_R + \phi(0^+) + \int_{0^+}^{\infty} e^{-i\omega s} d\phi + \eta_0 i\omega,$$

or, integrating by parts,

$$\mu^*(\omega) = G_R + \phi(0^+) - \phi(0^+) + i\omega \int_{0^+}^{\infty} e^{-i\omega s} \phi(s) ds + \eta_0 i\omega$$

$$= G_R + i\omega \int_{0^+}^{\infty} e^{-i\omega s} \phi(s) ds + \eta_0 i\omega \tag{7.50}$$

Substituting Eqs. 7.24 into Eq. 7.43 yields $\mu_*(\omega)$ in the form

$$\mu_*(\omega) = G_R \left[\pi\delta(\omega) + \frac{1}{i\omega} \right] + \int_{0^+}^{\infty} e^{-i\omega s} \phi(s) ds + \eta_0 \tag{7.51}$$

or

$$i\omega\mu_*(\omega) = G_R[i\pi\omega\delta(\omega) + 1] + i\omega\int_{0^+}^{\infty} e^{-i\omega s}\phi(s)ds + \eta_0 i\omega,$$

or, due to the fact (Eq. 1.110b) that $\omega\delta(\omega) = 0$,

$$i\omega\mu_*(\omega) = G_R + i\omega\int_{0^+}^{\infty} e^{-i\omega s}\phi(s)ds + \eta_0 i\omega \tag{7.52}$$

Comparing the equation above with Eq. 7.50, we can see that Eq. 7.42 is verified. However, the relationship between $\mu_*(\omega)$ and $\tilde{\mu}(i\omega)$ (Eq. 7.44) must be modified, due to the term $G_R H(t)$ in the function $\mu(t)$, into the following (see Eq. 1.144a)

$$\mu_*(\omega) = \tilde{\mu}(i\omega) + G_R\pi\delta(\omega) \tag{7.53}$$

Multiplying the equation above by $i\omega$ and observing the fact that $\omega\delta(\omega) = 0$ we find

$$i\omega\mu_*(\omega) = i\omega\tilde{\mu}(i\omega) \tag{7.54}$$

Hence Eq. 7.45 still holds, though Eq. 7.44 does not. The real and imaginary parts of Eq. 7.50 are, respectively,

$$\mu_1(\omega) = G_R + \omega\int_{0^+}^{\infty} \phi(s)\sin\omega s\, ds \tag{7.55a}$$

$$\mu_2(\omega) = \eta_0\omega + \omega\int_{0^+}^{\infty} \phi(s)\cos\omega s\, ds \tag{7.55b}$$

Since

$$\mu_1(0) = G_R \tag{7.56}$$

the inverse Fourier transforms of Eqs. 7.55a and b give, respectively,

$$\mu(t) = \left\{\mu_1(0) + \frac{2}{\pi}\int_0^{\infty}\left[\frac{\mu_1(\omega) - \mu_1(0)}{\omega}\right]\sin\omega t\, d\omega\right\}H(t) + \eta_0\delta(t) \tag{7.57a}$$

$$\mu(t) = \left\{\mu_1(0) + \frac{2}{\pi}\int_0^{\infty}\left[\frac{\mu_2(\omega) - \eta_0\omega}{\omega}\right]\cos\omega t\, d\omega\right\}H(t) + \eta_0\delta(t) \tag{7.57b}$$

The infinite integrals involve only absolutely integrable functions. While Eq. 7.50 gives a way of finding $\mu^*(\omega)$ from $\mu(t)$, Eqs. 7.57a and b give a way of finding $\mu(t)$ from $\mu^*(\omega)$. Explicitly, the complex modulus of the generalized Maxwell model can be expressed as

$$\mu^*(\omega) = G_R + \sum_{i=1}^{N}\frac{\omega^2\eta_i\tau_i}{1 + \omega^2\tau_i^2} + i\left[\omega\eta_0 + \sum_{i=1}^{N}\frac{\omega\eta_i}{1 + \omega^2\tau_i^2}\right] \tag{7.58}$$

For the generalized Kelvin model, the creep compliance are given in Eq. 7.23a, or

$$J(t) = \left[J_R^c + \phi^c(t) + \frac{t}{\eta_0}\right]H(t) \tag{7.59}$$

where

$$J_R^c = J_g + \psi(\infty) \tag{7.60a}$$

$$\phi^c(t) = \psi(t) - \psi(\infty) \tag{7.60b}$$

Note $\phi^c(t)$ monotonically approaches zero as t tends to infinity. Substituting Eq. 7.59 into Eq. 7.48a yields

$$J^*(\omega) = \int_{0^-}^{\infty} e^{-i\omega s}[J_R^c + \phi^c(s)]\,\delta(s)ds + \int_{0^+}^{\infty} e^{-i\omega s}d\phi^c + \frac{1}{\eta_0}\int_{0^-}^{\infty} s\,\delta(s)ds + \frac{1}{\eta_0}\int_{0^-}^{\infty} e^{-i\omega s}H(s)ds,$$

or, due to the properties of $\delta(t)$ and the Fourier transform of $H(t)$ (Eq. 1.144a),

$$J^*(\omega) = J_R^c + \phi^c(0) + \int_{0^+}^{\infty} e^{-i\omega s}d\phi^c + \frac{1}{\eta_0}\left[\pi\delta(\omega) + \frac{1}{i\omega}\right],$$

or, integrating by parts,

$$J^*(\omega) = J_R^c + i\omega\int_{0^+}^{\infty} e^{-i\omega s}d\phi^c(s)ds + \frac{1}{\eta_0}\left[\pi\delta(\omega) + \frac{1}{i\omega}\right] \tag{7.61}$$

On the other hand, the Fourier transform of $J(t)$ is

$$J_*(\omega) = \int_{0^-}^{\infty} e^{-i\omega s}J(s)ds$$

$$= J_R^c\int_{0^-}^{\infty} e^{-i\omega s}H(s)ds + \int_{0^+}^{\infty} e^{-i\omega s}\phi^c(s)ds + \frac{1}{\eta_0}\int_{0^-}^{\infty} e^{-i\omega s}sH(s)ds,$$

or, due to Eqs. 1.144,

$$J_*(\omega) = J_R^c\left[\pi\delta(\omega) + \frac{1}{i\omega}\right] + \int_{0^+}^{\infty} e^{-i\omega s}\phi^c(s)ds + \frac{1}{\eta_0}\left[\pi i\delta^{(1)}(\omega) - \frac{1}{\omega^2}\right],$$

or

$$i\omega J_*(\omega) = J_R^c[\pi i\omega\delta(\omega) + 1] + i\omega\int_{0^+}^{\infty} e^{-i\omega s}\phi^c(s)ds + \frac{1}{\eta_0}\left[-\pi\omega\delta^{(1)}(\omega) + \frac{1}{i\omega}\right],$$

or, due to the fact (Eqs. 1.110) that $\omega\delta(\omega) = 0$ and $\omega\delta^{(1)}(\omega) = -\delta(\omega)$,

$$i\omega J_*(\omega) = J_R^c + i\omega\int_{0^+}^{\infty} e^{-i\omega s}\phi^c(s)ds + \frac{1}{\eta_0}\left[\pi\delta(\omega) + \frac{1}{i\omega}\right] \tag{7.62}$$

Comparing the equation above with Eq. 7.61, we can see that Eq. 7.48b is verified. However, Eq. 7.48d must be modified, due to the term $(J_R^c + t/\eta_0)H(t)$ in the function $J(t)$, so that it reads

$$J_*(\omega) = \bar{J}(i\omega) + J_R^c\pi\delta(\omega) + \frac{\pi i\delta^{(1)}(\omega)}{\eta_0} \tag{7.63}$$

Multiplying the equation above by $i\omega$ and observing that $\omega\delta(\omega) = 0$ and $\omega\delta^{(1)}(\omega) = -\delta(\omega)$, we find

$$i\omega J_*(\omega) = i\omega\bar{J}(i\omega) + \frac{\pi\delta(\omega)}{\eta_0} \tag{7.64}$$

Hence Eq. 7.48e must be modified to

$$J^*(\omega) = \mathrm{i}\omega\bar{J}(\mathrm{i}\omega) + \frac{\pi\delta(\omega)}{\eta_0} \tag{7.65}$$

It should be noted that Eq. 7.39b can be used to obtain correct $\mu^*(\omega)$ from a known $J^*(\omega)$, but in reverse the same equation cannot yield the term $\delta(\omega)$ in obtaining $J^*(\omega)$ from a known $\mu^*(\omega)$. This complication does not arise if there is no dashpot in series with other units in the model, i.e. when $\eta_0 \to \infty$. The real and imaginary parts of Eq. 7.61 are, respectively,

$$J_1(\omega) = J_R^c + \omega\int_{0^+}^{\infty} \phi^c(s)\sin\omega s\,ds + \frac{\pi\delta(\omega)}{\eta_0} \tag{7.66a}$$

$$J_2(\omega) = \omega\int_{0^+}^{\infty} \phi^c(s)\cos\omega s\,ds - (\eta_0\omega)^{-1} \tag{7.66b}$$

Because

$$J_1(0^+) = J_R^c \tag{7.67}$$

the inverse Fourier transforms of Eqs. 7.66a and b give, respectively,

$$J(t) = \left\{J_1(0^+) + \frac{2}{\pi}\int_0^{\infty}\left[\frac{J_1(\omega) - J_1(0^+) - \pi\delta(\omega)/\eta_0}{\omega}\right]\sin\omega t\,d\omega + \frac{t}{\eta_0}\right\}H(t) \tag{7.68a}$$

$$J(t) = \left\{J_1(0^+) + \frac{2}{\pi}\int_0^{\infty}\left[\frac{J_2(\omega) + (\eta_0\omega)^{-1}}{\omega}\right]\cos\omega t\,d\omega + \frac{t}{\eta_0}\right\}H(t) \tag{7.68b}$$

Again the infinite integrals involve only absolutely integrable functions. While Eq. 7.61 provides a way of finding $J^*(\omega)$ from $J(t)$, Eqs. 7.68a and b give a way of finding $J(t)$ from $J^*(\omega)$. Explicitly, the complex compliance of the generalized Kelvin model can be expressed as

$$J^*(\omega) = (G_0)^{-1} + \frac{\pi\delta(\omega)}{\eta_0} + \sum_{i=1}^{N} \tau_i^2[\eta_i\tau_i(1+\omega^2\tau_i^2)]^{-1} - \mathrm{i}\left\{(\eta_0\omega)^{-1} + \sum_{i=1}^{N} \omega\tau_i^2[\eta_i(1+\omega^2\tau_i^2)]^{-1}\right\} \tag{7.69}$$

Next, let us consider the dissipation of energy in steady state harmonic oscillation. For a one-dimensional mechanical model, the work done in the time interval from t_0 to any t is

$$\Delta U = \int_{t_0}^{t} \Re[\sigma(s)]\,\Re(d\varepsilon) \tag{7.70}$$

Substituting Eqs. 7.37 into Eq. 7.70, in view of Eq. 7.39a, we find

$$\Delta U = -\varepsilon_0^2\omega\Re\left[\mu^*(\omega)\int_{t_0}^{t} e^{\mathrm{i}(\omega s - \delta)}\sin(\omega s - \delta)ds\right] \tag{7.71}$$

For the time interval equivalent to one cycle, the equation above becomes

$$\Delta U = -\varepsilon_0^2\omega\Re\left[\mu^*(\omega)\int_{t_0}^{t_0+2\pi/\omega} e^{i(\omega s-\delta)}\sin(\omega s-\delta)ds\right]$$

$$= -\varepsilon_0^2\omega\Re\left[\frac{\mu^*(\omega)i\pi}{\omega}\right],$$

or, due to Eq. 7.40,

$$\Delta U = \pi\varepsilon_0^2\mu_2(\omega) = \frac{\pi\sigma_0^2\mu_2(\omega)}{|\mu^*(\omega)|^2} \tag{7.72}$$

which is the amount of *energy dissipation per cycle*. Due to Eqs. 7.39c and d, this equation can be rewritten

$$\Delta U = \pi\varepsilon_0^2|\mu^*(\omega)|\sin\delta \tag{7.73}$$

Defining U_{max} as

$$U_{max} = \frac{1}{2}|\operatorname{Max}\sigma(t)|\,|\operatorname{Max}\varepsilon(t)| \tag{7.74a}$$

$$= \frac{1}{2}|\mu^*(\omega)|\,\varepsilon_0^2 \tag{7.74b}$$

we find the *specific loss* to be

$$\frac{\Delta U}{U_{max}} = 2\pi\sin\delta \tag{7.75}$$

If δ is small, the energy dissipation per cycle and the specific loss assume the approximate forms

$$\Delta U \approx \pi\varepsilon_0^2|\mu^*(\omega)|\tan\delta \tag{7.76a}$$

$$\frac{\Delta U}{U_{max}} \approx 2\pi\tan\delta \tag{7.76b}$$

Appropriately, $\mu_2(\omega)$ is called the *loss modulus*, and δ the *loss angle*. For elastic models, these two quantities are zero, and there is no energy dissipation.

7.6 THERMORHEOLOGICALLY SIMPLE SOLIDS

For a wide variety of materials known as *thermorheologically simple solids*, the mechanical properties depend on the temperature in a simple analytical fashion. This temperature dependence was first proposed by Leadermann (1943) and Ferry (1970), and applied by Schwarzl and Staverman (1952). The present derivation follows the work of Morland and Lee (1960). For this type of materials, the temperature dependence of the relaxation modulus is as shown in Fig. 7.10, in which

$$F(\log t) = \mu(t) \tag{7.77}$$

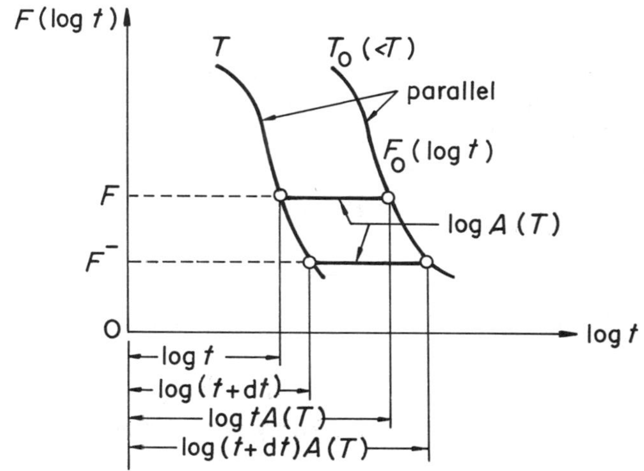

Fig. 7.10 Relaxation modulus of thermorheologically simple solids.

Note in the figure that the curve for a temperature T has a constant 'shift' $\log A(T)$ away from that for the reference temperature T_0, i.e.

$$F(\log t) = F_0[\log t + \log A(t)]$$

$$= F_0[\log tA(t)],$$

or

$$\mu(t) = \mu_0(\xi) \tag{7.78}$$

where the subscript ($_0$) denotes the reference temperature, and

$$\xi = tA(T) \tag{7.79}$$

which is known as the *pseudo* or *reduced time*. Naturally $A(T_0)$ is equal to unity. It will be shown later that the creep compliance at an elevated temperature can be obtained from that at the reference temperature in the same manner as Eq. 7.78, i.e.

$$\mu(t) = \mu_0(t)|_{t=tA(T)} \tag{7.80a}$$

$$J(t) = J_0(t)|_{t=tA(T)} \tag{7.80b}$$

In the process of getting $\mu(t)$ and $J(t)$ by Eqs. 7.80, one should note the following Heaviside step function, the Dirac-delta function and its derivatives

$$H[tA(T)] = H(t) \tag{7.81a}$$

$$\delta[tA(T)] = \frac{\delta(t)}{A(T)} \tag{7.81b}$$

$$\delta^{(m)}[tA(T)] = \frac{\delta^{(m)}(t)}{A^{m+1}(T)} \tag{7.81c}$$

In an isothermal process, T is kept independent of t, so the stress-strain relationship assumes the usual form, i.e. the creep law or the relaxation law, Eq. 7.7 or 7.15a respectively, while temperature enters the equations only as a parameter.

For *isothermal harmonic vibration*, Eqs. 7.37a and b become respectively

$$\sigma(t) = \sigma_0 e^{i\omega^*\xi} \tag{7.82a}$$

$$\varepsilon(t) = \varepsilon_0 e^{i\omega^*\xi} \tag{7.82b}$$

where ξ is given in Eq. 7.79, and ω^* is the *pseudo* or *reduced frequency*, i.e.

$$\omega^* = \frac{\omega}{A(T)} \tag{7.83}$$

The complex amplitudes σ_0 and ε_0 are related by the equation

$$\sigma_0 = [\mu_1(\omega^*) + i\mu_2(\omega^*)]\,\varepsilon_0 \tag{7.84a}$$

or

$$\varepsilon_0 = [J_1(\omega^*) + iJ_2(\omega^*)]\,\sigma_0 \tag{7.84b}$$

where $\mu_1(\omega)$ and $\mu_2(\omega)$ are real and imaginary parts of the complex modulus $\mu^*(\omega)$ at the reference temperature T_0, and $J_1(\omega)$ and $J_2(\omega)$ are those of the complex compliance $J^*(\omega)$. The dissipation of energy per cycle, Eq. 7.72, is

$$\Delta U = \pi|\varepsilon_0|^2\mu_2(\omega^*) = -\pi|\sigma_0|^2 J_2(\omega^*) \tag{7.85a,b}$$

Figure 7.10 also shows that if the time taken for μ to decrease from some value μ to μ^- is dt, then the time taken for μ_0 to decrease the same amount is $d\xi = A(T)dt$. If ξ corresponds to varying temperature $T(t)$, then

$$\mu(t) = \hat{\mu}(\xi) = \mu_0(t)|_{t=\xi} \tag{7.86a}$$

$$A(T)dt < d\xi < \left[A(T) + \frac{dA(T)}{dT}\frac{dT(t)}{dt}dt\right]dt \tag{7.86b}$$

Neglecting the $(dt)^2$ term in the equation above, we get

$$d\xi = A(T)dt \tag{7.87a}$$

or

$$\frac{d\xi}{dt} = A(T) \tag{7.87b}$$

Equations 7.87 hold for both isothermal and non-isothermal cases. Integrating Eq. 7.87b and using the initial condition $\xi(0) = 0$ we find

$$\xi(t) = \int_0^t A[T(p)]\,dp \tag{7.88}$$

Applying Leibnitz's rule for differentiating an integral, Eq. 1.137, to this equation we recover Eq. 7.87b. Putting $t = t - s$ in Eq. 7.88 we find

$$\xi(t-s) = \int_0^{t-s} A(T)\mathrm{d}p.$$

This integral is the shaded area in Fig. 7.11, in which the identity

$$q = p + s$$

should be noted. Using q instead of p as the integrating variable we find

$$\xi(t-s) = \int_s^t A(T)\mathrm{d}q$$

$$= \int_0^t A(T)\mathrm{d}q - \int_0^s A(T)\mathrm{d}q,$$

or

$$\xi(t-s) = \xi(t) - \xi(s) \tag{7.89}$$

Thus if $f(t) = \hat{f}(\xi)$, then we can write

$$f(t-s) = \hat{f}(\xi - \zeta) \tag{7.90}$$

where ξ is as defined by Eq. 7.88, and

$$\zeta = \xi(s) = \int_0^s A(T)\mathrm{d}p \tag{7.91}$$

The relationship between the relaxation modulus and the creep compliance, Eq. 7.14a, is

$$\tilde{\bar{J}}(p) = \left[p^2 \tilde{\bar{\mu}}(p)\right]^{-1} \tag{7.92a}$$

$$\bar{J}_0(p) = [p^2 \tilde{\mu}_0(p)]^{-1} \tag{7.92b}$$

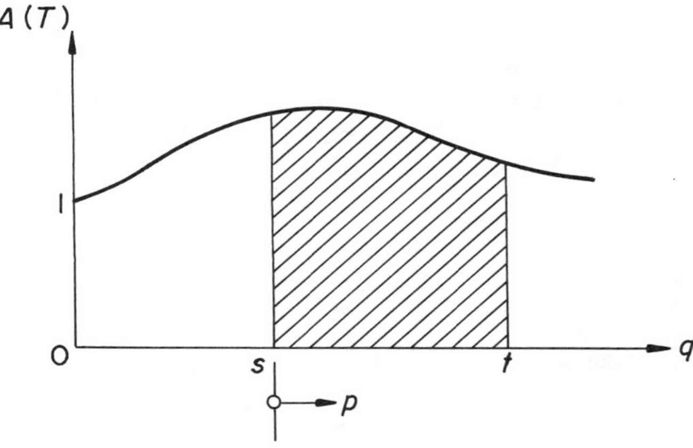

Fig. 7.11 Graphical representation of $\xi(t-s)$.

where $\tilde{\hat{J}}(p)$ and $\tilde{J}_0(p)$ are Laplace transforms of $\hat{J}(\xi)$ and $J_0(t)$, respectively. Note that we can, without loss of generality, use the same Laplace transform parameter p against both ξ and t. Due to Eq. 7.86a,

$$\tilde{\hat{\mu}}(p) = \tilde{\mu}_0(p).$$

Thus Eqs. 7.92 give

$$J(t) = \hat{J}(\xi) = J_0(t)|_{t=\xi} \tag{7.93}$$

Equations 7.86a and 7.93 are identical to Eqs. 7.80a and b, respectively. Taking the initial temperature, i.e. $T(0) = T_0$, we have $A[T(0)] = A(T_0) = 1$. So the Heaviside step function and the Dirac-delta function assume identical forms with the argument t or ξ, i.e.

$$H(t) = H(\xi), \qquad \delta(t) = \delta(\xi) \tag{7.94a,b}$$

but the derivatives of the Dirac-delta function follow the following recurrence formula

$$\delta^{(m)}(t) = A(T)\frac{\mathrm{d}}{\mathrm{d}\xi}[\delta^{(m-1)}(t)], \quad (m = 1,2,3,\ldots..) \tag{7.94c}$$

or, reversely,

$$\delta^{(m)}(\xi) = \frac{1}{A(T)}\frac{\mathrm{d}}{\mathrm{d}t}[\delta^{(m-1)}(\xi)], \quad (m = 1,2,3,\ldots..) \tag{7.94d}$$

As an example, both Eqs. 7.94c and d give $\delta^{(1)}(t) = A(T)\delta^{(1)}(\xi)$. Incorporating the temperature effects, the stress-strain relationship, Eq. 7.7 or 7.15a, can be written

$$\varepsilon(t) = \alpha(t) + \int_{0^-}^{t} J(t-s)\mathrm{d}\sigma(s) \tag{7.95a}$$

$$\sigma(t) = \int_{0^-}^{t} \mu(t-s)\mathrm{d}[\varepsilon(s) - \alpha(s)] \tag{7.95b}$$

where $\alpha(t)$ is the *strain due to the temperature change in a stress-free state.* As functions of the pseudo time ξ, $\hat{\mu}(\xi)$ and $\hat{J}(\xi)$ are given by Eqs. 7.86a and 7.93 respectively, while the others can be introduced as the following

$$\hat{\sigma}(\xi) = \sigma(t)|_{t=t(\xi)} \tag{7.96a}$$

$$\hat{\varepsilon}(\xi) = \varepsilon(t)|_{t=t(\xi)} \tag{7.96b}$$

$$\hat{\alpha}(\xi) = \alpha(t)|_{t=t(\xi)} \tag{7.96c}$$

and their time differentials take the form

$$\mathrm{d}f(s) = \mathrm{d}\hat{f}(\zeta) = \frac{\mathrm{d}\hat{f}(\zeta)}{\mathrm{d}\zeta}\mathrm{d}\zeta \tag{7.96d}$$

Substituting Eqs. 7.96 into Eqs. 7.95, and using Eq. 7.90, we obtain

$$\hat{\varepsilon}(\xi) = \hat{\alpha}(\xi) + \int_{0^-}^{\xi} \hat{J}(\xi - \zeta)\mathrm{d}\hat{\sigma}(\zeta) \tag{7.97a}$$

$$\hat{\sigma}(\xi) = \int_{0^-}^{\xi} \hat{\mu}(\xi - \zeta)\mathrm{d}[\hat{\varepsilon}(\zeta) - \hat{\alpha}(\zeta)] \tag{7.97b}$$

The final results are then obtained as functions of t by simply noting that

$$f(t) = \hat{f}(\xi)|_{\xi = \xi(t)} \tag{7.98}$$

and incorporating the special identities given in Eqs. 7.94.

7.7 THREE-DIMENSIONAL THEORY

The constitutive law for three-dimensional continua may be generalized from that already described for one-dimensional mechanical models. We state it in the differential form

$$P_{ijkl}(D)\sigma_{kl} = Q_{ijkl}(D)\varepsilon_{kl} \tag{7.99a}$$

or the corresponding integral form

$$\sigma_{ij}(t) = C_{ijkl}{*}\mathrm{d}\varepsilon_{kl} = \varepsilon_{kl}{*}\mathrm{d}C_{ijkl} \tag{7.99b}$$

The convolution integral symbol are as defined in Eq. 7.12. The latter form (Eq. 7.99b) is more convenient for describing various aspects of the theory, and we will use it exclusively. As in linear elasticity, C_{ijkl} has the symmetric property given in Eq. 3.14 and is positive definite at any value of t. Unless specified otherwise the discussion which follows is restricted to isothermal conditions and thermorheologically simple solids; the temperature enters only as a parameter, as in Eq. 7.79.

In the material domain V, the equation of motion and the strain-displacement relationship are the same as those for any solids under small deformation, i.e. respectively

$$\sigma_{ji,j} + X_i = \rho \frac{\partial^2 u_i}{\partial t^2} \tag{7.100}$$

$$\varepsilon_{ij} = \frac{1}{2}(u_{i,j} + u_{j,i}) \tag{7.101}$$

For isotropic materials, the stress-strain relationship, Eq. 7.99b, can be reduced to

$$\sigma_{ij}(t) = \delta_{ij}\lambda{*}\mathrm{d}\varepsilon_{kk} + 2\mu{*}\mathrm{d}\varepsilon_{ij} = \delta_{ij}\varepsilon_{kk}{*}\mathrm{d}\lambda + 2\varepsilon_{ij}{*}\mathrm{d}\mu \tag{7.102}$$

which is analogous to Eq. 3.32 in linear elasticity. The introduction of the stress deviator s_{ij} (Eq. 2.34) and the strain deviator e_{ij} (Eq. 2.47) makes it possible to state separate relationships for shear and bulk:

$$s_{ij} = 2\mu{*}\mathrm{d}e_{ij} \quad \text{or} \quad e_{ij} = \frac{1}{2}J{*}\mathrm{d}s_{ij} \tag{7.103a}$$

$$\sigma_{kk} = 3K*d\varepsilon_{kk} \quad \text{or} \quad \varepsilon_{kk} = \frac{1}{3}B*d\sigma_{kk} \tag{7.103b}$$

Here $\mu(t)$ and $K(t)$ are relaxation moduli for shear and bulk respectively, $J(t)$ and $B(t)$ are their respective creep compliances. The relationship between $\lambda(t)$, $\mu(t)$ and $K(t)$ is the same as in elasticity, i.e. $K(t) = \lambda(t) + 2\mu(t)/3$. The relationship between a relaxation modulus and its corresponding compliance is given in Eqs. 7.14. Equations 7.103 are analogous to Eqs. 3.34 and 3.35 in isotropic elasticity.

For uniaxial stress in the x_1-direction (only the stress σ_{11} is non-zero), *Young's modulus* $E(t)$ and *Poisson's ratio* $\nu(t)$ can be introduced through the relationship

$$\sigma_{11}(t) = E*d\varepsilon_{11} \tag{7.104a}$$

and

$$\varepsilon_{22}(t) = \varepsilon_{33}(t) = -\nu*d\varepsilon_{11} \tag{7.104b}$$

or, in Laplace transforms,

$$\tilde{\sigma}_{11}(p) = p\tilde{E}(p)\tilde{\varepsilon}_{11}(p) \tag{7.104c}$$

and

$$\tilde{\varepsilon}_{22}(p) = \tilde{\varepsilon}_{33}(p) = -p\tilde{\nu}(p)\tilde{\varepsilon}_{11}(p) \tag{7.104d}$$

We can obtain the following relationship between Laplace transforms of material functions as:

$$\tilde{E} = \frac{\tilde{\mu}(3\tilde{\lambda}+2\tilde{\mu})}{\tilde{\lambda}+\tilde{\mu}} = \frac{9\tilde{\mu}\tilde{K}}{\tilde{\mu}+3\tilde{K}} \tag{7.105a,b}$$

$$\tilde{\nu} = \frac{\tilde{\lambda}}{2p(\tilde{\lambda}+\tilde{\mu})} = \frac{3\tilde{K}-2\tilde{\mu}}{2p(\tilde{\mu}+3\tilde{K})} \tag{7.105c,d}$$

Table 7.1

	$\tilde{\lambda},\tilde{\mu}$	$\tilde{E},\tilde{\nu}$	$\tilde{\mu},\tilde{K}$	$\tilde{\mu},\tilde{\nu}$
$\tilde{\lambda}$	$\tilde{\lambda}$	$\frac{\tilde{E}p\tilde{\nu}}{(1+p\tilde{\nu})(1-2p\tilde{\nu})}$	$\frac{3\tilde{K}-2\tilde{\mu}}{3}$	$\frac{2\tilde{\mu}p\tilde{\nu}}{1-2p\tilde{\nu}}$
$\tilde{\mu}$	$\tilde{\mu}$	$\frac{\tilde{E}}{2(1+p\tilde{\nu})}$	$\tilde{\mu}$	$\tilde{\mu}$
$\tilde{K}$	$\tilde{\lambda}+\frac{2}{3}\tilde{\mu}$	$\frac{\tilde{E}}{3(1-2p\tilde{\nu})}$	$\tilde{K}$	$\frac{2\tilde{\mu}(1+p\tilde{\nu})}{3(1-2p\tilde{\nu})}$
$\tilde{E}$	$\frac{\tilde{\mu}(3\tilde{\lambda}+2\tilde{\mu})}{\tilde{\lambda}+\tilde{\mu}}$	$\tilde{E}$	$\frac{9\tilde{K}\tilde{\mu}}{3\tilde{K}+\tilde{\mu}}$	$2\tilde{\mu}(1+p\tilde{\nu})$
$\tilde{\nu}$	$\frac{\tilde{\lambda}}{2p(\tilde{\lambda}+\tilde{\mu})}$	$\tilde{\nu}$	$\frac{3\tilde{K}-2\tilde{\mu}}{2p(3\tilde{K}+\tilde{\mu})}$	$\tilde{\nu}$

A number of useful relations between the Laplace transforms of these isotropic relaxation moduli are shown in Table 7.1, which is the analogue of Table 3.1 in elasticity theory.

Combining Eqs. 7.100, 7.101 and 7.102 we obtain Navier equations for viscoelasticity, namely

$$\mu * \mathrm{d}u_{i,jj} + (\lambda+\mu) * \mathrm{d}u_{k,ki} + X_i = \rho \frac{\partial^2 u_i}{\partial t^2} \tag{7.106}$$

which is the analogue of Eq. 3.66 in isotropic elasticity.

For *quasi-static* problems, the inertial term $\partial^2 u_i/\partial t^2$ is neglected, so that the equations of equilibrium become

$$\sigma_{ij,j} + X_i = 0 \tag{7.107}$$

In terms of stress components, the compatibility conditions, Eq. 2.62, for isotropic materials are

$$\sigma_{ij,kk} * \mathrm{d}J + \sigma_{kk,ij} * \mathrm{d}\Delta + \delta_{ij} X_{k,k} * \mathrm{d}\chi + (X_{i,j} + X_{j,i}) * \mathrm{d}J = 0 \tag{7.108}$$

where

$$\Delta(t) = 2[3J(t) + B(t)]/9 \tag{7.109a}$$

$$\tilde{\chi}(p) = \tilde{J}(3\tilde{J} - 2\tilde{B})/(3\tilde{J} + 4\tilde{B}) \tag{7.109b}$$

Equation 7.108 can be called the Beltrami-Michell compatibility conditions for viscoelasticity. The *energy identity* is defined as

$$I(t) = \int_V X_i * \mathrm{d}u_i \mathrm{d}V + \int_S \overline{X}_i * \mathrm{d}u_i \mathrm{d}S \tag{7.110a}$$

where S is the surface completely enclosing the domain V, and $\overline{X}_i$ is the surface force (Eq. 2.6). This can be written

$$I(t) = \int_V \sigma_{ij} * \mathrm{d}\varepsilon_{ij} \mathrm{d}V \tag{7.110b}$$

$$= \int_V \left(s_{ij} * \mathrm{d}e_{ij} + \frac{1}{3}\sigma_{kk} * \mathrm{d}\varepsilon_{mm} \right) \mathrm{d}V \tag{7.110c}$$

or, for isotropic materials,

$$I(t) = \int_V [(2\mu * \mathrm{d}e_{ij}) * \mathrm{d}e_{ij} + (K * \mathrm{d}\varepsilon_{kk}) * \mathrm{d}\varepsilon_{mm}] \,\mathrm{d}V \tag{7.111a}$$

$$= \int_V \left[\left(\frac{1}{2} J * \mathrm{d}s_{ij} \right) * \mathrm{d}s_{ij} + \left(\frac{1}{9} B * \mathrm{d}\sigma_{kk} \right) * \mathrm{d}\sigma_{mm} \right] \mathrm{d}V \tag{7.111b}$$

The Maxwell-Betti reciprocal theorem linking two states 1 and 2 can be written

$$\int_V X_i^{(1)} * \mathrm{d}u_i^{(2)} \mathrm{d}V + \int_S \overline{X}_i^{(1)} * \mathrm{d}u_i^{(2)} \mathrm{d}S = \int_V X_i^{(2)} * \mathrm{d}u_i^{(1)} \mathrm{d}V + \int_S \overline{X}_i^{(2)} * \mathrm{d}u_i^{(1)} \mathrm{d}S$$

$$= \int_V \sigma_{ij}^{(1)} * \mathrm{d}\varepsilon_{ij}^{(2)} \mathrm{d}V$$

$$= \int_V \sigma_{ij}^{(2)} * \mathrm{d}\varepsilon_{ij}^{(1)} \mathrm{d}V \tag{7.112}$$

Equation 7.112 holds for anisotropy as well as isotropy. The *volume change*, i.e. $\int_V \varepsilon_{kk} \mathrm{d}V$, can be evaluated by using Eq. 7.103b:

$$\int_V \varepsilon_{kk} \mathrm{d}V = \frac{1}{3}\int_V \sigma_{kk} * \mathrm{d}B\, \mathrm{d}V$$

$$= \frac{1}{3}\int_V \delta_{kj}\sigma_{kj} * \mathrm{d}B\, \mathrm{d}V$$

$$= \frac{1}{3}\int_V \frac{\partial x_k}{\partial x_j}\sigma_{kj} * \mathrm{d}B\, \mathrm{d}V$$

$$= \frac{1}{3}\int_V [(x_k\sigma_{kj})_{,j} - x_k\sigma_{kj,j}] * \mathrm{d}B\, \mathrm{d}V.$$

If $B(t)$ is assumed to be spatially uniform, then Gauss's divergence theorem applied to the first term on the right-hand side yields

$$\int_V \varepsilon_{kk} \mathrm{d}V = \frac{1}{3}\left[\int_S x_k n_j \sigma_{kj} * \mathrm{d}B\, \mathrm{d}S - \int_V x_k \sigma_{kj,j} * \mathrm{d}B\, \mathrm{d}V\right].$$

Using the stress vector formula (Eq. 2.6) and the equation of equilibrium (Eq. 7.107), we obtain

$$\int_V \varepsilon_{kk} \mathrm{d}V = \frac{1}{3}\left[\int_S x_k \overline{X}_k * \mathrm{d}B\, \mathrm{d}S + \int_V x_k X_k * \mathrm{d}B\, \mathrm{d}V\right] \tag{7.113}$$

This equation expresses the volume change of a homogeneous domain in terms of integrals of the prescribed body and surface forces.

7.8 QUASI-STATIC SOLUTION BY SEPARATION OF VARIABLES

Following Alfrey (1944) for incompressible solids ($\nu = 1/2$) and Tsien (1950) for a constant Poisson's ratio, i.e.

$$\nu(t) = \nu H(t) \tag{7.114a}$$

we obtain separable solutions for the case in which the body force, boundary tractions and prescribed boundary displacements have the form

$$X_i(x_i, t) = X_i^0(x_i) X(t), \quad (\text{in } V) \tag{7.114b}$$

$$\overline{X}_i(x_i, t) = \overline{X}_i^0(x_i) X(t), \quad (\text{on } S_\sigma) \tag{7.114c}$$

$$u_i(x_i, t) = \overline{u}_i^0(x_i) u(t), \quad (\text{on } S_u) \tag{7.114d}$$

where S_σ and S_u are parts of the surface S. We assume a separable solution

$$u_i(x_i, t) = u_i^0(x_i) u(t) \tag{7.115a}$$

$$\sigma_{ij}(x_i,t)=\sigma^0_{ij}(x_i)X(t) \tag{7.115b}$$

Note that the displacement time function is the same as that prescribed on the surface S_u, and the stress time function is the same as that of the surface force on the surface S_σ and the same as that of the body force in the domain V.

In addition, the spatial functions in the equations correspond to the elastostatic problem with unit shear modulus, i.e. they satisfy the equilibrium equations

$$\sigma^0_{ji,j}(x_i)+X^0_i(x_i)=0 \tag{7.116a}$$

the strain-displacement relationship

$$\varepsilon^0_{ij}(x_i)=\frac{1}{2}[u^0_{i,j}(x_i)+u^0_{j,i}(x_i)] \tag{7.116b}$$

and the stress-strain relationship (for $\mu=1$)

$$\sigma^0_{ij}(x_i)=\frac{2\nu}{(1-2\nu)}\delta_{ij}\varepsilon^0_{kk}(x_i)+2\varepsilon^0_{ij}(x_i) \tag{7.116c}$$

If the stresses are the basic unknowns in the solution scheme, then we must incorporate the Beltrami-Michell compatibility conditions (Eq. 3.77)

$$\sigma^0_{ij,kk}(x_i)+\frac{1}{1+\nu}\sigma^0_{kk,ij}(x_i)+\frac{\nu}{1-\nu}\delta_{ij}X^0_{k,k}(x_i)+X^0_{i,j}(x_i)+X^0_{j,i}(x_i)=0 \tag{7.116d}$$

The boundary conditions are

$$\overline{X}^0_i(x_i)=\sigma^0_{ji}(x_i)n_j, \qquad (\text{on } S_\sigma) \tag{7.116e}$$

$$u^0_i(x_i)=\overline{u}^0_i(x_i), \qquad (\text{on } S_u) \tag{7.116f}$$

The viscoelastic stress-strain relationship, Eq. 7.102 or 7.103, gives

$$X(t)=\mu^*\mathrm{d}u \tag{7.117a}$$

$$u(t)=J^*\mathrm{d}X \tag{7.117b}$$

The displacements and stresses given in Eqs. 7.115 can be shown to satisfy all the quasi-static viscoelastic conditions if the time functions for displacements and forces are related by one of Eqs. 7.117. Note that the boundary surfaces must not move, i.e. S_σ and S_u must not change with time.

7.9 STEADY STATE HARMONIC SOLUTION SCHEME

If the boundary conditions and body forces are specified as steady state harmonic functions of time, the solution will have the same form

$$u_i(x_i,t)=u^0_i(x_i)e^{i\omega t} \tag{7.118a}$$

$$\varepsilon_{ij}(x_i,t)=\varepsilon^0_{ij}(x_i)e^{i\omega t} \tag{7.118b}$$

$$\sigma_{ij}(x_i,t)=\sigma^0_{ij}(x_i)e^{i\omega t} \tag{7.118c}$$

This solution is a special separable solution. The steady state harmonic stress-strain relations for three-dimensional isotropic continua may be written

$$\sigma_{ij}^0(x_i) = \delta_{ij}\lambda^*(\omega)\varepsilon_{kk}^0(x_i) + 2\mu^*(\omega)\varepsilon_{ij}^0(x_i) \tag{7.119}$$

or, for shear and bulk separately,

$$s_{ij}^0(x_i) = 2\mu^*(\omega)e_{ij}^0(x_i) \tag{7.120a}$$

$$\sigma_{kk}^0(x_i) = 3K^*(\omega)\varepsilon_{kk}^0(x_i) \tag{7.120b}$$

The stress-strain relationship has exactly the same form as in elasticity, except that it involves complex quantities. This means that we may solve the elasticity equations to find the spatial parts $u_i^0(x_i)$, $\varepsilon_{ij}^0(x_i)$ and $\sigma_{ij}^0(x_i)$, and then replace μ and K in the elasticity solution by $\mu^*(\omega)$ and $K^*(\omega)$ respectively.

Gottenberg and Christensen (1964) studied the torsional oscillation of a hollow circular cylinder. They included inertial effects, and found analytical relationship between the applied torque and the angle of twist. They used this relationship, together with experimental measurements of the applied torque and the angular response, to determine the complex shear modulus.

Note that the general treatment by separation of variables given in the previous section can be applied only when inertial terms are neglected. By contrast, the solution scheme for steady state harmonic oscillation discussed in the present section can include inertial terms with no difficulty.

7.10 INTEGRAL TRANSFORM METHODS AND THEIR LIMITATIONS

The Laplace transforms of the equations of quasi-static viscoelasticity are identical to those of linear elasto-quasi-statics. In particular, for the isotropic case (Table 7.2), the Laplace transformed viscoelastic solution is obtained directly from the corresponding elastic solution by replacing λ, μ, K, E and ν with $p\tilde{\lambda}(p)$, $p\tilde{\mu}(p)$, $p\tilde{K}(p)$, $p\tilde{E}(p)$ and $p\tilde{\nu}(p)$, respectively. The actual solution is then realized by inverting the transformed solution. An entirely similar procedure also applies in the anisotropic case. If appropriate, we may use Fourier transforms instead of Laplace transforms. The association of the viscoelastic solution with the elastic solution is called the *elastic-viscoelastic correspondence principle*, or the *elastic-viscoelastic analogy*.

We can include inertial terms in any integral transform method, but then (Table 7.3) the correspondence is between the transformed elastic solution and the transformed viscoelastic solution. This correspondence is much less useful than that for the quasi-static case, and there is in general no advantage in constructing the dynamic-viscoelastic solution by this means. The dynamic-viscoelastic solution can be approached directly by using an appropriate integral transform. The inversion of a transformed dynamic solution is naturally much more difficult than that in the quasi-

Table 7.2

Elasto-quasi-statics	Laplace transformed viscoelasto-quasi-statics
Equilibrium:	
$\sigma_{ji,j} + X_i = 0$	$\tilde{\sigma}_{ji,j} + \tilde{X}_i = 0$
Strain-displacement relationship:	
$\varepsilon_{ij} = \frac{1}{2}(u_{i,j} + u_{j,i})$	$\tilde{\varepsilon}_{ij} = \frac{1}{2}(\tilde{u}_{i,j} + \tilde{u}_{j,i})$
Stress-strain relationship:	
$s_{ij} = 2\mu e_{ij}$	$\tilde{s}_{ij} = 2p\tilde{\mu}\tilde{e}_{ij}$
$\sigma_{kk} = 3K\varepsilon_{kk}$	$\tilde{\sigma}_{kk} = 3p\tilde{K}\tilde{\varepsilon}_{kk}$
Boundary conditions:	
$\sigma_{ij}n_j = \overline{X}_i$ on S_σ	$\tilde{\sigma}_{ij}n_j = \tilde{\overline{X}}_i$ on S_σ
$u_i = \overline{u}_i$ on S_u	$\tilde{u}_i = \tilde{\overline{u}}_i$ on S_u

static case. Again note that, in any integral transform method, the boundary surfaces S_σ and S_u must not change with time.

Gottenberg and Christensen (1964) applied the elastic-viscoelastic correspondence principle to the quasi-static torsional oscillation of a hollow circular cylinder. Naturally, the solution was found to consist of transient and steady-state parts. The transient part is an exponentially decaying function of time.

In the book of Christensen (1971), there is a problem of a long cylinder being subjected to an internal pressure, but restrained by an elastic outer case. Inertial effects are included and the problem was solved by means of the Laplace transform. Axisymmetry of the problem and the incompressibility of the material make the mathematics tractable. Christensen and Schreiner (1965) studied the quasi-static problem for compressible materials. The Laplace transform was applied to all conditions involved. They also investigated a way of including the effect of a boundary which is changing due to burning or erosion of the material.

Lockett (1961) studied the pressurization of a spherical cavity in an infinite space using the Fourier transform. Inertial effects were included. They approximated the step function pressure history by the difference between two sine integrals of difference frequencies. This enabled them to use an inversion integral over a finite frequency range, rather than the usual infinite frequency range.

Table 7.3

Laplace transformed elastodynamics	Laplace transformed viscoelastodynamics
Equation of motion:	
$\tilde{\sigma}_{ji,j}+\bar{X}_i=\rho p^2\bar{u}_i$	$\tilde{\sigma}_{ji,j}+\bar{X}_i=\rho p^2\bar{u}_i$
Strain-displacement relationship:	
$\tilde{\varepsilon}_{ij}=\frac{1}{2}(\bar{u}_{i,j}+\bar{u}_{j,i})$	$\tilde{\varepsilon}_{ij}=\frac{1}{2}(\bar{u}_{i,j}+\bar{u}_{j,i})$
Stress-strain relationship:	
$\bar{s}_{ij}=2\mu\bar{e}_{ij}$	$\bar{s}_{ij}=2p\tilde{\mu}\bar{e}_{ij}$
$\tilde{\sigma}_{kk}=3K\tilde{\varepsilon}_{kk}$	$\tilde{\sigma}_{kk}=3p\tilde{K}\tilde{\varepsilon}_{kk}$
Boundary conditions:	
$\tilde{\sigma}_{ij}n_j=\tilde{\bar{X}}_i$ on S_σ	$\tilde{\sigma}_{ij}n_j=\tilde{\bar{X}}_i$ on S_σ
$\bar{u}_i=\tilde{\bar{u}}_i$ on S_u	$\bar{u}_i=\tilde{\bar{u}}_i$ on S_u

There are practical problems which involve variable boundary. One example is the deformation of a viscoelastic beam or a viscoelastic half space by a curved rigid indentor. As the indentor is pressed into the viscoelastic domain, there are some parts of the boundary of the domain which at first are traction free, but later they must conform to the geometry of the indentor in the contact region. For problems of this class, the use of integral methods fails, and there is no generally available alternative. Such problems are to be expected to be more difficult than those with non-moving boundaries. Some problems of this class have been solved by particular methods.

In his book, Christensen (1971) studied the simplest problem of this type, i.e. the deformation of a uniform viscoelastic beam by a curve rigid indentor. It should be possible to extend his procedure to dropping indentors, elastic beams with viscoelastic coatings, culp indentors, viscoelastic plates, and viscoelastic beams and plates including shear deformation effects. This would comprise a generalization of the elastic contact analysis given by Essenburg (1962).

The problem of the indentation of a viscoelastic half space by a rigid spherical indentor was studied by Lee and Radok (1960) and Hunter (1960) and later by Graham (1965). Christensen (1971) took the outline of the solution approach from Graham (1965). More recently Graham (1967) extended his procedure to include the case where the contact area has several relative maxima rather than just a single one. Ting (1968) has also treated this more general case. Graham (1968) and Ting (1969) have outlined restricted classes of viscoelastic contact problems which do admit a direct application of the elastic viscoelastic correspondence principle.

A similar moving boundary problem, but even more complicated, is that posed by a cylinder rolling over a viscoelastic half space. Here, again the elastic viscoelastic correspondence principle does not apply, and other techniques of solution must be developed. Such studies have been given by Hunter (1961) and Morland (1962, 1967). See also Kalker (1990).

7.11 THREE-DIMENSIONAL THERMOVISCOELASTICITY

For thermorheologically simple solids, the stress-strain relationship for three-dimensional domains with time varying temperature can be obtained by generalizing that for one-dimensional models described in Section 7.6:

$$\hat{\sigma}_{ij}(\xi) = \int_{0^-}^{\xi} \hat{C}_{ijkl}(\xi - \zeta)\mathrm{d}[\hat{\varepsilon}_{ij}(\zeta) - \hat{\alpha}_{ij}(\zeta)] \tag{7.121}$$

Here ξ and ζ are as defined in Eqs. 7.88 and 7.91, and $\alpha_{ij}(t)$ is the strain tensor in the stress-free state due to the temperature change from the initial temperature T_0 to the current temperature $T(t)$. The isotropic form of Eq. 7.121 is

$$\hat{s}_{ij}(\xi) = 2\int_{0^-}^{\xi} \hat{\mu}(\xi - \zeta)\mathrm{d}\hat{e}_{ij}(\zeta) \tag{7.122a}$$

$$\hat{\sigma}_{kk}(\xi) = 3\int_{0^-}^{\xi} \hat{K}(\xi - \zeta)\mathrm{d}[\hat{\varepsilon}_{kk}(\zeta) - \hat{\alpha}_{kk}(\zeta)] \tag{7.122b}$$

Specializing Eqs. 7.122 for the uniaxial stress case (Eqs. 7.104 with varying temperature) we find

$$\hat{\sigma}_{11}(\xi) = \int_{0^-}^{\xi} \hat{E}(\xi - \zeta)\mathrm{d}[\hat{\varepsilon}_{11}(\zeta) - \hat{\alpha}(\zeta)] \tag{7.123a}$$

$$\hat{\varepsilon}_{22}(\xi) = \hat{\varepsilon}_{33}(\xi) = \hat{\alpha}(\xi) - \int_{0^-}^{\xi} \hat{\nu}(\xi - \zeta)\mathrm{d}[\hat{\varepsilon}_{11}(\zeta) - \hat{\alpha}(\zeta)] \tag{7.123b}$$

where $\alpha(t)$ is the stress-free normal strain component in any direction due to the temperature change. The convolution integrals with respect to the pseudo time ξ in the equations above seem to suggest that the Laplace transform might conveniently be used to solve quasi-static boundary value problems. However, in terms of ξ, the equilibrium equations become

$$\hat{\sigma}_{ij,j} + \frac{\partial \hat{\sigma}_{ij}}{\partial \xi}\xi_{,j} + \hat{X}_i(\xi) = 0 \tag{7.124}$$

and the strain-displacement relations are

$$\hat{\varepsilon}_{ij}(\xi)=\frac{1}{2}\left(\hat{u}_{i,j}+\frac{\partial \hat{u}_i}{\partial \xi}\xi_{,j}+\hat{u}_{j,i}+\frac{\partial \hat{u}_j}{\partial \xi}\xi_{,i}\right) \tag{7.125}$$

The presence of the terms $\xi_{,i}$ means that we cannot apply the Laplace transform with respect to ξ, nor construct a correspondence between elastic solutions and viscoelastic solutions. When the temperature is spatially uniform but time dependent, such terms $\xi_{,i}$ vanish, and then we can apply integral transform methods with respect to ξ.

Another situation which is amenable to the use of integral transform methods occurs when the temperature distribution $T(x_i,t)$ is known, whether from experimental observations or by some analytical means. In such a case, the pseudo time ξ and the stress free dilatation α are known functions of the coordinate x_i as well as the time variable t. Muki and Sternberg (1961) solved a problem of this type using the Laplace transform. The problem is that of an infinite isotropic viscoelastic slab subjected to heating conditions which produce a known temperature variation through the thickness. Lee and Rogers (1963) pointed out that this problem may be solved without using the Laplace transform, and the problem was further discussed by Sternberg (1964). A related but more difficult problem, involving thermal stresses in a cylinder, was studied by Lockett and Morland (1967).

In a more general situation of *coupled thermoviscoelasticity*, the temperature upon which the mechanical properties depend is an unknown function of x_i and t. These problems are necessarily nonlinear, and must be expected to be very difficult to solve analytically.

7.12 PROBLEMS

With permission of the publisher, Academic Press, Problems 7.12.4 to 7.12.6 and 7.12.10 are materials adapted from the book by Christensen (1971).

7.12.1 *Reciprocal Theorem for Harmonic Oscillation*

For steady-state harmonic oscillation, the Maxwell-Betti reciprocal theorem can be written

$$\begin{aligned}
&\int_V X_{0i}^{(1)}u_{0i}^{(2)}\mathrm{d}V+\int_S \overline{X}_{0i}^{(1)}u_{0i}^{(2)}\mathrm{d}S \\
&=\int_V X_{0i}^{(2)}u_{0i}^{(1)}\mathrm{d}V+\int_S \overline{X}_{0i}^{(2)}u_{0i}^{(1)}\mathrm{d}S \\
&=\int_V[-\rho\omega^2u_{0i}^{(1)}u_{0i}^{(2)}+\sigma_{0ij}^{(1)}\varepsilon_{0ij}^{(2)}]\mathrm{d}V \\
&=\int_V[-\rho\omega^2u_{0i}^{(2)}u_{0i}^{(1)}+\sigma_{0ij}^{(2)}\varepsilon_{0ij}^{(1)}]\mathrm{d}V
\end{aligned} \tag{7.126}$$

where a subscript ($_0$) denotes the amplitude of a steady-state harmonic function, and a superscript in parentheses denotes a loading system as adopted in Section 3.4.

7.12.2 *Vibration of a bar with viscoelastic support*

A viscoelastic slender bar of length L, cross sectional area A and mass density ρ is supported viscoelastically and loaded axially as shown in Fig. 7.12. The bar material has a constant Poisson's ratio but behaves like a standard linear solid in shear. The whole system is in steady-state harmonic vibration due to the applied force $P^0 \cos \omega t$. Equations 7.19e and 7.45 give the complex modulus of the support as

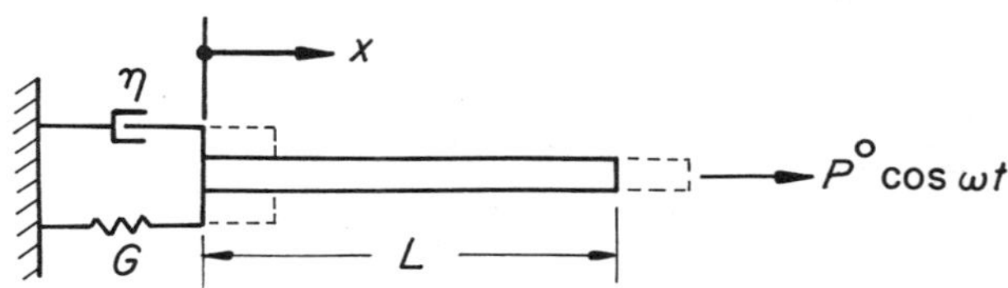

Fig. 7.12

$$G^*(\omega) = G(1 + i\omega\tau) \tag{7.127}$$

where $\tau = \eta/G$. With the same symbols as in Fig. 7.3, the complex Young's modulus of the bar is obtained as

$$E^*(\omega) = \frac{2(1+\nu)[G_0 + (G_0 + G_1)i\omega\tau_1]}{(1 + i\omega\tau_1)} \tag{7.128}$$

where ν is the constant Poisson's ratio of the bar. In the bar, the pertinent stress-strain relationship is

$$\sigma^0_{xx}(x) = E^*(\omega)\frac{du^0(x)}{dx} \tag{7.129}$$

and the equation of motion is

$$\frac{d\sigma^0_{xx}(x)}{dx} = -\rho\omega^2 u^0(x) \tag{7.130}$$

where a superscript (^{0}) denotes the amplitude of a steady-state harmonic function. Substituting Eq. 7.129 into Eq. 7.130 yields

$$\frac{d^2u^0(x)}{dx^2} + \Omega^2 u^0(x) = 0 \tag{7.131}$$

where

$$\Omega = \omega\sqrt{\frac{\rho}{E^*(\omega)}} \tag{7.132a}$$

$$= \omega\left(\frac{\rho}{|E^*(\omega)|}\right)^{1/2}\left(\cos\frac{\gamma}{2} + \mathrm{i}\sin\frac{\gamma}{2}\right) \tag{7.132b}$$

$$\tan\gamma = -\frac{E_2(\omega)}{E_1(\omega)} \tag{7.132c}$$

$$E_1(\omega) = \Re[E^*(\omega)] \tag{7.132d}$$

$$E_2(\omega) = \Im[E^*(\omega)] \tag{7.132e}$$

The solution to Eq. 7.131 takes the form

$$u^0(x) = P^0[C_1(\omega)\cos\Omega x + C_2(\omega)\sin\Omega x] \tag{7.133}$$

where $C_1(\omega)$ and $C_2(\omega)$ are to be determined from the boundary conditions

$$A\sigma^0_{xx}(0) = G^*(\omega)u^0(0), \quad A\sigma^0_{xx}(L) = P^0 \tag{7.134a,b}$$

The results are

$$C_1(\omega) = [G^*(\omega)\cos\Omega L - A\Omega E^*(\omega)\sin\Omega L]^{-1} \tag{7.135a}$$

$$C_2(\omega) = \frac{G^*(\omega)C_1(\omega)}{A\Omega E^*(\omega)} \tag{7.135b}$$

Since the applied force is

$$P^0\cos\omega t = \Re(P^0 e^{i\omega t}) \tag{7.136}$$

the displacement of the end of the bar is

$$\Re[u^0(L)e^{i\omega t}].$$

Due to Eq. 7.133, we can write

$$P^0 = \mu^*(\omega)u^0(L) \tag{7.137}$$

where

$$\mu^*(\omega) = [C_1(\omega)\cos\Omega L + C_2(\omega)\sin\Omega L]^{-1} \tag{7.138}$$

Thus the loss angle (Eq. 7.39c) is

$$\delta = \arctan\left[\frac{\mu_2(\omega)}{\mu_1(\omega)}\right] \tag{7.139a}$$

and the energy loss per cycle (Eq. 7.72) is

$$\Delta U = \frac{\pi(P^0)^2\mu_2(\omega)}{|\mu^*(\omega)|^2} \tag{7.139b}$$

where

$$\mu_1(\omega) = \Re[\mu^*(\omega)], \quad \mu_2(\omega) = \Im[\mu^*(\omega)] \tag{7.140a,b}$$

7.12.3 *Indentation on a Viscoelastic Half Space*

If a normal indentation is made on the surface of a homogeneous isotropic elastic half space by a smooth rigid circular plate, the stress at the contact between the rigid plate and the half space is given by Eq. 6.78a. The half space is $0 \le r,\ z < \infty$; the half space surface is $z = 0$; the radius of the rigid plate is a; the contact surface is $z = 0$ and $0 \le r \le a$; and w_0 is the prescribed magnitude of the statical indentation. For quasi-statics and a homogeneous isotropic viscoelastic half space with a constant Poisson's ratio (Eq. 7.114a), we suppose that the prescribed indentation is

$$u_z(r,\theta,0;t) = w_0 u(t), \quad (0 \le r \le a) \tag{7.141}$$

The contact normal stress can be obtained by separation of variables (Section 7.8) as

$$\sigma_{zz}(r,\theta,0;t) = -\frac{2w_0 X(t)}{\pi(1-\nu)(a^2-r^2)^{1/2}}, \quad (0 \le r < a) \tag{7.142}$$

where $X(t)$ is as given by Eq. 7.117a.

Alternatively, we can obtain the result by the elastic-viscoelastic correspondence principle as

$$\tilde{\sigma}_{zz}(r,\theta,0;p) = -\frac{2w_0 p\tilde{\mu}(p)\tilde{u}(p)}{\pi(1-\nu)(a^2-r^2)^{1/2}}, \quad (0 \le r < a) \tag{7.143}$$

which is the same as Eq. 7.142.

7.12.4 *Torsional Oscillation of a Hollow Cylinder*

In this problem, we consider the torsional vibration of a hollow cylinder under steady-state harmonic conditions. As in any axisymmetric torsion problems, the only nonzero displacement is u_θ, and the relevant equation of motion for an elastic cylinder is

$$\mu\left(\frac{\partial^2 u_\theta}{\partial r^2} - \frac{u_\theta}{r^2} + \frac{1}{r}\frac{\partial u_\theta}{\partial r} + \frac{\partial^2 u_\theta}{\partial z^2}\right) = \rho\frac{\partial^2 u_\theta}{\partial t^2} \tag{7.144}$$

where u_θ is the displacement in θ-direction, ρ the mass density, and r, θ and z the cylindrical coordinates. The nonvanishing stresses are $\sigma_{\theta z}$ and $\sigma_{r\theta}$ which are related to u_θ through the equations

$$\sigma_{\theta z} = \mu\frac{\partial u_\theta}{\partial z} \tag{7.145a}$$

$$\sigma_{r\theta} = \mu\left(\frac{\partial u_\theta}{\partial r} - \frac{u_\theta}{r}\right) \tag{7.145b}$$

The function u_θ may be assumed in the form

$$u_\theta = r\Phi(z)e^{i\omega t} \tag{7.146}$$

where $\Phi(z)$ is the amplitude of the angle of twist. Substituting Eq. 7.146 into Eq. 7.144, we find

$$\Phi(z) = A\sin\frac{\Omega z}{h} + B\cos\frac{\Omega z}{h} \tag{7.147}$$

where

$$\Omega = \omega h\sqrt{\frac{\rho}{\mu}} \tag{7.148}$$

h is the length of the cylinder, and A and B are arbitrary constants. Note that $\sigma_{r\theta}$ is identically zero. The remaining nonzero stress $\sigma_{\theta z}$ may be integrated over the cross-section of the cylinder to obtain the total twisting moment

$$M(z,t) = \frac{\pi(b^4 - a^4)\rho\omega^2 h}{2\Omega}\left(A\cos\frac{\Omega z}{h} - B\sin\frac{\Omega z}{h}\right)e^{i\omega t} \tag{7.149}$$

where a and b are the inner and outer radii, respectively. If the end $z = h$ is fixed while the end $z = 0$ has the applied torque $\overline{M}e^{i\omega t}$, we can find the values of A and B and further find that the ratio of the applied torque and the angle of twist at the end $z = 0$ is given by

$$\frac{\overline{M}}{\Phi(0)} = \frac{\pi(b^4 - a^4)\rho\omega^2 h\cot\Omega}{2\Omega} \tag{7.150}$$

Replacing μ in the equation above by $\mu^*(\omega)$ yields the required steady state harmonic viscoelastic solution as

$$\frac{\overline{M}}{\Phi(0)} = \frac{\pi(b^4 - a^4)\rho\omega^2 h\cot\Omega^*}{2\Omega^*} \tag{7.151}$$

where

$$\Omega^* = \omega h\sqrt{\frac{\rho}{\mu^*(\omega)}} \tag{7.152}$$

If we know $\mu^*(\omega)$, then Eq. 7.151 allows us to determine the angle of twist at $z = 0$ for a specified value of the applied torque. Alternatively, we may measure the applied torque and the angular response and determine the complex shear modulus from Eq. 7.151, as did Gottenberg and Christensen (1964).

7.12.5 *Quasi-Static Torsional Oscillation of a Hollow Cylinder*

We consider the same Problem 7.12.4, but now include transient starting effects. While the cylinder end at $z = h$ is fixed, the other end at $z = 0$ has a prescribed angle of rotation given

$$\Phi(0,t) = 0, \qquad (t < 0) \tag{7.153a}$$

$$\Phi(0,t) = k \sin \omega t, \quad (0 < t < \infty) \tag{7.153b}$$

Here k is the given amplitude of oscillation, and ω is the angular frequency taken to be sufficiently small that inertial terms may be neglected. The corresponding elastostatic relationship between the twisting moment $M(0,t)$ and the angle of twist $\Phi(0,t)$ is given by

$$M(0,t) = \frac{\pi(b^4 - a^4)\mu\Phi(0,t)}{2h} \tag{7.154}$$

or, due to Eq. 7.153b,

$$M(0,t) = \frac{\pi(b^4 - a^4)\mu k \sin \omega t}{2h} \tag{7.155}$$

Applying the elastic-viscoelastic correspondence principle, i.e. replacing μ by $p\tilde{\mu}(p)$, $M(0,t)$ by $\tilde{M}(0,p)$ and $\Phi(0,t)$ by $\tilde{\Phi}(0,p)$ in the equation above, we find the Laplace transform of the viscoelastic torque to be

$$\tilde{M}(0,p) = \frac{\pi(b^4 - a^4)p\tilde{\mu}(p)k\omega}{2h(p^2 + \omega^2)} \tag{7.156}$$

Assume the shear relaxation modulus in the form

$$\mu(t) = \left[G_0 + \sum_{i=1}^{N} G_i e^{-t/\tau_i}\right] H(t) \tag{7.157}$$

Its Laplace transform can be put in the form (Eq. 7.22f)

$$\tilde{\mu}(p) = \left[G_0 + \sum_{i=1}^{N} \frac{G_i p}{p + \tau_i^{-1}}\right] p^{-1} \tag{7.158}$$

or, with a common denominator,

$$p\tilde{\mu}(p) = \frac{A(p)}{\prod_{i=1}^{N}(p + \tau_i^{-1})} \tag{7.159}$$

in which $A(p)$ is an Nth degree polynomial in p. Substituting Eq. 7.159 into Eq. 7.156 and inverting the results by means of the Heaviside expansion formula (Spiegel, 1965), we find

$$M(0,t) = \frac{\pi(b^4 - a^4)}{2h} k\omega \left\{ \frac{-e^{-i\omega t} A(-i\omega)}{2i\omega \prod_{j=1}^{N}(-i\omega + \tau_j^{-1})} + \frac{e^{i\omega t} A(i\omega)}{2i\omega \prod_{j=1}^{N}(i\omega + \tau_j^{-1})} \right.$$

$$\left. + \sum_{n=1}^{N} \lim_{p \to -\tau_n^{-1}} \left[\frac{A(p)(p + \tau_n^{-1})}{\prod_{j=1}^{N}(p + \tau_j^{-1})} \right] \frac{e^{-t/\tau_n}}{(\tau_n^{-2} + \omega^2)} \right\} \tag{7.160}$$

The first two terms in this equation represent the steady state response, while the last term describes the transient response. Gottenberg and Christensen (1964) used this equation to predict the starting transient of the specimen, for which the complex shear modulus was determined in the preceding section.

7.12.6 *Dynamic Response of an Incompressible Cylinder*

Consider the problem of a cylinder subjected to an internal pressure and restrained from expanding freely by an elastic outer case. The problem is axisymmetric, the cylinder length is sufficiently large, and the cylinder material is incompressible. Thus the zero dilatation condition can be written

$$\frac{\partial u_r(r,t)}{\partial r}+\frac{u_r(r,t)}{r}=0 \tag{7.161}$$

where $u_r(r,t)$ is the radial displacement in the cylindrical coordinate (r,θ,z) system. The solution of this equation is

$$u_r(r,t)=\frac{C(t)}{r} \tag{7.162}$$

Hooke's law for the present problem is given by

$$\sigma_{rr}=2\mu\frac{\partial u_r}{\partial r}+\frac{I_1}{3} \tag{7.163a}$$

$$\sigma_{\theta\theta}=2\mu\frac{u_r}{r}+\frac{I_1}{3} \tag{7.163b}$$

$$\sigma_{zz}=\frac{I_1}{3} \tag{7.163c}$$

where $I_1(r)$ is the first stress invariant, i.e. $I_1=\sigma_{rr}+\sigma_{\theta\theta}+\sigma_{zz}$. Applying the correspondence principle to the elastic solution in Eqs. 7.163, we find the Laplace transforms of the viscoelastic stresses as

$$\bar{\sigma}_{rr}(r,p)=-\frac{2p\bar{\mu}(p)\bar{C}(p)}{r^2}+\frac{\bar{I}_1(r,p)}{3} \tag{7.164a}$$

$$\bar{\sigma}_{\theta\theta}(r,p)=\frac{2p\bar{\mu}(p)\bar{C}(p)}{r^2}+\frac{\bar{I}_1(r,p)}{3} \tag{7.164b}$$

$$\bar{\sigma}_{zz}(r,p)=\frac{\bar{I}_1(r,p)}{3} \tag{7.164c}$$

The single non-trivial equation of motion (in the r-direction) has the form

$$\frac{\partial\sigma_{rr}}{\partial r}+\frac{\sigma_{rr}-\sigma_{\theta\theta}}{r}=\rho\frac{\partial^2 u_r}{\partial t^2} \tag{7.165}$$

or, in the form of its Laplace transform,

$$\frac{\partial\tilde{\sigma}_{rr}}{\partial r}+\frac{\tilde{\sigma}_{rr}-\tilde{\sigma}_{\theta\theta}}{r}=\rho p^2\tilde{u}_r \tag{7.166}$$

Substituting Eqs. 7.164 into this equation, we obtain

$$\frac{\partial\tilde{I}_1(r,p)}{3\partial r}=\frac{\rho p^2\tilde{C}(p)}{r} \tag{7.167}$$

or, integrating,

$$\tilde{I}_1(r,p)=3\rho\log r p^2\tilde{C}(p)+3\tilde{D}(p) \tag{7.168}$$

Substituting Eq. 7.168 into Eqs. 7.164 we find

$$\tilde{\sigma}_{rr}(r,p)=-\frac{2p\tilde{\mu}\tilde{C}}{r^2}+\rho p^2\tilde{C}\log r+\tilde{D} \tag{7.169a}$$

$$\tilde{\sigma}_{\theta\theta}(r,p)=\frac{2p\tilde{\mu}\tilde{C}}{r^2}+\rho p^2\tilde{C}\log r+\tilde{D} \tag{7.169b}$$

$$\tilde{\sigma}_{zz}(r,p)=\rho p^2\tilde{C}\log r+\tilde{D} \tag{7.169c}$$

The boundary conditions of this problem are

$$\sigma_{rr}(a,t)=-q(t) \tag{7.170a}$$

$$\sigma_{rr}(b,t)=-ku_r(b,t) \tag{7.170b}$$

where $q(t)$ is the pressure prescribed inside the cylinder, and k is the stiffness of the outer case restraining the cylinder. For a restraining case with a small thickness h,

$$k=\frac{E_c h}{b^2(1-\nu_c^2)} \tag{7.171}$$

where E_c and ν_c are the Young's modulus and Poisson's ratio of the case. We express the relaxation function in shear as

$$p\tilde{\mu}(p)=\frac{A(p)}{B(p)} \tag{7.172}$$

where $A(p)$ and $B(p)$ are polynomials in p. The satisfaction of the Laplace transform of the boundary conditions (Eqs. 7.170) determines $\tilde{C}(p)$ and $\tilde{D}(p)$ as

$$\tilde{C}(p)=-\frac{p\tilde{q}(p)B(p)}{F(p)} \tag{7.173a}$$

$$\tilde{D}(p)=-\frac{p\tilde{q}(p)E(p)}{F(p)} \tag{7.173b}$$

where

$$F(p)=-2pA(p)\left(\frac{1}{a^2}-\frac{1}{b^2}\right)+\rho p^3B(p)\log\frac{a}{b}-\frac{pkB(p)}{b} \tag{7.174a}$$

$$E(p) = \frac{A(p)}{b^2} - \rho p^2 B(p) \log b - \frac{kB(p)}{b} \tag{7.174b}$$

If, for example, the applied pressure

$$q(t) = q_0 H(t) \tag{7.175}$$

then its Laplace transform is

$$\bar{q}(p) = \frac{q_0}{p} \tag{7.176}$$

The denominator $F(p)$ is a polynomial in p and can be written

$$F(p) = c(p - a_1)(p - a_2)\ldots(p - a_m) \tag{7.177}$$

Here c is a real constant and a_j are roots of $F(p) = 0$; they can be determined by standard computer programs. Substituting Eqs. 7.173 into Eqs. 7.169 and inverting the results, by using the Heaviside expansion formula (Spiegel, 1965), we find

$$\sigma_{rr}(r,t) = -q_0 \sum_{j=1}^{m} F_j(t)\left[-\frac{2}{r^2} A(a_j) + B_j(r)\right] \tag{7.178a}$$

$$\sigma_{\theta\theta}(r,t) = -q_0 \sum_{j=1}^{m} F_j(t)\left[\frac{2}{r^2} A(a_j) + B_j(r)\right] \tag{7.178b}$$

$$\sigma_{zz}(r,t) = -q_0 \sum_{j=1}^{m} F_j(t) B_j(r) \tag{7.178c}$$

where

$$F_j(t) = \frac{e^{a_j t}}{\lim\limits_{p \to a_j} [F(p)/(p - a_j)]} \tag{7.179a}$$

$$B_j(r) = \rho a_j^2 \log r B(a_j) + E(a_j) \tag{7.179b}$$

7.12.7 *Isothermal Harmonic Vibration*

Some materials have real and imaginary parts of the complex compliance of the form

$$J_1(\omega) = k_1 \omega^{-n} \tag{7.180a}$$

$$J_2(\omega) = k_2 \omega^{-n} \tag{7.180b}$$

where k_1 and k_2 are constants, and n is approximately 0.21. Moreover, the following 'shift factor' $A(T)$ has been observed in a vicinity of a certain reference temperature T_0;

$$A(T) = 10^{b_0(T - T_0)} = \exp[2.303 b_0 (T - T_0)] \tag{7.181}$$

where b_0 is a constant and T is the temperature being kept constant all the time. According to Eq. 7.83, the pseudo or reduced frequency becomes

$$\omega^* = \omega \exp[-2.303 b_0 (T - T_0)] \tag{7.182}$$

The real and imaginary parts of the complex compliance at the constant temperature T are

$$J_1(\omega^*) = k_1 (\omega^*)^{-n} \tag{7.183a}$$

$$J_2(\omega^*) = k_2 (\omega^*)^{-n} \tag{7.183b}$$

The dissipation of energy per cycle, Eq. 7.85b, is

$$\Delta U = -\pi |\sigma_0|^2 k_2 \omega^{-n} \exp[2.303 n b_0 (T - T_0)] \tag{7.184}$$

7.12.8 *Isothermal Effects on Stretched String*

Consider a viscoelastic string stretched between two fixed points. The initial tensile stress is σ^0, the properties of the string are uniform, the string material behaves as a Maxwell solid in shear but as a Hookean solid in bulk, and the temperature T is constant in space and time. The 'shift factor' $A(T)$ is given as

$$A(T) = e^{b_0 (T - T_0)} \tag{7.185}$$

where b_0 is a known positive constant. Assume quasi-statics.

At the reference temperature T_0, the Laplace transforms of the bulk modulus $K_0(t)$ and the shear modulus $\mu_0(t)$ are given, respectively, by Eqs. 7.16c and 7.18h as

$$\bar{K}_0(p) = K p^{-1} \tag{7.186a}$$

$$\bar{\mu}_0(p) = G (p + \tau^{-1})^{-1} \tag{7.186b}$$

Here K is the elastic spring constant in bulk, and τ is given by Eq. 7.18f in which G and η are, respectively, the elastic spring constant and the dashpot coefficient in the Maxwell model as depicted in Fig. 7.2(a). With the help of Table 7.1, we can write the Laplace transforms of the Young's modulus and Poisson's ratio, respectively, at the reference temperature as

$$\bar{E}_0(p) = C (p + \Gamma^{-1})^{-1} \tag{7.187a}$$

$$\bar{\nu}_0(p) = (2p)^{-1} + D (p + \Gamma^{-1})^{-1} \tag{7.187b}$$

where

$$C = \frac{9KG}{3K + G}, \quad D = -\frac{3G}{2(3K + G)} \tag{7.188a,b}$$

$$\Gamma = \frac{3K + G}{3K} \tau \tag{7.188c}$$

The inverse Laplace transforms of Eqs. 7.187 give

$$E_0(t)=Ce^{-t/\Gamma}H(t) \tag{7.189a}$$

$$\nu_0(t)=\left(\frac{1}{2}+De^{-t/\Gamma}\right)H(t) \tag{7.189b}$$

The Young's modulus and Poisson's ratio at the temperature T, due to Eqs. 7.80 and 7.81, are

$$E(t)=\hat{E}(\xi)=Ce^{-\xi/\Gamma}H(\xi) \tag{7.190a}$$

$$\nu(t)=\hat{\nu}(\xi)=\left(\frac{1}{2}+De^{-\xi/\Gamma}\right)H(\xi) \tag{7.190b}$$

where $\xi=tA(T)$ as in Eq. 7.79 and $H(\xi)=H(t)$ as in Eq. 7.81a.

Since the string has fixed ends, the longitudinal normal strain $\varepsilon(t)$ is a constant function of t, i.e.

$$\varepsilon(t)=\varepsilon^0 H(t) \tag{7.191a}$$

or

$$\varepsilon(t)=\hat{\varepsilon}(\xi)=\varepsilon^0 H(\xi) \tag{7.191b}$$

Here ε^0 is the initial longitudinal normal strain, and can be related to the initial tensile stress σ^0 by means of Eq. 7.6, i.e.

$$\varepsilon^0=\frac{\sigma^0}{E(0^+)} \tag{7.192}$$

where $E(0^+)$ is the initial Young's modulus. Due to Eqs. 7.190,

$$E(0^+)=C,\quad \nu(0^+)=\frac{1}{2}+D \tag{7.193a,b}$$

The initial value of a function can be obtained from its Laplace transform by means of Eq. 1.139. Substituting Eqs. 7.190a and 7.191a into Eq. 7.104a we find the longitudinal normal stress

$$\sigma(t)=C\varepsilon^0 e^{-tA(T)/\Gamma}H(t)=\varepsilon^0 E(t) \tag{7.194}$$

For the lateral normal strains, Eqs. 7.104b and 7.190b should be employed to obtain

$$\varepsilon_{22}(t)=\varepsilon_{33}(t)=-\left(\frac{1}{2}+De^{-tA(T)/\Gamma}\right)\varepsilon^0 H(t)=-\varepsilon^0\nu(t) \tag{7.195}$$

7.12.9 *Varying Temperature Effects on a Stretched String*

Consider the same viscoelastic string as in the preceding section, but allow the temperature T to vary linearly with time t, i.e.

$$T(t)=T_0+a_0 t \tag{7.196}$$

where a_0 is a known positive constant, and T_0 is the initial as well as the reference temperature. The thermal coefficient (normal strain per unit temperature change) is known as c.

The Young's modulus and Poisson's ratio at any temperature T are given by Eqs. 7.190a and b respectively, but the pseudo or reduced time as given by Eq. 7.88, due to Eqs. 7.185 and 7.196, is

$$\xi = \frac{\exp(a_0 b_0 t) - 1}{a_0 b_0} \tag{7.197}$$

Hence

$$t = \frac{\log(a_0 b_0 \xi + 1)}{a_0 b_0} \tag{7.198}$$

Equations 7.189 and 7.190 give the initial values

$$E_0(0^+) = E(0^+) = C \tag{7.199a}$$

$$\nu_0(0^+) = \nu(0^+) = \frac{1}{2} + D \tag{7.199b}$$

The normal strain due to the temperature change in a stress-free state is

$$\alpha(t) = c(T - T_0) \tag{7.200a}$$

or, due to Eq. 7.196,

$$\alpha(t) = a_0 c t H(t) \tag{7.200b}$$

or, due to Eqs. 7.94, 7.96 and 7.198,

$$\alpha(t) = \hat{\alpha}(\xi) = \frac{c \log(a_0 b_0 \xi + 1)}{b_0} H(\xi) \tag{7.200c}$$

Substituting Eqs. 7.190a, 7.191b and 7.200c into Eq. 7.123a we find the longitudinal normal stress

$$\sigma(t) = \varepsilon^0 \hat{E}(\xi) - C c a_0 e^{-\xi/\Gamma} \int_0^{\xi} \frac{e^{\zeta/\Gamma}}{a_0 b_0 \zeta + 1} d\zeta$$

$$= \varepsilon^0 \hat{E}(\xi) - \frac{C c e^{-(a + \xi/\Gamma)}}{b_0} [f(aQ) - f(a)] H(\xi) \tag{7.201}$$

where

$$a = (a_0 b_0 \Gamma)^{-1}, \quad Q = a_0 b_0 \xi + 1 \tag{7.202a,b}$$

$$f(Q) = \log Q + \frac{Q^1}{1!} + \frac{Q^2}{2 \cdot 2!} + \frac{Q^3}{3 \cdot 3!} + \dots \tag{7.202c}$$

Substituting Eqs. 7.190b, 7.191b and 7.200c into Eq. 7.123b we find the lateral normal strains

$$\varepsilon_{22}(t) = \varepsilon_{33}(t) = \alpha(t) - \varepsilon^0 \hat{\nu}(\xi) + \left\{ \frac{c \log Q}{2 b_0} + \frac{D c e^{-(a + \xi/\Gamma)}}{b_0} [f(aQ) - f(a)] \right\} H(\xi) \tag{7.203}$$

Functions of ξ in Eqs. 7.201 and 7.203 can be express as functions of t by recalling Eqs. 7.94 and 7.98, i.e.

$$\xi = \frac{Q-1}{a_0 b_0} \tag{7.204a}$$

$$\hat{E}(\xi) = E(t) = C \exp\left(\frac{1-Q}{a_0 b_0 \Gamma}\right) H(t) \tag{7.204b}$$

$$\hat{\nu}(\xi) = \nu(t) = \left[\frac{1}{2} + D \exp\left(\frac{1-Q}{a_0 b_0 \Gamma}\right)\right] H(t) \tag{7.204c}$$

where

$$Q = e^{a_0 b_0 t} \tag{7.205}$$

7.12.10 *Heating of an Infinite Slab*

The problem is that of an infinite homogeneous isotropic viscoelastic slab of thickness $2a$ being subjected to heating conditions which produce a known temperature variation through the thickness. Rectangular coordinates are used such that the stress free boundaries are at $z = \pm a$. The displacements are zero except $u_z(z,t)$, the strains are zero except $\varepsilon_{zz}(z,t)$, and the nonzero stresses are $\sigma_{xx}(z,t)$, $\sigma_{yy}(z,t)$ and $\sigma_{zz}(z,t)$ with

$$\sigma_{xx}(z,t) = \sigma_{yy}(z,t) \tag{7.206}$$

The nontrivial equation of equilibrium is

$$\frac{\partial \sigma_{zz}(z,t)}{\partial z} = 0 \tag{7.207}$$

The pertinent boundary conditions are

$$\sigma_{zz}(a,t) = 0, \quad \sigma_{zz}(-a,t) = 0 \tag{7.208a,b}$$

Thus the only possible solution for $\sigma_{zz}(z,t)$ is trivial, i.e.

$$\sigma_{zz}(z,t) = 0 \tag{7.209}$$

With these conditions, the Laplace transforms with respect to ξ of Eqs. 7.122 give, respectively

$$\tilde{\hat{\sigma}}_{xx} = -2p\tilde{\hat{\mu}}\tilde{\hat{\varepsilon}}_{zz} \tag{7.210a}$$

$$\tilde{\hat{\sigma}}_{xx} = \frac{3}{2} p \tilde{\hat{K}} \left(\tilde{\hat{\varepsilon}}_{zz} - \tilde{\hat{\alpha}}\right) \tag{7.210b}$$

in which p is the transform parameter, and $\alpha(z,t)$ is the only nonzero strain due to the temperature change in a stress-free state. Eliminating $\tilde{\hat{\varepsilon}}_{zz}$ from the pair of equations above we obtain

$$\tilde{\hat{\sigma}}_{xx} = -p \tilde{\hat{S}} \tilde{\hat{a}} \tag{7.211}$$

where

$$\tilde{\hat{S}} = \frac{6\tilde{\hat{\mu}}\tilde{\hat{K}}}{4\tilde{\hat{\mu}}+3\tilde{\hat{K}}} \tag{7.212}$$

The inverse Laplace transform of Eq. 7.211 is

$$\hat{\sigma}_{xx}(z,\xi) = \hat{\sigma}_{yy}(z,\xi) = -\int_{0^-}^{\xi} \hat{S}(\xi-\zeta)\mathrm{d}\hat{\alpha}(z,\zeta) \tag{7.213}$$

The integration above can be carried out first with respect to the pseudo time ξ, then the final result as a function of the physical time t can be obtained by means of Eq. 7.98. Alternatively, Eq. 7.213 can be expressed in terms of t, so that

$$\sigma_{xx}(z,t) = \sigma_{yy}(z,t) = -\int_{0^-}^{t} S(t-s)\mathrm{d}\alpha(z,s) \tag{7.214}$$

When the temperature T is a known function of z and t, $\alpha(z,t)$ can be derived, ξ can be found from Eq. 7.88, $\hat{S}(\xi)$ from Eq. 7.212, $S(t)$ from Eq. 7.98, and thence the stresses from Eq. 7.214. A more general treatment of this problem was given by Muki and Sternberg (1961) with detailed numerical results.

CHAPTER VIII

WAVE PROPAGATION

8.1 TERMINOLOGY IN WAVE PROPAGATION

When a portion of a deformable medium is disturbed, it is deformed; the deformation disturbs the neighboring parts of the medium, and so propagates the disturbance through the medium. Such a disturbance is termed a *wave* and its progress through the medium is called *wave propagation*. As the wave propagates it carries energy in the form of kinetic and potential energies. The transmission of these energies is not by any bulk motion of the medium, but is passed on from one particle to the next. In other words, waves in deformable bodies are characterized by the transport of energy through motions of particles about an equilibrium position. Deformability and inertia of a medium are essential for the occurrence of waves. If the medium were not deformable, any part of the medium would immediately experience a disturbance upon application of a disturbance to a certain part of the medium, thus the response of the medium to the disturbance would be immediate, not gradual as in wave propagation. Similarly, if the medium had no inertia there would be no delay in the displacement of particles, and the transmission of the disturbance from one particle to another would be instantaneous. When a wave propagates through a three-dimensional medium, we can at a certain instant of time draw a surface through all points undergoing an identical disturbance. As time goes on, such a surface, which is called a *wavefront,* moves along. The normals to the wavefront, defining the direction of wave propagation, are called *rays*.

In one-dimensional problem a wave may be defined by a function $f(x-ct)$, where f, a function of the spatial coordinate x and the time t, represents a physical quantity such as a displacement, a velocity or a stress component. Figure 8.1 shows two identical forms of the function f at two instants, t and $t+\Delta t$ for a sufficiently small Δt , thus

$$f(x-ct) \equiv f(x+\Delta x-ct-c\Delta t),$$

hence

$$x-ct = x+\Delta x-ct-c\Delta t,$$

or

$$c = \Delta x/\Delta t.$$

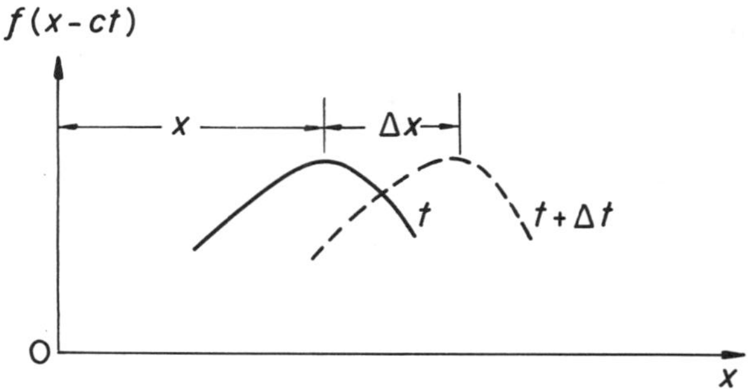

Fig. 8.1 An outgoing simple wave.

Appropriately, c is termed the *phase velocity*, and $f(x-ct)$ a *simple wave function* since it propagates though the medium without change of shape and magnitude. If f is the displacement of the particle at position x and time t, then $\partial f/\partial t$ is the particle velocity. This is not the same as the phase velocity. An observer at the origin 0 will 'see' that the wave $f(x-ct)$ shown in Fig 8.1 is moving away from him/her with the speed c. Such a wave is called *outgoing*. An *incoming* simple wave is described by a function $f(x+ct)$. If the shape of a wave does not travel, it is called a *standing wave*. A standing wave can be obtained by superposition of an outgoing simple wave and an incoming simple wave of the same characteristics, $f(x-ct)+f(x+ct)$, e.g.

$$2\sin\omega x\cos\omega t = \sin\omega(x-ct)+\sin\omega(x+ct).$$

Some travelling waves are not 'simple' but *attenuate*. Figure 8.2 shows the wave with

$$f(x,t) = e^{-ax} f_1(x-ct),$$

where a is a positive real constant. The plots of the function f at two separate instants, t and $t+\Delta t$, show that the function is unchanged in shape but decreases in magnitude as time increases.

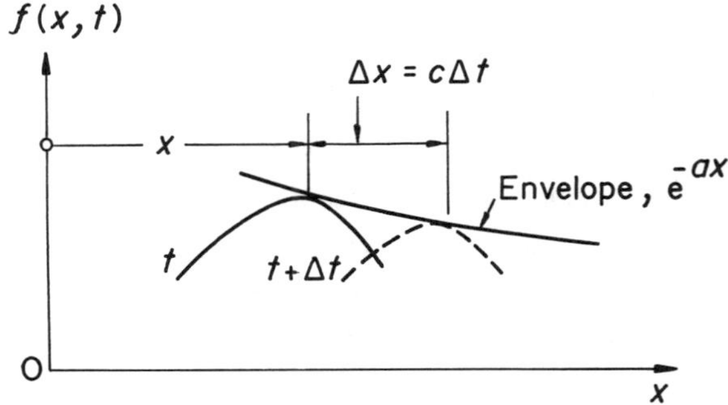

Fig. 8.2 An outgoing attenuating wave.

The three-dimensional wave equation is Eq. 3.111 or 3.112. In one dimension this reduces to the *simple wave equation.*

$$\frac{\partial^2 f}{\partial x^2}=c^{-2}\frac{\partial^2 f}{\partial t^2} \tag{8.1}$$

In oder to find the general solution to this equation we introduce two new variables

$$y=x-ct, \quad z=x+ct \tag{8.2a,b}$$

so that

$$x=(y+z)/2, \quad t=(z-y)/(2c) \tag{8.2c,d}$$

Now

$$\frac{\partial}{\partial y}=\frac{\partial}{\partial x}\frac{\partial x}{\partial y}+\frac{\partial}{\partial t}\frac{\partial t}{\partial y}=\frac{1}{2}\left(\frac{\partial}{\partial x}-\frac{1}{c}\frac{\partial}{\partial t}\right) \tag{8.2e}$$

and

$$\frac{\partial}{\partial z}=\frac{\partial}{\partial x}\frac{\partial x}{\partial z}+\frac{\partial}{\partial t}\frac{\partial t}{\partial z}=\frac{1}{2}\left(\frac{\partial}{\partial x}+\frac{1}{c}\frac{\partial}{\partial t}\right) \tag{8.2f}$$

Thus

$$\left(\frac{\partial^2 f}{\partial x^2}-\frac{1}{c^2}\frac{\partial^2 f}{\partial t^2}\right)=\left(\frac{\partial}{\partial x}-\frac{1}{c}\frac{\partial}{\partial t}\right)\left(\frac{\partial f}{\partial x}+\frac{1}{c}\frac{\partial f}{\partial t}\right)=4\frac{\partial^2 f}{\partial y\partial z}=0 \tag{8.2g}$$

which has the general solution

$$f(x,t)=f(y)+f(z)=f(x-ct)+f(x+ct) \tag{8.3}$$

Consider the problem of an elastic slender bar under axial excitation. If the x_1-axis runs along the bar, the equation of motion of the bar in the absence of any body force is specialized from Eq. 2.10 to be

$$\sigma_{11,1}=\rho\frac{\partial^2 u_1}{\partial t^2}.$$

Since $\sigma_{11}=Eu_{1,1}$ (Eqs. 2.43a and 3.36a), this is

$$\frac{\partial^2 u_1}{\partial x_1^2}=\frac{\rho}{E}\frac{\partial^2 u_1}{\partial t^2} \tag{8.4}$$

Thus, the *axial wave speed* is

$$c_a=\sqrt{E/\rho} \tag{8.5}$$

Consider a one-dimensional *plane harmonic wave*

$$u(x,t)=A\exp[\mathrm{i}(kx-\omega t)] \tag{8.6}$$

where k and ω are real constants independent of the spatial variable x and the time t. The term A is known as the *amplitude,* and ω as the *angular frequency*. The wave has phase velocity c given by

$$c=\frac{\omega}{k} \tag{8.7}$$

Figure 8.3(a) shows a plot of the wave at a particular time, while Fig. 8.3(b) shows that at a particular position. If x increases by $2\pi/k$ or t by $2\pi/\omega$, the argument $kx-\omega t$ changes by 2π so that the exponential function returns to its original value. This shows that the wave has a *wavelength* Λ and a *period* T given by

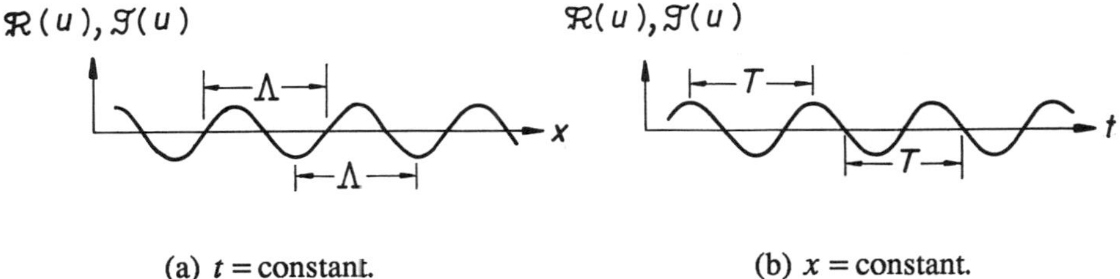

(a) $t=$ constant. (b) $x=$ constant.

Fig. 8.3 Wavelength and period of a harmonic wave.

$$\Lambda=\frac{2\pi}{k}, \quad T=\frac{2\pi}{\omega} \tag{8.8a,b}$$

The quantity k $(=2\pi/\Lambda)$ which counts the number of wavelength in 2π, is termed the *wavenumber*. A system is said to be *dispersive*, if its wavenumber k depends on the frequency ω, or, in other words, its phase velocity c depends on the wavelength Λ.

Let us consider the superposition of two plane harmonic waves of the same amplitude but slightly different wavenumbers and frequencies. The resultant of the superposed waves is

$$\begin{aligned} u+u' &= A\{\exp[i(kx-\omega t)]+\exp[i(k'x-\omega' t)]\} \\ &= 2A\exp\frac{i}{2}[(k+k')x-(\omega+\omega')t]\cos\frac{1}{2}[(k-k')x-(\omega-\omega')t] \end{aligned} \tag{8.9}$$

which has the characteristics of 'beats'. The exponential factor represents a *carrier wave* whose wavenumber and frequency are equal to the means of the superposed waves. The cosine factor, which varies slowly when the difference in characteristics of the two

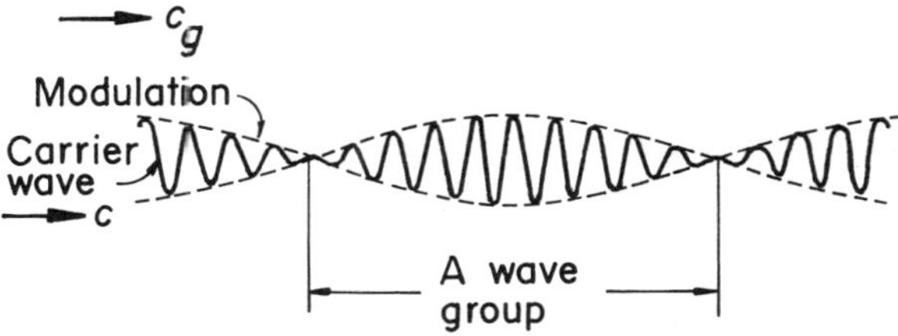

Fig. 8.4 An illustration of a wave group.

waves is small, represents the *modulation* and may be regarded as a varying amplitude. The situation is depicted in Fig. 8.4. The '*wave group*' ends wherever the cosine becomes zero. The velocity of the advance of these points is called the *group velocity* with the value

$$c_g = \frac{\omega - \omega'}{k - k'}.$$

For long groups (or slow beats), the expression above may be rewritten with sufficient accuracy as

$$c_g = \frac{\mathrm{d}\omega}{\mathrm{d}k} \tag{8.10a}$$

$$= c + k\frac{\mathrm{d}c}{\mathrm{d}k} \tag{8.10b}$$

$$= c - \Lambda\frac{\mathrm{d}c}{\mathrm{d}\Lambda} \tag{8.10c}$$

and another useful relation is

$$\frac{1}{c_g} = \frac{1}{c} - \frac{\omega}{c^2}\frac{\mathrm{d}c}{\mathrm{d}\omega} \tag{8.10d}$$

In general $c_g < c$, and thus $\mathrm{d}c/\mathrm{d}k < 0$. The group velocity is the same as the *velocity of energy transport*. This is defined as the ratio of the time average of the power per cross section and the time average of the total energy per unit length of the wave. Some verification of this point may be seen in the book by Achenbach (1973). If the system is nondispersive, ω is not a function of k (hence Λ as well), and c_g becomes identical to c.

If the only non-zero displacement component is u_1 and it is a function of x_1 only, then the governing equation for an isotropic elastic domain can be obtained by specializing Eq. 3.66 into

$$\frac{\partial^2 u_1}{\partial x_1^2} = \frac{\rho}{\lambda + 2\mu}\frac{\partial^2 u_1}{\partial t^2} \tag{8.11}$$

which is a special case of the three-dimensional pressure wave equation, Eq. 3.111. The equation above implies that the motion is in the same direction as the propagation, therefore the wave is called a *longitudinal wave, compressive wave, pressure wave* or a *P-wave* (P standing for pressure or primary), and the wave speed is

$$c_p = \sqrt{\frac{\lambda + 2\mu}{\rho}} \tag{8.12}$$

For incompressible solids, c_p becomes infinity following the value of λ.

For the same type of domain but the only non-zero displacement component u_1 being a function of x_2 only, the governing equation is

$$\frac{\partial^2 u_1}{\partial x_2^2}=\frac{\rho}{\mu}\frac{\partial^2 u_1}{\partial t^2} \tag{8.13}$$

which is a special case of the three-dimensional shear wave equation, Eq. 3.112. The equation above implies that the motion is normal to the direction of the propagation, therefore the wave is called a *transverse wave, rotational wave*, *shear wave*, or an *S-wave* (S standing for shear or secondary), and the wave speed is

$$c_s = \sqrt{\mu/\rho} \tag{8.14a}$$

$$= \sqrt{\frac{\mu}{\lambda+2\mu}}c_p \tag{8.14b}$$

$$= \sqrt{\frac{1-2\nu}{2(1-\nu)}}c_p \tag{8.14c}$$

Thus c_s is always smaller than c_p.

If in such a domain the dilatation is zero or uniform, i.e.

$$\varepsilon_{kk}=0 \tag{8.15a}$$

or

$$\varepsilon_{kk,i}=0 \tag{8.15b}$$

then the governing equation, Eq. 3.66, becomes

$$c_s^2\nabla^2 u_i=\frac{\partial^2 u_i}{\partial t^2} \tag{8.16}$$

Comparing this equation with Eq. 3.112, we can say that it is an S-wave equation governing each displacement component individually. Recall that the radial spherical coordinate R is related to the Cartesian x_i as

$$R^2=x_i x_i \tag{8.17}$$

The derivative of the equation above with respect to x_k is

$$2R\frac{\partial R}{\partial x_k}=2x_i\delta_{ik},$$

or

$$\frac{\partial R}{\partial x_k}=\frac{x_k}{R},$$

or, by squaring,

$$\left(\frac{\partial R}{\partial x_k}\right)^2=\frac{x_k x_k}{R^2},$$

or, due to Eq. 8.17,

$$\left(\frac{\partial R}{\partial x_k}\right)^2=1 \tag{8.18}$$

Taking derivative of the equation above with respect to R we find

$$\frac{\partial R}{\partial x_k}\frac{\partial}{\partial R}\left(\frac{\partial R}{\partial x_k}\right)=\frac{\partial^2 R}{\partial x_k^2}=0 \tag{8.19}$$

If u_i is independent of any angular spherical coordinates, then

$$u_{i,k}=\frac{\partial u_i}{\partial R}\frac{\partial R}{\partial x_k},$$

and

$$\begin{aligned} u_{i,kk} &= \frac{\partial}{\partial R}\left(\frac{\partial u_i}{\partial R}\frac{\partial R}{\partial x_k}\right)\frac{\partial R}{\partial x_k} \\ &= \frac{\partial^2 u_i}{\partial R^2}\left(\frac{\partial R}{\partial x_k}\right)^2+\frac{\partial u_i}{\partial R}\frac{\partial}{\partial R}\left(\frac{\partial R}{\partial x_k}\right)\frac{\partial R}{\partial x_k} \end{aligned} \tag{8.20}$$

or, due to Eqs. 8.18 and 8.19,

$$u_{i,kk}=\frac{\partial^2 u_i}{\partial R^2} \tag{8.21}$$

Substituting Eq. 8.21 into Eq. 8.16 we obtain the simple wave equation

$$\frac{\partial^2 u_i}{\partial R^2}=c_s^{-2}\frac{\partial^2 u_i}{\partial t^2} \tag{8.22}$$

The wave is called an *equivoluminal wave,* and has a *spherical wavefront* and a wave speed equal to c_s. Note a consequence that any wave equation is transformable into a simple wave equation if the spherical radial coordinate R is the only spatial variable involved.

If the deformation in such domain is irrotational, i.e.

$$\omega_{ik}=0 \tag{8.23}$$

or, by definition of the rotation tensor (Eq. 2.48),

$$\frac{1}{2}(u_{i,k}-u_{k,i})=0,$$

then the tensor $u_{i,k}$ is symmetric, i.e.

$$u_{k,i}=u_{i,k} \tag{8.24}$$

Accordingly,

$$u_{k,ki}=u_{i,kk},$$

or

$$\varepsilon_{kk,i}=u_{i,kk} \tag{8.25}$$

Substituting this equation into Eq. 3.66 we find

$$c_p^2\nabla^2 u_i=\frac{\partial^2 u_i}{\partial t^2} \tag{8.26}$$

Comparing this equation with Eq. 3.111, we can say that it is a P-wave equation governing each displacement component individually. In the same way as obtaining Eq. 8.22, we can change this equation into the simple wave equation

$$\frac{\partial^2 u_i}{\partial R^2} = c_p^{-2} \frac{\partial^2 u_i}{\partial t^2} \tag{8.27}$$

Therefore the wave is called an *irrotational wave* or a *dilatational wave*, has a spherical wavefront, and the wave speed is c_p.

When each displacement component assumes the form

$$u_i = d_i f(x_k n_k - ct) \tag{8.28}$$

where d_i and n_i are unit vectors defining the directions of motion and propagation, respectively, the wave is a *plane wave* since the quantity $x_i n_i$ = *constant* describes a plane with n_i as its unit normal vector. The wave is propagating outward with the phase velocity c. Introduce a new scalar

$$y = x_k n_k - ct \tag{8.29}$$

Accordingly,

$$\frac{\partial y}{\partial x_j} = \delta_{jk} n_k = n_j \tag{8.30a}$$

$$u_{i,j} = \frac{\partial f}{\partial y}\frac{\partial y}{\partial x_j} d_i = n_j d_i \frac{\partial f}{\partial y} \tag{8.30b}$$

$$u_{j,i} = n_i d_j \frac{\partial f}{\partial y} \tag{8.30c}$$

$$u_{k,k} = n_k d_k \frac{\partial f}{\partial y},$$

or

$$\varepsilon_{kk} = n_k d_k \frac{\partial f}{\partial y} \tag{8.30d}$$

$$\varepsilon_{kk,i} = n_k d_k \frac{\partial^2 f}{\partial y^2}\frac{\partial y}{\partial x_i}$$

$$= n_k d_k n_i \frac{\partial^2 f}{\partial y^2} \tag{8.30e}$$

If

$$n_i = \pm d_i \tag{8.31}$$

then Eqs. 8.30b and c become identical, i.e.

$$u_{i,j} = u_{j,i},$$

or, by definition of the rotation tensor (Eq. 2.48),

$$\omega_{ij} = 0 \tag{8.32}$$

meaning that when the directions of motion and propagation are the same, the wave is 'irrotational'. On the other hand, if

$$n_k d_k = 0 \tag{8.33}$$

then Eqs. 8.30d and e give, respectively,

$$\varepsilon_{kk} = 0 \tag{8.34a}$$

$$\varepsilon_{kk,i} = 0 \tag{8.34b}$$

meaning that the wave becomes 'equivoluminal' when the directions of motion and propagation are orthogonal.

In the following three sections (Sections 8.2, 8.3 and 8.4) we shall derive a theory for the propagation of waves in continua. The general conditions for wavefronts, dynamic and kinematic jumps, and rays in any type of solids will be considered first. The derivation of a wavefront from a preceding wavefront will be made. Nonstationary and stationary surfaces of discontinuity will be defined. Then inhomogeneous isotropic elastic media will be considered. Pressure and shear waves will be distinguished. The relationship between a velocity jump at a wavefront and that at a preceding wavefront will be derived, and subsequently specialized for homogeneous isotropic elastic solids. Then the problem of reflection and transmission of waves from an arbitrary smooth interface of two media will be considered. Finally, we shall treat the reflection of waves from the bounding surface of a body for various types of boundary conditions. The derivation (in Sections 8.2, 8.3 and 8.4) is based upon the work by Keller (1964).

8.2 WAVEFRONT AND JUMPS

Let V be an arbitrary domain with a piecewise boundary S in the four-dimensional x_ν-space in which the position vector components are

$$\{x_\nu\} = \begin{Bmatrix} x_1 \\ x_2 \\ x_3 \\ t \end{Bmatrix} \tag{8.35a}$$

and let a four-dimensional vector be defined as

$$\{\tau_{i\nu}\} = \begin{Bmatrix} \tau_{i1} \\ \tau_{i2} \\ \tau_{i3} \\ \tau_{i4} \end{Bmatrix} = \begin{Bmatrix} \sigma_{i1} \\ \sigma_{i2} \\ \sigma_{i3} \\ -\rho v_i \end{Bmatrix} \tag{8.35b}$$

Note that the subscript ν runs from 1 to 4, i from 1 to 3, and $v_i \equiv \partial u_i/\partial t$. The equations of motion (Eq. 2.10) can be written

$$\tau_{i\nu,\nu} = -X_i \tag{8.36}$$

in which the comma-subscript for a derivative and Einstein summation conventions are employed. Integrating the equation above over the *hyperdomain* V and applying Gauss's divergence theorem (Eq. 1.88), we get

$$\int_S \tau_{i\nu}\zeta_\nu \mathrm{d}S = -\int_V X_i \mathrm{d}V \tag{8.37}$$

where ζ_ν is the outward unit normal to the *hypersurface* boundary S. This equation is one of the most general forms of the equations of motion.

Consider two portions of V, i.e. V_1 and V_2 which are so chosen that S_{1-2} is their common boundary (see Fig. 8.5). The disturbance is taken as being in V_1 only. If ζ_ν is the unit vector normal to S_{1-2} and outwards from V_1, then $-\zeta_\nu$ is the unit vector normal to S_{1-2} and outwards from V_2. For domains V_1 and V_2, Eq. 8.37 gives

$$\int_{S_1} \tau_{i\nu}\zeta_\nu \mathrm{d}S + \int_{S_{1-2}} \tau_{i\nu}\zeta_\nu \mathrm{d}S = -\int_{V_1} X_i \mathrm{d}V \tag{8.38a}$$

$$\int_{S_2} \tau_{i\nu}\zeta_\nu \mathrm{d}S - \int_{S_{1-2}} \tau_{i\nu}\zeta_\nu \mathrm{d}S = -\int_{V_2} X_i \mathrm{d}V \tag{8.38b}$$

Noting that $\tau_{i\nu}$ is discontinuous on the *hyperwavefront* S_{1-2} and that the integrals in Eq. 8.37 are just summation of those for V_1 and V_2, the sum of Eqs. 8.38 gives

$$\int_{S_{1-2}} [\tau_{i\nu}]\zeta_\nu \mathrm{d}S = 0 \tag{8.39a}$$

where $[\tau_{i\nu}]$ denotes the *jump* in $\tau_{i\nu}$ across S_{1-2}, i.e.

$$[\tau_{i\nu}] = \tau_{i\nu}^{+} - \tau_{i\nu}^{-} \tag{8.39b}$$

The superminus ($^-$) and superplus ($^+$) signs denote the quantity in the disturbed domain

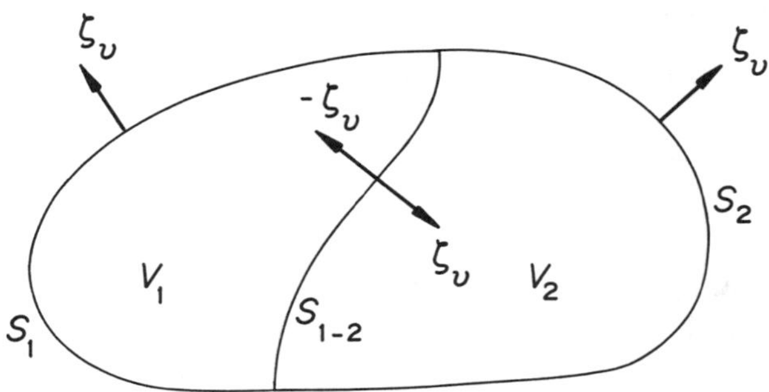

Fig. 8.5 Hypersurface S_{1-2} separating disturbed hyperdomain V_1 from undisturbed hyperdomain V_2.

behind the wavefront, and that in the undisturbed domain in front of the wavefront, respectively. Since Eq. 8.39a must hold for an arbitrary section of S_{1-2}, we can conclude that

$$[\tau_{i\nu}]\zeta_\nu = 0$$

at all points of the surface of discontinuity. Equation 8.35b shows that this equation is equivalent to

$$[\sigma_{ij}]\zeta_j = [\rho v_i]\zeta_4 \tag{8.40}$$

where $i,j = 1$ to 3. If the hypersurface of the discontinuity is represented by

$$\phi(x_i, t) = 0 \tag{8.41}$$

then we have (Eq. 1.99)

$$\begin{Bmatrix} \zeta_1 \\ \zeta_2 \\ \zeta_3 \\ \zeta_4 \end{Bmatrix} = (\phi_{,i}\phi_{,i} + \phi_{,t}^2)^{-1/2} \begin{Bmatrix} \phi_{,1} \\ \phi_{,2} \\ \phi_{,3} \\ \phi_{,t} \end{Bmatrix} \tag{8.42}$$

where

$$\phi_{,t} \equiv \frac{\partial \phi}{\partial t}.$$

There can be no jump in displacements even at the hyperwavefront S_{1-2} and the same can be said for a tangential derivative, i.e.

$$\left[\frac{\partial u_i}{\partial s}\right] = 0 \tag{8.43}$$

where s is an arc length of a space curve tangent to S_{1-2}. Now,

$$\frac{\partial u_i}{\partial s} = u_{i,\nu}\frac{\partial x_\nu}{\partial s} \tag{8.44}$$

and, as discussed in Section 1.8,

$$\frac{\partial x_\nu}{\partial s} = N_\nu \tag{8.45}$$

where N_ν is a unit vector tangent to s thus also to S_{1-2}. Equations 8.43 to 8.45 show that

$$[u_{i,\nu}]N_\nu = 0 \tag{8.46}$$

which implies that the vector $[u_{i,\nu}]$ in the four-dimensional space is normal to any N_ν. Thus it must be normal to ζ_ν, i.e.

$$[u_{i,\nu}] = \alpha_i \zeta_\nu,$$

where α_i is a scalar (in the four-dimensional space). The equation above can be rewritten more explicitly as

$$[u_{i,j}] = \alpha_i \zeta_j,$$

$$[u_{i,t}] = \alpha_i \zeta_4.$$

But $u_{i,t}$ is the velocity v_i, so that

$$[u_{i,j}] = \alpha_i(\phi_{,k}\phi_{,k} + \phi_{,t}^2)^{-1/2}\phi_{,j} \tag{8.47a}$$

$$[v_i] = \alpha_i(\phi_{,k}\phi_{,k} + \phi_{,t}^2)^{-1/2}\phi_{,t} \tag{8.47b}$$

where i, j, $k = 1$ to 3. Determining α_i from Eq. 8.47b and substituting the result in Eq. 8.47a we find that, if $\phi_{,t} \neq 0$, then

$$[u_{i,j}] = \frac{[v_i]\phi_{,j}}{\phi_{,t}} \tag{8.48}$$

Suppose that we have solved Eq. 8.41 for t as a function of x_i. Then we may write

$$\phi(x_i, t) \equiv \psi(x_i) - t \tag{8.49a}$$

So that the surface of discontinuity is given by

$$\psi(x_i) = t \tag{8.49b}$$

The surface $\psi(x_i)$ = *constant* (in the physical x_i-space) is a *wavefront* across which the stress may be discontinuous at an appropriate instant of time. The function $\psi(x_i)$ is known as the *wavefront function.* Accordingly, we can write

$$\phi_{,i} = \psi_{,i} \tag{8.50a}$$

$$\phi_{,t} = -1 \tag{8.50b}$$

$$\frac{d\psi}{dt} = 1 \tag{8.50c}$$

Figure 8.6 shows wavefronts at times t and $t + \Delta t$ spaced apart by a distance Δs, i.e.

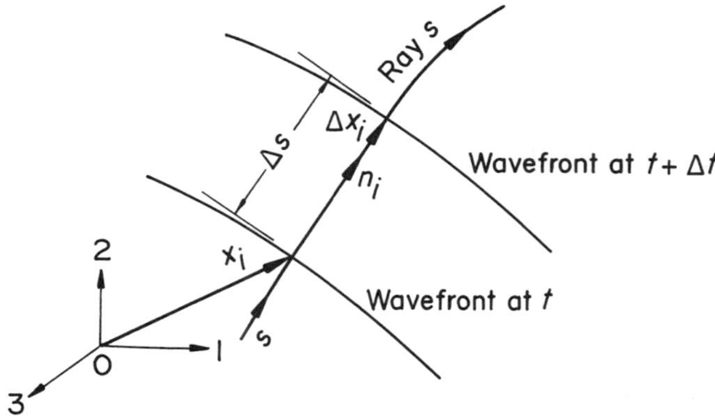

Fig. 8.6 Wavefronts at times t and $t + \Delta t$.

$$\Delta s = c\Delta t \tag{8.51a}$$

$$\Delta x_i = \Delta s\, n_i \tag{8.51b}$$

$$= c\Delta t n_i \tag{8.51c}$$

$$\frac{dx_i}{dt} = c n_i \tag{8.51d}$$

where c is the wave speed, and n_i the unit vector normal to the wavefront, i.e.

$$n_i = \frac{\psi_{,i}}{\sqrt{\psi_{,j}\psi_{,j}}} \tag{8.52}$$

Equation 8.51a shows that

$$\frac{d}{ds} = c^{-1}\frac{d}{dt} \tag{8.53}$$

Equation 8.49b shows that

$$\psi(x_i + \Delta x_i) - \psi(x_i) = t + \Delta t - t,$$

or

$$\Delta\psi = \Delta t.$$

If we divide this equation by Δx_i, we get

$$\psi_{,i} = \Delta t/\Delta x_i \tag{8.54a}$$

or,

$$\frac{\psi_{,i}}{\Delta t} = \frac{1}{\Delta x_i} \tag{8.54b}$$

Multiplying this equation by $A(x_i + \Delta x_i) - A(x_i)$, we get

$$\psi_{,i}\frac{dA}{dt} = A_{,i} \tag{8.55a}$$

or

$$\frac{dA}{dt} = \frac{A_{,i}\psi_{,i}}{\psi_{,j}\psi_{,j}} \tag{8.55b}$$

But, by the chain rule of differentiation

$$\frac{dA}{dt} = A_{,i}\frac{dx_i}{dt},$$

or, due to Eq. 8.51d,

$$\frac{dA}{dt} = c n_i A_{,i}.$$

Equating this equation to Eq. 8.55b we find

$$c n_i A_{,i} = \frac{A_{,i}\psi_{,i}}{\psi_{,j}\psi_{,j}}.$$

Since the equation above holds for any $A(x_i)$, we can cancel $A_{,i}$ from both sides of it, and obtain

$$cn_i = \frac{\psi_{,i}}{\psi_{,j}\psi_{,j}},$$

or, due to Eq. 8.52,

$$\psi_{,j}\psi_{,j} = c^{-2} \tag{8.56}$$

This equation, known as the *Eiconal equation* in geometrical optics, is a nonlinear first order partial differential equation, the solutions of which yield possible wavefronts. Using this equation in Eqs. 8.52, 8.54a and 8.55b, we find

$$n_i = c\psi_{,i} \tag{8.57a}$$

$$\frac{\Delta t}{\Delta x_i} = c^{-1} n_i \tag{8.57b}$$

$$\frac{\mathrm{d}A}{\mathrm{d}t} = c^2 \psi_{,i} A_{,i} \tag{8.57c}$$

or

$$\frac{\mathrm{d}A}{\mathrm{d}t} = c n_i A_{,i} \tag{8.57d}$$

or

$$\frac{\mathrm{d}A}{\mathrm{d}t} = A_{,i} \frac{\mathrm{d}x_i}{\mathrm{d}t} \tag{8.57e}$$

Incorporating the results above in Eqs. 8.40 and 8.48, we get, respectively

$$[\sigma_{ij}]n_j = -c[\rho v_i] \tag{8.58a}$$

$$[u_{i,j}] = -c^{-1}[v_i]n_j \tag{8.58b}$$

Equation 8.58a is known as the *dynamic jump condition,* and characterizes the momentum conservation across the wavefront, i.e. the jump in surface traction is proportional to the jump in momentum. Equation 8.58b is known as the *kinematic jump condition.*

Differentiating Eq. 8.57a with respect to t, we get

$$\frac{\mathrm{d}n_i}{\mathrm{d}t} = \frac{\mathrm{d}c}{\mathrm{d}t}\psi_{,i} + c\frac{\mathrm{d}\psi_{,i}}{\mathrm{d}t}$$

$$= \frac{\mathrm{d}c}{\mathrm{d}t}c^{-1}n_i + c\psi_{,ij}\frac{\mathrm{d}x_j}{\mathrm{d}t},$$

or, due to Eq. 8.51d,

$$\frac{\mathrm{d}n_i}{\mathrm{d}t} = c^{-1}\frac{\mathrm{d}c}{\mathrm{d}t}n_i + c^3\psi_{,ij}\psi_{,j}$$

$$= c^{-1}\frac{dc}{dt}n_i + \frac{c^3}{2}(\psi_{,j}\psi_{,j})_{,i},$$

or, due to Eq. 8.56,

$$\frac{dn_i}{dt} = c^{-1}\frac{dc}{dt}n_i - c_{,i} \tag{8.59a}$$

or, due to Eq. 8.53,

$$c_{,i} = c^{-1}\frac{dc}{dt}n_i - c\frac{dn_i}{ds},$$

or, due to Eq. 1.91a,

$$c_{,i} = c^{-1}\frac{dc}{dt}n_i - cKN_i \tag{8.59b}$$

where K and N_i are curvature and unit normal, respectively, of the ray. When c is constant, Eqs. 8.59 give a constant n_i and zero curvature meaning that the ray is straight. Differentiating Eq. 8.57a with respect to x_i we find

$$c\psi_{,ii} = n_{i,i} - c_{,i}\psi_{,i} = n_{i,i} - c_{,i}c^{-1}n_i,$$

or, due to Eq. 8.59b,

$$c\psi_{,ii} = n_{i,i} - c^{-2}\frac{dc}{dt} \tag{8.60}$$

By Gauss's divergence theorem (Eq. 1.87), we may write the term $n_{i,i}$ as

$$n_{i,i} = \lim_{V\to 0}\frac{1}{V}\int_S n_i B_i dS \tag{8.61}$$

in which V is the volume of the space containing a 'bundle' of rays, S the surface

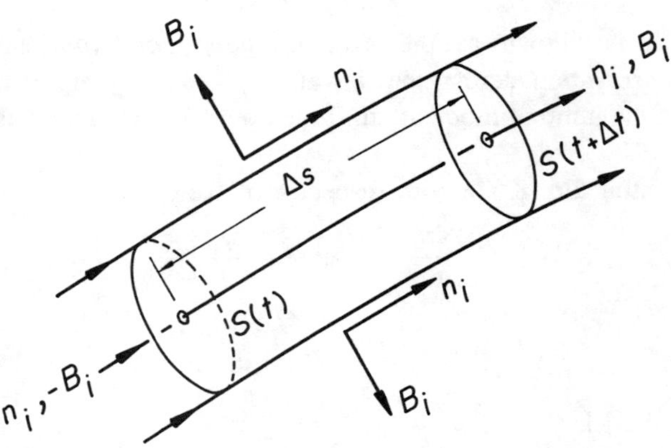

Fig. 8.7 Same bundle of rays normal to wavefronts at times t and $t+\Delta t$.

completely enclosing V, and B_i the unit vector normal to S and outward from V. The situation is depicted more clearly in Fig. 8.7. Note that the scalar product n_iB_i on the two ends and curved surface of S is

$$n_iB_i = -1\,, \quad \text{on} \quad S(t) \tag{8.62a}$$

$$= 1\,, \quad \text{on} \quad S(t+\Delta t) \tag{8.62b}$$

$$= 0\,, \quad \text{elsewhere} \tag{8.62c}$$

in which $S(t)$ denotes the area of the wavefront at the time t. Incorporating Eqs. 8.62, we can rewrite Eq. 8.61 as

$$n_{i,i} = \operatorname*{Lim}_{V \to 0} \frac{S(t+\Delta t) - S(t)}{V}$$

$$= \operatorname*{Lim}_{\Delta t \to 0} \frac{S(t+\Delta t) - S(t)}{S(t)c\Delta t}$$

$$= \frac{c^{-1}}{S(t)} \frac{\mathrm{d}S(t)}{\mathrm{d}t} \tag{8.63}$$

The relationship between small areas of wavefronts at times t_0 and t can be put as

$$S(t) = J(t_0, t)S(t_0) \tag{8.64}$$

where $J(t_0, t)$ is the Jacobian of the mapping of areas of wavefronts normal to the same group of rays, i.e.

$$J(t_0, t) = \frac{\varepsilon_{ijk} n_i (\partial x_j/\partial\alpha)(\partial x_k/\partial\beta)}{\varepsilon_{ijk} n_i^0 (\partial x_j^0/\partial\alpha)(\partial x_k^0/\partial\beta)} \tag{8.65}$$

where the superscript (^{0}) denotes a quantity at the time t_0, and α and β are the parameters of any wavefront. A point on the wavefront at the time t is located by the position vector $x_i(\alpha, \beta)$. In fact, we have already described this topic by Eqs. 1.100 and 1.101. Substituting Eq. 8.64 into Eq. 8.63 yields

$$n_{i,i} = c^{-1} J^{-1} \frac{\mathrm{d}J}{\mathrm{d}t} \tag{8.66}$$

Consequently, Eq. 8.60 becomes

$$c\psi_{,ii} = c^{-1}\left(J^{-1}\frac{\mathrm{d}J}{\mathrm{d}t} - c^{-1}\frac{\mathrm{d}c}{\mathrm{d}t} \right) \tag{8.67a}$$

$$= c^{-1}\frac{\mathrm{d}\log(J/c)}{\mathrm{d}t} \tag{8.67b}$$

Moreover, due to Eqs. 8.57c and 8.67b, we can show that

$$\frac{c^2}{A}(A\psi_{,j})_{,j} = \frac{\mathrm{d}\log(AJ/c)}{\mathrm{d}t} \tag{8.67c}$$

If all conditions are known at a wavefront for the time t, a subsequent wavefront for $t+\Delta t$ can be located easily by Eq. 8.51c, and rays normal to the subsequent wavefront can be determined by solving Eq. 8.59a. Such a solution exists wherever the wavefronts are smooth. Besides its location, we may be interested also in magnitudes of various jumps at the subsequent wavefront. Now we shall formulate equations governing such magnitudes. We know that the displacement jump is always zero, i.e.

$$[u_i]=0 \tag{8.68}$$

The derivative of any jump can be derived as follows

$$\begin{aligned}\frac{\partial}{\partial x_j}[f(x_i,t)] &= \frac{\partial}{\partial x_j}[f(x_i,\psi)] \\ &= \left[\frac{\partial}{\partial x_j}f(x_i,\psi)\right]^+ - \left[\frac{\partial}{\partial x_j}f(x_i,\psi)\right]^- \\ &= (f_{,j}+f_{,t}\psi_{,j})^+ - (f_{,j}+f_{,t}\psi_{,j})^- \\ &= [f_{,j}]+[f_{,t}]\psi_{,j}\end{aligned} \tag{8.69}$$

or, due to Eq. 8.57a,

$$\frac{\partial}{\partial x_j}[f(x_i,t)] = [f_{,j}]+[f_{,t}]n_j c^{-1} \tag{8.70}$$

Differentiating Eq. 8.68 with respect to x_j, and using Eq. 8.69, we find

$$[u_{i,j}] = -[u_{i,t}]\psi_{,j} = -[v_i]\psi_{,j} = -V_i\psi_{,j} \tag{8.71}$$

where V_i denotes the *velocity jump* as a continuously differentiable function of x_i, i.e.

$$V_i(x_i) = [v_i(x_i,\psi)] \tag{8.72}$$

The kinematic jump obtained earlier as Eq. 8.58b is identical to Eq. 8.71. The derivative of the latter equation with respect to x_k gives

$$[u_{i,jk}] = -[v_{i,j}]\psi_{,k} - (V_i\psi_{,j})_{,k} \tag{8.73a}$$

and in particular this implies

$$[u_{i,kk}] = -[v_{i,k}]\psi_{,k} - (V_i\psi_{,k})_{,k} \tag{8.73b}$$

$$[u_{k,ki}] = -[v_{k,k}]\psi_{,i} - (V_k\psi_{,k})_{,i} \tag{8.73c}$$

Differentiating Eq. 8.72 with respect to x_j we find

$$V_{i,j} = [v_{i,j}] + [v_{i,t}]\psi_{,j} \tag{8.74a}$$

or

$$[v_{i,t}]\psi_{,j} = V_{i,j} - [v_{i,j}] \tag{8.74b}$$

Multiplying the equation above by $\psi_{,j}$ and incorporating the Eiconal equation, we obtain

$$[v_{i,t}] = c^2(V_{i,j}\psi_{,j} - [v_{i,j}]\psi_{,j}) \tag{8.75}$$

Due to Eq. 8.74b, it can be shown that

$$[v_{i,j}]\psi_{,i}\psi_{,j} = V_{i,j}\psi_{,i}\psi_{,j} - [v_{i,t}]\psi_{,j}\psi_{,i}\psi_{,j},$$

and

$$-[v_{k,k}]\psi_{,i}\psi_{,i} = -V_{k,k}\psi_{,i}\psi_{,i} + [v_{k,t}]\psi_{,k}\psi_{,i}\psi_{,i}.$$

Combining the last two equations, we get

$$[v_{i,j}]\psi_{,i}\psi_{,j} - [v_{k,k}]\psi_{,i}\psi_{,i} = V_{i,j}\psi_{,i}\psi_{,j} - V_{k,k}\psi_{,i}\psi_{,i} \tag{8.76}$$

Substituting Eq. 8.75 into Eq. 8.74a yields for $i=j$,

$$c^2(V_{i,j}\psi_{,i}\psi_{,j} - [v_{i,j}]\psi_{,i}\psi_{,j}) = V_{i,i} - [v_{i,i}] \tag{8.77}$$

With the new definition of the velocity jump, we can write the dynamic and kinematic jumps (Eqs. 8.58), respectively, as

$$[\sigma_{ij}]n_j = -c\rho V_i \tag{8.78a}$$

$$[u_{i,j}] = -c^{-1}V_i n_j \tag{8.78b}$$

and the dilatation jump is

$$[u_{k,k}] = -c^{-1}V_k n_k \tag{8.78c}$$

The derivation so far is based on a *nonstationary surface of discontinuity,* i.e. $\phi_{,t} \neq 0$. If $\phi_{,t} = 0$, the surface is fixed in space, i.e. *stationary,* then Eq. 8.47b becomes

$$V_i = 0 \tag{8.79}$$

and, accordingly, the jump conditions Eqs. 8.78a and 8.47a become

$$[\sigma_{ij}]n_j = 0 \tag{8.80a}$$

$$[u_{i,j}] = \alpha_i n_j \tag{8.80b}$$

and

$$[u_{i,i}] = \alpha_i n_i \tag{8.80c}$$

for some $\alpha_i \neq 0$.

8.3 VELOCITY JUMPS IN ISOTROPIC ELASTIC DOMAINS

While Section 8.2 is independent of any particular constitutive laws, the present section is restricted to isotropic domains which may be inhomogeneous but continuous, i.e. the mass density ρ and the elastic constants λ and μ are continuous and diferentiable functions of x_i. Due to the kinematic jump conditions, we can write the constitutive relationship at the wavefront in the form

$$[\sigma_{ij}] = -\lambda\delta_{ij}V_k n_k c^{-1} - \mu V_i n_j c^{-1} - \mu V_j n_i c^{-1} \tag{8.81a}$$

or, due to the dynamic jump condition,

$$[(c^2 - c_s^2)\delta_{ij} - (c_p^2 - c_s^2)n_i n_j]V_j = 0 \tag{8.81b}$$

where c_p and c_s are as defined in Eqs. 8.12 and 8.14, respectively. The equation above is an eigenequation with c^2 and V_j being eigenvalues and eigenvectors, respectively. There are three eigensolutions. The first solution can be easily identified by observation, i.e.

$$c^2 = c_p^2 \tag{8.82a}$$

$$V_j = \frac{a n_j}{c_p} = a\psi_{,j} \tag{8.82b}$$

where a is an arbitrary scalar. The second solution should have the eigenvector orthogonal to the first one (Eq. 8.82b), i.e.

$$V_j \psi_{,j} = 0 \tag{8.83a}$$

which makes Eq. 8.81b yield the eigenvalue as

$$c^2 = c_s^2 \tag{8.83b}$$

The third eigenvalue can be shown to be a duplicate of the second one. In case of Eqs. 8.82, the velocity jump is normal to the wavefronts, thus it is called a *longitudinal* or *compressive discontinuity* and the speed c_p is just that of P-waves, Eq. 8.12. In case of Eqs. 8.83, the velocity jump is tangential to the surface of discontinuity, thus it is called a *transverse* or *shear discontinuity* and the speed c_s is just that of the S-waves (Eq. 8.14). Moreover, the latter case also has a zero jump in dilatation (Eq. 8.78c) which is another common feature in the classical theory of elastic S-wave propagation.

The equations of motion, Eq. 2.14, may be written in terms of displacements for inhomogeneous isotropic elastic solids in the form

$$v_{i,t} = (c_p^2 - c_s^2)u_{k,ki} + c_s^2 u_{i,kk} + \frac{\lambda_{,i}}{\rho} u_{k,k} + \frac{\mu_{,k}}{\rho}(u_{i,k} + u_{k,i}) \tag{8.84}$$

Substituting Eqs. 8.71, 8.73 and 8.75 into the equation above yields

$$c^2(V_{i,j}\psi_{,j} - [v_{i,j}]\psi_{,j}) + (c_p^2 - c_s^2)\,[[v_{k,k}]\psi_{,i} + (V_k\psi_{,k})_{,i}]$$

$$+ c_s^2[[v_{i,k}]\psi_{,k} + (V_i\psi_{,k})_{,k}] + \frac{\lambda_{,i}}{\rho} V_k\psi_{,k} + \frac{\mu_{,k}}{\rho}(V_i\psi_{,k} + V_k\psi_{,i}) = 0 \tag{8.85}$$

For P-waves, the problem is characterized by Eqs. 8.82, thus

$$V_{i,j} = a_{,j}\psi_{,i} + a\psi_{,ij} \tag{8.86}$$

Setting $c^2 = c_p^2$ in Eq. 8.85 and multiplying the result by $\psi_{,i}$, then incorporating Eq. 8.77 we find

$$V_{k,k} + c_p^2(V_k\psi_{,k})_{,i}\psi_{,i} + c_s^2([v_{i,k}]\psi_{,k} - [v_{k,k}]\psi_{,i})\psi_{,i} + c_s^2[(V_i\psi_{,k})_{,k}$$

$$-(V_k\psi_{,k})_{,i}]\psi_{,i} + \frac{\lambda_{,i}}{\rho}(V_k\psi_{,k})\psi_{,i} + \frac{\mu_{,k}}{\rho}(V_i\psi_{,k}\psi_{,i} + V_k\psi_{,i}\psi_{,i}) = 0.$$

Substituting Eq. 8.76 into the equation above, then incorporating Eq. 8.86, we get after an extensive manipulation

$$\psi_{,k}a_{,k}+\frac{a}{2\rho}(\rho\psi_{,k})_{,k}=0 \tag{8.87}$$

Substituting Eqs. 8.57c and 8.67c into the equation above we find

$$\frac{\mathrm{d}a}{\mathrm{d}t}+\frac{a\,\mathrm{d}\log(\rho J/c_p)}{2\mathrm{d}t}=0 \tag{8.88}$$

The solution to the equation above is

$$a=a^0\left(\frac{\rho^0 c_p}{\rho c_p^0}J^{-1}\right)^{1/2} \tag{8.89}$$

where the superscript (^{0}) denotes a quantity at an initial time t_0, and the terms without any superscripts belong to a subsequent time t. Substituting Eq. 8.89 into Eq. 8.82b, we can get the velocity jump at a subsequent wavefront as

$$V_i=n_i\left(V_j^0V_j^0J^{-1}\frac{\rho^0c_p^0}{\rho c_p}\right)^{1/2} \tag{8.90}$$

For S-waves, the problem is characterized by Eqs. 8.83, thus

$$V_{k,j}\psi_{,k}=-V_k\psi_{,kj}.$$

Multiplying the equation above by $\psi_{,j}$ we find

$$V_{k,j}\psi_{,k}\psi_{,j}=-\frac{1}{2}V_k(\psi_{,j}\psi_{,j})_{,k},$$

or, due to the Eiconal equation, Eq. 8.56,

$$V_{k,j}\psi_{,k}\psi_{,j}=V_kc_s^{-3}c_{s,k} \tag{8.91}$$

In Eq. 8.85, setting $c^2=c_s^2$ and incorporating Eq. 8.83a we find

$$c_s^2V_{i,j}\psi_{,j}+(c_p^2-c_s^2)[v_{k,k}]\psi_{,i}+c_s^2(V_i\psi_{,k})_{,k}+\frac{\mu_{,k}}{\rho}(V_i\psi_{,k}+V_k\psi_{,i})=0 \tag{8.92}$$

Multiplying the equation above by $\psi_{,j}$, we can solve for $[v_{k,k}]$ as

$$[v_{k,k}]=-\frac{c_s^2}{(c_p^2-c_s^2)}\left(2c_s^2V_{i,j}\psi_{,j}\psi_{,i}+c_s^{-2}\frac{\mu_{,k}}{\rho}V_k\right),$$

or, due to Eq. 8.91,

$$[v_{k,k}]=-\frac{c_s^2}{(c_p^2-c_s^2)}\left(2c_s^{-1}c_{s,k}V_k+c_s^{-2}\frac{\mu_{,k}}{\rho}V_k\right) \tag{8.93}$$

Substituting Eq. 8.93 into Eq. 8.92, we can get

$$\psi_{,k}V_{i,k}+\frac{1}{2\mu}(\mu\psi_{,k})_{,k}V_i-\frac{c_{s,k}}{c_s}V_k\psi_{,i}=0 \tag{8.94}$$

Substituting Eqs. 8.57a, 8.57c and 8.67c into the equation above we find

$$\frac{\mathrm{d}V_i}{\mathrm{d}t}+\frac{V_i\mathrm{d}\log(\mu J/c_s)}{2\mathrm{d}t}-c_{s,k}V_kn_i=0 \tag{8.95}$$

In order to satisfy Eq. 8.83a, a general form of V_i is

$$V_i = aN_i + bB_i \tag{8.96}$$

where N_i and B_i are unit normal and unit binormal to the ray, respectively, and a and b are scalars. Substituting Eq. 8.96 into Eq. 8.95, then incorporating Eq. 8.59b and the Frenet-Serret formulas, Eqs 1.91 and 1.92, we get

$$\frac{d\gamma}{dt} + \left(\frac{d\log(\mu J/c_s)}{2dt} + ic_s\tau\right)\gamma = 0 \tag{8.97}$$

where τ is the torsion of the ray, and

$$\gamma = a + ib \tag{8.98}$$

in which i is the imaginary quantity, i.e. $i^2 = -1$. Note that $|\gamma|$ and $|V_i|$ are identical. The solution to Eq. 8.97 can be shown to be

$$\gamma = |\gamma^0| \left(\frac{c_s\mu^0 J^{-1}}{c_s^0\mu}\right)^{1/2} e^{i\delta} \tag{8.99}$$

where

$$\delta = \delta^0 - \int_{t_0}^{t} c_s\tau dt' \tag{8.100}$$

Substituting Eq. 8.99 into Eq. 8.98 we find the velocity jump at a subsequent wavefront for propagating shear waves, Eq. 8.96, to be

$$V_i = \left(V_k^0 V_k^0 \frac{c_s\mu^0}{c_s^0\mu} J^{-1}\right)^{1/2} (N_i\cos\delta + B_i\sin\delta) \tag{8.101a}$$

or, since $\mu = \rho c_s^2$,

$$V_i = \left(V_k^0 V_k^0 \frac{\rho^0 c_s^0}{\rho c_s} J^{-1}\right)^{1/2} (N_i\cos\delta + B_i\sin\delta) \tag{8.101b}$$

In fact, δ^0 is also defined by the equation above with $t = t_0$, i.e.,

$$\tan\delta^0 = \frac{V_i^0 B_i^0}{V_j^0 N_j^0} \tag{8.102}$$

For a homogeneous domain, the rays are straight, thus a wavefront at a time t can be constructed by advancing from the wavefront at a time t_0 a distance $c(t-t_0)$ along a ray. Such construction of wavefronts, known as the *Huyghen's principle,* is illustrated in Fig. 8.8, in which r_1 and r_2 are measured along rays, and r_0 along radii perpendicular to the 'axis of revolution'. Essential geometric relations involved are

$$r_0^0 = r_2^0\sin\phi' \tag{8.103a}$$

$$r_0 = r_2\sin\phi' \tag{8.103b}$$

$$r_1 - r_1^0 = r_2 - r_2^0 = c(t-t_0) \tag{8.103c}$$

The terms r_1 and r_2 are also known as the principal radii of curvatures of wavefronts.

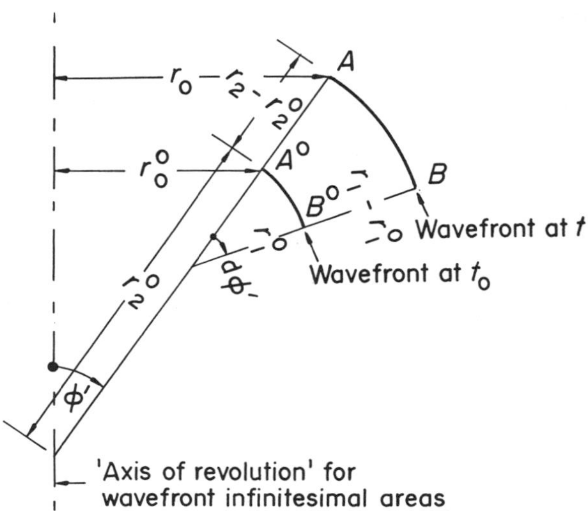

Fig. 8.8 Construction of initial and subsequent wavefronts for homogeneous medium.

Corresponding infinitesimal areas on the initial and subsequent wavefronts are generated by 'revolving' arcs A^0B^0 and AB about the axis of revolution through a small angle $\mathrm{d}\theta$, thus they can be put, respectively, as

$$S(t_0) = r_1^0 r_2^0 \sin\phi' \mathrm{d}\phi' \mathrm{d}\theta \tag{8.104a}$$

$$S(t) = r_1 r_2 \sin\phi' \mathrm{d}\phi' \mathrm{d}\theta \tag{8.104b}$$

Hence the Jacobian of area transformation (Eq. 8.64) becomes

$$J(t_0, t) = \frac{r_1 r_2}{r_1^0 r_2^0} \tag{8.105}$$

Moreover, it can be stated for the homogeneous case that $\rho^0 = \rho$, $c_p^0 = c_p$, $c_s^0 = c_s$, $n_i^0 = n_i$, $N_i^0 = N_i$, $B_i^0 = B_i$, $\tau = 0$ and $\delta^0 = \delta$, thus Eqs 8.90 and 8.101 give

$$V_i V_i = \frac{r_1^0 r_2^0}{r_1 r_2} V_i^0 V_i^0 \tag{8.106}$$

This shows that the magnitude of a field variable at the wavefront decays as R^{-1} for spherical waves (with radius R) and as $r^{-1/2}$ for cylindrical waves (with radius r).

8.4 REFLECTION AND TRANSMISSION AT INTERFACES AND BOUNDARIES

In Sections 8.2 and 8.3, we have considered the propagation of waves in continuous media. We shall now study the situation at an interface between two media with different properties. Let an interface S separating media A and B be represented by $x_i = x_i(\alpha, \beta)$ where α and β are the parameters of the surface S. We assume that the surface S is smooth, thus $\partial x_i/\partial\alpha$ and $\partial x_i/\partial\beta$ are continuous in α and β, and $\varepsilon_{ijk}(\partial x_j/\partial\alpha)(\partial x_k/\partial\beta) \neq 0$. At each point on S, the unique unit normal to S is given by

$$T_i = \frac{\varepsilon_{ijk}(\partial x_j/\partial\alpha)(\partial x_k/\partial\beta)}{[\varepsilon_{imn}\varepsilon_{iop}(\partial x_m/\partial\alpha)(\partial x_n/\partial\beta)(\partial x_o/\partial\alpha)(\partial x_p/\partial\beta)]^{1/2}} \tag{8.107}$$

In fact, $\partial x_i/\partial\alpha$ and $\partial x_i/\partial\beta$ are two unique tangents to S. The positive direction of T_i is chosen as pointing from medium A into medium B.

Any 'incident' signal in medium A is a linear combination of compressive and shear signals. Thus we specify the incident family of wavefronts as $\psi^{(0)}(x_i) = t$, where (Eq. 8.56)

$$\psi^{(0)}_{,i}\psi^{(0)}_{,i} = c_0^{-2} \tag{8.108}$$

The unit normal to a wavefront, in the direction of propagation, of the signal is (Eq. 8.57a)

$$n_i^{(0)} = c_0\psi^{(0)}_{,i} \tag{8.109}$$

The magnitude of the signal is specified by $V_i^{(0)}(x_i)$, the jump in particle velocity across a wavefront. For an incident S-wave we must have

$$n_i^{(0)}V_i^{(0)} = 0, \quad c_0 \equiv c_{sA} \tag{8.110a}$$

while for an incident P-wave we have

$$\varepsilon_{ijk}n_j^{(0)}V_k^{(0)} = 0, \quad c_0 = c_{pA} \tag{8.110b}$$

Here c_{sA} and c_{pA} are, respectively, the speeds of shear and compressive waves in medium A.

When the incident wavefront or rays meet the surface S, we can no longer be assured that only the original shear or compressive wave is present in medium A, so must allow both types of waves to be present in both media. If we denote the wavefronts of these waves by

$$\psi^{(\nu)}(x_i) = t, \quad (\nu = 1, 2, 3, 4) \tag{8.111}$$

then the wave functions $\psi^{(\nu)}$ and unit normals to the fronts $n_i^{(\nu)}$ are determined by

$$\psi^{(\nu)}_{,i}\psi^{(\nu)}_{,i} = c_\nu^{-2}, \quad n_i^{(\nu)} = c_\nu\psi^{(\nu)}_{,i}, \quad (\nu = 1, 2, 3, 4) \tag{8.112}$$

The type of each of these waves is specified by requiring that

$$c_1 = c_{pA}, \qquad \varepsilon_{ijk}n_j^{(1)}V_k^{(1)} = 0 \tag{8.113a}$$

$$c_2 = c_{sA}, \qquad n_i^{(2)}V_i^{(2)} = 0 \tag{8.113b}$$

$$c_3 = c_{pB}, \qquad \varepsilon_{ijk}n_j^{(3)}V_k^{(3)} = 0 \tag{8.113c}$$

$$c_4 = c_{sB}, \qquad n_i^{(4)}V_i^{(4)} = 0 \tag{8.113d}$$

Thus $V^{(1)}$ and $V^{(3)}$ are the magnitudes of compressive signals in media A and B, respectively, while $V^{(2)}$ and $V^{(4)}$ are the magnitudes of shear signals in those respective media. The waves for $\nu = 1$ to 2 in medium A are called *reflected,* and the waves for $\nu = 3$ and 4 in medium B called *transmitted.*

At each instant of time the reflected and transmitted wavefronts emanate from the intersection of the incident wavefront and S. We must then have, on such intersection which is assumed sufficiently small,

$$\psi^{(\nu)} = \psi^{(0)}, \quad (\nu = 1, 2, 3, 4) \tag{8.114a}$$

and the derivatives of every ψ with respect to α or β must be equal, i.e.

$$\psi_{,i}^{(\nu)}\frac{\partial x_i}{\partial \alpha} = \psi_{,i}^{(0)}\frac{\partial x_i}{\partial \alpha} \tag{8.114b}$$

$$\psi_{,i}^{(\nu)}\frac{\partial x_i}{\partial \beta} = \psi_{,i}^{(0)}\frac{\partial x_i}{\partial \beta} \tag{8.114c}$$

Equations 8.114b and 8.114c are equivalent, respectively, to

$$(c_\nu n_i^{(0)} - c_0 n_i^{(\nu)})\frac{\partial x_i}{\partial \alpha} = 0,$$

$$(c_\nu n_i^{(0)} - c_0 n_i^{(\nu)})\frac{\partial x_i}{\partial \beta} = 0.$$

Thus the vector in parentheses in the equations above must be in the direction T_i and so there exist scalars a_ν such that

$$n_i^{(\nu)} = \frac{c_\nu}{c_0} n_i^{(0)} + a_\nu T_i, \quad (\nu = 1, 2, 3, 4) \tag{8.115}$$

The last equation implies that the five wave normals $n_i^{(\nu)}$, $\nu = 0$, 1, 2, 3, 4 and the interface normal T_i are in the same plane, known as the *plane of incidence*. Introduce a unit vector N_i, tangent to S in the plane of incidence, in such a way that the angle θ_0 between $n_i^{(0)}$ and T_i can be put in the form

$$n_i^{(0)}T_i = \cos\theta_0 \tag{8.116a}$$

$$n_i^{(0)}N_i = \sin\theta_0 \tag{8.116b}$$

with

$$0 \le \theta_0 \le \frac{\pi}{2} \tag{8.116c}$$

Similarly the angle θ_ν between $n_i^{(\nu)}$ and T_i can be put in the form

$$n_i^{(\nu)}T_i = \cos\theta_\nu \tag{8.117a}$$

$$n_i^{(\nu)}N_i = \sin\theta_\nu \tag{8.117b}$$

Taking the scalar product of each equation in Eq. 8.115 with N_i and T_i, we find that

$$\frac{\sin\theta_\nu}{c_\nu}=\frac{\sin\theta_0}{c_0} \tag{8.118a}$$

$$a_\nu=\frac{\sin(\theta_0-\theta_\nu)}{\sin\theta_0} \tag{8.118b}$$

Since all c_ν are greater than zero, Eq. 8.118a implies $0\le\theta_\nu\le\pi$. Further considering the media in which each wave belongs, we can deduce

$$\frac{\pi}{2}\le\theta_1,\theta_2\le\pi \tag{8.119a}$$

$$0\le\theta_3,\theta_4\le\frac{\pi}{2} \tag{8.119b}$$

These angles and the relevant normal vectors are illustrated in Fig. 8.9. Equations 8.115 and 8.118 determine the normals $n_i^{(\nu)}$, in terms of the given incident wave normal $n_i^{(0)}$, and the surface normal T_i.

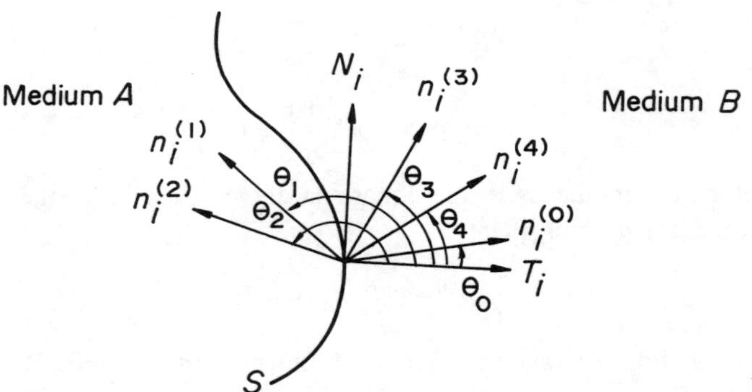

Fig. 8.9 Plane of incidence.

Since $c_0=c_1$ or c_2, then by Eq. 8.118a θ_1 or $\theta_2=\pi-\theta_0$ meaning that the reflected ray of the same type as the incident ray suffers 'equal-angle' reflection. If for any c_ν $(\nu=1,2,3,4)$ and incident angle θ_0,

$$\frac{c_\nu}{c_0}\sin\theta_0>1 \tag{8.120}$$

then Eq. 8.118a implies that there can be wave (reflected or transmitted) corresponding to that value of ν. For example, if the incident signal is an S-wave then no reflected P-wave is possible for any incident angle θ_0 if

$$\Theta<\theta_0\le\frac{\pi}{2} \tag{8.121}$$

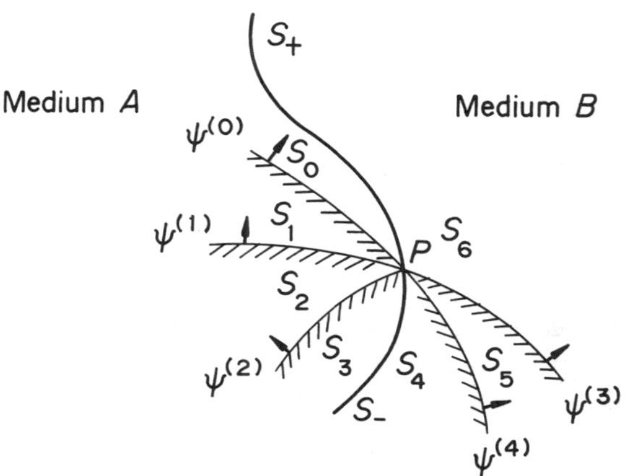

Fig. 8.10 Plane of incidence being partitioned, by five wavefronts $\psi^{(\nu)}$ and interfaces S_+ and S_-, into seven 'angular' sectors S_n.

in which the critical angle Θ is defined by

$$\sin\Theta = \frac{c_{sA}}{c_{pA}}, \quad \left(0 < \Theta < \frac{\pi}{2}\right) \tag{8.122}$$

It is clear that a reflected shear wave is possible for all angles of incidence of a compressive signal. The various possibilities for transmitted waves can be deduced easily by the reader.

Let P be a point at (x_i^0, t_0) where S and the wavefronts meet. A neighborhood of P is divided into seven 'angular' sectors S_n by the five wavefronts, and interface S which may be considered as two parts, S_+ and S_-, corresponding to $t > t_0$ and $t < t_0$, respectively (see Fig. 8.10). In the interior of each such sector the deformation and stress fields are continuous. Denote by $v_i^{(n)}$ and $\sigma_{ij}^{(n)}$ the limits of v_i and σ_{ij} as $(x_i, t) \to P$ while remaining interior to the sector S_n. Then in obvious notations;

$$\begin{aligned} &V_i^{(0)} = v_i^{(0)} - v_i^{(1)}, \quad V_i^{(3)} = v_i^{(6)} - v_i^{(5)}, \\ &V_i^{(1)} = v_i^{(1)} - v_i^{(2)}, \quad V_i^{(4)} = v_i^{(5)} - v_i^{(4)}, \\ &V_i^{(2)} = v_i^{(2)} - v_i^{(3)}, \quad V_i^{(-)} = v_i^{(4)} - v_i^{(3)}, \\ &V_i^{(+)} = v_i^{(6)} - v_i^{(0)} \end{aligned} \tag{8.123}$$

Summing the above we find the following identity

$$V_i^{(0)} + V_i^{(1)} + V_i^{(2)} + V_i^{(+)} = V_i^{(3)} + V_i^{(4)} + V_i^{(-)} \tag{8.124}$$

But a velocity jump at a stationary surface must vanish (Eq. 8.79), i.e.

$$V_i^{(+)} = V_i^{(-)} = 0 \tag{8.125}$$

hence the identity becomes

$$\sum_{\nu=0}^{2} V_i^{(\nu)} = \sum_{\nu=3}^{4} V_i^{(\nu)} \tag{8.126}$$

In the same manner, it can be shown using Eq. 8.80a that

$$\sum_{\nu=0}^{2} [\sigma_{ij}^{(\nu)}]T_j = \sum_{\nu=3}^{4} [\sigma_{ij}^{(\nu)}]T_j \tag{8.127}$$

Note that $[\sigma_{ij}^{(\nu)}]$ in the equation above can be put in terms of $V_j^{(\nu)}$ due to the constitutive relationship at the wavefronts, Eq. 8.81a.

In addition to the orthonormals T_i and N_i, we may recall the *binormal* (Eq. 1.90),

$$B_i = \varepsilon_{ijk} T_j N_k \tag{8.128}$$

which is tangent to S but normal to the plane of incidence. Now all wave normals can be represented in terms of these unit base vectors as

$$n_i^{(\nu)} = \cos\theta_\nu T_i + \sin\theta_\nu N_i + (0)B_i \tag{8.129}$$

Using this representation in Eqs. 8.113, we find that the most general form for the velocity jumps at various wavefronts is

$$V_i^{(1)} = \alpha_1 \cos\theta_1 T_i + \alpha_1 \sin\theta_1 N_i \tag{8.130a}$$

$$V_i^{(2)} = \alpha_2 \sin\theta_2 T_i - \alpha_2 \cos\theta_2 N_i + \beta_2 B_i \tag{8.130b}$$

$$V_i^{(3)} = \alpha_3 \cos\theta_3 T_i + \alpha_3 \sin\theta_3 N_i \tag{8.130c}$$

$$V_i^{(4)} = \alpha_4 \sin\theta_4 T_i - \alpha_4 \cos\theta_4 N_i + \beta_4 B_i \tag{8.130d}$$

and for the incident signal (Eqs. 8.110), which is assumed known, is

$$V_i^{(0)} = \alpha_0 \cos\theta_0 T_i + \alpha_0 \sin\theta_0 N_i, \quad (c_0 = c_{pA}) \tag{8.130e}$$

or

$$V_i^{(0)} = \alpha_0 \sin\theta_0 T_i - \alpha_0 \cos\theta_0 N_i + \beta_0 B_i, \quad (c_0 = c_{sA}) \tag{8.130f}$$

At this stage one may note that the task is to determine six unknowns α_1, α_2, α_3, α_4, β_2 and β_4 from six linear equations, Eqs. 8.126 and 8.127. The task can be organized by rewriting Eqs. 8.129 and 8.130 as

$$n_i^{(\nu)} = a_{\nu 1} T_i + a_{\nu 2} N_i + a_{\nu 3} B_i \tag{8.131}$$

$$V_i^{(\nu)} = b_{\nu 1} T_i + b_{\nu 2} N_i + b_{\nu 3} B_i \tag{8.132}$$

in which $a_{\nu j}$ and $b_{\nu j}$ are as defined in Table 8.1. Accordingly and due to Eq.8.81a, we can write the following

$$[\sigma_{ij}^{(\nu)}]T_j = T_{\nu 1} T_i + T_{\nu 2} N_i + T_{\nu 3} B_i \tag{8.133}$$

where

Table 8.1

$a_{\nu j}$		$j=1$	$j=2$	$j=3$	
$\nu=$	0	$\cos\theta_0$	$\sin\theta_0$	0	
	1	$\cos\theta_1$	$\sin\theta_1$	0	
	2	$\cos\theta_2$	$\sin\theta_2$	0	
	3	$\cos\theta_3$	$\sin\theta_3$	0	
	4	$\cos\theta_4$	$\sin\theta_4$	0	
$b_{\nu j}$		$j=1$	$j=2$	$j=3$	c_ν
$\nu=$	0 (P-wave)	$\alpha_0\cos\theta_0$	$\alpha_0\sin\theta_0$	0	c_{pA}
	0 (S-wave)	$\alpha_0\sin\theta_0$	$-\alpha_0\cos\theta_0$	β_0	c_{sA}
	1	$\alpha_1\cos\theta_1$	$\alpha_1\sin\theta_1$	0	c_{pA}
	2	$\alpha_2\sin\theta_2$	$-\alpha_2\cos\theta_2$	β_2	c_{sA}
	3	$\alpha_3\cos\theta_3$	$\alpha_3\sin\theta_3$	0	c_{pB}
	4	$\alpha_4\sin\theta_4$	$-\alpha_4\cos\theta_4$	β_4	c_{sB}

$$T_{\nu 1}=-\frac{1}{c_\nu}[(\lambda+2\mu)a_{\nu 1}b_{\nu 1}+\lambda a_{\nu 2}b_{\nu 2}+\lambda a_{\nu 3}b_{\nu 3}] \tag{8.134a}$$

$$T_{\nu 2}=-\frac{\mu}{c_\nu}(a_{\nu 1}b_{\nu 2}+a_{\nu 2}b_{\nu 1}) \tag{8.134b}$$

$$T_{\nu 3}=-\frac{\mu}{c_\nu}(a_{\nu 1}b_{\nu 3}+a_{\nu 3}b_{\nu 1}) \tag{8.134c}$$

in which it should be understood that $\lambda=\lambda_A$ and $\mu=\mu_A$ for $\nu=0,1,2$; and $\lambda=\lambda_B$ and $\mu=\mu_B$ for $\nu=3,4$. Equations 8.126 and 8.127 become, respectively

$$b_{1j}+b_{2j}-b_{3j}-b_{4j}=-b_{0j} \tag{8.135a}$$

$$T_{1j}+T_{2j}-T_{3j}-T_{4j}=-T_{0j} \tag{8.135b}$$

Since $b_{13}=b_{33}=0$ and $a_{\nu 3}=0$ for all ν, Eqs. 8.135a and 8.135b for $j=3$ assume simple respective forms as

$$b_{23}-b_{43}=-b_{03} \tag{8.136a}$$

$$T_{23}-T_{43}=-T_{03} \tag{8.136b}$$

while other four equations remain as

$$b_{1j}+b_{2j}-b_{3j}-b_{4j}=-b_{0j}, \quad (j=1,2) \tag{8.136c}$$

$$T_{1j}+T_{2j}-T_{3j}-T_{4j}=-T_{0j}, \quad (j=1,2) \tag{8.136d}$$

Equations 8.136a and 8.136b yield the values of β_2 and β_4, which exist only for an incident S-wave, as

$$\beta_2 = \beta_0 \frac{[(\mu_A/c_0)\cos\theta_0 - (\mu_B/c_4)\cos\theta_4]}{\Delta} \tag{8.137a}$$

$$\beta_4 = \beta_0 \frac{[(\mu_A/c_0)\cos\theta_0 - (\mu_A/c_2)\cos\theta_2]}{\Delta} \tag{8.137b}$$

where

$$\Delta \equiv \frac{\mu_B}{c_4}\cos\theta_4 - \frac{\mu_A}{c_2}\cos\theta_2 \tag{8.138}$$

In the meantime, Eqs. 8.136c and 8.136d form a set of four linear equations for determining α_1 to α_4 as

$$\begin{bmatrix} \cos\theta_1 & \sin\theta_2 & -\cos\theta_3 & -\sin\theta_4 \\ \sin\theta_1 & -\cos\theta_2 & -\sin\theta_3 & \cos\theta_4 \\ -\rho_A c_1 \cos 2\theta_2 & -\rho_A c_2 \sin 2\theta_2 & \rho_B c_3 \cos 2\theta_4 & \rho_B c_4 \sin 2\theta_4 \\ -\frac{\mu_A}{c_1}\sin 2\theta_1 & \rho_A c_2 \cos 2\theta_2 & \frac{\mu_B}{c_3}\sin 2\theta_3 & -\rho_B c_4 \cos 2\theta_4 \end{bmatrix} \begin{Bmatrix} \alpha_1 \\ \alpha_2 \\ \alpha_3 \\ \alpha_4 \end{Bmatrix} = \alpha_0 \begin{Bmatrix} K_1 \\ K_2 \\ K_3 \\ K_4 \end{Bmatrix}$$

Here K_ν are given for an incident P-wave ($c_0 = c_1$, $\theta_0 = \pi - \theta_1$) as

$$K_1 = \cos\theta_1, \qquad K_2 = -\sin\theta_1 \tag{8.139a,b}$$

$$K_3 = \rho_A c_1 \cos 2\theta_2, \qquad K_4 = -\frac{\mu_A}{c_1}\sin 2\theta_1 \tag{8.139c,d}$$

and for an incident S-wave ($c_0 = c_2$, $\theta_0 = \pi - \theta_2$) as

$$K_1 = -\sin\theta_2, \qquad K_2 = -\cos\theta_2 \tag{8.140a,b}$$

$$K_3 = -\rho_A c_2 \sin 2\theta_2, \quad K_4 = -\rho_A c_2 \cos 2\theta_2 \tag{8.140c,d}$$

These results agree formally with the appropriate special cases in the book by Ewing, Jardetzky and Press (1957) where the case of plane stress waves incident on plane interfaces is treated.

We shall now consider reflection at boundaries. The bounding surface of a medium can be considered as a special case of an interface. Now, however, there are no transmitted signals and specific conditions are to be imposed on the displacements and/or stresses at the surface. These conditions serve to replace the jumps specified on an interface.

If all three displacement components are specified at the boundary,

$$v_i^{(0)} = v_i^{(3)} = \text{prescribed quantity},$$

then Eqs. 8.123 give

$$\sum_{\nu=0}^{2} V_i^{(\nu)} = 0 \tag{8.141}$$

It follows that for an incident P-wave with $c_0 = c_1$ and $\theta_0 = \pi - \theta_1$,

$$\alpha_1 = \alpha_0 \frac{\cos(\theta_2 - \theta_0)}{\cos(\theta_2 + \theta_0)}, \quad \alpha_2 = \alpha_0 \frac{\sin 2\theta_0}{\cos(\theta_2 + \theta_0)}, \quad \beta_2 = 0 \tag{8.142a}$$

Similarly for an incident shear wave, with $c_0 = c_2$ and $\theta_0 = \pi - \theta_2$, we find that

$$\alpha_1 = -\alpha_0 \frac{\sin 2\theta_0}{\cos(\theta_1 + \theta_0)}, \quad \alpha_2 = \alpha_0 \frac{\cos(\theta_1 - \theta_0)}{\cos(\theta_1 + \theta_0)}, \quad \beta_2 = -\beta_0 \tag{8.142b}$$

If all three stress vector components are specified at the boundary,

$$\sigma_{ij}^{(0)} T_j = \sigma_{ij}^{(3)} T_j = \text{prescribed quantity.}$$

then

$$\sum_{\nu=0}^{2} [\sigma_{ij}^{(\nu)}] T_j = 0.$$

It follows that, for an incident P-wave,

$$\alpha_1 = -\alpha_0 \left(\frac{c_1^2 \cos^2 2\theta_2 + c_2^2 \sin 2\theta_2 \sin 2\theta_0}{c_1^2 \cos^2 2\theta_2 - c_2^2 \sin 2\theta_2 \sin 2\theta_0} \right),$$

$$\alpha_2 = \alpha_0 \left(\frac{2 c_1 c_2 \cos 2\theta_2 \sin 2\theta_0}{c_1^2 \cos^2 2\theta_2 - c_2^2 \sin 2\theta_2 \sin 2\theta_0} \right),$$

$$\beta_2 = 0 \tag{8.143a}$$

and for an incident shear wave,

$$\alpha_1 = -\alpha_0 \left(\frac{2 c_1 c_2 \cos 2\theta_0 \sin 2\theta_0}{c_1^2 \cos^2 2\theta_0 - c_2^2 \sin 2\theta_1 \sin 2\theta_0} \right),$$

$$\alpha_2 = \alpha_0 \left(\frac{-c_1^2 \cos^2 2\theta_0 + c_2^2 \sin 2\theta_1 \sin 2\theta_0}{c_1^2 \cos^2 2\theta_0 - c_2^2 \sin 2\theta_1 \sin 2\theta_0} \right),$$

$$\beta_2 = \beta_0 \tag{8.143b}$$

If the normal stress component and two tangential displacement components are specified on the boundary,

$$\sigma_{ij}^{(0)} T_j T_i = \sigma_{ij}^{(3)} T_j T_i = \text{prescribed quantity,}$$

$$v_i^{(0)} N_i = v_i^{(3)} N_i = \text{prescribed quantity,}$$

$$v_i^{(0)} B_i = v_i^{(3)} B_i = \text{prescribed quantity ,}$$

then

$$\sum_{\nu=0}^{2} [\sigma_{ij}^{(\nu)}] T_j T_i = \sum_{\nu=0}^{2} V_i^{(\nu)} N_i = \sum_{\nu=0}^{2} V_i^{(\nu)} B_i = 0.$$

It follows that for an incident P-wave,

$$\alpha_1 = -\alpha_0, \quad \alpha_2 = 0, \quad \beta_2 = 0 \tag{8.144a}$$

and for an incident shear wave,

$$\alpha_1 = 0, \quad \alpha_2 = \alpha_0, \quad \beta_2 = -\beta_0 \tag{8.144b}$$

Finally, if the normal displacement component and the two tangential stress components are specified,

$$v_i^{(0)}T_i = v_i^{(3)}T_i = \text{prescribed quantity},$$

$$\sigma_{ij}^{(0)}T_jN_i = \sigma_{ij}^{(3)}T_jN_i = \text{prescribed quantity},$$

$$\sigma_{ij}^{(0)}T_jB_i = \sigma_{ij}^{(3)}T_jB_i = \text{prescribed quantity},$$

then

$$\sum_{\nu=0}^{2} V_i^{(\nu)}T_i = \sum_{\nu=0}^{2} [\sigma_{ij}^{(\nu)}]T_jN_i = \sum_{\nu=0}^{2} [\sigma_{ij}^{(\nu)}]T_jB_i = 0.$$

It follows that for an incident P-wave,

$$\alpha_1 = \alpha_0, \quad \alpha_2 = 0, \quad \beta_2 = 0 \tag{8.145a}$$

and for an incident shear signal,

$$\alpha_1 = 0, \quad \alpha_2 = -\alpha_0, \quad \beta_2 = -\beta_0 \tag{8.145b}$$

8.5 WAVES IN ISOTROPIC VISCOELASTIC MEDIA

The main purpose of this section is to modify the subject matters of Sections 8.3 and 8.4 to account for isotropic viscoelastic media. For a material point that has been disturbed for a duration t, the viscoelastic constitutive law, Eq. 7.102, can be written explicitly as

$$\sigma_{ij} = \delta_{ij}\lambda\varepsilon_{kk0} + 2\mu\varepsilon_{ij0} + \delta_{ij}\int_{0^+}^{t} \lambda(t-s)\mathrm{d}\varepsilon_{kk}(s) + 2\int_{0^+}^{t} \mu(t-s)\mathrm{d}\varepsilon_{ij}(s) \tag{8.146}$$

in which the subscript ($_0$) denotes $t = 0$. Implicitly, it has been assumed in Eq. 8.146 that the functions $\lambda(t)$ and $\mu(t)$ are not singular at $t = 0$. Thus this equation does not hold if the constitutive relationship is characterized by a mechanical model which contains a 'free' dashpot in parallel with other mechanical elements. In other words, the case which is characterized by a generalized Maxwell model (Fig. 7.5) with $\eta_0 \neq 0$ is not applicable to Eq. 8.146 nor the derivation to follow. For a point at a wavefront, t in Eq. 8.146 is zero since the material there has been disturbed for a zero duration. Then the same equation gives a relation among jumps as follows

$$[\sigma_{ij}] = \delta_{ij}\lambda_0[\varepsilon_{kk}] + 2\mu_0[\varepsilon_{ij}] \tag{8.147}$$

or, due to Eqs. 8.78b and c,

$$[\sigma_{ij}] = -\delta_{ij}\lambda_0 V_k n_k c^{-1} - \mu_0 V_i n_j c^{-1} - \mu_0 V_j n_i c^{-1} \tag{8.148}$$

which is the same as Eq. 8.81a if initial values of Lamé's functions are used there. Thus like isotropic elastic solids, there can be two types of waves propagating in isotropic viscoelastic solids, i.e. the P-waves characterized by Eqs. 8.82 and S-waves by Eqs. 8.83, while the wave speeds are

$$c_p = \sqrt{\frac{\lambda_0 + 2\mu_0}{\rho}} \tag{8.149}$$

$$c_s = \sqrt{\mu_0/\rho} \tag{8.150}$$

Recalling the Leibnitz's rule, Eq. 1.137, the time derivative of Eq. 8.146 is obtained as

$$\sigma_{ij,t} = \delta_{ij}\lambda'\varepsilon_{kk0} + 2\mu'\varepsilon_{ij0} + \delta_{ij}\lambda_0\varepsilon_{kk,t} + 2\mu_0\varepsilon_{ij,t}$$

$$+ \delta_{ij}\int_{0^+}^{t} \lambda'(t-s)\mathrm{d}\varepsilon_{kk}(s) + 2\int_{0^+}^{t} \mu'(t-s)\mathrm{d}\varepsilon_{ij}(s) \tag{8.151}$$

where the prime ($'$) denotes a derivative with respect to t. At a wavefront, t in Eq. 8.151 is zero, so the same equation gives a relation among jumps as

$$[\sigma_{ij,t}] = \delta_{ij}\lambda_0'[\varepsilon_{kk}] + 2\mu_0'[\varepsilon_{ij}] + \delta_{ij}\lambda_0[\varepsilon_{kk,t}] + 2\mu_0[\varepsilon_{ij,t}] \tag{8.152}$$

The derivative of Eq. 8.147 with respect to x_m, in view of the formula for derivatives of jumps (Eq. 8.70), can be put as

$$[\sigma_{ij,m}] + [\sigma_{ij,t}]n_m c^{-1} = \delta_{ij}\lambda_0[\varepsilon_{kk,m}] + \delta_{ij}\lambda_0[\varepsilon_{kk,t}]n_m c^{-1} + \delta_{ij}\lambda_{0,m}[\varepsilon_{kk}]$$

$$+ 2\mu_0[\varepsilon_{ij,m}] + 2\mu_0[\varepsilon_{ij,t}]n_m c^{-1} + 2\mu_{0,m}[\varepsilon_{ij}] \tag{8.153}$$

Substituting Eq. 8.152 into Eq. 8.153 we find

$$[\sigma_{ij,m}] = \delta_{ij}\lambda_0[\varepsilon_{kk,m}] + 2\mu_0[\varepsilon_{ij,m}] + \delta_{ij}\lambda_{0,m}[\varepsilon_{kk}] + 2\mu_{0,m}[\varepsilon_{ij}]$$

$$- \delta_{ij}\lambda_0'[\varepsilon_{kk}]n_m c^{-1} - 2\mu'[\varepsilon_{ij}]n_m c^{-1} \tag{8.154}$$

Setting m equal to j in the last equation yields

$$[\sigma_{ij,j}] = \lambda_0[\varepsilon_{kk,i}] + 2\mu_0[\varepsilon_{ij,j}] + \lambda_{0,i}[\varepsilon_{kk}] + 2\mu_{0,j}[\varepsilon_{ij}]$$

$$- \lambda_0'[\varepsilon_{kk}]n_i c^{-1} - 2\mu_0'[\varepsilon_{ij}]n_j c^{-1} \tag{8.155}$$

Due to the equation of motion, $\sigma_{ij,j} = \rho v_{i,t}$, and the strain definition, $\varepsilon_{ij} = (u_{i,j} + u_{j,i})/2$, the equation above can be rewritten as

$$[v_{i,t}] = (c_p^2 - c_s^2)[u_{k,ki}] + c_s^2[u_{i,kk}] + \frac{\lambda_{0,i}}{\rho}[u_{k,k}] + \frac{\mu_{0,k}}{\rho}[u_{i,k} + u_{k,i}]$$

$$- \lambda_0'(c\rho)^{-1}[u_{k,k}]n_i - \mu_0'(c\rho)^{-1}[u_{i,k} + u_{k,i}]n_k \tag{8.156}$$

Substituting Eqs. 8.71, 8.73 and 8.75 into this equation we find

$$c^2(V_{i,j}\psi_{,j}-[v_{i,j}]\psi_{,j})+(c_p^2-c_s^2)\,[[v_{k,k}]\psi_{,i}+(V_k\psi_{,k})_{,i}]$$

$$+c_s^2[[v_{i,k}]\psi_{,k}+(V_i\psi_{,k})_{,k}]+\frac{\lambda_{0,i}}{\rho}V_k\psi_{,k}+\frac{\mu_{0,k}}{\rho}(V_i\psi_{,k}+V_k\psi_{,i})$$

$$-\lambda_0'(c\rho)^{-1}V_k\psi_{,k}n_i-\mu_0'(c\rho)^{-1}(V_i\psi_{,k}+V_k\psi_{,i})n_k=0 \tag{8.157}$$

The resemblance of the last two equations with those for elastic solids (Eqs. 8.84 and 8.85) should be noted.

Proceeding in the same manner as already used to obtain subsequent jump magnitudes for elastic solids (Eqs. 8.90 and 8.101), we can get

$$V_i=V_i^e\exp\left(\frac{1}{2}\int_{t_0}^{t}\Omega dt'\right) \tag{8.158}$$

in which V_i^e is the velocity jump for elastic solids (Eq. 8.90 or Eq. 8.101b depending on the wave type), and

$$\Omega=\frac{\lambda_0'+2\mu_0'}{\lambda_0+2\mu_0},\quad \text{for P-waves} \tag{8.159a}$$

$$=\frac{\mu_0'}{\mu_0},\quad \text{for S-waves} \tag{8.159b}$$

Normally, Ω is negative, hence the decay of the velocity jump as the propagation proceeds is more than that in elastic solids by an exponential factor. For a homogeneous domain, it can be shown that

$$V_iV_i=\frac{r_1^0r_2^0}{r_1r_2}\exp[\Omega(t-t_0)]\,V_i^0V_i^0 \tag{8.160}$$

in which the superscript (^{0}) denotes a quantity at the time t_0, and r_1, r_2 are principal radii of wavefronts as depicted in Fig. 8.8 and defined in Eqs. 8.103.

For the reflection and transmission at interfaces and boundaries, the derivation in Section 8.4 is also applicable to isotropic viscoelastic solids if initial ($t=0$) values of λ and μ are used instead in all equations involved.

8.6 IN-PLANE HARMONIC SURFACE WAVES

In this section, we shall consider the harmonic vibration of isotropic elastic planes of the same geometry as Section 6.2 where statics of planes with $-\infty<x<\infty$ was considered. Knowing that the solution of a plane stress problem with a Poisson's ratio equal to $\nu/(1-\nu)$ can be obtained from its corresponding plane strain problem with a Poisson's ratio ν, we shall derive in this section only for plane strain cases. The equations of motion, in terms of displacements and in the absence of the real body force, can be specialized from Eq. 6.1 by taking X_i there as the D'Alembert's force. The resulting equations in explicit symbols are

$$(\lambda+2\mu)\frac{\partial^2 u}{\partial x^2}+\mu\frac{\partial^2 u}{\partial y^2}+(\lambda+\mu)\frac{\partial^2 v}{\partial x\partial y}=\rho\frac{\partial^2 u}{\partial t^2} \tag{8.161a}$$

$$(\lambda+2\mu)\frac{\partial^2 v}{\partial y^2}+\mu\frac{\partial^2 v}{\partial x^2}+(\lambda+\mu)\frac{\partial^2 u}{\partial x\partial y}=\rho\frac{\partial^2 v}{\partial t^2} \tag{8.161b}$$

where ρ is the mass density. For a 'discrete' harmonic wave which is outgoing in the x-direction parallel to the surfaces and the stratification of the plane, the displacement functions can be put in the form

$$u(x,y,t)=u(y)\mathrm{e}^{\mathrm{i}(\omega t-kx)} \tag{8.162a}$$

$$v(x,y,t)=v(y)\mathrm{e}^{\mathrm{i}(\omega t-kx)} \tag{8.162b}$$

where ω and k are angular frequency and wavenumber, respectively. The derivation to follow is mainly concerned with the determination of the wavenumber while there is no applied traction in the plane. In other words, we are looking for what is commonly known as a *harmonic surface wave of Rayleigh type* (J. W. Strutt, Lord Rayleigh, 1887). Substituting Eqs. 8.162 into Eqs. 8.161 we find

$$-\frac{\eta^2}{\eta_p^2}u+\frac{1}{\eta_s^2}\frac{\mathrm{d}^2 u}{\mathrm{d}y^2}-\mathrm{i}\eta\left(\frac{1}{\eta_p^2}-\frac{1}{\eta_s^2}\right)\frac{\mathrm{d}v}{\mathrm{d}y}=-u \tag{8.163a}$$

$$\frac{1}{\eta_p^2}\frac{\mathrm{d}^2 v}{\mathrm{d}y^2}-\frac{\eta^2}{\eta_s^2}v-\mathrm{i}\eta\left(\frac{1}{\eta_p^2}-\frac{1}{\eta_s^2}\right)\frac{\mathrm{d}u}{\mathrm{d}y}=-v \tag{8.163b}$$

Here a constant a of length dimension is used to nondimensionalize the independent variables x and y; η is the dimensionless surface wavenumber; η_p and η_s are dimensionless frequencies of P and S-waves, respectively, i.e.

$$\eta=ak,\quad \eta_p=\frac{a\omega}{c_p},\quad \eta_s=\frac{a\omega}{c_s} \tag{8.164a,b,c}$$

We call η_p and η_s dimensionless wavenumbers of *body waves*. The solution to Eqs. 8.163 takes the form

$$u(y)=A\mathrm{e}^{\alpha y},\quad v(y)=B\mathrm{e}^{\alpha y} \tag{8.165a,b}$$

Substituting Eqs. 8.165 into Eqs. 8.163 we find

$$\begin{bmatrix}(\eta_p^2\eta_s^2-\eta^2\eta_s^2+\eta_p^2\alpha^2) & -\mathrm{i}\eta\alpha(\eta_s^2-\eta_p^2)\\ -\mathrm{i}\eta\alpha(\eta_s^2-\eta_p^2) & (\eta_p^2\eta_s^2-\eta^2\eta_p^2+\eta_s^2\alpha^2)\end{bmatrix}\begin{Bmatrix}A\\B\end{Bmatrix}=\begin{Bmatrix}0\\0\end{Bmatrix} \tag{8.166}$$

A non-trivial solution of this system of equations exists if the determinant vanishes, leading to the condition

$$(\alpha^2-\alpha_p^2)(\alpha^2-\alpha_s^2)=0,$$

where

$$\alpha_p(\eta) = +(\eta^2 - \eta_p^2)^{1/2}, \quad \alpha_s(\eta) = +(\eta^2 - \eta_s^2)^{1/2} \tag{8.167a,b}$$

Thus

$$\alpha = \alpha_p, -\alpha_p, \alpha_s, -\alpha_s \tag{8.168}$$

For each α, the eigenvector involved can be determined from Eq. 8.166, and Eqs. 8.165 become

$$u(y) = Ae^{-\alpha_p y} + Ce^{\alpha_p y} + Be^{-\alpha_s y} + De^{\alpha_s y} \tag{8.169a}$$

$$v(y) = -\frac{i\alpha_p}{\eta}\left(Ae^{-\alpha_p y} - Ce^{\alpha_p y}\right) - \frac{i\eta}{\alpha_s}\left(Be^{-\alpha_s y} - De^{\alpha_s y}\right) \tag{8.169b}$$

where A, B, C and D are arbitrary constants. If $\eta_p < \eta < \eta_s$, it is more convenient to write the equations above as

$$u(y) = Ae^{-\alpha_p y} + Ce^{\alpha_p y} + B\cos\bar{\alpha}_s y + D\sin\bar{\alpha}_s y \tag{8.170a}$$

$$v(y) = -\frac{i\alpha_p}{\eta}\left(Ae^{-\alpha_p y} - Ce^{\alpha_p y}\right) - \frac{i\eta}{\bar{\alpha}_s}(D\cos\bar{\alpha}_s y - B\sin\bar{\alpha}_s y) \tag{8.170b}$$

where

$$\bar{\alpha}_s(\eta) = i\alpha_s(\eta) = +(\eta_s^2 - \eta^2)^{1/2} \tag{8.171}$$

If $\eta < \eta_p < \eta_s$, the solution form should be modified further for the sake of convenience as

$$u(y) = A\cos\bar{\alpha}_p y + C\sin\bar{\alpha}_p y + B\cos\bar{\alpha}_s y + D\sin\bar{\alpha}_s y \tag{8.172a}$$

$$v(y) = \frac{i\bar{\alpha}_p}{\eta}(C\cos\bar{\alpha}_p y - A\sin\bar{\alpha}_p y) - \frac{i\eta}{\bar{\alpha}_s}(D\cos\bar{\alpha}_s y - B\sin\bar{\alpha}_s y) \tag{8.172b}$$

where

$$\bar{\alpha}_p(\eta) = i\alpha_p(\eta) = +(\eta_p^2 - \eta^2)^{1/2} \tag{8.173}$$

The arbitrary constants A, B, C and D in the various solutions are to be determined from boundary conditions in the same manner as described for statics in Section 6.2. However, the present case is an eigenvalue problem, since we are investigating the possible existence of harmonic vibration without any applied traction. For a plane of N layers each having a finite dimension in the y-direction, the formulation leads to a set of $4N$ linear homogeneous equations in $4N$ unknown arbitrary constants; A_1 to A_N, B_1 to B_N, C_1 to C_N, and D_1 to D_N. The eigenvalue η of this problem is obtained in the usual way, i.e. by setting the determinant of the set of equations to zero, leading to the following equation

$$F_R(\eta) = 0 \tag{8.174}$$

Roots of the equation above are commonly known as dimensionless *Rayleigh wave-numbers* and will be denoted by η_R.

The simplest Rayleigh wave problem is probably the case of a homogeneous half plane with $-\infty < x < \infty$, $0 \le y < \infty$. The most convenient solution form is given by Eqs. 8.169, in which C and D must vanish for boundedness of the solution of the wave as y tends to infinity. The boundary conditions for this case are

$$\sigma_{yy}(x,0)=0, \quad \sigma_{xy}(x,0)=0 \tag{8.175}$$

which leads to the following set of homogeneous equations

$$\begin{bmatrix} 2\alpha_p & [2-(\eta_s/\eta)^2]\eta^2/\alpha_s \\ [2-(\eta_s/\eta)^2] & 2 \end{bmatrix} \begin{Bmatrix} A \\ B \end{Bmatrix} = \begin{Bmatrix} 0 \\ 0 \end{Bmatrix} \tag{8.176}$$

Consequently, the famous Rayleigh wave equation for a homogeneous isotropic elastic half plane is obtained as

$$F_R(\eta) \equiv (2\eta^2 - \eta_s^2)^2 - 4\eta^2\alpha_p(\eta)\alpha_s(\eta) = 0 \tag{8.177}$$

The equation above can be further nondimensionalized by introducing

$$\eta_R = s\eta_s \tag{8.178a}$$

$$\gamma = \frac{\eta_p}{\eta_s} = \frac{c_s}{c_p} = \sqrt{\frac{1-2\nu}{2(1-\nu)}} \tag{8.178b}$$

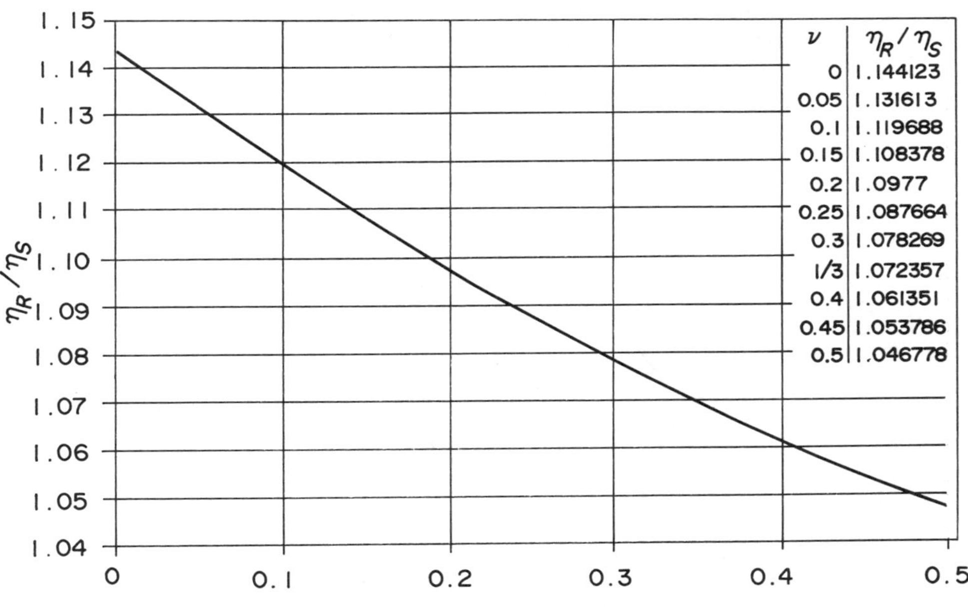

ν	η_R/η_S
0	1.144123
0.05	1.131613
0.1	1.119688
0.15	1.108378
0.2	1.0977
0.25	1.087664
0.3	1.078269
1/3	1.072357
0.4	1.061351
0.45	1.053786
0.5	1.046778

Fig. 8.11 Rayleigh wavenumber for homogeneous half plane.

where η_R is the root of Eq. 8.177, i.e. the dimensionless Rayleigh wavenumber for a homogeneous isotropic elastic half plane. Substituting Eqs. 8.178 into Eq. 8.177 we find

$$(2s^2-1)^2-4s^2(s^2-\gamma^2)^{1/2}(s^2-1)^{1/2}=0 \tag{8.179}$$

Due to the absence of the frequency ω in Eq. 8.179, we can say that the Rayleigh wave in this case is nondispersive. Rationalizing Eq. 8.179 by squaring we find

$$16(1-\gamma^2)s^6-8(3-2\gamma^2)s^4+8s^2-1=0 \tag{8.180}$$

which is a cubic equation in s^2, with three roots. However, only one of these satisfies the original equation, Eq. 8.179, while the other two values are spurious, arising from the squaring. Moreover, since the wave must not be incoming as y tends to infinity, the correct value of s^2 must be a positive real quantity and greater than unity, so that

$$\eta_p<\eta_s<\eta_R \tag{8.181a}$$

$$c_R<c_s<c_p \tag{8.181b}$$

where c_R is the *Rayleigh wave velocity* $=a\omega/\eta_R$. Results for various values of the Poisson's ratio ν are presented in Fig. 8.11.

Next to be considered is the occurrence of another special type of Rayleigh waves. The material domain in this case consists of two different half planes, perfectly bonded together as shown in Fig. 8.12. The displacement functions given for a general case should be modified to suit the present case as

$$u_1(y)=C_1e^{\alpha_{p1}y}+D_1e^{\alpha_{s1}y} \tag{8.182a}$$

$$v_1(y)=\frac{i\alpha_{p1}}{\eta}C_1e^{\alpha_{p1}y}+\frac{i\eta}{\alpha_{s1}}D_1e^{\alpha_{s1}y} \tag{8.182b}$$

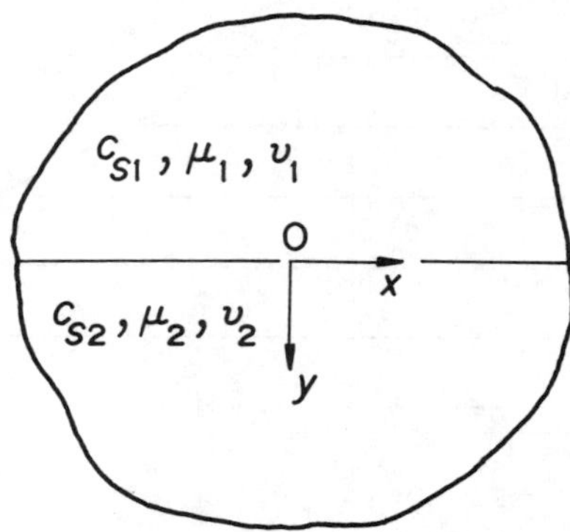

Fig. 8.12 Two different half planes perfectly bonded together.

$$u_2(y) = A_2 e^{-\alpha_{p2} y} + B_2 e^{-\alpha_{s2} y} \tag{8.182c}$$

$$v_2(y) = -\frac{i\alpha_{p2}}{\eta} A_2 e^{-\alpha_{p2} y} - \frac{i\eta}{\alpha_{s2}} B_2 e^{-\alpha_{s2} y} \tag{8.182d}$$

in which the subscript 1 denotes the material domain where $-\infty < y \le 0$, and the subscript 2 for $0 \le y < \infty$. The boundary conditions are

$$u_1(0) = u_2(0), \quad v_1(0) = v_2(0) \tag{8.183a,b}$$

$$\sigma_{yy1}(x,0) = \sigma_{yy2}(x,0), \quad \sigma_{xy1}(x,0) = \sigma_{xy2}(x,0) \tag{8.183c,d}$$

Accordingly, the Rayleigh wave equation can be obtained in the form

$$F_R(\eta) \equiv \eta^2(\eta^2 - \beta_1\eta_{s1}^2 - \beta_2\eta_{s2}^2)^2 + \eta^2\alpha_{p1}\alpha_{p2}\alpha_{s1}\alpha_{s2} - \alpha_{p1}\alpha_{s1}(\eta^2 - \beta_2\eta_{s2}^2)^2$$

$$- \alpha_{p2}\alpha_{s2}(\eta^2 - \beta_1\eta_{s1}^2)^2 + \beta_1\beta_2\eta_{s1}^2\eta_{s2}^2(\alpha_{p1}\alpha_{s2} + \alpha_{p2}\alpha_{s1}) = 0 \tag{8.184a}$$

or

$$s^2(s^2 - \beta_1\beta_3 - \beta_2)^2 + s^2[(s^2 - \beta_3\gamma_1^2)(s^2 - \gamma_2^2)(s^2 - \beta_3)(s^2 - 1)]^{1/2}$$

$$-[(s^2 - \beta_3\gamma_1^2)(s^2 - \beta_3)]^{1/2}(s^2 - \beta_2)^2 - [(s^2 - \gamma_2^2)(s^2 - 1)]^{1/2}(s^2 - \beta_1\beta_3)^2$$

$$+\beta_1\beta_2\beta_3\left\{[(s^2 - \beta_3\gamma_1^2)(s^2 - 1)]^{1/2} + [(s^2 - \gamma_2^2)(s^2 - \beta_3)]^{1/2}\right\} = 0 \tag{8.184b}$$

in which the dimensionless quantities defined by Eqs. 8.164 and 8.178b have been used in the process, and

$$s = \frac{\eta_R}{\eta_{s2}}, \quad \beta_1 = \frac{1}{2(1-\beta_4)} \tag{8.185a,b}$$

$$\beta_2 = \frac{\beta_4}{2(\beta_4 - 1)}, \quad \beta_3 = \left(\frac{c_{s2}}{c_{s1}}\right)^2, \quad \beta_4 = \frac{\mu_2}{\mu_1} \tag{8.185c,d,e}$$

Observing the absence of the frequency in Eq. 8.184b, we can say that this particular surface wave is nondispersive. Stoneley (1924) made a thorough investigation on the subject, thus gained the recognition in having it named as the *Stoneley wave*. Koppe (1948) solved Eq. 8.184b numerically and concluded that

$$\eta_{s1}, \eta_{s2} < \eta_R < \eta_{Ri} \tag{8.186}$$

where η_{Ri} is the greater Rayleigh wavenumber among those obtained separately (by solving Eq. 8.177) for each of the constituent homogeneous half planes. Other analyses were given by Love (1911, 1926), Sezawa and Kanai (1939), Scholte (1947) and Cagniard (1962). There can be at most one Stoneley wave. While Love showed that this wave could exist under the stringent condition that the S-wave velocities of the two constituent half planes were nearly equal, Cagniard found a more general condition for such existence. Without loss of generality, that condition (by Cagniard) can be restricted to the case where $\beta_3 < 1$, then put in the following form

$$-(\beta_3+\beta_4-2)^2+(\beta_4-2)^2[(1-\beta_3\gamma_1^2)(1-\beta_3)]^{1/2}+\beta_3\beta_4[(1-\gamma_2^2)(1-\beta_3)]^{1/2}<0 \quad (8.187)$$

Equation 8.187 should serve as a quick check for the existence, while Eq. 8.186 as a good guide for quick computation of the Stoneley wavenumber from Eq. 8.184b.

8.7 ANTIPLANE HARMONIC SURFACE WAVES

Now consider the possible existence of harmonic surface waves in antiplane problems; these are commonly known as *Love waves*. We suppose the plane is infinitely extended in the x-direction; displacement components are independent of z; u and v vanish; waves propagate in x-direction; and the equation of motion is

$$c_s^2\nabla^2 w=\frac{\partial^2 w}{\partial t^2} \quad (8.188)$$

This equation has Eq. 6.9 as its statical counterpart, and has a solution of the form

$$w(x,y,t)=w(y)e^{i(\omega t-\eta x)} \quad (8.189)$$

where η is the dimensionless wavenumber as defined in Eq. 8.164a, and x and y have been nondimensionalized by a constant a of length dimension. Proceeding in the same manner as the in-plane problem in the preceding section, we find

$$w(y)=Be^{-\alpha_s y}+De^{\alpha_s y} \quad (8.190)$$

where $\alpha_s(\eta)$ is as defined in Eq. 8.167b. If $\eta<\eta_s$, it is more convenient to rewrite Eq. 8.190 as

$$w(y)=B\cos\bar{\alpha}_s y+D\sin\bar{\alpha}_s y \quad (8.191)$$

where $\bar{\alpha}_s(\eta)$ is as defined in Eq. 8.171. In either form, the constants B and D are to be determined from boundary conditions, as in Section 8.6. For a plane of N layers each having a finite dimension in the y-direction, the formulation leads to a set of $2N$ homogeneous equations in $2N$ unknown arbitrary constants; B_1 to B_N, and D_1 to D_N. Setting the determinant of that set of equations to zero, we find the *Love wave equation*

$$F_L(\eta)=0 \quad (8.192)$$

Roots of the equation above are commonly known as dimensionless *Love wavenumbers* and will be denoted by η_L.

It can be shown that there is no Love wave in a homogeneous half plane, nor in a full plane consisting of two different half planes as shown in Fig. 8.12. The simplest Love wave problem is probably the case of a layered half plane as shown in Fig. 8.13. Employing a subscript 1 for the upper layer where $-h/a\le y\le 0$, and subscript 2 for the underlying half plane where $0\le y<\infty$; we find that the solution assumes the following form

$$w_1(y)=B_1\cos\bar{\alpha}_{s1}y+D_1\sin\bar{\alpha}_{s1}y \quad (8.193a)$$

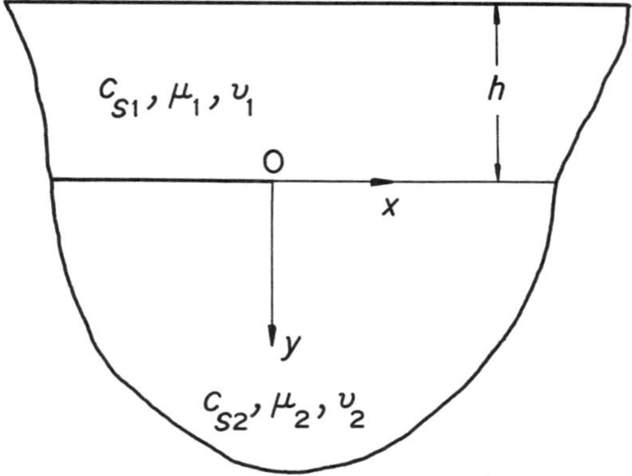

Fig. 8.13 Layered half plane.

$$w_2(y) = B_2 e^{-\alpha_{s2} y} \tag{8.193b}$$

The boundary conditions are

$$w_1(0) = w_2(0), \quad \sigma_{yz1}(x,0) = \sigma_{yz2}(x,0) \tag{8.194a,b}$$

$$\sigma_{yz1}(x,-h/a) = 0 \tag{8.194c}$$

The first boundary condition, Eq. 8.194a, readily gives

$$B_2 = B_1 \tag{8.195a}$$

Then the remaining boundary conditions lead to

$$\begin{bmatrix} \mu_2\alpha_{s2} & \mu_1\bar{\alpha}_{s1} \\ \bar{\alpha}_{s1}\sin(\bar{\alpha}_{s1}h/a) & \bar{\alpha}_{s1}\cos(\bar{\alpha}_{s1}h/a) \end{bmatrix} \begin{Bmatrix} B_1 \\ D_1 \end{Bmatrix} = \begin{Bmatrix} 0 \\ 0 \end{Bmatrix} \tag{8.195b}$$

Consequently, the Love equation for this case is

$$F_L(\eta) \equiv \tan(\bar{\alpha}_{s1}h/a) - \frac{\mu_2\alpha_{s2}}{\mu_1\bar{\alpha}_{s1}} = 0 \tag{8.196}$$

Since the equation above cannot be nondimensionalized further to be completely independent of both η_{s1} and η_{s2}, its roots are dependent upon the frequency ω, thus surface waves in this case are dispersive. Equation 8.196 can have several roots, but all of them must be real and positive and fall within the range

$$\eta_{s2} < \eta_L < \eta_{s1}, \quad \text{or} \quad c_{s1} < c_L < c_{s2} \tag{8.197a,b}$$

where c_L is the *Love wave velocity* $= a\omega/\eta_L$. This shows that there is no Love wave in a layered half plane if the shear wave velocity in the upper layer is greater than that

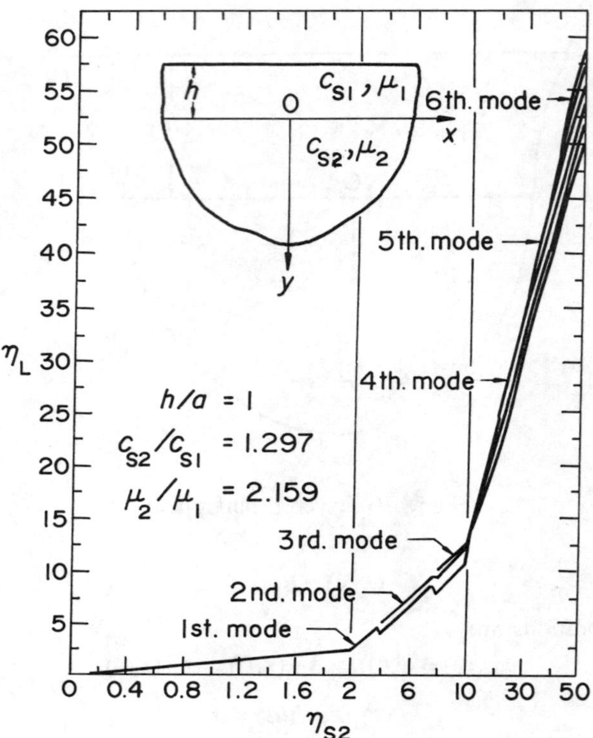

(a) First six modes of vibration.

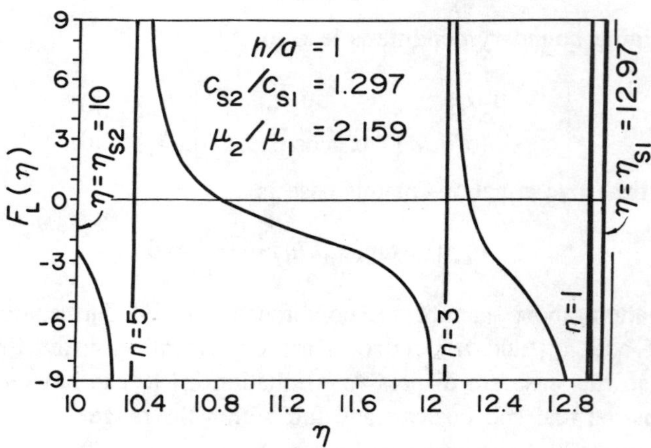

(b) Example plotting of $F_L(\eta)$ v.s η.

Fig. 8.14 Love wavenumbers for a layered half plane.

in the underlying half plane. In fact, the Love wave function $F_L(\eta)$ in this case is discontinuous and singular, if

$$\bar{\alpha}_{s1}h/a = n\pi/2, \quad (n = 1,3,5,\ldots),$$

or

$$\eta = \left[\eta_{s1}^2 - \left(\frac{n\pi a}{2h}\right)^2\right]^{1/2}, \quad [F_L(\eta) = \pm\infty, \quad n = 1,3,5,\ldots] \tag{8.198}$$

Conditions stipulated by Eqs. 8.197a and 8.198 should serve as good guides for quick computation of values of η_L from Eq. 8.196. An example problem has been solved, and its Love wavenumbers of the first six *modes of vibration* are presented in Fig. 8.14. The higher the mode, the closer the derivative of $F_L(\eta)$ with respect to η tends to infinity. Extensive quantitative information on the phase and group velocities of this type of Love wave can be seen in the book by Ewing, Jardetzky and Press (1957).

8.8 VIBRATION OF MULTILAYERED SPACES

In this section, we shall consider three-dimensional problems of harmonic vibration of isotropic elastic spaces of the same geometry as Section 6.3 where statics of spaces with $0 \le r < \infty$ is considered. The displacement functions of the present case take the form

$$u_r(r,\theta,z,t) = u_r(r,\theta,z)e^{i\omega t} \tag{8.199a}$$

$$u_\theta(r,\theta,z,t) = u_\theta(r,\theta,z)e^{i\omega t} \tag{8.199b}$$

$$u_z(r,\theta,z,t) = u_z(r,\theta,z)e^{i\omega t} \tag{8.199c}$$

The equations of motion, in terms of the displacements and in the absence of real body force, are given by Eqs. 3.67, namely

$$(\lambda+\mu)\frac{\partial}{\partial r}(\varepsilon_{kk}) + \mu\left(\nabla^2 u_r - \frac{u_r}{r^2} - \frac{2}{r^2}\frac{\partial u_\theta}{\partial\theta}\right) = -\omega^2\rho u_r \tag{8.200a}$$

$$(\lambda+\mu)\frac{1}{r}\frac{\partial}{\partial\theta}(\varepsilon_{kk}) + \mu\left(\nabla^2 u_\theta + \frac{2}{r^2}\frac{\partial u_r}{\partial\theta} - \frac{u_\theta}{r^2}\right) = -\omega^2\rho u_\theta \tag{8.200b}$$

$$(\lambda+\mu)\frac{\partial}{\partial z}(\varepsilon_{kk}) + \mu\nabla^2 u_z = -\omega^2\rho u_z \tag{8.200c}$$

where $\varepsilon_{kk}(r,\theta,z)$ is the shape function of the dilatation, i.e. Eq. 2.72. Harkrider (1964) introduced potentials Φ, Ψ and χ so that

$$a^{-1}u_r(r,\theta,z) = \frac{\partial\Phi}{\partial r} + \frac{\partial^2\Psi}{\partial r\partial z} + \frac{1}{r}\frac{\partial\chi}{\partial\theta} \tag{8.201a}$$

$$a^{-1}u_\theta(r,\theta,z) = \frac{1}{r}\frac{\partial\Phi}{\partial\theta} + \frac{1}{r}\frac{\partial^2\Psi}{\partial z\partial\theta} - \frac{\partial\chi}{\partial r} \tag{8.201b}$$

$$a^{-1}u_z(r,\theta,z)=\frac{\partial\Phi}{\partial z}-\nabla^2\Psi+\frac{\partial^2\Psi}{\partial z^2} \tag{8.201c}$$

and so transformed Eqs. 8.200 into

$$\nabla^2\Phi=-\eta_p^2\Phi \tag{8.202a}$$

$$\nabla^2\Psi=-\eta_s^2\Psi \tag{8.202b}$$

$$\nabla^2\chi=-\eta_s^2\chi \tag{8.202c}$$

here a constant a of length dimension has been used to nondimensionalize the spatial variables r and z, and η_p and η_s are dimensionless frequencies or body wavenumbers as defined in Eqs. 8.164b and c. Using Hooke's law, we can express the stress components in terms of these potentials as

$$\sigma_{zz}(r,\theta,z)=2\mu\left(\frac{\partial^2\Phi}{\partial z^2}+\frac{\partial^3\Psi}{\partial z^3}+\eta_s^2\frac{\partial\Psi}{\partial z}+\frac{2\eta_p^2-\eta_s^2}{2}\Phi\right) \tag{8.203a}$$

$$\sigma_{\theta z}(r,\theta,z)=\mu\left(\frac{2}{r}\frac{\partial^2\Phi}{\partial z\partial\theta}+\frac{2}{r}\frac{\partial^3\Psi}{\partial z^2\partial\theta}+\frac{\eta_s^2}{r}\frac{\partial\Psi}{\partial\theta}-\frac{\partial^2\chi}{\partial r\partial z}\right) \tag{8.203b}$$

$$\sigma_{rz}(r,\theta,z)=\mu\left(2\frac{\partial^2\Phi}{\partial z\partial r}+2\frac{\partial^3\Psi}{\partial z^2\partial r}+\eta_s^2\frac{\partial\Psi}{\partial r}+\frac{1}{r}\frac{\partial^2\chi}{\partial z\partial\theta}\right) \tag{8.203c}$$

The other three stress components can also be expressed in terms of these potentials. As in the elastostatics of layered spaces in Chapter VI, the potentials can be expressed in the form of Fourier series with respect to θ and Hankel transforms with respect to r as

$$\Phi(r,\theta,z)=\sum_m\cos m\theta\int_0^\infty\tilde{\Phi}(z,\eta)J_m(\eta r)d\eta \tag{8.204a}$$

$$\Psi(r,\theta,z)=\sum_m\cos m\theta\int_0^\infty\tilde{\Psi}(z,\eta)J_m(\eta r)d\eta \tag{8.204b}$$

$$\chi(r,\theta,z)=\sum_m\sin m\theta\int_0^\infty\tilde{\chi}(z,\eta)J_m(\eta r)d\eta \tag{8.204c}$$

Substituting Eqs. 8.204 into Eqs. 8.201 and 8.203 we find

$$a^{-1}u_r(r,\theta,z)=\sum_m\cos m\theta\int_0^\infty\tilde{u}_r(r,z,\eta)d\eta \tag{8.205a}$$

$$a^{-1}u_\theta(r,\theta,z)=\sum_m\sin m\theta\int_0^\infty\tilde{u}_\theta(r,z,\eta)d\eta \tag{8.205b}$$

$$a^{-1}u_z(r,\theta,z)=\sum_m\cos m\theta\int_0^\infty\tilde{u}_z(r,z,\eta)d\eta \tag{8.205c}$$

$$\sigma_{zz}(r,\theta,z)=\sum_m\cos m\theta\int_0^\infty\tilde{\sigma}_{zz}(r,z,\eta)d\eta \tag{8.205d}$$

$$\sigma_{\theta z}(r,\theta,z)=\sum_m \sin m\theta\int_0^\infty \tilde{\sigma}_{\theta z}(r,z,\eta)\mathrm{d}\eta \tag{8.205e}$$

$$\sigma_{rz}(r,\theta,z)=\sum_m \cos m\theta\int_0^\infty \tilde{\sigma}_{rz}(r,z,\eta)\mathrm{d}\eta \tag{8.205f}$$

where

$$\tilde{u}_r(r,z,\eta)=\eta\left[u_R(z,\eta)\frac{\mathrm{d}J_m(\eta r)}{\mathrm{d}(\eta r)}+mv_L(z,\eta)\frac{J_m(\eta r)}{\eta r}\right] \tag{8.206a}$$

$$\tilde{u}_\theta(r,z,\eta)=\eta\left[-mu_R(z,\eta)\frac{J_m(\eta r)}{\eta r}-v_L(z,\eta)\frac{\mathrm{d}J_m(\eta r)}{\mathrm{d}(\eta r)}\right] \tag{8.206b}$$

$$\tilde{u}_z(r,z,\eta)=w_R(z,\eta)J_m(\eta r) \tag{8.206c}$$

$$\tilde{\sigma}_{zz}(r,z,\eta)=\sigma_R(z,\eta)J_m(\eta r) \tag{8.206d}$$

$$\tilde{\sigma}_{\theta z}(r,z,\eta)=\eta\left[-m\tau_R(z,\eta)\frac{J_m(\eta r)}{\eta r}-\tau_L(z,\eta)\frac{\mathrm{d}J_m(\eta r)}{\mathrm{d}(\eta r)}\right] \tag{8.206e}$$

$$\tilde{\sigma}_{rz}(r,z,\eta)=\eta\left[\tau_R(z,\eta)\frac{\mathrm{d}J_m(\eta r)}{\mathrm{d}(\eta r)}+m\tau_L(z,\eta)\frac{J_m(\eta r)}{\eta r}\right] \tag{8.206f}$$

in which

$$u_R(z,\eta)=\tilde{\Phi}+\frac{\mathrm{d}\tilde{\Psi}}{\mathrm{d}z} \tag{8.207a}$$

$$w_R(z,\eta)=\frac{\mathrm{d}\tilde{\Phi}}{\mathrm{d}z}+\frac{\mathrm{d}^2\tilde{\Psi}}{\mathrm{d}z^2}+\eta_s^2\tilde{\Psi} \tag{8.207b}$$

$$\sigma_R(z,\eta)=2\mu\left[\frac{\mathrm{d}^2\tilde{\Phi}}{\mathrm{d}z^2}+\frac{\mathrm{d}^3\tilde{\Psi}}{\mathrm{d}z^3}+\eta_s^2\frac{\mathrm{d}\tilde{\Psi}}{\mathrm{d}z}+(2\eta_p^2-\eta_s^2)\frac{\tilde{\Phi}}{2}\right] \tag{8.207c}$$

$$\tau_R(z,\eta)=\mu\left(2\frac{\mathrm{d}\tilde{\Phi}}{\mathrm{d}z}+2\frac{\mathrm{d}^2\tilde{\Psi}}{\mathrm{d}z^2}+\eta_s^2\tilde{\Psi}\right) \tag{8.207d}$$

$$v_L(z,\eta)=\tilde{\chi} \tag{8.208a}$$

$$\tau_L(z,\eta)=\mu\frac{\mathrm{d}\tilde{\chi}}{\mathrm{d}z} \tag{8.208b}$$

The equations governing $\tilde{\Phi}(z,\eta)$, $\tilde{\Psi}(z,\eta)$ and $\tilde{\chi}(z,\eta)$ can be obtained by substituting Eqs. 8.204 into Eqs. 8.202, i.e.

$$\frac{\mathrm{d}^2\tilde{\Phi}}{\mathrm{d}z^2}=\alpha_p^2(\eta)\tilde{\Phi} \tag{8.209a}$$

$$\frac{\mathrm{d}^2\tilde{\Psi}}{\mathrm{d}z^2}=\alpha_s^2(\eta)\tilde{\Psi} \tag{8.209b}$$

$$\frac{d^2\tilde{\chi}}{dz^2}=\alpha_s^2(\eta)\tilde{\chi} \tag{8.209c}$$

where $\alpha_p(\eta)$ and $\alpha_s(\eta)$ are as defined in Eqs. 8.167. The solution to Eqs. 8.209, which are ordinary differential equations, is simple, i.e.

$$\tilde{\Phi}=A_\Phi e^{-\alpha_p z}+B_\Phi e^{\alpha_p z} \tag{8.210a}$$

$$\tilde{\Psi}=A_\Psi e^{-\alpha_s z}+B_\Psi e^{\alpha_s z} \tag{8.210b}$$

$$\tilde{\chi}=A_\chi e^{-\alpha_s z}+B_\chi e^{\alpha_s z} \tag{8.210c}$$

where A_Φ, B_Φ, A_Ψ, B_Ψ, A_χ and B_χ are arbitrary constants. If $\eta_p<\eta<\eta_s$, it is more convenient to write Eqs. 8.210b and c as

$$\tilde{\Psi}=A_\Psi \cos\bar{\alpha}_s z+B_\Psi \sin\bar{\alpha}_s z \tag{8.211a}$$

$$\tilde{\chi}=A_\chi \cos\bar{\alpha}_s z+B_\chi \sin\bar{\alpha}_s z \tag{8.211b}$$

where $\bar{\alpha}_s(\eta)$ is as defined in Eq. 8.171. If $\eta<\eta_p<\eta_s$, Eq. 8.210a should be modified also for the sake of convenience as

$$\tilde{\Phi}=A_\Phi \cos\bar{\alpha}_p z+B_\Phi \sin\bar{\alpha}_p z \tag{8.212}$$

where $\bar{\alpha}_p(\eta)$ is as defined in Eq. 8.173. The arbitrary constants are to be determined from boundary conditions. For a space of N layers each having a finite dimension in the z-direction, the formulation leads to a set of $6N$ linear equations in $6N$ unknown arbitrary constants; $A_{\Phi 1}$ to $A_{\Phi N}$, $B_{\Phi 1}$ to $B_{\Phi N}$, $A_{\Psi 1}$ to $A_{\Psi N}$, $B_{\Psi 1}$ to $B_{\Psi N}$, $A_{\chi 1}$ to $A_{\chi N}$, and $B_{\chi 1}$ to $B_{\chi N}$. As an example, consider the interface at $z=z_{n-1}$ between the layers $n-1$ and n. If there are no any applied tractions at this interface, the relevant conditions are the continuity of

$$u_r,\ u_\theta,\ u_z,\ \sigma_{zz},\ \sigma_{\theta z} \text{ and } \sigma_{rz}.$$

Noting, in Eqs. 8.206, that

$$\frac{dJ_m(\eta r)}{d(\eta r)} \quad \text{and} \quad \frac{J_m(\eta r)}{\eta r}$$

are independent of each other we see that these conditions can be expressed (for $z=z_{n-1}$) as

$$\begin{Bmatrix} u_{Rn}(z,\eta) \\ w_{Rn}(z,\eta) \\ \sigma_{Rn}(z,\eta) \\ \tau_{Rn}(z,\eta) \end{Bmatrix} = \begin{Bmatrix} u_{Rn-1}(z,\eta) \\ w_{Rn-1}(z,\eta) \\ \sigma_{Rn-1}(z,\eta) \\ \tau_{Rn-1}(z,\eta) \end{Bmatrix} \tag{8.213}$$

$$\begin{Bmatrix} v_{Ln}(z,\eta) \\ \tau_{Ln}(z,\eta) \end{Bmatrix} = \begin{Bmatrix} v_{Ln-1}(z,\eta) \\ \tau_{Ln-1}(z,\eta) \end{Bmatrix} \tag{8.214}$$

in which the subscripts $n-1$ and n denote the respective layers. Deliberately, the last two equations are put separately to show that Φ and Ψ appear together in a set of four equations, while χ appears separately in another set of two equations. Deserving notice is that these equations are independent of Fourier series harmonic m. Only the R-set of equations, which is for Φ and Ψ, exists in the axisymmetric (torsion free) case where $u_\theta = 0$ and all functions are independent of θ.

For torsion about the z-axis, u_r and u_z must vanish, all functions are independent of θ, and the only non-trivial equation of motion is

$$\nabla_1^2 u_\theta(r,z) = -\eta_s^2 u_\theta(r,z) \tag{8.215}$$

in which ∇_1^2 is defined by Eq. 6.47 with $m = 1$. Thus $u_\theta(r,z)$ assumes the form

$$a^{-1}u_\theta(r,z) = \int_0^\infty \bar{u}_\theta(z,\eta) J_1(\eta r) d\eta \tag{8.216}$$

and we can write, according to Hooke's law,

$$\sigma_{\theta z}(r,z) = \int_0^\infty \tilde{\sigma}_{\theta z}(z,\eta) J_1(\eta r) d\eta \tag{8.217}$$

where

$$\tilde{\sigma}_{\theta z}(r,z) = \mu \frac{d\bar{u}_\theta(z,\eta)}{dz} \tag{8.218}$$

Now, if we change symbols as follows: $\bar{u}_\theta$ into $\tilde{\chi}(\equiv v_L)$, and $\tilde{\sigma}_{\theta z}$ into τ_L, we can write boundary conditions in the identical form as the L-set of equations of the general case. In other words, only the L-set of equations exists in this 'pure torsion' case. Substituting Eq. 8.216 into Eq. 8.215 we obtain an ordinary differential equation of the second order with respect to z, with the solution

$$\bar{u}_\theta(z,\eta) = \eta\left(Be^{-\alpha_s z} + De^{\alpha_s z}\right) \tag{8.219}$$

or

$$\bar{u}_\theta(z,\eta) = \eta(B\cos\bar{\alpha}_s z + D\sin\bar{\alpha}_s z) \tag{8.220}$$

where B and D are arbitrary constants to be determined from boundary conditions.

For a sufficiently large r, the approximate formulas for Bessel functions (Eqs. 1.121) can be used to reduce the amount of computation, since Eqs. 8.206a,b,e and f are simplified into

$$\bar{u}_r(r,z,\eta) = \eta u_R(z,\eta) J_{m-1}(\eta r), \quad (r \to \infty) \tag{8.221a}$$

$$\bar{u}_\theta(r,z,\eta) = -\eta v_L(z,\eta) J_{m-1}(\eta r), \quad (r \to \infty) \tag{8.221b}$$

$$\tilde{\sigma}_{\theta z}(r,z,\eta) = -\eta \tau_L(z,\eta) J_{m-1}(\eta r), \quad (r \to \infty) \tag{8.221c}$$

$$\tilde{\sigma}_{rz}(r,z,\eta) = \eta \tau_R(z,\eta) J_{m-1}(\eta r), \quad (r \to \infty) \tag{8.221d}$$

These equations show the separability of boundary conditions into R and L sets of equations. Moreover, we can treat the problem as two separate problems; one being associated with potentials Φ and Ψ, and another with χ only.

When we are investigating the possible existence of surface waves, we seek to determine discrete values of η. In other words, we investigate the possible existence of the harmonic vibration when there is no applied traction on the surfaces nor on any interfaces between layers. The R-set of $4N$ linear homogeneous equations gives the Rayleigh wave equation identical to that for the in-plane case, i.e. Eq. 8.174. On the other hand, the L-set of $2N$ linear homogeneous equations gives the Love wave equation identical to that for the antiplane case, i.e. Eq. 8.192. In general, Rayleigh waves and Love waves are dispersive, but their wavenumbers must be real, as concluded by Ewing, Jardetzky and Press (1957). The solution corresponding to a surface wavenumber η can be constructed simply by removing the integration symbol in all expressions derived so far, and replacing $J_m(\eta r)d\eta$ in those expressions by

$$J_m(\eta r) \quad \text{or} \quad Y_m(\eta r) \quad \text{or} \quad H_m^{(1)}(\eta r) \quad \text{or} \quad H_m^{(2)}(\eta r).$$

Which function of r should be selected is determined by the requirement that each surface wave must be attenuating and outgoing as r tends to infinity. Thus shape functions for far field Rayleigh waves and Love waves assume the forms of

$$H_m^{(2)}(\eta_R r) \quad \text{and} \quad H_m^{(2)}(\eta_L r),$$

respectively. It should be noted that the effects of the Love waves on 'axisymmetric' functions (associated with potentials Φ and Ψ), e.g.

$$u_r, \quad u_z, \quad \sigma_{zz} \quad \text{and} \quad \sigma_{rz}$$

are negligible for $r \to \infty$. The same can be said for the Rayleigh waves for the 'pure torsion' functions (associated with potential χ), e.g.

$$u_\theta \quad \text{and} \quad \sigma_{\theta z}.$$

The complete solution to problems of forced vibration consists of the surface wave functions already discussed and also the *Cauchy's principal values* (Spiegel, 1963) of the infinite integrals, Eqs. 8.204. The integrands of these infinite integrals are naturally singular at the values of the surface wavenumbers. Unfortunately, none of the existing solutions for multilayered spaces are in such a form that a detailed investigation of the associated infinite integrals is possible. Jardetzky (1953) made an attempt to answer some of the important questions associated with these integrals without obtaining a complete solution to the problem. But the problems considered by Jardetzky were axisymmetric, so only Rayleigh waves were discussed. Later, Press, Harkrider and Seafeldt (1961) developed, for multilayered half spaces, a computer algorithm to determine the roots of Rayleigh and Love wave equations, Eqs. 8.174 and 8.192, respectively. The number and values of these roots play an important role in the solution approach, since they are *poles* of the integrands when evaluating the infinite integrals by means of appropriate contours in a complex plane. Furthermore, such a process involves *branch line integrals*. Jardetzky ventured to make a general conclusion that, in all cases of wave propagation from a point source in a half space formed by parallel layers displaying different elastic properties, all expected branch line integrals vanished except the one corresponding to the *branch points* (η_p and η_s) of the underlying half space. Thus each potential was obtained in the form of a sum of two parts, of which the first part represented a discrete spectrum of surface waves and the second a continuous spectrum given in the form of a branch line integral. This important point (by Jardetzky) will be scrutinized further.

Formally, the C.P.V. (Cauchy's principal value) of each infinite integral in Eqs. 8.204 can be obtained by means of appropriate *contours in the plane of the complex variable* ζ (Churchill, 1960), i.e.

$$\zeta = \eta + i\xi.$$

These contours bypass branch points and poles on the real axis η. Branch points are at $\zeta = \pm\eta_s$ and $\pm\eta_p$, while poles are at $\zeta = \pm\eta_R$ or $\pm\eta_L$. The C.P.V. obtained in such a way contains continuous spectra of body waves and discrete surface waves. Each surface wave is directly associated with the *residues* at a particular surface wavenumber (or two opposite poles). However, these surface waves do not assume the outgoing nature for large r. Thus a superposition upon such a surface wave by another surface wave of the same wavenumber is needed to make that particular wave to be in the form of

$$H_m^{(2)}(\eta_R r) \quad \text{or} \quad H_m^{(2)}(\eta_L r), \quad (\text{for large } r).$$

In summary, the complete solution to problems of forced vibration can be taken as a summation of a *body wave solution* and a *surface wave solution,* and each of these solutions individually satisfies all governing conditions outside the loaded domain. The body wave solution consists of continuous spectra of waves, while the surface wave solution consists of discrete waves. In addition, we can derive a unique relationship among all functions of surface waves, of the same as well as of different wavenumbers, but of the same wave type — (Rayleigh type or Love type). Various important conclusions made in this paragraph, on the composition of the complete solution and the individual natures of body and surface wave solutions, are to be verified in the next two sections where two simple cases are considered.

8.9 ASYMMETRIC VIBRATION OF A HOMOGENEOUS HALF SPACE

Consider the harmonic vibration of a homogeneous isotropic elastic half space occupying the domain $0 \le z < \infty$ and subjected to a uniform shear traction on its surface in a circular area of radius a, such that

$$\sigma_{zz}(r,\theta,0) = 0, \quad (0 \le r < \infty) \tag{8.222}$$

$$\sigma_{xz}(r,\theta,0) = \mu, \quad (0 \le r < 1) \tag{8.223a}$$

$$= 0, \quad (1 < r < \infty) \tag{8.223b}$$

Equations 8.223 can be expressed as the single equation

$$\sigma_{xz}(r,\theta,0) = \mu \int_0^\infty J_1(\eta) J_0(\eta r) d\eta \tag{8.224}$$

or, due to a recurrence formula of Bessel functions, Eq. 1.112c,

$$\sigma_{xz}(r,\theta,0)=\mu\int_0^\infty J_1(\eta)\left[\frac{\mathrm{d}J_1(\eta r)}{\mathrm{d}(\eta r)}+\frac{J_1(\eta r)}{\eta r}\right]\mathrm{d}\eta \tag{8.225}$$

The associated cylindrical stress components are

$$\sigma_{\theta z}(r,\theta,0)=-\sigma_{xz}(r,\theta,0)\sin\theta \tag{8.226}$$

$$\sigma_{rz}(r,\theta,0)=\sigma_{xz}(r,\theta,0)\cos\theta \tag{8.227}$$

In view of Eqs. 8.205 to 8.208, 8.222, 8.223 and 8.225 to 8.227, we can set $m=1$, and

$$\sigma_R(0,\eta)=0 \tag{8.228a}$$

$$\tau_R(0,\eta)=\mu\eta^{-1}J_1(\eta) \tag{8.228b}$$

$$\tau_L(0,\eta)=\mu\eta^{-1}J_1(\eta) \tag{8.229}$$

Substituting Eqs. 8.207c and d into Eqs. 8.228, and using Eqs. 8.210a and b (with $B_\Phi=B_\Psi=0$ for boundedness of the solutions as z tends to infinity), we find

$$\begin{bmatrix}2\eta^2-\eta_s^2 & -2\eta^2\alpha_s\\ -2\alpha_p & 2\eta^2-\eta_s^2\end{bmatrix}\begin{Bmatrix}A_\Phi\\ A_\Psi\end{Bmatrix}=\begin{Bmatrix}0\\ \eta^{-1}J_1(\eta)\end{Bmatrix} \tag{8.230}$$

On the other hand, using Eqs. 8.208b and 8.210c (with $B_\chi=0$ for boundedness of the solution as z tends to infinity), in Eq. 8.229, we get

$$-\alpha_s A_\chi=\eta^{-1}J_1(\eta) \tag{8.231}$$

Solving the preceding two equations yields

$$A_\Phi=\frac{2\eta\alpha_s(\eta)J_1(\eta)}{F_R(\eta)} \tag{8.232a}$$

$$A_\Psi=\frac{(2\eta^2-\eta_s^2)J_1(\eta)}{\eta F_R(\eta)} \tag{8.232b}$$

$$A_\chi=-\frac{J_1(\eta)}{\eta\alpha_s(\eta)} \tag{8.233}$$

where $F_R(\eta)$ is the Rayleigh wave function identical to that of the in-plane problem, i.e. Eq. 8.177. We can, consequently, say again that there exists a Rayleigh wave but not a Love wave for the vibration of a homogeneous half space. Subsequently, we can write Eqs. 8.204 in the form

$$\Phi(r,\theta,z)=\cos\theta\int_0^\infty\frac{2\eta\alpha_s(\eta)J_1(\eta)J_1(\eta r)e^{-\alpha_p(\eta)z}}{F_R(\eta)}\mathrm{d}\eta \tag{8.234a}$$

$$\Psi(r,\theta,z)=\cos\theta\int_0^\infty\frac{(2\eta^2-\eta_s^2)J_1(\eta)J_1(\eta r)e^{-\alpha_s(\eta)z}}{\eta F_R(\eta)}\mathrm{d}\eta \tag{8.234b}$$

$$\chi(r,\theta,z) = -\sin\theta \int_0^{\infty} \frac{J_1(\eta)J_1(\eta r)e^{-\alpha_s(\eta)z}}{\eta\alpha_s(\eta)} d\eta \tag{8.235}$$

Now, our immediate task is to determine the Cauchy's principal value (C.P.V.) of each infinite integral in the equations above. For that in Eq. 8.234a and $r > 1$, consider the complex integrand

$$\frac{2\zeta\alpha_s(\zeta)J_1(\zeta)H_1^{(1)}(\zeta r)e^{-\alpha_p(\zeta)z}}{F_R(\zeta)} d\zeta \tag{8.236}$$

and use the dual integral contours recommended by Lamb (1904) as depicted in Fig. 8.15, where ζ is the complex variable, i.e.

$$\zeta = \eta + i\xi \tag{8.237}$$

Consistent values of radical quantities $\alpha_p(\zeta)$ and $\alpha_s(\zeta)$ are also indicated on various parts of the contours. The integrand in question vanishes on the integral paths where $|\zeta| \rightarrow \infty$. Thus, according to the *residue theorem of complex variables,* the contour C_1 gives

$$\int_0^{\infty} \frac{2\eta\alpha_s(\eta)J_1(\eta)H_1^{(1)}(\eta r)e^{-\alpha_p(\eta)z}}{F_R(\eta)} d\eta = \frac{\pi i 2\eta_R\alpha_s(\eta_R)J_1(\eta_R)H_1^{(1)}(\eta_R r)e^{-\alpha_p(\eta_R)z}}{F_R'(\eta_R)}$$
$$+ i\int_0^{\infty} \frac{2i\xi\alpha_s(i\xi)J_1(i\xi)H_1^{(1)}(i\xi r)e^{-\alpha_p(i\xi)z}}{F_R(i\xi)} d\xi \tag{8.238}$$

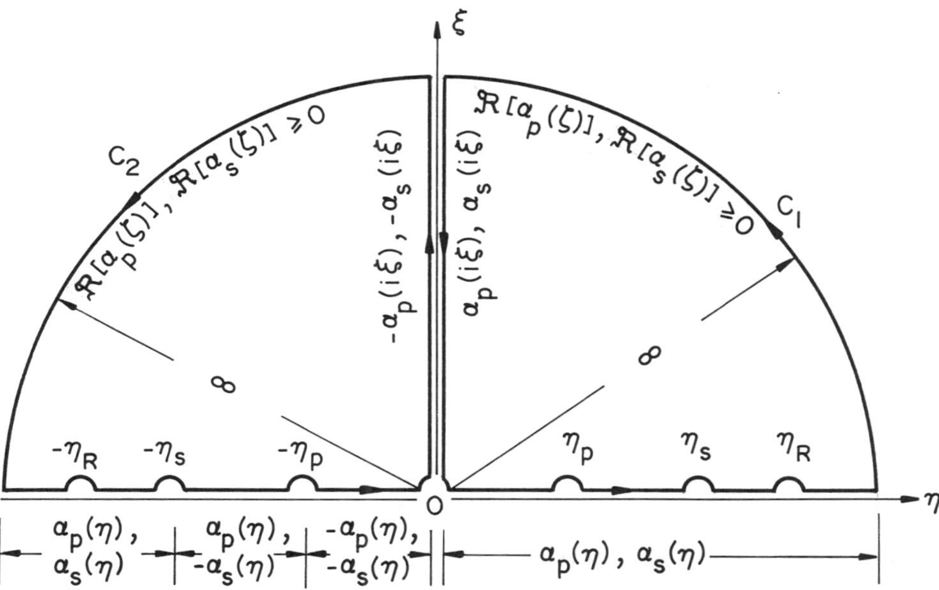

Fig. 8.15 Integral contours for $\Phi(r,\theta,z)$ and $\Psi(r,\theta,z)$, Eqs. 8.234.

where

$$F_R'(\eta_R) = \left[\frac{F_R(\eta)}{(\eta - \eta_R)}\right]_{\eta=\eta_R} = \left.\frac{dF_R(\eta)}{d\eta}\right|_{\eta=\eta_R} \tag{8.239}$$

On the other hand, the contour C_2 gives

$$\int_{-\infty}^{-\eta_s} \frac{2\eta\alpha_s(\eta)J_1(\eta)H_1^{(1)}(\eta r)e^{-\alpha_p(\eta)z}}{F_R(\eta)} d\eta - \int_{-\eta_s}^{-\eta_p} \frac{2\eta\alpha_s(\eta)J_1(\eta)H_1^{(1)}(\eta r)e^{-\alpha_p(\eta)z}}{\overline{F}_R(\eta)} d\eta$$

$$- \int_{-\eta_p}^{0} \frac{2\eta\alpha_s(\eta)J_1(\eta)H_1^{(1)}(\eta r)e^{\alpha_p(\eta)z}}{F_R(\eta)} d\eta = \frac{\pi i 2(-\eta_R)\alpha_s(-\eta_R)J_1(-\eta_R)H_1^{(1)}(-\eta_R r)e^{-\alpha_p(-\eta_R)z}}{F'_R(-\eta_R)}$$

$$+ i\int_0^{\infty} \frac{2i\xi\alpha_s(i\xi)J_1(i\xi)H_1^{(1)}(i\xi r)e^{\alpha_p(i\xi)z}}{F_R(i\xi)} d\xi \tag{8.240}$$

where

$$\overline{F}_R(\eta) = (2\eta^2 - \eta_s^2)^2 + 4\eta^2\alpha_p(\eta)\alpha_s(\eta) \tag{8.241}$$

Due to the analytic continuation, Eq. 1.114e, we can have identities involving Hankel functions of negative arguments as follows

$$H_m^{(1)}(-\eta) = -H_m^{(2)}(\eta), \quad (m = 0, 2, 4, 6, \ldots) \tag{8.242a}$$

$$= H_m^{(2)}(\eta), \quad (m = 1, 3, 5, 7, \ldots) \tag{8.242b}$$

Subsequently, rewriting Eq. 8.240 to involve only positive limits of integration and positive arguments of functions we find

$$\int_0^{\infty} \frac{2\eta\alpha_s(\eta)J_1(\eta)H_1^{(2)}(\eta r)e^{-\alpha_p(\eta)z}}{F_R(\eta)} d\eta = -\frac{\pi i 2\eta_R\alpha_s(\eta_R)J_1(\eta_R)H_1^{(2)}(\eta_R r)e^{-\alpha_p(\eta_R)z}}{F'_R(\eta_R)}$$

$$+ \int_0^{\eta_p} \frac{2\eta\alpha_s(\eta)J_1(\eta)H_1^{(2)}(\eta r)\,[e^{\alpha_p(\eta)z} + e^{-\alpha_p(\eta)z}]}{F_R(\eta)} d\eta$$

$$+ \int_{\eta_p}^{\eta_s} \frac{4(2\eta^2 - \eta_s^2)^2\eta\alpha_s(\eta)J_1(\eta)H_1^{(2)}(\eta r)e^{-\alpha_p(\eta)z}}{F_R(\eta)\overline{F}_R(\eta)} d\eta$$

$$+ i\int_0^{\infty} \frac{2i\xi\alpha_s(i\xi)J_1(i\xi)H_1^{(1)}(i\xi r)e^{\alpha_p(i\xi)z}}{F_R(i\xi)} d\xi \tag{8.243}$$

Adding Eqs. 8.238 and 8.243 together, and using Eqs. 1.113a and b, we get

$$\int_0^{\infty} \frac{2\eta\alpha_s(\eta)J_1(\eta)J_1(\eta r)e^{-\alpha_p(\eta)z}}{F_R(\eta)} d\eta = -\frac{\pi 2\eta_R\alpha_s(\eta_R)J_1(\eta_R)Y_1(\eta_R r)e^{-\alpha_p(\eta_R)z}}{F_R'(\eta_R)}$$

$$
+\int_0^{\eta_p}\frac{2\eta\alpha_s(\eta)J_1(\eta)H_1^{(2)}(\eta r)\cos\left[(\eta_p^2-\eta^2)^{1/2}z\right]}{F_R(\eta)}\mathrm{d}\eta
$$

$$
+\int_{\eta_p}^{\eta_s}\frac{2(2\eta^2-\eta_s^2)^2\eta\alpha_s(\eta)J_1(\eta)H_1^{(2)}(\eta r)e^{-\alpha_p(\eta)z}}{F_R(\eta)\overline{F}_R(\eta)}\mathrm{d}\eta
$$

$$
-\int_0^{\infty}\frac{2\xi\alpha_s(i\xi)J_1(i\xi)H_1^{(1)}(i\xi r)\cos\left[(\xi^2+\eta_p^2)^{1/2}z\right]}{F_R(i\xi)}\mathrm{d}\xi \tag{8.244}
$$

It should be noted that every term on the right-hand side of the equation above, except the first one, characterizes a wave which is outgoing and/or attenuating as r increases. Due to the presence of $H_1^{(1)}(i\xi r)$, the last term of that equation decays exponentially. To make the first term on the right-hand side of Eq. 8.244 to represent an outgoing wave for large r as well, we should superimpose upon it a surface wave function equal to

$$
-\frac{\pi i2\eta_R\alpha_s(\eta_R)J_1(\eta_R)J_1(\eta_R r)e^{-\alpha_p(\eta_R)z}}{F_R'(\eta_R)} \tag{8.245}
$$

Accordingly, we have for $r>1$,

$$
\Phi(r,\theta,z)=\cos\theta\left\{-\frac{\pi i2\eta_R\alpha_s(\eta_R)J_1(\eta_R)H_1^{(2)}(\eta_R r)e^{-\alpha_p(\eta_R)z}}{F_R'(\eta_R)}\right.
$$

$$
+\int_0^{\eta_p}\frac{2\eta\alpha_s(\eta)J_1(\eta)H_1^{(2)}(\eta r)\cos\left[(\eta_p^2-\eta^2)^{1/2}z\right]}{F_R(\eta)}\mathrm{d}\eta
$$

$$
+\int_{\eta_p}^{\eta_s}\frac{2(2\eta^2-\eta_s^2)^2\eta\alpha_s(\eta)J_1(\eta)H_1^{(2)}(\eta r)e^{-\alpha_p(\eta)z}}{F_R(\eta)\overline{F}_R(\eta)}\mathrm{d}\eta
$$

$$
\left.-\int_0^{\infty}\frac{2\xi\alpha_s(i\xi)J_1(i\xi)H_1^{(1)}(i\xi r)\cos\left[(\xi^2+\eta_p^2)^{1/2}z\right]}{F_R(i\xi)}\mathrm{d}\xi\right\} \tag{8.246}
$$

The expressions derived so far are not valid for a case of small values of r which is our next task. To determine the Cauchy's principal value (C.P.V.) of the infinite integral in Eq. 8.234a for $r<1$, the complex integrand to be used is

$$
\frac{2\zeta\alpha_s(\zeta)H_1^{(1)}(\zeta)J_1(\zeta r)e^{-\alpha_p(\zeta)z}}{F_R(\zeta)}\mathrm{d}\zeta \tag{8.247}
$$

but, for continuity of solution at $r=1$, the same superimposing surface wave function, Eq. 8.245, must be used. Proceeding in the same manner as having obtained Eq. 8.246 we can eventually show that this equation holds as well for $r<1$, if in that equation we replace

$$\begin{aligned} J_1(\eta) &\quad \text{by} \quad H_1^{(2)}(\eta) \\ H_1^{(2)}(\eta r) &\quad \text{by} \quad J_1(\eta r) \\ J_1(\mathrm{i}\xi) &\quad \text{by} \quad H_1^{(1)}(\mathrm{i}\xi) \\ H_1^{(1)}(\mathrm{i}\xi r) &\quad \text{by} \quad J_1(\mathrm{i}\xi r) \end{aligned} \tag{8.248}$$

However, the presence of the terms $\cos\left[(\eta_p^2-\eta^2)^{1/2}z\right]$ and $\cos\left[(\xi^2+\eta_p^2)^{1/2}z\right]$ in the expressions derived so far hints that they are not suitable for large values of z. For large z, the integration in Eq. 8.234a can be truncated at η^*, which is sufficiently larger than η_p, and we have

$$\Phi(r,\theta,z)=\cos\theta\int_0^{\eta^*}\frac{2\eta\alpha_s(\eta)J_1(\eta)J_1(\eta r)e^{-\alpha_p(\eta)z}}{F_R(\eta)}\mathrm{d}\eta \tag{8.249}$$

which properly contains spectra of outgoing and exponentially decaying waves as z tends to infinity. At this point, we may conclude that any physical quantity given by infinite integrals as follows

$$\Lambda(r,z)=\int_0^{\infty}\frac{f(\eta,z)J_m(\eta r)}{F_R(\eta)}\mathrm{d}\eta \tag{8.250}$$

is equal to, for small and intermediate values of z,

$$\Lambda(r,z)=\text{C.P.V.}\left[\int_0^{\infty}\frac{f(\eta,z)J_m(\eta r)}{F_R(\eta)}\mathrm{d}\eta\right]-\frac{\pi i f(\eta_R,z)J_m(\eta_R r)}{F_R'(\eta_R)} \tag{8.251}$$

In addition, Eq. 8.251 implies that the C.P.V. of an infinite integral alone implicitly contains discrete non-outgoing waves for large r, thus does not give the correct value for a physical quantity.

At this stage, it should be quite obvious that the evaluation of $\Psi(r,\theta,z)$ defined in Eq. 8.234b can follow the same procedures as used for $\Phi(r,\theta,z)$. The results are the following: for small to intermediate z,

$$\begin{aligned} \Psi(r,\theta,z)=\cos\theta\Bigg\{\text{C.P.V.}&\left[\int_0^{\infty}\frac{(2\eta^2-\eta_s^2)J_1(\eta)J_1(\eta r)e^{-\alpha_s(\eta)z}}{\eta F_R(\eta)}\mathrm{d}\eta\right] \\ &-\frac{\pi i(2\eta_R^2-\eta_s^2)J_1(\eta_R)J_1(\eta_R r)e^{-\alpha_s(\eta_R)z}}{\eta_R F_R'(\eta_R)}\Bigg\} \end{aligned} \tag{8.252}$$

and for large z,

$$\Psi(r,\theta,z)=\cos\theta\left\{\text{C.P.V.}\left[\int_0^{\eta^*}\frac{(2\eta^2-\eta_s^2)J_1(\eta)J_1(\eta r)e^{-\alpha_s(\eta)z}}{\eta F_R(\eta)}\mathrm{d}\eta\right]\right\} \tag{8.253}$$

where η^* is sufficiently larger than η_s. In detail, Eq. 8.252 can be written for $r>1$ and small to intermediate z as,

$$\Psi(r,\theta,z)=\cos\theta\left\{-\frac{\pi i(2\eta_R^2-\eta_s^2)J_1(\eta_R)H_1^{(2)}(\eta_R r)e^{-\alpha_s(\eta_R)z}}{\eta_R F_R'(\eta_R)}-\frac{\cos(\eta_s z)}{2\eta_s^2 r}\right.$$

$$-i\int_0^{\eta_p}\frac{(2\eta^2-\eta_s^2)J_1(\eta)H_1^{(2)}(\eta r)\sin\left[(\eta_s^2-\eta^2)^{1/2}z\right]}{\eta F_R(\eta)}d\eta$$

$$+\int_{\eta_p}^{\eta_s}\frac{(2\eta^2-\eta_s^2)J_1(\eta)H_1^{(2)}(\eta r)}{2\eta}\left[\frac{e^{-\alpha_s(\eta)z}}{F_R(\eta)}-\frac{e^{\alpha_s(\eta)z}}{\overline{F}_R(\eta)}\right]d\eta$$

$$\left.+i\int_0^{\infty}\frac{(2\xi^2+\eta_s^2)J_1(i\xi)H_1^{(1)}(i\xi r)\sin\left[(\xi^2+\eta_s^2)^{1/2}z\right]}{\xi F_R(i\xi)}d\xi\right\} \qquad (8.254)$$

The equation above also holds for $r<1$ and small to intermediate z, if we replace various terms in the equation as stipulated in Eq. 8.248 and replace r^{-1} (in the second term of Eq. 8.254) by r. In fact, the second term of Eq. 8.254 is the residue of the integral at $\zeta=0$, and the singular integrands at the lower limit of integration in the third and fifth terms of Eq. 8.254 nullify each other.

Now, there remains the task of determining $\chi(r,\theta,z)$ defined in Eq. 8.235, which does not have any surface wave characteristics, hence the task is much simpler than for $\Phi(r,\theta,z)$ and $\Psi(r,\theta,z)$. The dual integral contours to be used are as depicted in

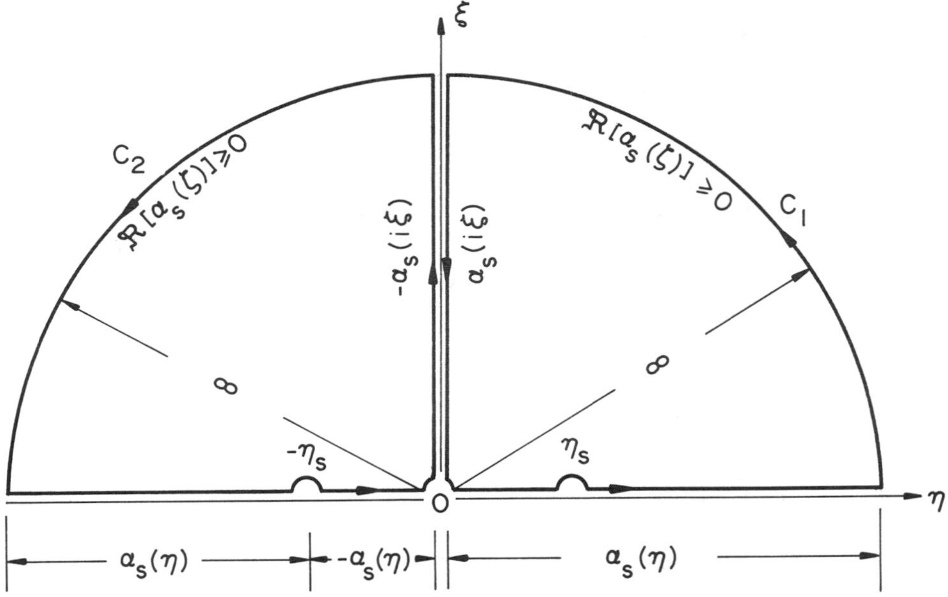

Fig. 8.16 integral contours for $\chi(r,\theta,z)$, Eq. 8.235.

Fig. 8.16. The results contain only the Cauchy's principal value (C.P.V.) of the infinite integral involved, i.e. for small to intermediate z,

$$\chi(r,\theta,z)=-\sin\theta\left\{\text{C.P.V.}\left[\int_0^{\infty}\frac{J_1(\eta)J_1(\eta r)e^{-\alpha_s(\eta)z}}{\eta\alpha_s(\eta)}d\eta\right]\right\} \tag{8.255}$$

and for large z,

$$\chi(r,\theta,z)=-\sin\theta\left\{\text{C.P.V.}\left[\int_0^{\eta^*}\frac{J_1(\eta)J_1(\eta r)e^{-\alpha_s(\eta)z}}{\eta\alpha_s(\eta)}d\eta\right]\right\} \tag{8.256}$$

where η^* is sufficiently larger than η_s. In detail, Eq. 8.255 can be written for $r>1$ and small to intermediate z as,

$$\chi(r,\theta,z)=-\sin\theta\left\{-\frac{\sin(\eta_s z)}{2r\eta_s}+\int_0^{\eta_s}\frac{J_1(\eta)H_1^{(2)}(\eta r)\cos\left[(\eta_s^2-\eta^2)^{1/2}z\right]}{\eta\alpha_s(\eta)}d\eta\right.$$

$$\left.+\int_0^{\infty}\frac{J_1(i\xi)H_1^{(1)}(i\xi r)\cos\left[(\xi^2+\eta_s^2)^{1/2}z\right]}{\xi\alpha_s(i\xi)}d\xi\right\} \tag{8.257}$$

The equation above also holds for $r<1$ and small to intermediate z, if we replace various terms in the equation as stipulated in Eq. 8.248 and replacing r^{-1} (in the first term of Eq. 8.257) by r. Again, it should be noted that the singular integrands at the lower limit of integration in the second and third terms of Eq. 8.257 nullify each other.

In summary, the complete solution to this problem can be expressed as a sum of a surface wave solution and a body wave solution, i.e.

$$\Phi(r,\theta,z)=\Phi^s(r,\theta,z)+\Phi^b(r,\theta,z) \tag{8.258a}$$

$$\Psi(r,\theta,z)=\Psi^s(r,\theta,z)+\Psi^b(r,\theta,z) \tag{8.258b}$$

$$\chi(r,\theta,z)=\chi^s(r,\theta,z)+\chi^b(r,\theta,z) \tag{8.259}$$

in which the superscripts s and b denote a surface wave and a body wave, respectively. Explicitly, the surface wave solution functions, for $r>1$ and z not large, are

$$\Phi^s(r,\theta,z)=-\frac{2\pi i\eta_R\alpha_s(\eta_R)J_1(\eta_R)H_1^{(2)}(\eta_R r)e^{-\alpha_p(\eta_R)z}\cos\theta}{F_R'(\eta_R)} \tag{8.260a}$$

$$\Psi^s(r,\theta,z)=-\frac{\pi i(2\eta_R^2-\eta_s^2)J_1(\eta_R)H_1^{(2)}(\eta_R r)e^{-\alpha_s(\eta_R)z}\cos\theta}{\eta_R F_R'(\eta_R)} \tag{8.260b}$$

$$\chi^s(r,\theta,z)=0 \tag{8.261}$$

Comparing Eqs. 8.258a,b and 8.259 with Eqs. 8.246, 8.254 and 8.257, respectively, and noting Eqs. 8.260 and 8.261, we can identify explicitly the body wave solution functions

$$\Phi^b(r,\theta,z),\quad \Psi^b(r,\theta,z)\quad \text{and}\quad \chi^b(r,\theta,z).$$

With a brief reflection on the results obtained in this section, one should be able to appreciate some conclusions made in the last paragraph of the preceding section for the general case.

8.10 AXISYMMETRIC TORSION OF A LAYERED HALF SPACE

Consider the harmonic torsional vibration of a layered half space consisting of an upper layer (1) where $-h/a \le z \le 0$, and an underlying half space (2) where $0 \le z < \infty$. There will be Love waves provided that the inequality Eq. 8.197b holds. The half space is subjected to a shear traction on its top surface in a circular area of radius a such that

$$\sigma_{z\theta 1}(r,-h/a) = \mu_1 r, \quad (r<1) \tag{8.262a}$$

$$= 0, \quad (r>1) \tag{8.262b}$$

Equations 8.262 can be written as the single equation (Eq. 6.112g),

$$\sigma_{z\theta 1}(r,-h/a) = \mu_1 \int_0^\infty J_2(\eta) J_1(\eta r)\mathrm{d}\eta, \quad (0 \le r < \infty) \tag{8.263}$$

Relevant boundary conditions for the present case are

$$u_{\theta 1}(r,0) = u_{\theta 2}(r,0), \quad \sigma_{z\theta 1}(r,0) = \sigma_{z\theta 2}(r,0) \tag{8.264a,b}$$

$$\sigma_{z\theta 1}(r,-h/a) = \mu_1 \int_0^\infty J_2(\eta) J_1(\eta r)\mathrm{d}\eta \tag{8.264c}$$

The solution is chosen to take the form of Eq. 8.220 for the upper layer, and Eq. 8.219 with $D_2 = 0$ for the underlying half space, i.e.

$$u_{\theta 1}(r,z) = \int_0^\infty \eta(B_1 \cos\bar{\alpha}_{s1} z + D_1 \sin\bar{\alpha}_{s1} z) J_1(\eta r)\mathrm{d}\eta \tag{8.265}$$

$$u_{\theta 2}(r,z) = \int_0^\infty \eta B_2 e^{-\alpha_{s2} z} J_1(\eta r)\mathrm{d}\eta \tag{8.266}$$

The first boundary condition, Eq. 8.264a, readily gives

$$B_2 = B_1 \tag{8.267a}$$

then the remaining boundary conditions can be put as

$$\begin{bmatrix} \mu_2 \alpha_{s2} & \mu_1 \bar{\alpha}_{s1} \\ \bar{\alpha}_{s1} \sin(\bar{\alpha}_{s1} h/a) & \bar{\alpha}_{s1} \cos(\bar{\alpha}_{s1} h/a) \end{bmatrix} \begin{Bmatrix} B_1 \\ D_1 \end{Bmatrix} = \begin{Bmatrix} 0 \\ J_2(\eta)/\eta \end{Bmatrix} \tag{8.267b}$$

This equation can be solved for B_1 and D_1. Substituting the results in Eqs. 8.265 and 8.266 yields

$$u_{\theta 1}(r,z) = \int_0^\infty \frac{[A(\eta,z) + C(\eta,z)] J_2(\eta) J_1(\eta r)}{F_L(\eta)}\mathrm{d}\eta \tag{8.268a}$$

$$u_{\theta 2}(r,z) = \int_0^\infty \frac{A_2(\eta,z) J_2(\eta) J_1(\eta r)}{F_L(\eta)}\mathrm{d}\eta \tag{8.268b}$$

where $F_L(\eta)$ is the Love wave function, i.e. Eq. 8.196, and

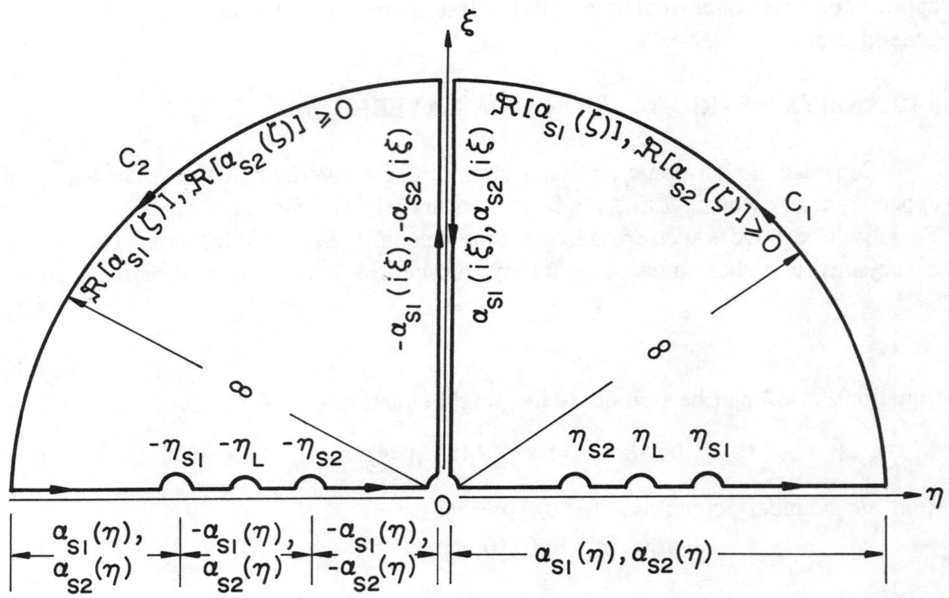

Fig. 8.17 Integral contours for $u_\theta(r,z)$ in Eqs. 8.268.

$$A(\eta,z)=\frac{\cos\bar{\alpha}_{s1}z}{\bar{\alpha}_{s1}\cos(\bar{\alpha}_{s1}h/a)} \tag{8.269a}$$

$$C(\eta,z)=-\frac{\mu_2\alpha_{s2}\sin\bar{\alpha}_{s1}z}{\mu_1\bar{\alpha}_{s1}^2\cos(\bar{\alpha}_{s1}h/a)} \tag{8.269b}$$

$$A_2(\eta,z)=\frac{e^{-\alpha_{s2}z}}{\bar{\alpha}_{s1}\cos(\bar{\alpha}_{s1}h/a)} \tag{8.269c}$$

The Cauchy's principal values of infinite integrals in Eqs. 8.268 can be determined as in the previous section using dual integral contours as shown in Fig. 8.17. For simplicity, only one surface wavenumber is explicitly shown in the figure; i.e. one pole at $\zeta=\eta_L$ lying in between branch points at $\zeta=\eta_{s1}$ and η_{s2}, and another pole $-\eta_L$ in between branch points $-\eta_{s1}$ and $-\eta_{s2}$. For the present case, we find that for $r>1$,

$$u_{\theta1}(r,z)=-\sum\frac{\pi i[A(\eta_L,z)+C(\eta_L,z)]J_2(\eta_L)H_1^{(2)}(\eta_L r)}{F_L'(\eta_L)}$$
$$+\int_0^{\eta_{s2}}A_3(\eta,z)J_2(\eta)H_1^{(2)}(\eta r)d\eta+i\int_0^{\infty}A_3(i\xi,z)J_2(i\xi)H_1^{(1)}(i\xi r)d\xi \tag{8.270a}$$

$$u_{\theta 2}(r,z) = -\sum \frac{\pi i A(\eta_L, z) J_2(\eta_L) H_1^{(2)}(\eta_L r)}{F_L'(\eta_L)} + \int_0^{\eta_{s2}} A_4(\eta, z) J_2(\eta) H_1^{(2)}(\eta r) d\eta$$

$$+ i \int_0^{\infty} A_4(i\xi, z) J_2(i\xi) H_1^{(1)}(i\xi r) d\xi, \qquad (z \text{ not large}) \qquad (8.270b)$$

Here Σ denotes summation of all surface waves, and

$$F_L'(\eta_L) = \left[\frac{F_L(\eta)}{\eta - \eta_L}\right]_{\eta = \eta_L} = \left.\frac{dF_L(\eta)}{d\eta}\right|_{\eta = \eta_L} \qquad (8.271a)$$

$$A_3(\eta, z) = \frac{A(\eta, z) + C(\eta, z)}{2F_L(\eta)} - \frac{A(\eta, z) - C(\eta, z)}{2\overline{F}_L(\eta)} \qquad (8.271b)$$

$$A_4(\eta, z) = \frac{A_2(\eta, z)}{2F_L(\eta)} - \frac{A_2(\eta, -z)}{2\overline{F}_L(\eta)} \qquad (8.271c)$$

in which

$$\overline{F}_L(\eta) = \tan(\overline{\alpha}_{s1} h/a) + \frac{\mu_2 \alpha_{s2}}{\mu_1 \overline{\alpha}_{s1}} \qquad (8.272)$$

The higher the surface wave mode, the less significant it is, due to the closer to infinity its $F_L'(\eta_L)$ is. It should be noted that Eqs. 8.270 do not contain any integral with respect to η in the range from η_{s2} to η_{s1}. For $r > 1$ and large z, Eq. 8.270b does not hold, so $u_{\theta 2}(r,z)$ should be calculated anew by truncating the integration in Eq. 8.268b at η^*, which is sufficiently larger than η_{s2}, i.e.

$$u_{\theta 2}(r,z) = \text{C.P.V.}\left[\int_0^{\eta^*} \frac{A_2(\eta, z) J_2(\eta) J_1(\eta r)}{F_L(\eta)} d\eta\right] \qquad (8.273)$$

The equation above holds for $r \leq 1$ also as long as z is sufficiently large. For small and intermediate z and $r < 1$, we can show that Eqs. 8.270 hold, provided that we replace

$$\begin{array}{lcl} J_2(\eta) & \text{by} & H_2^{(2)}(\eta) \\ H_1^{(2)}(\eta r) & \text{by} & J_1(\eta r) \\ J_2(i\xi) & \text{by} & H_2^{(1)}(i\xi) \\ H_1^{(1)}(i\xi r) & \text{by} & J_1(i\xi r) \end{array} \qquad (8.274)$$

and add to Eqs. 8.270a and b, respectively, the following terms

$$r\left[\frac{A(0,z) + C(0,z)}{2F_L(0)} + \frac{A(0,z) - C(0,z)}{2\overline{F}_L(0)}\right] \qquad (8.275a)$$

and

$$r\left[\frac{A_2(0,z)}{2F_L(0)}+\frac{A_2(0,-z)}{2\overline{F}_L(0)}\right] \tag{8.275b}$$

which are residues of integrals at the pole $\zeta=0$.

The complete solution to this problem can be put as a sum of a surface wave solution and a body wave solution, i.e.

$$u_{\theta 1}(r,z)=u^s_{\theta 1}(r,z)+u^b_{\theta 1}(r,z) \tag{8.276a}$$

$$u_{\theta 2}(r,z)=u^s_{\theta 2}(r,z)+u^b_{\theta 2}(r,z) \tag{8.276b}$$

The surface wave functions

$$u^s_{\theta 1}(r,z) \quad \text{and} \quad u^s_{\theta 2}(r,z),$$

for $r>1$ and z not large, are the first terms on the right-hand side of Eqs. 8.270a and b, respectively.

8.11 TOTAL SOLUTION TO VIBRATION OF HALF PLANES

For in-plane problems of harmonic vibration of isotropic elastic half planes, any physical quantity is in the form of an infinite integral as follows

$$\Lambda(x,y)=\int_0^\infty \frac{f(\eta,y)\cos\eta x}{F_R(\eta)}\,\mathrm{d}\eta \tag{8.277a}$$

or

$$\Lambda(x,y)=\int_0^\infty \frac{f(\eta,y)\sin\eta x}{F_R(\eta)}\,\mathrm{d}\eta \tag{8.277b}$$

where $F_R(\eta)$ is the Rayleigh wave function, Eq. 8.174. For antiplane problems, any physical quantity takes the same forms as above with $F_R(\eta)$ being replaced by the Love wave function $F_L(\eta)$, Eq. 8.192. Any surface wavenumber, η_R or η_L, if it exists, must be real positive and greater than the shear wavenumber (η_s) of the underlying half plane, and greater than that of the overlying half plane for a multilayered full plane. For the underlying half plane, $f(\eta,y)$ in Eqs. 8.277 is exponentially decaying, i.e.

$$f(\eta,y)=f_1(\eta)e^{-\alpha(\eta)y} \tag{8.278}$$

where $\alpha(\eta)$ is a radical function for the underlying half plane given by Eq. 8.167a or b. For the overlying half plane, $\alpha(\eta)$ should be replaced by $-\alpha(\eta)$.

After going through the same arguments as carried out for three dimensional problems in the preceding sections, one should be able to arrive at the following conclusions: Any physical quantity given by Eq. 8.277a is, for small and intermediate values of y,

$$\Lambda(x,y)=\text{C.P.V.}\left[\int_0^\infty \frac{f(\eta,y)\cos\eta x}{F_R(\eta)}\,\mathrm{d}\eta\right]-\Sigma\frac{\pi i f(\eta_R,y)\cos\eta_R x}{F_R'(\eta_R)} \tag{8.279}$$

where Σ denotes summation of all Rayleigh waves involved in the problem being considered, and $F_R'(\eta_R)$ is the derivative of the Rayleigh wave function at a Rayleigh wavenumber η_R. For sufficiently large values of y in the underlying half plane, the integration in Eq. 8.277a can be truncated at η^*, i.e.

$$\Lambda(x,y) = \text{C.P.V.}\left[\int_0^{\eta^*} \frac{f(\eta,y)\cos\eta x}{F_R(\eta)} \mathrm{d}\eta\right] \tag{8.280}$$

where η^* is sufficiently larger than η_p or η_s depending on which is involved in $\alpha(\eta)$ in Eq. 8.278. The Cauchy principal value (C.P.V.) in Eq. 8.279 can be put in a more explicit form by integrating a complex integrand along appropriate contours. For $x > 1$, the complex integrand to be used is

$$\frac{f(\zeta,y)e^{i\zeta x}}{F_R(\zeta)} \mathrm{d}\zeta \tag{8.281}$$

where $\Re(\zeta) = \eta$. For a simple case such as a homogeneous half plane, the dual contours depicted in Figs. 8.15 and 8.16 can be used. However, for a more complicated case such as a layered or multilayered half plane, we follow the procedure: formulate the surface wave function and its derivatives, and Eq. 8.277a symbolically; evaluate the surface wavenumbers, the derivative of the surface wave function at each and every surface wavenumber; evaluate Eqs. 8.279 and 8.280 numerically.

All these schemes are applicable to any physical quantity given by an infinite sine integral of the form Eq. 8.277b.

8.12 VIBRATION OF VISCOELASTIC HALF SPACES

The solution to the vibration of an isotropic viscoelastic solid can be obtained from that of the corresponding elastic one by simply replacing the Lamé's constants λ and μ by the complex moduli $\lambda^*(\omega)$ and $\mu^*(\omega)$, respectively. These moduli have positive real parts and positive imaginary parts. So the real and imaginary parts of the wave velocities, c_p and c_s given by Eqs. 8.12 and 8.14 respectively, are also positive. The dimensionless body wavenumbers, ζ_p and ζ_s ($\zeta = \eta + i\xi$) determined from Eqs. 8.164b and c, will have positive real parts but negative imaginary parts. The surface Rayleigh and Love wavenumbers, ζ_R and ζ_L determined from Eqs. 8.174 and 8.192 respectively, will have the same signs as the body wavenumbers.

Then any physical quantity in the vibration of a multilayered viscoelastic half plane can be evaluated with the same procedures as those described for the elastic one in the preceding section. However, each physical quantity in the present case is given solely by the C.P.V. of its infinite integral. With reference to the dual contours as in Fig. 8.15 (or 8.16 or 8.17), branch points and poles (if any) will not be on the contours like the elastic case. The 'positive' branch points and poles (ζ_p, ζ_s, ζ_R or ζ_L) are in the fourth quadrant of the complex plane, while the 'negative' branch points and poles ($-\zeta_p$,

$-\zeta_s$, $-\zeta_R$ or $-\zeta_L$) are in the second quadrant inside the contour C_2. The residue at each pertinent pole, $-\zeta_p$ or $-\zeta_L$, has an outgoing wave form as x tends to infinity. Hence, the C.P.V. of an infinite integral representing a physical quantity already contains (though not explicitly) discrete outgoing surface wave functions, and it is not necessary to superimpose as the second term on the right-hand side of Eq. 8.279.

This scheme is applicable to three-dimensional viscoelastic cases.

8.13 INFINITE ELEMENTS FOR A HOMOGENEOUS HALF SPACE

Analytical methods such as that described in Sections 8.9 and 8.10 become impractical if not impossible when extended to load transfer problems and/or multi-layered half spaces. Several numerical methods have been developed to bypass such difficulties. At this stage of development, the most useful numerical algorithm seems to be the one that involves infinite elements for the far field and finite elements for the near field, as shown in Fig. 8.18(a). At present this algorithm is restricted to a homogeneous isotropic elastic half space. Research on algorithms for layered and multilayered half spaces is in progress. The present section describes infinite elements which have been developed by Rajapakse and Karasudhi (1986) and Karasudhi (1986) for load transfer problems in a homogeneous isotropic elastic half space.

The whole material domain is separated into a near field and a far field as shown in Fig. 8.18. The near field, consisting of the partially embedded bar and a finite region of the half space around it, is discretized into conventional finite elements. The far field covering the rest of the half space is descretized into infinite elements. Near the surface of the half space are horizontal infinite elements (HE), while the remainder of the far field is occupied by radiating infinite elements (RE). Every infinite element has point nodes only on the interface between the near field and the far field. Displacement functions of an elastodynamic infinite element should decrease with distance, and represent outgoing waves. A logical way to obtain such displacement functions is to investigate analytically the far field behavior of the half space.

Axisymmetric torsional oscillation of a homogeneous half space is one of the simplest wave propagation problems, since there is only one body wave and a single displacement component u_θ governed by Eq. 8.215. For sufficiently large r, the solution is

$$u_\theta \sim H_m^{(2)}(\eta_s r) \tag{8.282}$$

Instead, if the spherical Navier equations (Eqs. 3.68) are considered as the governing equations, then for sufficiently large R

$$u_\theta \sim h_m^{(2)}(\eta_s R) \tag{8.283}$$

A verification of Eqs. 8.282 and 8.283 can be achieved with the help of the approximate formulas for Bessel functions for large argument, Eqs. 1.121. These solutions are rather

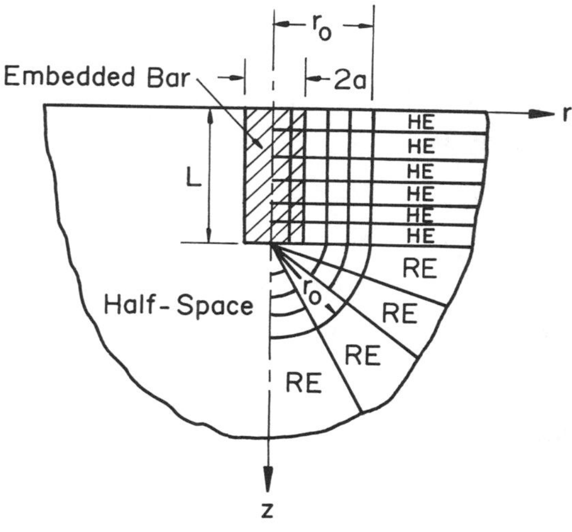

(a) Geometry.

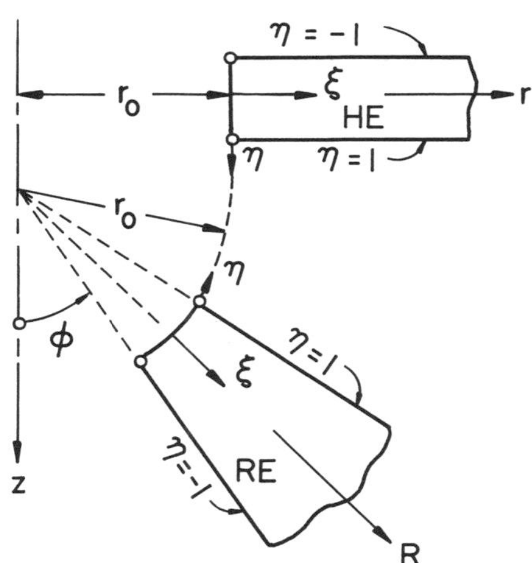

(b) Infinite elements, *HE* and *RE*.

Fig. 8.18 Discretization of material domains.

special and 'particular'. While they satisfy the equations of motion and the attenuating and outgoing wave requirements, they cannot satisfy the boundary conditions at the half space surface, since each is a function of a radial coordinate (r or R) only. Weighing advantages and disadvantages of these solutions against those of continuous spectra of waves (see Eqs. 8.270 and 8.273), Karasudhi *et al* favored the former. They used Eq. 8.282 for horizontal infinite elements and Eq. 8.283 for radial infinite elements. These equations imply that cylindrical wavefronts exist in a horizontal infinite element, and spherical wavefronts in a radiating infinite element. The waves in these respective infinite elements attenuate as $r^{-1/2}$ and R^{-1}, i.e. for a horizontal infinite element,

$$u_\theta^b \sim (\eta_s r)^{-1/2} e^{-i\eta_s r} \tag{8.284}$$

and for a radiating infinite element,

$$u_\theta^b \sim (\eta_s R)^{-1} e^{-i\eta_s R} \tag{8.285}$$

The superscript b denotes the body wave solution, which is also the complete solution for the axisymmetric torsion of a homogeneous half space. In order to enable simple analytical evaluation of infinite integrals of impedance and mass matrices of infinite elements, the displacement function for a horizontal infinite element was taken in the form

$$u_\theta^b \sim e^{-(1+i)\eta_s r} \tag{8.286a}$$

and for a radiating infinite element,

$$u_\theta^b \sim e^{-(1+i)\eta_s R} \tag{8.286b}$$

In general vibration of a homogeneous half space, each displacement component consists of a body wave part and a surface wave part. Following the scheme adopted for torsional vibration, we find the body wave displacement functions, for a horizontal infinite element,

$$u_r^b \sim e^{-(1+i)\eta_p r} \tag{8.287a}$$

$$u_\theta^b \sim e^{-(1+i)\eta_s r} \tag{8.287b}$$

$$u_z^b \sim e^{-(1+i)\eta_s r} \tag{8.287c}$$

and for a radiating infinite element,

$$u_R^b \sim e^{-(1+i)\eta_p R} \tag{8.288a}$$

$$u_\phi^b \sim e^{-(1+i)\eta_s R} \tag{8.288b}$$

$$u_\theta^b \sim e^{-(1+i)\eta_s R} \tag{8.288c}$$

Analytically, the surface wave displacement components can be obtained as

$$u_{rm}^s = -A_m(e^{-pz} - a_1 e^{-qz}) \frac{\partial H_m^{(2)}(\eta_R r)}{\partial(\eta_R r)} \tag{8.289a}$$

$$u_{\theta m}^s = m A_m(e^{-pz} - a_1 e^{-qz}) \frac{H_m^{(2)}(\eta_R r)}{\eta_R r} \tag{8.289b}$$

$$u^s_{zm} = \eta_R^{-1} A_m p (e^{-pz} - a_2 e^{-qz}) H_m^{(2)}(\eta_R r) \qquad (8.289c)$$

where the superscript s denotes the surface wave, m the harmonic in the Fourier series with respect to θ, A_m an arbitrary constant, η_R the Rayleigh surface wavenumber of the homogeneous half space, and

$$p = \alpha_s(\eta_R), \quad q = \alpha_p(\eta_R) \qquad (8.290a,b)$$

$$a_1 = 2pq/\bar{\alpha}_p(i\eta_R), \quad a_2 = 2\eta_R^2/\bar{\alpha}_p(i\eta_R) \qquad (8.290c,d)$$

The radical quantities in the equations above are as defined in Eqs. 8.167, 8.171 and 8.173. It should be noted that the origin in the coordinate system for Eqs. 8.289 is on the surface of the half space. Equations 8.289 may be assumed to take the following approximate form

$$u^s_{rm} = iA_m b_1(z) e^{-(1+i)\eta_R r} \qquad (8.291a)$$

$$u^s_{\theta m} \sim b_1(z) e^{-(1+i)\eta_R r} \qquad (8.291b)$$

$$u^s_{zm} = A_m b_2(z) e^{-(1+i)\eta_R r} \qquad (8.291c)$$

where

$$b_1(z) = (e^{-pz} - a_1 e^{-qz}) \qquad (8.292a)$$

$$b_2(z) = \eta_R^{-1} p (e^{-pz} - a_2 e^{-qz}) \qquad (8.292b)$$

Corresponding spherical displacement components for this surface wave can be obtained by a standard coordinate transformation as

$$u^s_{Rm} = A_m[ib_1(z)\sin\phi + b_2(z)\cos\phi]\, e^{-(1+i)\eta_R r} \qquad (8.293a)$$

$$u^s_{\phi m} = A_m[ib_1(z)\cos\phi - b_2(z)\sin\phi]\, e^{-(1+i)\eta_R r} \qquad (8.293b)$$

$$u^s_{\theta m} \sim b_1(z) e^{-(1+i)\eta_R r} \qquad (8.293c)$$

Assuming that the displacement components in Eqs. 8.291 and 8.293 are independent of each other, we obtain surface wave displacement components for a horizontal infinite element as

$$u^s_{rm} \sim b_1(z) e^{-(1+i)\eta_R r} \qquad (8.294a)$$

$$u^s_{\theta m} \sim b_1(z) e^{-(1+i)\eta_R r} \qquad (8.294b)$$

$$u^s_{zm} \sim b_2(z) e^{-(1+i)\eta_R r} \qquad (8.294c)$$

and for a radiating infinite element as

$$u^s_{Rm} \sim [ib_1(z)\sin\phi + b_2(z)\cos\phi]\, e^{-(1+i)\eta_R r} \qquad (8.295a)$$

$$u^s_{\phi m} \sim [ib_1(z)\cos\phi - b_2(z)\sin\phi]e^{-(1+i)\eta_R r} \tag{8.295b}$$

$$u^s_{\theta m} \sim b_1(z)e^{-(1+i)\eta_R r} \tag{8.295c}$$

Total displacement components for a horizontal infinite element can be written

$$u_{rm} = A^b_{rm}u^b_{rm} + A^s_{rm}u^s_{rm} \tag{8.296a}$$

$$u_{\theta m} = A^b_{\theta m}u^b_{\theta m} + A^s_{\theta m}u^s_{\theta m} \tag{8.296b}$$

$$u_{zm} = A^b_{zm}u^b_{zm} + A^s_{zm}u^s_{zm} \tag{8.296c}$$

and for a radiating infinite element

$$u_{Rm} = A^b_{Rm}u^b_{Rm} + A^s_{Rm}u^s_{Rm} \tag{8.297a}$$

$$u_{\phi m} = A^b_{\phi m}u^b_{\phi m} + A^s_{\phi m}u^s_{\phi m} \tag{8.297b}$$

$$u_{\theta m} = A^b_{\theta m}u^b_{\theta m} + A^s_{\theta m}u^s_{\theta m} \tag{8.297c}$$

Here

$$A^b_{rm}, A^s_{rm}, \ldots, A^s_{\theta m}$$

are arbitrary constants, and

$$u^b_{rm} \text{ to } u^b_{zm}, \qquad u^b_{Rm} \text{ to } u^b_{\theta m}$$

$$u^s_{rm} \text{ to } u^s_{zm}, \text{ and } u^s_{Rm} \text{ to } u^s_{\theta m}$$

are given by Eqs. 8.287, 8.288, 8.294 and 8.295, respectively. Considering each infinite element to have two nodes as shown in Fig. 8.18(b), we can express the unknown constants,

$$A^b_{rm} \text{ to } A^s_{zm} \text{ or } A^b_{Rm} \text{ to } A^s_{\theta m},$$

in terms of the displacement components and coordinates of the nodes.

8.14 EXERCISE PROBLEMS

8.14.1 Show detailed derivation for velocity jumps in isotropic elastic domains, from Eq. 8.84 to Eq. 8.102.

8.14.2 Show detailed derivation for waves in isotropic viscoelastic media of Section 8.5.

8.14.3 A homogeneous isotropic elastic half plane $-\infty < x < \infty$, $0 \le y < \infty$, is subjected to the following time harmonic tractions on its surface $y = 0$, $-a < x < a$: (a) a uniformly distributed normal force, (b) a normal force distributed in such a way that the displacement component in the y-direction on the contact area is uniform in the statical case (Eq. 6.66a). Find the relationship between the applied force and the contact normal displacement, and compare the results with those by Quinlan (1953).

8.14.4 With the same geometry and time harmonic functions as in the preceding problem, the displacement in the y-direction is prescribed uniformly in the region $-a < x < a$, while the shear traction is assumed to vanish everywhere on the half plane surface. Moreover, the normal traction vanishes outside the loaded area. Find the distribution of the contact normal stress in terms of the prescribed indention, and the relationship between the applied force and the indentation. Reference: Karasudhi, Keer and Lee (1968).

8.14.5 Solve the shear counterpart of the preceding problem, i.e. when the displacement in the x-direction is prescribed uniformly in the region $-a < x < a$, while the normal traction is assumed to vanish everywhere on the half plane surface.

8.14.6 Solve the rocking counterpart of Problem 8.14.4, i.e. when a uniform rocking (rotation about the z-axis) is prescribed in the region $-a < x < a$, while the shear traction is assumed to vanish everywhere on the half plane surface.

8.14.7 A homogeneous isotropic elastic half space $0 \leq r < \infty$, $0 \leq z < \infty$, is subjected to the following time harmonic tractions on its surface $z = 0$, $0 \leq r < a$: (a) a uniformly distributed normal force, (b) a normal force distributed in such a way that the displacement component in the z-direction on the contact area is uniform in the statical case (Eq. 6.78a). Find the relationship between the applied force and the contact normal displacement, and compare the results with those by Quinlan (1953) and Sung (1953).

8.14.8 With the same geometry and time harmonic functions as in the preceding problem, the displacement in the z-direction is prescribed uniformly in the region $0 \leq r < a$, while the shear tractions in the x and y-directions are assumed to vanish everywhere on the half space surface. Moreover, the normal traction vanishes outside the loaded area. Find the distribution of the contact normal stress in terms of the prescribed displacement, and the relationship between the applied force and the prescribed displacement. In other words, this is a three dimensional counterpart of Problem 8.14.4. Reference: Robertson (1966).

8.14.9 Formulate an analytical scheme for solution to general vibrations under few loading types of two homogeneous elastic half spaces (and half planes) perfectly bonded together.

8.14.10 Formulate an analytical scheme for solution to axisymmetric torsion and antiplane vibrations under few loading types of a layered elastic half space.

8.14.11 Formulate symbolically a solution scheme for axisymmetric (torsion free) and in-plane vibrations under few loading types of a layered elastic half space, then outline a practical algorithm to obtain results numerically.

8.14.12 Employ a few simple vibration cases of a homogeneous isotropic viscoelastic half space (or half plane) to verify various points made in Section 8.12.

CHAPTER IX

PLASTICITY

9.1 FACTS FROM SIMPLE TESTS

If a long metal rod is pulled in a testing machine, the normal stress in the direction of the applied load (σ) may be plotted against the normal strain in the direction of the elongation (ε) as shown in Fig. 9.1, in which OA^* denotes the *elastic range*, AB the *ideal plastic range*, BD the nonlinear *work-hardening range*, and DE 'necking' range. There is a small unstable region A^*A, but it will be ignored in this study, since no well accepted theories have been proposed as yet for unstable materials. In the ideal plastic range AB, the strain ε increases while the stress σ remains constant. There is a nonlinear rise of the curve beyond the point B until the *ultimate load* at the point D. A test

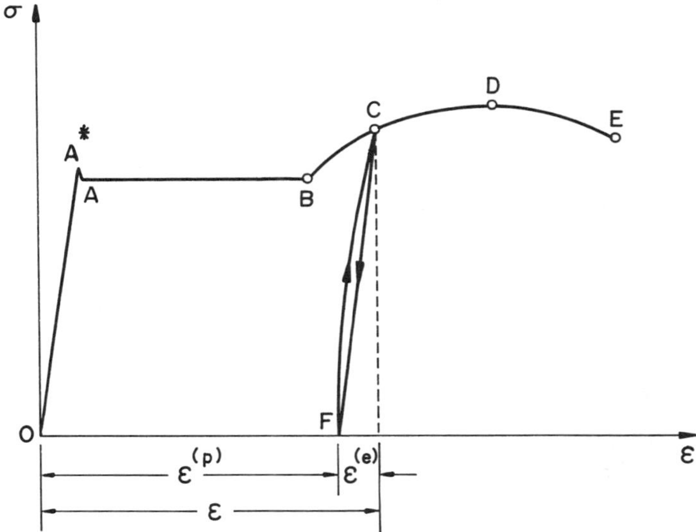

Fig. 9.1 Typical stress-strain curve in simple tension tests of elastic-ideal plastic-work hardening metals.

specimen may 'neck' at a certain strain, so that its cross-sectional area is reduced in a small region. When necking occurs under continued elongation, the load reaches a maximum (ultimate load) and then decreases, although the actual average stress at the necked cross-section continues to increase. The specimen metal flows beyond the point D and the specimen breaks at the point E. For mild steel, the flat yield region AB is large due to many microscopic slips along slip planes of the material crystal. For many other solids, the stress-strain curve may not have such a flat region (Fig. 9.2), or even not have an elastic range (Fig. 9.3). Materials like cast iron or any rock materials, which allow very little plastic deformation before reaching the breaking point, are called *brittle* materials. However, rocks tend to become ductile when subjected to large hydrostatic pressure (large negative mean stress σ_{ii}). This point of view was discovered by von Karman (1911) in his experiments with marbles.

Upon unloading at any stage in the deformation, the strain is reduced along an elastic path and reloading retraces the unloading path with minor deviation, and then further plastic deformation is produced when approximately the previously applied stress is exceeded. For better understanding of this point, we may refer to the complete unloading-reloading cycle CFC in Figs. 9.1 to 9.3. For all practical purposes, the unloading path CF and the reloading path FC are identical straight lines; parallel to the elastic line OA for Figs. 9.1 and 9.2, and parallel to the tangent at zero stress for Fig. 9.3. Thus we can see that, at any stage of deformation such as C in the figures, the total strain ε can be taken as

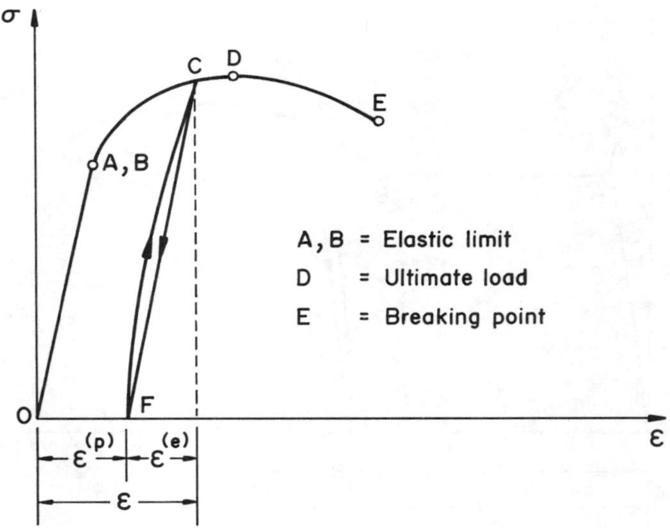

Fig. 9.2 Typical stress-strain curve in simple tension test of elastic-work hardening materials.

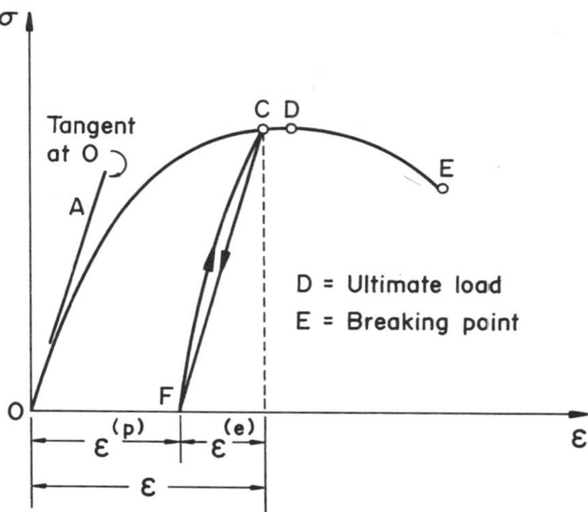

Fig. 9.3 Typical stress-strain curve in simple tension test of work hardening materials.

$$\varepsilon = \varepsilon^{(p)} + \varepsilon^{(e)} \tag{9.1}$$

where $\varepsilon^{(p)}$ is the *plastic strain* and *non-recoverable*; while $\varepsilon^{(e)}$ is the *elastic strain* and can be computed from the stress at that state by the Hooke's law. If we accept this fact (that the line CF is parallel to the elastic line OA), Eq. 9.1 holds also for a state C_1 below its yielding state C (Fig. 9.4), and the plastic strain does not change with the position C_1. However, such a state like C_2 which is above its yielding state C is *inadmissible*.

Tests of specimens subjected to simple compression and simple shear give diagrams similar to those of Figs. 9.1 to 9.4. Simple experiments on specimens under combined stresses have been mostly carried out on thin-walled circular cylinders; subjected to a combination of an axial tension, twisting moment about the cylinder axis, and an internal or external pressure. It is found that the shear stress, among other things, is by far the major cause of yielding. Moreover, experiments by Bridgman (1923) have shown that hydrostatic pressures of the order of the yield stress have practically no influence on yielding of metals. If a tensile test or shear test is run at atmospheric pressure in the standard manner, the stress-strain curve recorded will hardly differ in the small strain range from those obtained in immersion of the whole apparatus in a pressurized chamber. The major effect of the hydrostatic pressure is to increase greatly the ductility of the material, i.e. to permit much larger deformation before fracture. Conversely, very small changes in material density are found when a metal is subjected to repeated plastic deformation, indicating that the plastic volume change $\varepsilon_{ii}^{(p)}$ is small.

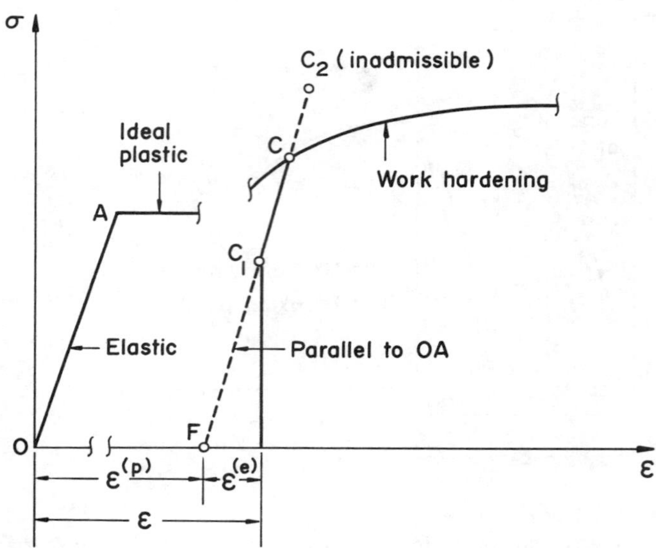

Fig. 9.4 No change in plastic strain below yielding, and inadmissible above-yielding state.

9.2 BASIC ASSUMPTIONS AND COMMON CHARACTERISTICS OF VARIOUS THEORIES

There are several theories of plasticity, but their common basic assumption, based on results of simple tests reviewed in the preceding section, can be listed for the general three-dimensional case as the following: the strain tensor can be expressed as the sum of the *plastic strain tensor* $\varepsilon_{ij}^{(p)}$ and the *elastic strain tensor* $\varepsilon_{ij}^{(e)}$, i.e.

$$\varepsilon_{ij}=\varepsilon_{ij}^{(p)}+\varepsilon_{ij}^{(e)} \tag{9.2}$$

By its meaning, $\varepsilon_{ij}^{(e)}$ is related to the stress tensor σ_{ij} by Hooke's law, i.e.

$$\sigma_{ij}=C_{ijkl}\varepsilon_{kl}^{(e)} \tag{9.3}$$

Following Eq. 9.2, we may write the dilatation and the strain deviation tensor, respectively, as

$$\varepsilon_{ii}=\varepsilon_{ii}^{(p)}+\varepsilon_{ii}^{(e)} \tag{9.4}$$

$$e_{ij}=e_{ij}^{(p)}+e_{ij}^{(e)} \tag{9.5}$$

With the assumption that the plastic volume change is negligible, we have

$$\varepsilon_{ii}^{(p)}=0 \tag{9.6}$$

and Eq. 9.4 becomes

$$\varepsilon_{ii} = \varepsilon_{ii}^{(e)} \tag{9.7}$$

Equations 9.3 and 9.7 imply that Hooke's law holds for the mean stress and the mean strain at all times, provided the material is isotropic, i.e.

$$\sigma_{ii} = 3K\varepsilon_{ii} \tag{9.8}$$

where K is the bulk modulus. However, Eq. 9.8 does not exist for the general case of anisotropic materials. By the definition of the strain deviator, Eq. 2.47, we can write

$$e_{ij}^{(p)} = \varepsilon_{ij}^{(p)} - \frac{1}{3}\varepsilon_{kk}^{(p)}\delta_{ij} \tag{9.9}$$

Substituting Eq. 9.6 into Eq. 9.9 we find

$$e_{ij}^{(p)} = \varepsilon_{ij}^{(p)} \tag{9.10}$$

which allows us to use the notation $\varepsilon_{ij}^{(p)}$ in places where we should write $e_{ij}^{(p)}$ to indicate a strain deviation.

A common essential characteristics of all theories of plasticity is that there exists a *yield function*

$$f(\sigma_{ij}, \varepsilon_{ij}^{(p)}, \kappa) \tag{9.11}$$

which depends on the state of stress and strain and on the history of loading denoted by the *work-hardening parameter* κ. The parameter κ is a function of the plastic strain tensor $\varepsilon_{ij}^{(p)}$. The equation $f = 0$ represents a closed surface in the nine-dimensional *stress space*, with axes σ_{ij} $(i, j = 1, 2, 3)$. When a state of stress can be represented by a point on the yield surface $f = 0$, change in plastic deformation occurs. For a state of stress with $f < 0$, the material is elastic (governed by Eq. 9.3), and there is no change in plastic deformation, i.e.

$$\frac{d\varepsilon_{ij}^{(p)}}{dt} \equiv \dot{\varepsilon}_{ij}^{(p)} = 0, \quad f < 0 \tag{9.12}$$

The state of stress where $f > 0$ has no meaning, i.e. it is *inadmissible*. The states $f = 0$, $f < 0$ and $f > 0$ correspond respectively to points C, C_1 and C_2 in Fig. 9.4, which is an illustration of one-dimensional stress.

At this stage, it is appropriate to consider the meaning of loading and unloading in a plastic state, where the *yield condition* governs, i.e.

$$f(\sigma_{ij}, \varepsilon_{ij}^{(p)}, \kappa) = 0 \tag{9.13}$$

The time derivative of the yield function is

$$\dot{f} = \frac{\partial f}{\partial \sigma_{ij}}\dot{\sigma}_{ij} + \frac{\partial f}{\partial \varepsilon_{ij}^{(p)}}\dot{\varepsilon}_{ij}^{(p)} + \frac{\partial f}{\partial \kappa}\dot{\kappa} \tag{9.14}$$

where $\dot{\kappa}$ is the time rate of κ, which is a function of $\varepsilon_{ij}^{(p)}$. Thus

$$\dot{\kappa} = \frac{\partial \kappa}{\partial \varepsilon_{ij}^{(p)}}\dot{\varepsilon}_{ij}^{(p)} \tag{9.15}$$

Obviously, $f=0$ and $\dot{f}<0$ at a time t would imply $f<0$ the next instant $t+dt$. Such a change leads to an elastic state and corresponds to *unloading*. However, for such a state, Eq. 9.12 applies, i.e. $\dot{\varepsilon}_{ij}^{(p)}=0$. Hence, due to Eq. 9.15, $\dot{\kappa}$ must also vanish. Thus, for *unloading*, $f=0$ at its onset, and

$$\dot{f}=\frac{\partial f}{\partial \sigma_{ij}}\dot{\sigma}_{ij}<0 \tag{9.16}$$

Otherwise, *loading* occurs, if

$$\frac{\partial f}{\partial \sigma_{ij}}\dot{\sigma}_{ij}>0, \quad f=0 \tag{9.17}$$

or *neutral loading*, if

$$\frac{\partial f}{\partial \sigma_{ij}}\dot{\sigma}_{ij}=0, \quad f=0 \tag{9.18}$$

The yield function f is sometime called a *loading function* because of its direct association with these loading criteria. Before making a geometric interpretation of these criteria (and for reasons to be discussed later), we have to choose a sign convention in writing Eq. 9.13 such that $\partial f/\partial \sigma_{ij}$ be directed along the outer normal to 'hypersurface' $f=0$. Then, for a state of stress on the yield surface; loading, unloading, or neutral loading takes place, according to whether the stress increment vector is directed outward, inward, or along a tangent to the yield surface. Since loading from one plastic state must lead to another plastic state, f remains zero and so does its time rate

$$\dot{f}\equiv\frac{\partial f}{\partial \sigma_{ij}}\dot{\sigma}_{ij}+\frac{\partial f}{\partial \varepsilon_{ij}^{(p)}}\dot{\varepsilon}_{ij}^{(p)}+\frac{\partial f}{\partial \kappa}\frac{\partial \kappa}{\partial \varepsilon_{ij}^{(p)}}\dot{\varepsilon}_{ij}^{(p)}=0 \tag{9.19}$$

For this reason, Eq. 9.13 is called *consistency condition* by Prager (1948).

For *ideal plastic* solids, f is not affected by $\varepsilon_{ij}^{(p)}$, thus during loading from a plastic state, Eq. 9.19 gives

$$\frac{\partial f}{\partial \sigma_{ij}}\dot{\sigma}_{ij}=0, \quad f=0 \tag{9.20}$$

A comparison of this equation with Eqs. 9.17 and 9.18 shows that loading from a plastic state for ideal plastic solids is always *neutral*.

A material is said to be *isotropic* if the yield function f depends only on the *invariants* (with respect to orientation of the reference coordinate system) of stress, strain and strain history. If f is an isotropic function of the stress alone, the theory is called an *isotropic stress theory* for ideal plastic solids. In such theories

$$f=f(I_1,I_2,I_3) \tag{9.21}$$

where I_1, I_2 and I_3 are the three invariants of stress tensor, i.e.

$$I_1=\sigma_{ii}=\sigma_1+\sigma_2+\sigma_3 \tag{9.22}$$

$$I_2=\frac{1}{2}(I_1^2-\sigma_{ij}\sigma_{ji})=\sigma_1\sigma_2+\sigma_2\sigma_3+\sigma_3\sigma_1 \tag{9.23}$$

$$I_3 = \frac{1}{6}(2\sigma_{ij}\sigma_{jk}\sigma_{ki} - 3I_1\sigma_{ij}\sigma_{ji} + I_1^3) = \sigma_1\sigma_2\sigma_3 \tag{9.24}$$

in which σ_1, σ_2 and σ_3 are principal stresses.

As mentioned in Section 9.1, it has been found by Bridgman (1923) that the plastic deformation of metals is essentially independent of hydrostatic pressure. The result has been verified by Crossland (1954) and many others. If this finding is accepted, the yield condition will be independent of I_1 and can be written as

$$f = f(J_2, J_3) \tag{9.25}$$

where J_2 and J_3 are the second and third deviatoric stress invariants (Eqs. 2.39 to 2.41), i.e.

$$J_2 = \frac{1}{2}s_{ij}s_{ij} = \frac{I_1^2}{3} - I_2 \tag{9.26a}$$

$$J_3 = \frac{1}{3}s_{ij}s_{jk}s_{ki} = I_3 - \frac{1}{3}I_1I_2 + \frac{2}{27}I_1^3 \tag{9.26b}$$

or, written as functions of differences in principal stresses, they are

$$J_2 = \frac{1}{6}\{[(\sigma_1 - \sigma_3) - (\sigma_2 - \sigma_3)]^2 + (\sigma_1 - \sigma_3)^2 + (\sigma_2 - \sigma_3)^2\} \tag{9.27}$$

$$J_3 = \frac{1}{81}\{[2(\sigma_1 - \sigma_3) - (\sigma_2 - \sigma_3)]^3 + [2(\sigma_2 - \sigma_3) - (\sigma_1 - \sigma_3)]^3 - [(\sigma_1 - \sigma_3) + (\sigma_2 - \sigma_3)]^3\} \tag{9.28}$$

The yield surface $f(J_2, J_3) = 0$ can be represented by a cylinder in a three dimensional space with σ_1, σ_2, σ_3 as coordinate axes. The axis of the cylinder is equally inclined to the coordinate axes, i.e. perpendicular to the so-called *π-plane* where $\sigma_1 + \sigma_2 + \sigma_3 = 0$. If J_3 does not appear in f, the yield surface $f(J_2) = 0$ in the *principal stress space* will be a circular cylinder, i.e. its projection on the π-plane is a circle. If J_3 appears in f, then the cross-section of the yield surface $f(J_2, J_3) = 0$ on the π-plane will not be circular. A yield surface in the three dimensional principal stress space is shown in Fig. 9.5, while projection of two yield surfaces on the π-plane are shown in Fig. 9.6. Due to Eqs. 9.27 and 9.28, a two-dimensional plot of the surface $f(J_2, J_3) = 0$ with $\sigma_1 - \sigma_3$ and $\sigma_2 - \sigma_3$ as coordinate axes is also possible, as shown for example in Fig. 9.7.

9.3 VARIOUS YIELD FUNCTIONS

With permission of the Publisher, Prentice-Hall, a significant amount of material in this section was adapted from the book by Fung (1965).

H. Tresca (1868) stipulated that the maximum shear stress, during the plastic flow, must have the constant value k, the yield stress in simple shear. Thus the Tresca's yield function is

$$f = \sigma_1 - \sigma_3 - 2k \tag{9.29}$$

where σ_1 is the maximum principal stress and σ_3 is the minimum principal stress, i.e.

$$\sigma_1 \geq \sigma_2 \geq \sigma_3.$$

However, f in the form of Eq. 9.29 is not analytic; since it violates the rule that the labelling of the principal axes 1, 2, 3 should not affect the form of the yield function. To conform with this rule, we note that during plastic flow one of the differences $|\sigma_1 - \sigma_2|, |\sigma_2 - \sigma_3|, |\sigma_3 - \sigma_1|$ has the value $2k$. Hence, we may put f in a form symmetrical with respect to principal stresses as

$$f = [(\sigma_1 - \sigma_2)^2 - 4k^2]\,[(\sigma_2 - \sigma_3)^2 - 4k^2]\,[(\sigma_3 - \sigma_1)^2 - 4k^2] \tag{9.30}$$

or, due to Reuss, as

$$f = 4J_2^3 - 27J_3^2 - 36k^2J_2^2 + 96k^4J_2 - 64k^6 \tag{9.31}$$

When the direction and relative magnitudes of the principal stresses are known by symmetry consideration or by intuition, it is possible to use the simple form of Eq. 9.29. Otherwise, we are obliged to use the general form of Eq. 9.31.

In 1913, Richard von Mises assumed that the yield function f is independent of J_3 and takes the simple form

$$f = J_2 - k^2 \tag{9.32}$$

It can be shown that the yield criteria of von Mises and Tresca give the same yield stress in simple shear; but, in simple tension, the yield stress by von Mises' criterion is smaller than that predicted by Tresca's, by a factor $\sqrt{3}/2$. For no other type of stresses is the discrepancy between the predictions of the yield stresses by von Mises and by Tresca's criteria as large as that in simple tension (von Mises, 1913).

As illustrative examples of the yield surface $f = 0$ when f is not affected by hydrostatic pressure, the von Mises' criterion is plotted in the principal stress space as shown in Fig. 9.5, the von Mises' and Tresca's yield surfaces projected on π-plane in Fig. 9.6, and the two-dimensional plotting of these two yield surfaces in Fig. 9.7.

A more general form of f, in isotropic stress theory for ideal plastic solids which is unaffected by hydrostatic pressure, is

$$f = F(J_2, J_3) - k^2 \tag{9.33}$$

A simple example is

$$F = J_2^3 - CJ_3^2 \tag{9.34}$$

A good correlation with Osgood's experimental results (1947) is obtained by taking $C = 2.25$.

To reveal and preserve initial anisotropy, we may use

$$f = D_{ijkl}\sigma_{ij}\sigma_{kl} - k^2 \tag{9.35}$$

Here D_{ijkl} is composed of a set of 21 independent constants, due to the symmetry conditions

$$D_{ijkl} = D_{klij} = D_{jikl} = D_{ijlk}.$$

If incompressibility is assumed, the number of independent constants is reduced to 15.

When a metal rod is subjected to repeated tension-compression tests, the stress-strain curve sometimes appears as in Fig. 9.8. The tension stroke and the compression stroke are dissimilar. This is referred to as the *Bauschinger effect* (J. Bauschinger, 1886). Von Mises' and Tresca's yield functions discussed in the preceding paragraphs imply the initial isotropy, and the equality of tensile and compressive yield stresses at all stages of the deformation, i.e. no Bauschinger effect.

To include Bauschinger effect but preserve isotropy, some yield functions have been suggested for work hardening solids (by including $\varepsilon_{ij}^{(p)}$) as follows.

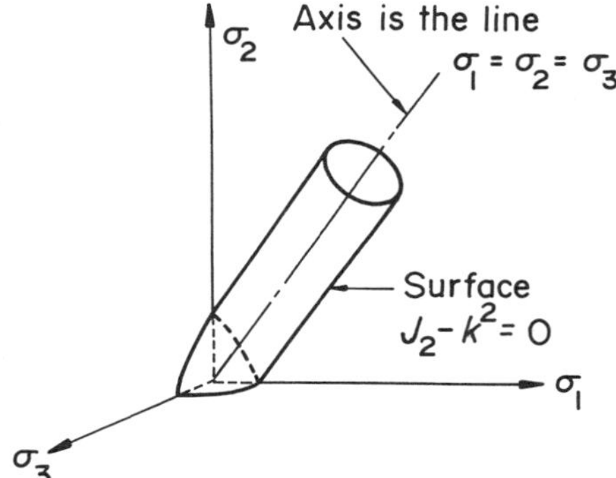

Fig. 9.5 A yield surface in the principal stress space.

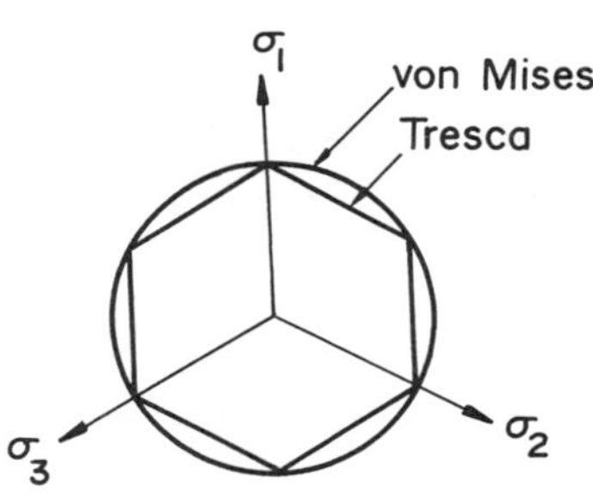

Fig. 9.6 Projection of yield surfaces on the π-plane.

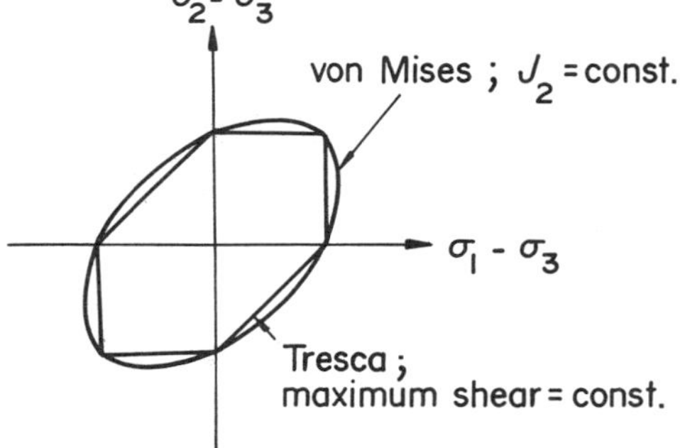

Fig. 9.7 Yield surface plotted on the plane of $(\sigma_1 - \sigma_3)$, $(\sigma_2 - \sigma_3)$.

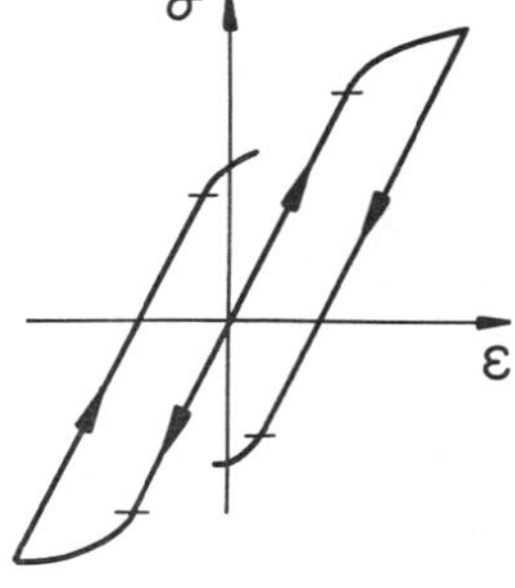

Fig. 9.8 Bauschinger effect in a simple tension-compression test.

$$f = F(J_2) - m s_{ij}\varepsilon_{ij}^{(p)} - k^2, \quad (m\text{=a constant}) \tag{9.36}$$

$$f = F(J_2, J_3) - m s_{ij}\varepsilon_{ij}^{(p)} - k^2 \tag{9.37}$$

$$f = F(J_2, J_3) - H(J_2, J_3) s_{ij}\varepsilon_{ij}^{(p)} - k^2 \tag{9.38}$$

$$f = F(J_2, J_3) - [P(J_2, J_3) s_{ij} + Q(J_2, J_3) t_{ij}]\varepsilon_{ij}^{(p)} - k^2 \tag{9.39}$$

where t_{ij} is deviator of the square of the stress deviator, i.e.

$$t_{ij} = s_{ik}s_{kj} - \frac{2}{3}J_2\delta_{ij} = \frac{\partial J_3}{\partial \sigma_{ij}} \tag{9.40}$$

Finally, the expression

$$f = D_{ijkl}(s_{ij} - m\varepsilon_{ij}^{(p)})(s_{kl} - m\varepsilon_{kl}^{(p)}) - k^2 \tag{9.41}$$

yields the Bauschinger effect, and does not preserve the initial anisotropy during deformation.

It should be noted that each of these suggested forms uses a single analytic function to represent the entire yield surface. If a yield surface is composed of piecewise smooth surfaces which meet to form corners, it would be convenient to use a separate expression for each of these piecewise smooth surfaces. This concept was generalized by Koiter (1953, 1960) and will be discussed in Section 9.4.

9.4 HARDENING AND FLOW RULES

With permission of the publisher, Prentice-Hall, a significant amount of material in this section was adapted from the book by Fung (1965).

Work hardening in a simple tension experiment means that the stress is a monotonically increasing function of increasing strain, as illustrated by points C and C' on the stress-strain curve in Fig. 9.9. In other words, when the stress σ increases by a small amount $d\sigma$, the plastic state $C(\sigma, \varepsilon)$ is transferred to another plastic state $C'(\sigma + d\sigma, \varepsilon + d\varepsilon)$, and

$$d\sigma d\varepsilon > 0 \tag{9.42}$$

On removing the stress $d\sigma$, the transfer will follow the elastic path $C'F'$, and the system will arrive at the elastic state $C''(\sigma, \varepsilon + d\varepsilon^{(p)})$. It is obvious from the figure that

$$d\sigma d\varepsilon^{(p)} > 0 \tag{9.43}$$

which shows that only the elastic part of the strain, $d\varepsilon^{(e)}$, is recovered upon the completion of a loading cycle. The concept has been generalized by Drucker (1951) and can be stated as: Suppose, at a point in the material domain with a state of stress σ_{ij} and strain ε_{ij}, an *external agency* increases the stress by $d\sigma_{ij}$ and produces a small strain increment $d\varepsilon_{ij}$. On the removal of the stress $d\sigma_{ij}$, an elastic strain $d\varepsilon_{ij}^{(e)}$ is recovered. Then, the material is said to be *work hardening* or *strain hardening* if the following

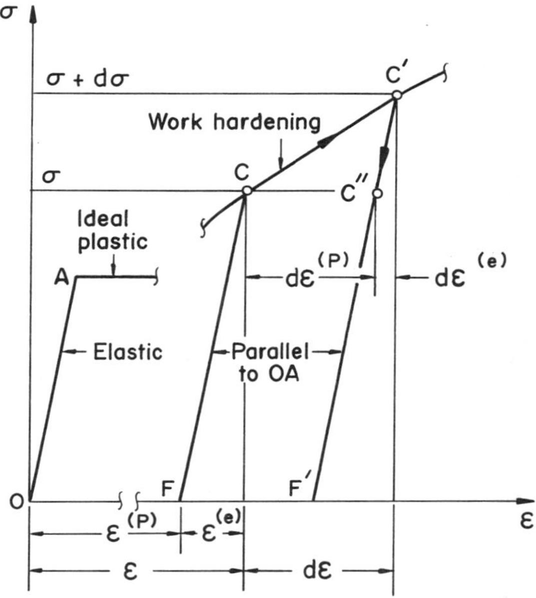

Fig. 9.9 Illustration of work-hardening by a simple tension test.

two conditions hold true.

$$d\sigma_{ij} d\varepsilon_{ij} > 0, \qquad \text{upon loading} \tag{9.44}$$

$$d\sigma_{ij}(d\varepsilon_{ij} - d\varepsilon_{ij}^{(e)}) \equiv d\sigma_{ij} d\varepsilon_{ij}^{(p)} > 0, \qquad \text{upon load removal} \tag{9.45}$$

If we accept Drucker's definition of strain hardening, we will say that there is *ideal plastic deformation* when the equality sign prevails in Eq. 9.45, i.e.

$$d\sigma_{ij} d\varepsilon_{ij}^{(p)} = 0 \tag{9.46}$$

and the yield function is independent of $\varepsilon_{ij}^{(p)}$. The differentials $d\sigma_{ij}$ and $d\varepsilon_{ij}^{(p)}$ must be interpreted as in Fig. 9.9.

In addition to having the yield function f to furnish a criterion when yielding occurs, we need further information on the increment or rate of deformation in order to complete our description of the material behavior. In other words, we need to derive the *plastic flow rule*. Equation 9.20 may be recalled for this purpose and rewritten as

$$\frac{\partial f}{\partial \sigma_{ij}} d\sigma_{ij} = 0 \tag{9.47}$$

A comparison of Eqs. 9.46 and 9.47 leads to

$$d\varepsilon_{ij}^{(p)} = \lambda \frac{\partial f}{\partial \sigma_{ij}} \tag{9.48}$$

where λ is an *arbitrary* constant of proportionality. Alternatively, Eq. 9.48 can be written in terms of the rate of strain as

$$\dot{\varepsilon}_{ij}^{(p)} = \Lambda \frac{\partial f}{\partial \sigma_{ij}} \tag{9.49}$$

The sign of λ or Λ is restricted by the condition that plastic flow must involve dissipation of energy, i.e.

$$\dot{W} \equiv \sigma_{ij}\dot{\varepsilon}_{ij}^{(p)} > 0 \tag{9.50}$$

Equation 9.48, the rule we are seeking for the plastic flow of ideal plastic solids, was first given by von Mises (1928), and is known as the *theory of the plastic potential.*

An interesting geometric interpretation of Eq. 9.49 has been given by Prager (1948). It is that the plastic strain rate $\dot{\varepsilon}_{ij}^{(p)}$ is a normal vector to the surface $f=0$ in the nine-dimensional stress space with $\sigma_{ij}(i,j=1,2,3)$ as coordinates.

It should be noted that a direct application of the flow rule in a one-dimensional stress experiment could be difficult, since the plastic flow will continue when the stress reaches its yielding limit, with no possibility of increasing the stress beyond that limit. To deduce some useful information from an experiment, it should be a combined-stress system, such as a thin-walled circular cylinder subjected to a twisting moment and to a tension or internal pressure. Such arrangement gives non-trivial changes of $d\sigma_{ij}$ and $d\varepsilon_{ij}$, to which the preceding derivation applies.

Koiter (1953, 1960) generalized the flow rule by allowing the yield criterion to be specified by several yield functions,

$$f_1(\sigma_{ij}), \quad f_2(\sigma_{ij}), \quad \ldots, \quad f_n(\sigma_{ij}) \tag{9.51}$$

We have an elastic state if all these functions are negative. At a plastic state, at least one yield function vanished. It is inadmissible to have any yield functions greater than zero. If functions $f_h = \ldots = f_m = 0$, whereas all other f's are negative, the Koiter's generalization of the flow rule (Eq. 9.48) is

$$\dot{\varepsilon}_{ij}^{(p)} = \Lambda_h \frac{\partial f_h}{\partial \sigma_{ij}} + \ldots + \Lambda_m \frac{\partial f_m}{\partial \sigma_{ij}} \tag{9.52}$$

where $\Lambda_h, \ldots, \Lambda_m$ are arbitrary nonnegative constants of proportionality. Thus, the basic concept leads at once to a general *incremental* stress-strain relationship in ideal plasticity.

For von Mises' yield function, Eq. 9.32, the flow rule, Eq. 9.49, gives

$$\dot{\varepsilon}_{ij}^{(p)} = \Lambda s_{ij} \tag{9.53}$$

For Tresca's yield condition, the multi-yield functions may be chosen as

$$f_1 = \sigma_2 - \sigma_3 - 2k, \qquad f_2 = \sigma_3 - \sigma_1 - 2k, \qquad f_3 = \sigma_1 - \sigma_2 - 2k,$$

$$f_4 = -(\sigma_2 - \sigma_3) - 2k, \quad f_5 = -(\sigma_3 - \sigma_1) - 2k, \quad f_6 = -(\sigma_1 - \sigma_2) - 2k \tag{9.54}$$

For example, if $f_1 = 0$, while all other f's are negative, then

$$\dot{\varepsilon}_1^{(p)} = 0, \quad \dot{\varepsilon}_2^{(p)} = \Lambda_1, \quad \dot{\varepsilon}_3^{(p)} = -\Lambda_1,$$

where $\Lambda_1 > 0$, and ε_1, ε_2, ε_3 are the principal strains, the directions of which coincide with those of the principal stresses under Tresca's condition of isotropy.

Following Drucker's definition of work hardening (or strain hardening) materials, we can prove the following:

(1) The yield (or loading) surface must be convex.

(2) The plastic strain increment must be a vector normal to the yield surface at a smooth point on the surface, and it must lie between adjacent normals to the yield surface at a corner of the surface.

(3) The plastic strain rate must be a linear function of the stress rate.

In a short statement; the above three points are often referred to as *convexity*, *normality*, and *linearity*, respectively. The statement is often interpreted as the requirement for stable materials. The yield function and the yield surface assume the general forms, Eqs. 9.11 and 9.13, respectively. Moreover, the yield surface should be taken as a *hypersurface* (in the nine-dimensional stress space). The proof of these points can be found in Naghdi (1960). Only a simple proof of linearity would be presented here, i.e. the *flow rule* will be derived.

The normality of the plastic strain rate at a smooth point of the yield surface implies that

$$\dot{\varepsilon}_{ij}^{(p)} = \Lambda \frac{\partial f}{\partial \sigma_{ij}} \tag{9.55}$$

where Λ is a function, in general, of stress, stress rate, strain, and strain history. Since $\dot{\varepsilon}_{ij}^{(p)}$ is associated with a positive work done during loading, we can show that Λ must be nonnegative. Substituting Eq. 9.55 into Eq. 9.19, we can solve for Λ. Eventually, Eq. 9.55 becomes

$$\dot{\varepsilon}_{ij}^{(p)} = \hat{G} \frac{\partial f}{\partial \sigma_{ij}} \frac{\partial f}{\partial \sigma_{kl}} \dot{\sigma}_{kl} \tag{9.56}$$

where

$$\hat{G} = -\left[\left(\frac{\partial f}{\partial \varepsilon_{mn}^{(p)}} + \frac{\partial f}{\partial \kappa} \frac{\partial \kappa}{\partial \varepsilon_{mn}^{(p)}} \right) \frac{\partial f}{\partial \sigma_{mn}} \right]^{-1} \tag{9.57}$$

Equation 9.56 proves the *linearity*. These results were first obtained by Prager (1948) and Drucker (1959).

For a yield surface being composed of a number (say n) of smooth surfaces f_p's which meet to form corners, Koiter's generalization (1953, 1960) gives the plastic strain rates as follows

$$\dot{\varepsilon}_{ij}^{(p)} = \sum_{p=1}^{n} C_p h_p \frac{\partial f_p}{\partial \sigma_{ij}} \frac{\partial f_p}{\partial \sigma_{kl}} \dot{\sigma}_{kl} \tag{9.58}$$

where

$$C_p = 0; \quad \text{if } f_p < 0, \text{ or } \frac{\partial f_p}{\partial \sigma_{ij}} \dot{\sigma}_{ij} < 0 \tag{9.59a}$$

$$C_p = 1; \quad \text{if } f_p = 0, \quad \text{and } \frac{\partial f_p}{\partial \sigma_{ij}} \dot{\sigma}_{ij} \geq 0 \tag{9.59b}$$

and h_p's are positive functions of stress, strain, and strain history. Conditions in Eqs. 9.59 specify, of course, whether yielding and/or loading occur as governed by Eqs. 9.16 to 9.18.

We may now wonder what possible forms the plastic strain $\varepsilon_{ij}^{(p)}$ can have in the yield function $f(\sigma_{ij}, \varepsilon_{ij}^{(p)}, \kappa)$. Laws governing this aspect are called *hardening rules*. If they are independent of the hydrostatic pressure, the yield surfaces in the principal stress space are cylinders of infinite length perpendicular to the π-plane $\sigma_1 + \sigma_2 + \sigma_3 = 0$, and characterized by their cross-section on this plane. We know also that these cross-sectional curves are closed, convex and piecewise smooth, but they could change in size and shape as plastic deformation proceeds. Several proposed hardening rules characterized by the π-plane cross-sectional curves are illustrated in Fig. 9.10. In Fig. 9.10(a) Tresca's initial yield surface is shown as a regular hexagon on the π-plane. In Fig. 9.10(b) is shown the so-called *isotropic hardening*, which assumes a uniform expansion of the initial yield surface. Isotropic hardening, though conveniently used in analysis, has little experimental support. Figure 9.10(c) illustrates Prager's *kinematic hardening*. The initial yield surface translates in the π-plane without rotation and without change in size. None of the flow rules deduced from Drucker's hypothesis is violated, the Bauschinger effect can be represented very simply, and the development of anisotropy due to plastic deformation appears most naturally. With some variation in the model, Prager (1954) was able to represent various models of plasticity: rigid perfectly plastic, rigid work-hardening, elastic, etc. Hodge (1956) pointed out that the concept of kinematic hardening must be applied in nine-dimensional stress space. Modified kinematic hardening by Ziegler (1959) is illustrated in Fig. 9.10(d). In Fig. 9.10(e) the plastic deformation causes a linear segment to move. In Fig. 9.10(f) the plane loading surface changes with plastic loading in some independent manner. More complicated hardening rules have been proposed from time to time, e.g. Hodge (1957) has extended the kinematic hardening to include an expansion of the yield surface simultaneously with its translation.

In contrast to the *incremental theories of plasticity* discussed so far, there is another group called *deformation theories of plasticity*. The difference lies in the flow rule. The characterizing feature of the deformation theory is the unique relationship between the plastic strain and the stress deviator in the plastic state; there is no such uniqueness in the incremental theories. The plastic strain as a unique function of the stress deviator given in Prager's form is

$$\varepsilon_{ij}^{(p)} = P s_{ij} + Q t_{ij},$$

where t_{ij} is as defined in Eq. 9.40. For isotropic materials P and Q are naturally functions of the invariants J_2 and J_3. It can be shown that Prager's form is the most

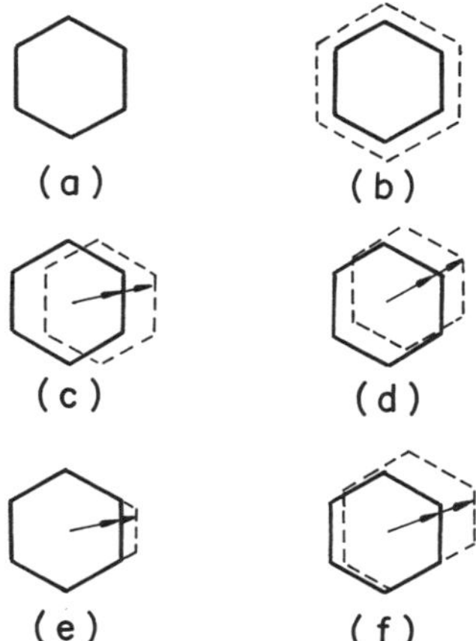

Fig. 9.10 Several hardening rules: (a) initial yield condition (Tresca), (b) isotropic hardening, (c) kinematic hardening (Prager), (d) kinematic hardening (Ziegler's modification), (e) independently acting plane loading surfaces, (f) interdependent plane loading surfaces. (From Naghdi, 1960).

general form under the assumption that the plastic strain can be expressed in powers of s_{ij}. Different choices of the yield function $f(s_{ij}, \varepsilon_{ij}^{(p)}, \kappa)$, functions P and Q, and the hardening parameter κ lead to different theories of plasticity. Deformation theories were developed by Hencky (1924) and Nadai (1950, 1963) and their followers. Handelman, Lin and Prager (1947) pointed out some conceptual difficulties of deformation theories. In some problems, a deformation theory yields the same answer as the corresponding incremental theory. Such is the case where in the loading process the ratio of the stress components

$$\sigma_{11} : \sigma_{22} : \sigma_{33} : \sigma_{12} : \sigma_{23} : \sigma_{31}$$

is kept constant at all times. For such problems the incremental flow rule can be integrated to yield the deformation rule. For a much wider class of loading, Budiansky (1959) and Kliushnikov (1958) independently showed that the deformation rule can be derived from the incremental rule provided that 'corners' develop in the loading surface as plastic deformation proceeds.

9.5 INCREMENTAL FORMULATION FOR ISOTROPIC HARDENING

The work increment per unit volume in the transfer from the plastic state of stress σ_{ij} and strain ε_{ij} to another plastic state of stress $\sigma_{ij} + d\sigma_{ij}$ and strain $\varepsilon_{ij} + d\varepsilon_{ij}$ (see the illustration of one-dimensional stress in Fig. 9.9) is

$$dW = \sigma_{ij} d\varepsilon_{ij} \tag{9.60}$$

which may be split into elastic and plastic parts

$$dW = dW^{(e)} + dW^{(p)} \tag{9.61}$$

where

$$dW^{(e)} = \sigma_{ij} d\varepsilon_{ij}^{(e)} \tag{9.62}$$

$$dW^{(p)} = \sigma_{ij} d\varepsilon_{ij}^{(p)} = s_{ij} d\varepsilon_{ij}^{(p)} \tag{9.63}$$

In writing the last equation, we have used the condition of plastic incompressibility, Eq. 9.6.

For *isotropic stress theory* of materials obeying *work-hardening hypothesis*, the yield function takes the form

$$f = f(J_2, J_3, W^{(p)}) \tag{9.64}$$

where $W^{(p)}$ is the total plastic work, i.e.

$$W^{(p)} = \int dW^{(p)} = \int s_{ij} d\varepsilon_{ij}^{(p)} \tag{9.65}$$

As the plastic flow proceeds, the yield surface f (in the nine-dimensional stress space) enlarges, as measured in the direction of its normal, by a distance

$$df = \frac{\partial f}{\partial \sigma_{ij}} d\sigma_{ij} \tag{9.66}$$

or, because of Eq. 9.55,

$$df = \Lambda^* d\varepsilon_{ij}^{(p)} d\sigma_{ij} \tag{9.67}$$

If the *Prandtl-Reuss equation* (Prandtl, 1924 and Reuss, 1930) holds, i.e.

$$d\varepsilon_{ij}^{(p)} = \chi s_{ij} \tag{9.68}$$

Eq. 9.67 can be rewritten as

$$df = \mathrm{T} dJ_2 \tag{9.69}$$

If T is a function of stress invariants, Eq. 9.69 shows that the yield function expands uniformly, i.e. there is an *isotropic hardening*.

Introducing the *equivalent* or *effective stress* $\bar{\sigma}$ as

$$\bar{\sigma} = (3J_2)^{1/2} = (3s_{ij}s_{ij}/2)^{1/2} \tag{9.70}$$

and, in a similar fashion, the *equivalent* or *effective plastic strain increment* $d\bar{\varepsilon}^{(p)}$ as

$$d\bar{\varepsilon}^{(p)} = (2d\varepsilon_{ij}^{(p)} d\varepsilon_{ij}^{(p)}/3)^{1/2} \tag{9.71}$$

we can have, due to Eq. 9.68,

$$d\bar{\varepsilon}^{(p)} = 2\chi\bar{\sigma}/3 \tag{9.72}$$

Furthermore, the *plastic work increment* $dW^{(p)}$ in Eq. 9.63 can be rewritten as

$$\mathrm{d}W^{(p)} = 2\chi(\bar{\sigma})^2/3 \tag{9.73}$$

$$= \bar{\sigma}\mathrm{d}\bar{\varepsilon}^{(p)} \tag{9.74}$$

With the assumption of the independence of f from J_3 and the existence of a *universal plastic stress-strain curve* relating $\bar{\sigma}$ and $\bar{\varepsilon}^{(p)}$, the total plastic work $W^{(p)}$ becomes a single-valued function of $\bar{\varepsilon}^{(p)}$, and Eq. 9.64 is simplified into

$$f = f[\bar{\sigma}, H(\bar{\varepsilon}^{(p)})] \tag{9.75}$$

which is known as the *strain-hardening hypothesis* for isotropic strain-hardening materials. Equations 9.56 and 9.57 for the present case can be written, respectively as

$$\mathrm{d}\varepsilon_{ij}^{(p)} = \hat{G}\left(\frac{\partial f}{\partial \bar{\sigma}}\right)^2 \frac{\partial \bar{\sigma}}{\partial \sigma_{ij}} \frac{\partial \bar{\sigma}}{\partial \sigma_{kl}} \mathrm{d}\sigma_{kl} \tag{9.76}$$

$$\hat{G} = -\left[\frac{\partial f}{\partial H} H' \frac{\partial \bar{\varepsilon}^{(p)}}{\partial \varepsilon_{mn}^{(p)}} \frac{\partial f}{\partial \bar{\sigma}} \frac{\partial \bar{\sigma}}{\partial \sigma_{mn}}\right]^{-1} \tag{9.77}$$

where

$$H' = \frac{\mathrm{d}H}{\mathrm{d}\bar{\varepsilon}^{(p)}} \tag{9.78}$$

Using Eq. 9.70, we get

$$\frac{\partial \bar{\sigma}}{\partial \sigma_{mn}} = \frac{3}{2} \frac{s_{mn}}{\bar{\sigma}} \tag{9.79}$$

Furthermore, dividing Eq. 9.72 by Eq. 9.68 we obtain

$$\frac{\partial \bar{\varepsilon}^{(p)}}{\partial \varepsilon_{mn}^{(p)}} = \frac{2}{3} \frac{\bar{\sigma}}{s_{mn}} \tag{9.80}$$

Thus Eq. 9.77 becomes

$$\hat{G} = -\left(\frac{\partial f}{\partial H} H' \frac{\partial f}{\partial \bar{\sigma}}\right)^{-1} \tag{9.81}$$

For materials obeying the Mises' yield condition, f may be taken as

$$f = \bar{\sigma} - H \tag{9.82}$$

Thus Eqs. 9.81 and 9.76 become, respectively,

$$\hat{G} = \frac{1}{H'} \tag{9.83}$$

$$\mathrm{d}\varepsilon_{ij}^{(p)} = \frac{9 s_{ij} s_{kl} \mathrm{d}s_{kl}}{H'(2\bar{\sigma})^2} \tag{9.84}$$

Due to Eq. 9.70, $s_{kl}\mathrm{d}s_{kl}$ can be written as

$$s_{kl}\mathrm{d}s_{kl} = \frac{2}{3}\bar{\sigma}\mathrm{d}\bar{\sigma} \tag{9.85}$$

Substituting Eq. 9.85 into Eq. 9.84 we find

$$d\varepsilon_{ij}^{(p)} = \frac{3}{2}\frac{s_{ij}d\bar{\sigma}}{H'\bar{\sigma}} \tag{9.86}$$

Numerical values of H' can be obtained easily as the slope of the stress versus plastic strain curve in uniaxial tension.

As expressed in Eq. 9.3, the stress σ_{ij} is related to the elastic strain $\varepsilon_{ij}^{(e)}$ according to Hooke's law, at all times. The same condition is applicable to their increments, i.e. for *isotropic materials*,

$$d\varepsilon_{ij}^{(e)} = \frac{1-2\nu}{3E}d\sigma_{kk}\delta_{ij} + \frac{ds_{ij}}{2\mu} \tag{9.87}$$

where E, μ and ν are Young's modulus, shear modulus, and Poisson's ratio, respectively. However, the total strain increment is the sum of elastic and plastic increments, i.e.

$$d\varepsilon_{ij} = d\varepsilon_{ij}^{(e)} + d\varepsilon_{ij}^{(p)} \tag{9.88}$$

Substituting Eqs. 9.86 and 9.87 into Eq. 9.88 we find the total strain increment tensor in terms of the stress increment tensor

$$d\varepsilon_{ij} = \frac{1-2\nu}{3E}d\sigma_{kk}\delta_{ij} + \frac{ds_{ij}}{2\mu} + \frac{3s_{ij}d\bar{\sigma}}{2H'\bar{\sigma}} \tag{9.89}$$

Multiplying Eq. 9.89 by s_{ij}, and using Eqs. 9.70 and 9.85, we find

$$s_{kl}d\varepsilon_{kl} = \bar{\sigma}d\bar{\sigma}\left(\frac{1}{H'} + \frac{1}{3\mu}\right)$$

or

$$\frac{d\bar{\sigma}}{H'\bar{\sigma}} = \frac{1}{(\bar{\sigma})^2}\frac{s_{kl}d\varepsilon_{kl}}{1+\frac{H'}{3\mu}} \tag{9.90}$$

Substituting Eq. 9.90 into Eqs. 9.86 and 9.89 leads to, respectively,

$$d\varepsilon_{ij}^{(p)} = \frac{3}{2(\bar{\sigma})^2}\frac{s_{ij}s_{kl}d\varepsilon_{kl}}{1+\frac{H'}{3\mu}} \tag{9.91}$$

$$d\varepsilon_{ij} = \frac{1-2\nu}{3E}d\sigma_{mm}\delta_{ij} + \frac{ds_{ij}}{2\mu} + \frac{3}{2(\bar{\sigma})^2}\frac{s_{ij}s_{kl}d\varepsilon_{kl}}{1+\frac{H'}{3\mu}} \tag{9.92}$$

Alternatively, the relationship between the stress increment tensor and the elastic strain increment tensor can be put in the form of the inverse of Eq. 9.87 as

$$d\sigma_{ij} = \frac{E}{3(1-2\nu)}d\varepsilon_{kk}^{(e)}\delta_{ij} + 2\mu de_{ij}^{(e)} \tag{9.93}$$

or, in view of Eqs. 9.7, 9.10 and 9.88,

$$d\sigma_{ij} = \frac{E}{3(1-2\nu)}d\varepsilon_{kk}\delta_{ij} + 2\mu de_{ij} - 2\mu d\varepsilon_{ij}^{(p)} \tag{9.94}$$

Substituting Eq. 9.91 into Eq. 9.94 we find the relationship between the stress increment tensor and the total strain increment tensor

$$d\sigma_{ij} = \frac{E}{3(1-2\nu)} d\varepsilon_{mm}\delta_{ij} + 2\mu de_{ij} - \frac{3\mu}{(\bar{\sigma})^2}\frac{s_{ij}s_{kl}d\varepsilon_{kl}}{1+\frac{H'}{3\mu}} \tag{9.95}$$

Equation 9.92 or its inverse, Eq. 9.95, can be conveniently used in the numerical analysis of boundary value problems, in which the loads (and displacements) are increased by small amounts at each step of the analysis.

For *ideal plastic* solids, the yield function is independent of the plastic strain tensor, thus Eq. 9.78 gives

$$H' = 0 \tag{9.96}$$

and, with the von Mises yield condition, the yield surface is

$$\bar{\sigma} - \sqrt{3}k = 0 \tag{9.97}$$

or

$$\bar{\sigma} = \sigma_Y \tag{9.98}$$

where k and σ_Y are yield stresses in simple shear and simple tension, respectively. Substituting Eqs. 9.96 and 9.98 into Eqs. 9.91, 9.92 and 9.95 leads to, respectively,

$$d\varepsilon_{ij}^{(p)} = \frac{3s_{ij}s_{kl}d\varepsilon_{kl}}{2\sigma_Y^2} \tag{9.99}$$

$$d\varepsilon_{ij} = \frac{1-2\nu}{3E} d\sigma_{mm}\delta_{ij} + \frac{ds_{ij}}{2\mu} + \frac{3s_{ij}s_{kl}d\varepsilon_{kl}}{2\sigma_Y^2} \tag{9.100}$$

$$d\sigma_{ij} = \frac{E}{3(1-2\nu)} d\varepsilon_{mm}\delta_{ij} + 2\mu de_{ij} - \frac{3\mu s_{ij}s_{kl}d\varepsilon_{kl}}{\sigma_Y^2} \tag{9.101}$$

Equation 9.100 or its inverse, Eq. 9.101, can be conveniently used in the numerical analysis of boundary value problems of ideal plastic solids.

For *rigid-strain hardening*, the elastic strain tensor is assumed to be negligible in comparison with the plastic strain tensor, thus E and μ tend to infinity,

$$\varepsilon_{ij} = \varepsilon_{ij}^{(p)} \tag{9.102}$$

and, due to Eq. 9.92,

$$d\varepsilon_{ij} = \frac{3}{2(\bar{\sigma})^2} s_{ij}s_{kl}d\varepsilon_{kl} \tag{9.103}$$

In addition, Eq. 9.63 states that the scalar quantity $s_{kl}d\varepsilon_{kl}$ in the equation above denotes the plastic work increment, which is exactly equal to $\bar{\sigma}d\bar{\varepsilon}^{(p)}$ as in Eq. 9.74. With the help of the definition of the equivalent strain increment in Eq. 9.71, we have

$$s_{kl}d\varepsilon_{kl} = \bar{\sigma}\left(\frac{2}{3}d\varepsilon_{kl}d\varepsilon_{kl}\right)^{1/2} \tag{9.104}$$

Substituting Eq. 9.104 into Eq. 9.103 we obtain the stress-strain increment relationship in the form

$$s_{ij} = \sqrt{\frac{2}{3}} \frac{\bar{\sigma} d\varepsilon_{ij}}{(d\varepsilon_{kl} d\varepsilon_{kl})^{1/2}} \tag{9.105}$$

in which $\bar{\sigma}$ is a function of the stress deviation tensor as in Eq. 9.70. The same type of relationship for *rigid-plastic solids* can be obtained by simply setting $\bar{\sigma} = \sigma_Y = \sqrt{3}k$ in Eq. 9.105.

9.6 VISCOPLASTICITY

Constitutive models of plasticity discussed so far can describe the plastic behavior of materials, but cannot be used to deal with rate sensitive yield materials due to their fundamental assumption of time independence. To overcome this difficulty, Bingham (1922) introduced the concept of a *viscoplastic model* through a stress-strain relation in a uniaxial context. Since then, many viscoplastic laws have been suggested. According to these laws, the viscous properties of these materials introduce a time dependence of the states of stress and strain, and the plastic properties make these states depend on the loading paths. Thus different results will be obtained for different loading paths and different durations of the process. Most existing constitutive laws of *viscoplasticity* can be described by stress-strain incremental relationships based, to a certain extent, on the elements of plasticity theory, i.e. the yield criterion, the flow rule and the hardening rule. An extensive review of these was made by Hill (1950).

Many metals when subjected to uniaxial tension or compression have been found to exhibit several rate effects. Among these the yield stress sensitivity to the rate of straining is the most pronounced one. As early as in 1909, Ludwik proposed an empirical constitutive relation to describe such metals under tension as

$$\sigma = \sigma_Y \log(\dot{\varepsilon}^{(vp)}/\dot{\varepsilon}_0^{(vp)}) \tag{9.106}$$

with $\dot{\varepsilon}^{(vp)}$ denoting the *viscoplastic strain rate*, the difference between the total strain rate and the elastic strain rate, i.e.

$$\dot{\varepsilon}^{(vp)} = \dot{\varepsilon} - \dot{\varepsilon}^{(e)} \tag{9.107}$$

The elastic strain rate obeys Hooke's law, i.e. (for the uniaxial tension case)

$$\dot{\varepsilon}^{(e)} = \frac{\dot{\sigma}}{E} \tag{9.108}$$

In addition, σ_Y is the yield stress in uniaxial tension, and $\dot{\varepsilon}_0^{(vp)}$ is a material constant. Obviously, Eq. 9.106 fails to describe the material behavior near the origin of the $(\sigma, \dot{\varepsilon}^{(vp)})$ plane. Later Prandtl (1928) proposed the hyperbolic sine speed law; this has the advantage of being valid at the origin and admissible in the compression region; it is

$$\dot{\varepsilon}^{(vp)} = \dot{\varepsilon}_0^{(vp)} \sinh \frac{\sigma}{\sigma_Y} \tag{9.109}$$

A general relation was proposed by Malvern (1951) in which the viscoplastic strain rate is assumed to be a function of the current stress and strain, i.e.

$$\dot{\varepsilon}^{(vp)} = f(\sigma, \varepsilon) \tag{9.110}$$

Many forms have been proposed for the function $f(\sigma, \varepsilon)$. For example, Gilman (1960) assumed

$$f(\sigma, \varepsilon) = E\dot{\varepsilon}_0 \exp(-A/\sigma),$$

where A and $\dot{\varepsilon}_0$ are material constants.

Sokolovsky (1948) in his attempt to separate dynamic effects from strain hardening, introduced a linear viscoplastic law in which the viscoplastic strain rate is a function of the dynamic overstress above the first yield point. Mathematically, it can be written as

$$\dot{\varepsilon}^{(vp)} = f(\sigma - \sigma_Y) \tag{9.111}$$

Perzyna (1963) likewise proposed the following relation for the viscoplastic strain rate

$$\dot{\varepsilon}^{(vp)} = f\left(\frac{\sigma}{\sigma_Y} - 1\right) \tag{9.112}$$

While all these constitutive laws provide useful explicit relations for computing viscoplastic strain rates, their applications generally come after the material stress state exceeds a specific yield condition. The total strain rate can then be split into an elastic part $\dot{\varepsilon}^{(e)}$ governed by Hooke's law and a viscoplastic part $\dot{\varepsilon}^{(vp)}$, given by one of these laws.

It should be mentioned that there exist several one-dimensional constitutive laws of viscoplasticity in the literature. An excellent survey made by Cormeau (1976) is recommended to all who are interested in the historical development.

Following the derivation adopted in the one dimensional cases, several models of three dimensional viscoplasticity have been developed by Perzyna (1966), Philips and Wu (1973), Nicholson and Philips (1974), and others. Simplified versions of Perzyna's model have been proposed by Zienkiewicz and Cormeau (1974), and Hughes and Taylor (1978). A brief review of Perzyna's model and these two modified models for isotropic solids will now be presented.

Perzyna's constitutive laws for elasto-viscoplastic materials under multi-axial deformation assume the existence of an initial yield surface beyond which viscoplastic flow will occur. The rate of increase of viscoplastic strain components is taken as a function of the extent to which the stress state exceeds the static yield criterion, which can be expressed by

$$f(\sigma_{ij}) = \frac{F(\sigma_m, J_2, J_3)}{\sigma_Y} - 1 \tag{9.113}$$

where f denotes the static yield surface, and

$$\sigma_m = \frac{1}{3}\sigma_{ii} \tag{9.114}$$

The sign of the yield functional F is taken such that f is negative for all elastic states, positive for viscoplastic states, and zero for the incipient plastic states.

The total strain rate $\dot{\varepsilon}_{ij}$ is resolved into elastic $\dot{\varepsilon}_{ij}^{(e)}$ and viscoplastic β_{ij} parts as

$$\dot{\varepsilon}_{ij} = \dot{\varepsilon}_{ij}^{(e)} + \beta_{ij} \tag{9.115}$$

where

$$\dot{\varepsilon}_{ij}^{(e)} = \frac{1}{2\mu}\dot{s}_{ij} + \frac{1}{3K}\dot{\sigma}_m \delta_{ij} \tag{9.116}$$

$$\beta_{ij} = \gamma < \phi(f) > \frac{\partial F}{\partial \sigma_{ij}} \tag{9.117}$$

in which μ, K and γ denote shear modulus, bulk modulus and a material dependent fluidity parameter, respectively; the symbol $< \phi(f) >$ defines a switch-on/off operator such that

$$< \phi(f) > = \phi(f), \quad \text{if } f > 0 \tag{9.118a}$$

$$= 0, \quad \text{if } f \le 0 \tag{9.118b}$$

For many practical cases $\phi(f)$ is assumed as f^n with n being a positive integer depending on the material under consideration. Larger values of n characterize slower 'creep' rates, but different values of n give the same final plastic solution. Using the chain rule of differentiation to evaluate $\partial F/\partial \sigma_{ij}$, we can transform Eq. 9.117 into

$$\beta_{ij} = \gamma < \phi(f) > \left(\frac{1}{3}\frac{\partial F}{\partial \sigma_m}\delta_{ij} + \frac{\partial F}{\partial J_2} s_{ij} + \frac{\partial F}{\partial J_3} t_{ij} \right) \tag{9.119}$$

where t_{ij} is as defined in Eq. 9.40. Equation 9.115 together with Eqs. 9.116 and 9.119 define the constitutive relations of Perzyna's model.

For von Mises materials, Zienkiewicz and Cormeau (1974) simplified Perzyna's model with the following viscoplastic strain rates

$$\beta_{ij} = \gamma < \phi > \frac{\partial f}{\partial \sigma_{ij}} \tag{9.120}$$

where

$$< \phi > = \left(\frac{f}{\sigma_Y} \right)^n, \quad \text{if } f > 0 \tag{9.121a}$$

$$= 0, \quad \text{if } f \le 0 \tag{9.121b}$$

$$f = \sqrt{3J_2} - \sigma_Y \tag{9.122}$$

Note that β_{ij} is a symmetric second order tensor, and thus can be expressed in the form of six distinct elements in a column matrix as

$$\{\beta\} = (\beta_{11} \quad \beta_{22} \quad \beta_{33} \quad 2\beta_{12} \quad 2\beta_{23} \quad 2\beta_{31})^T \tag{9.123}$$

where the superscript T denotes the transpose of a matrix.

Hughes and Taylor (1978), on the other hand, proposed

$$\{\beta\} = 3\gamma < \phi > \{s\}/\sigma_Y \tag{9.124}$$

where the individual terms are given by

$$\{s\} = [s_{11} \quad s_{22} \quad s_{33} \quad s_{12} \quad s_{23} \quad s_{31}]^T \tag{9.125}$$

$$< \phi > = f^n, \text{ if } f > 0 \tag{9.126a}$$

$$= 0, \quad \text{if } f \leq 0 \tag{9.126b}$$

and

$$f = \frac{3J_2}{\sigma_Y^2} - 1 \tag{9.127}$$

The basic characteristic of all these models is that when a stress state at a point falls outside the yield surface, they permit viscoplastic 'creep' at that point (Fig. 9.11) to occur, so the stresses are redistributed and relaxed till the stress state falls back on to the yield surface. Further, when the stress state at no point exceeds the yield surface, the solution by these models will correspond to the elastoplastic solution of the given problem.

In both strain rate models described above, n equal to one was used. It should be note that the relations proposed by Zienkiewicz and Cormeau (1974) have been well accepted for the whole path of the viscoplastic 'creep', while those introduced by Hughes and Taylor (1978) may not reflect the true behavior of such stress relaxation. However, both models should give the same steady state or final solution.

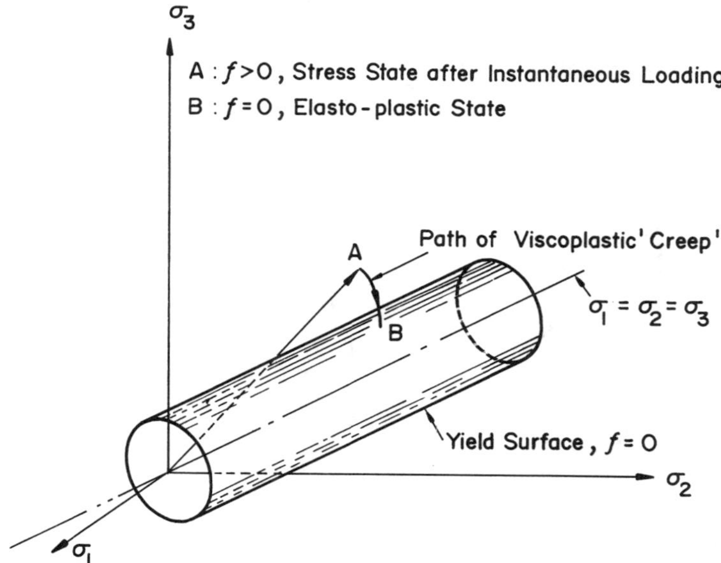

Fig. 9.11 Viscoplastic 'creep' on to three-dimensional von Mises yield surface.

CHAPTER X

FINITE DEFORMATION

10.1 DIFFERENT DESCRIPTIONS OF CHANGING CONFIGURATION

So far, we have restricted our analysis of deformation to problems involving small displacements only. Now we shall remove this restriction and consider finite displacements. The analysis involves the so called *geometrical nonlinearity*. Indicial notation rules using subscripts to denote Cartesian components of tensors will be followed from the present section until Section 10.6.

In Fig. 10.1, a particle at P^0 located by x_i in the original configuration is displaced to a new position P located by y_i in the deformed configuration. The *mapping* from x_i to y_i is *one-to-one* (Section 1.13), i.e. a particle originally at P^0 is displaced to a single position P only. In other words, the current location y_i of the particle which

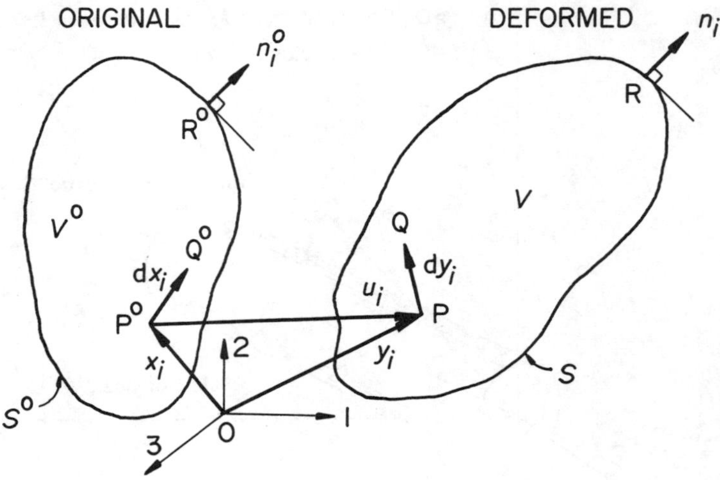

Fig. 10.1 Original and deformed configurations in the same ordinary physical space.

occupied the point x_i at time $t=0$ characterizes the mapping of the initial configuration into the current configuration (at time t) of the form

$$y_i = y_i(x_j, t) \tag{10.1}$$

In reverse, the mapping is in the form

$$x_i = x_i(y_j, t) \tag{10.2}$$

These two mappings are unique inverses of one another.

Coordinates x_i and y_i are normally called *Lagrangian coordinates* and *Eulerian coordinates*, respectively. A description which uses x_i and t as independent variables is called a *material* or *Lagrangian description*, while y_i and t are used as independent variables in what is called a *spatial* or *Eulerian description.* Truesdell (1952) found that the Lagrangian coordinates were introduced by L. Euler in 1762, and the Eulerian coordinates by J. le R. D'Alembert in 1752!

As shown in Fig. 10.1, the displacement from P^0 to P is denoted by u_i, i.e.

$$y_i = x_i + u_i \tag{10.3}$$

In addition, note that the vector $\mathrm{d}x_i$ defining the infinitesimal line element P^0Q^0 in the original configuration becomes $\mathrm{d}y_i$, defining the element PQ in the deformed configuration, thus

$$\mathrm{d}y_i = \frac{\partial y_i}{\partial x_j}\mathrm{d}x_j \tag{10.4}$$

or, due to Eq. 10.3,

$$\mathrm{d}y_i = \left(\delta_{ij} + \frac{\partial u_i}{\partial x_j}\right)\mathrm{d}x_j \tag{10.5}$$

and

$$\frac{\partial y_i}{\partial x_j} = \delta_{ij} + \frac{\partial u_i}{\partial x_j} \tag{10.6}$$

To be more complete it should be added that the displacement of point Q^0 is $u_i + \mathrm{d}u_i$, where

$$\mathrm{d}u_i = \frac{\partial u_i}{\partial x_j}\mathrm{d}x_j$$

All vectors involved in the transformation of P^0Q^0 into PQ are illustrated in Fig. 10.2.

In the preceding paragraph, y_i and u_i are regarded as functions of x_i and time t. Conversely, when x_i and u_i are regarded as functions of y_i and t,

$$\mathrm{d}x_i = \frac{\partial x_i}{\partial y_j}\mathrm{d}y_j \tag{10.7}$$

$$= \left(\delta_{ij} - \frac{\partial u_i}{\partial y_j}\right)\mathrm{d}y_j \tag{10.8}$$

$$\frac{\partial x_i}{\partial y_j} = \delta_{ij} - \frac{\partial u_i}{\partial y_j} \tag{10.9}$$

For a function $A(y_i, t)$, its partial derivative with respect to x_j is

$$\frac{\partial}{\partial x_j} A(y_i, t) = \frac{\partial A}{\partial y_k}\frac{\partial y_k}{\partial x_j} \tag{10.10}$$

If A in this equation is taken as x_i, we get the identity

$$\delta_{ij} = \frac{\partial x_i}{\partial y_k}\frac{\partial y_k}{\partial x_j} \tag{10.11}$$

Similarly, we can show that

$$\frac{\partial y_i}{\partial x_k}\frac{\partial x_k}{\partial y_j} = \delta_{ij} = \frac{\partial x_i}{\partial y_k}\frac{\partial y_k}{\partial x_j} \tag{10.12}$$

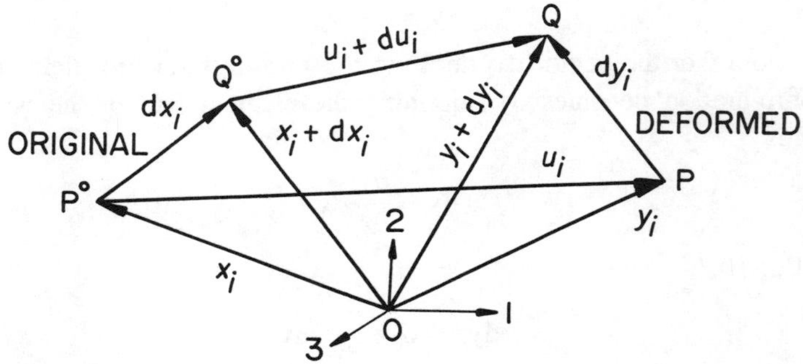

Fig. 10.2 One-to-one mapping of an infinitesimal line element between two configurations.

10.2 MATERIAL DERIVATIVE AND CONSERVATION OF MASS

In fact x_i is truly independent of t, but y_i is not so. For a function $A(y_j, t)$, its *material derivative* (L. Euler, 1770 and J. L. Lagrange, 1783) is

$$\frac{\mathrm{d}}{\mathrm{d}t} A(y_j, t) = \frac{\partial A}{\partial t} + \frac{\partial A}{\partial y_j} v_j \tag{10.13}$$

where v_i is the velocity, i.e.

$$v_i = \frac{\mathrm{d}y_i}{\mathrm{d}t} \tag{10.14}$$

or, due to Eqs. 10.3 and 10.13

$$v_i = \frac{\mathrm{d}u_i}{\mathrm{d}t} \tag{10.15}$$

$$= \frac{\partial u_i}{\partial t} + \frac{\partial u_i}{\partial y_j} v_j \tag{10.16}$$

The *acceleration* α_i is the material derivative of v_i and can be obtained according to the rule of Eq. 10.13 as

$$\alpha_i \equiv \frac{\mathrm{d}v_i}{\mathrm{d}t} = \frac{\partial v_i}{\partial t} + \frac{\partial v_i}{\partial y_j} v_j \tag{10.17}$$

The first and the second terms in this equation are called the *local* and the *convective* parts of the acceleration, respectively.

To lay a groundwork for the material derivative of the volume differential $\mathrm{d}V$, two very close points P and Q in the deformed configuration at time t are considered. Let y_i and $y_i + \mathrm{d}y_i$ denote the position vectors for P and Q, respectively as shown in Fig. 10.3. At time $t + \Delta t$ (where Δt is sufficiently small), these two points assume, respectively, the positions P' and Q'. Let the position vectors for P' and Q' be z_i and $z_i + \mathrm{d}z_i$, respectively. According to the concept of one-to-one mapping, we can state that

$$\mathrm{d}z_i = \frac{\partial z_i}{\partial y_j} \mathrm{d}y_j \tag{10.18}$$

If v_i is the velocity at P, then the position vector for P' can be put in the form

$$z_i = y_i + v_i \Delta t \tag{10.19}$$

Substituting the last equation into Eq. 10.18 yields

$$\mathrm{d}z_i = \mathrm{d}y_i + \frac{\partial v_i}{\partial y_j} \mathrm{d}y_j \Delta t \tag{10.20}$$

To be more complete, it should be added that the velocity at Q is $v_i + \mathrm{d}v_i$, where

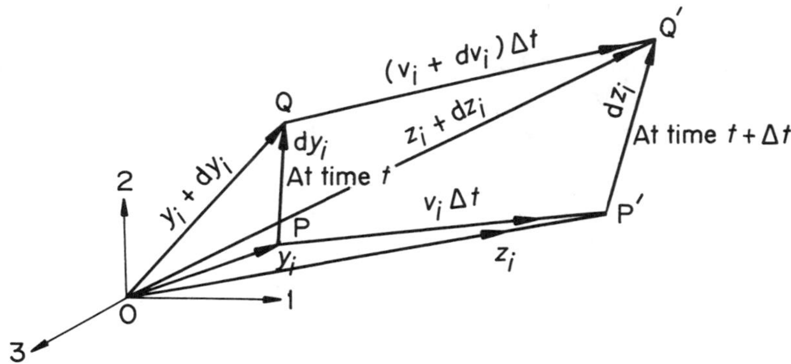

Fig. 10.3 Transformation of element $\mathrm{d}y_i$ at time t into element $\mathrm{d}z_i$ at time $t + \Delta t$.

$$dv_i = \frac{\partial v_i}{\partial y_j} dy_j \tag{10.21}$$

All vectors involved in the transformation of the line element PQ into the line element $P'Q'$ are illustrated in Fig. 10.3. Next, consider a volume differential dV at P, which we may choose for convenience to be a rectangular element as shown in Fig. 10.4, i.e.

$$dV = dy_1 dy_2 dy_3 \tag{10.22}$$

where dy_1, dy_2 and dy_3 are mutually orthogonal line elements characterized, respectively, by the following vectors

$$\begin{aligned} dy_i^{(1)} &= dy_1 \delta_{i1} \\ dy_i^{(2)} &= dy_2 \delta_{i2} \\ dy_i^{(3)} &= dy_3 \delta_{i3} \end{aligned} \tag{10.23}$$

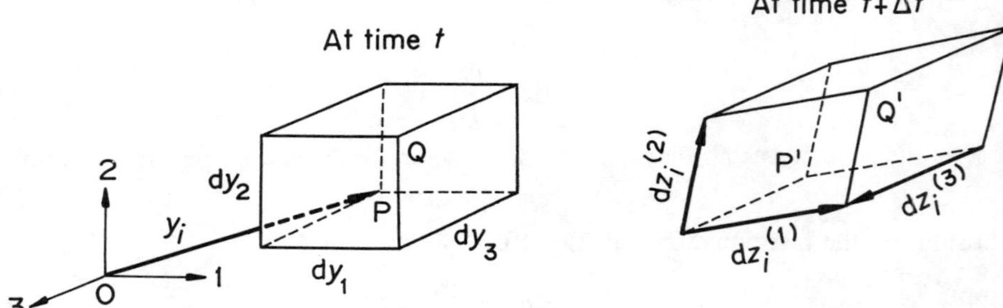

Fig. 10.4 Deformation of a volume differential dV at time t into dV' at time $t + \Delta t$.

Substituting Eqs. 10.23 into Eq. 10.20, we find the vectors characterizing the same line elements at time $t + \Delta t$

$$\begin{aligned} dz_i^{(1)} &= dy_1 \delta_{i1} + \frac{\partial v_i}{\partial y_1} dy_1 \Delta t \\ dz_i^{(2)} &= dy_2 \delta_{i2} + \frac{\partial v_i}{\partial y_2} dy_2 \Delta t \\ dz_i^{(3)} &= dy_3 \delta_{i3} + \frac{\partial v_i}{\partial y_3} dy_3 \Delta t \end{aligned} \tag{10.24}$$

In general, $dz_i^{(1)}$, $dz_i^{(2)}$ and $dz_i^{(3)}$ are no longer mutually orthogonal, but rather form a parallelepiped as shown in Fig. 10.4. The volume of the latter can be expressed, according to the law of products of vectors (Eqs. 1.5, 1.6, 1.31 and 1.32), in the form

$$dV' = \varepsilon_{ijk} dz_i^{(1)} dz_j^{(2)} dz_k^{(3)} \tag{10.25}$$

where ε_{ijk} is the permutation tensor defined in Eq. 1.33. On the other hand, the basic definition of the material derivative of dV is

$$\frac{d(dV)}{dt} = \lim_{\Delta t \to 0} \frac{dV' - dV}{\Delta t} \tag{10.26}$$

Substituting Eq. 10.25 into Eq. 10.26, using Eqs. 10.22 and 10.24, and neglecting the remaining terms of Δt we obtain

$$\frac{d(dV)}{dt} = \frac{\partial v_i}{\partial y_i} dV \tag{10.27}$$

For the material derivative of an integral in the form

$$\frac{d}{dt} \int_V A(y_j, t) dV,$$

where V is the domain of the continuum at time t, the operations of differentiation and integration may be interchanged, since the differentiation is with respect to a definite portion of the continuum (i.e. a specific mass system). Therefore

$$\frac{d}{dt} \int_V A(y_j, t) dV = \int_V \frac{d}{dt} [A(y_j, t) dV]$$

which, upon carrying out the differentiation and using Eq. 10.27, results in

$$\frac{d}{dt} \int_V A(y_j, t) dV = \int_V \left[\frac{dA(y_j, t)}{dt} + A(y_j, t) \frac{\partial v_i}{\partial y_i} \right] dV \tag{10.28}$$

Applying Gauss's divergence theorem (Eq. 1.87) into Eq. 10.28 we find

$$\frac{d}{dt} \int_V A(y_j, t) dV = \int_V \left[\frac{dA(y_j, t)}{dt} - \frac{\partial A(y_j, t)}{\partial y_i} v_i \right] dV + \int_S A(y_j, t) v_i n_i dS \tag{10.29}$$

where S is the surface enclosing V and n_i is the unit outward normal vector to S. With the help of Eq. 10.13, we can rewrite Eq. 10.29 in the form

$$\frac{d}{dt} \int_V A(y_j, t) dV = \int_V \frac{\partial A(y_j, t)}{\partial t} dV + \int_S A(y_j, t) v_i n_i dS \tag{10.30}$$

Without the assumption of the spatial differentiability of $A(x_j, t)$ and v_i, the material derivative above can be considered as

$$\frac{d}{dt} \int_V A(x_j, t) dV = \lim_{\Delta t \to 0} \frac{\Delta \left[\int_V A(x_j, t) dV \right]}{\Delta t} \tag{10.31}$$

The numerator on the right-hand side of the equation above can be considered as consisting of two terms; one over the region that $V(t)$ and $V(t + \Delta t)$ share in common, i.e.

$$\int_V \frac{\partial A}{\partial t} \Delta t dV,$$

and another the volume swept by particles occupying an element of area dS, i.e.

$$\int_S A(x_j, t) v_i \Delta t n_i dS.$$

Substituting these two terms into Eq. 10.31, we can readily obtain Eq. 10.30.

If $A(y_j, t)$ in Eqs. 10.28 to 10.30 is taken as a unit scalar quantity, we have the change of the material volume with respect to time, i.e. the material derivative of the volume, in the forms

$$\frac{d}{dt}\int_V dV = \int_V \frac{\partial v_i}{\partial y_i} dV \tag{10.32}$$

$$= \int_S v_i n_i dS \tag{10.33}$$

Note that Eq. 10.32 is equivalent to Eq. 10.27. If A in Eq. 10.30 is replaced by the material mass density ρ, then the left-hand side of that equation must vanish because of the *conservation of mass*, i.e.

$$\int_V \frac{\partial \rho}{\partial t} dV + \int_S \rho v_j n_j dS = 0 \tag{10.34}$$

Applying the same technique to Eq. 10.28, but dropping the integration symbols there (since the equation holds for any V), we find

$$\frac{d\rho}{dt} + \rho \frac{\partial v_j}{\partial y_j} = 0 \tag{10.35}$$

or, due to Eq. 10.13,

$$\frac{\partial \rho}{\partial t} + \frac{\partial(\rho v_j)}{\partial y_j} = 0 \tag{10.36}$$

Equations 10.34 to 10.36 are called the *equations of continuity* (J. le R. D'Alembert, 1752 and L. Euler, 1757). The integral form (Eq. 10.34) is useful when the differentiability of ρv_j cannot be assumed.

In the one-to-one correspondence between the original and the deformed configurations as depicted in Fig. 10.1, another basic form of the conservation of mass is

$$\rho^0 dV^0 = \rho dV \tag{10.37}$$

where ρ^0 and V^0 are the mass density and the material domain at the original state. For convenience, we may choose dV^0 as a volume of an infinitesimal rectangular element of the form

$$dV^0 = dx_1 dx_2 dx_3 \tag{10.38}$$

where dx_1, dx_2 and dx_3 are mutually orthogonal line elements in the original configuration, and characterized respectively by the following vectors

$$\begin{aligned} dx_i^{(1)} &= dx_1 \delta_{i1} \\ dx_i^{(2)} &= dx_2 \delta_{i2} \\ dx_i^{(3)} &= dx_3 \delta_{i3} \end{aligned} \tag{10.39}$$

Substituting Eqs. 10.39 into Eq. 10.4 yields, respectively, the vectors characterizing the same line elements in the deformed configuration as follows

$$\begin{aligned} dy_i^{(1)} &= \frac{\partial y_i}{\partial x_1} dx_1 \\ dy_i^{(2)} &= \frac{\partial y_i}{\partial x_2} dx_2 \\ dy_i^{(3)} &= \frac{\partial y_i}{\partial x_3} dx_3 \end{aligned} \tag{10.40}$$

In general, the line elements $dy_i^{(1)}$, $dy_i^{(2)}$ and $dy_i^{(3)}$ are not mutually orthogonal. In other words, an original rectangular element of volume dV^0 transforms into an element of a parallelepiped shape of volume dV. The latter can be put in the form

$$dV = \varepsilon_{ijk} dy_i^{(1)} dy_j^{(2)} dy_k^{(3)} \tag{10.41}$$

Due to Eqs. 10.38 and 10.40, Eqs. 10.41 and 10.37 become, respectively

$$dV = \varepsilon_{ijk} \frac{\partial y_i}{\partial x_1} \frac{\partial y_j}{\partial x_2} \frac{\partial y_k}{\partial x_3} dV^0 \tag{10.42}$$

$$\rho^0 dV^0 = \rho \varepsilon_{ijk} \frac{\partial y_i}{\partial x_1} \frac{\partial y_j}{\partial x_2} \frac{\partial y_k}{\partial x_3} dV^0,$$

or

$$\rho^0 = \varepsilon_{ijk} \frac{\partial y_i}{\partial x_1} \frac{\partial y_j}{\partial x_2} \frac{\partial y_k}{\partial x_3} \rho \tag{10.43}$$

Before proceeding further, we should recall the following properties of the determinant (Eqs. 1.46 and 1.48)

$$\left| \frac{\partial y_r}{\partial x_s} \right| = \varepsilon_{ijk} \frac{\partial y_i}{\partial x_1} \frac{\partial y_j}{\partial x_2} \frac{\partial y_k}{\partial x_3} \tag{10.44}$$

$$\varepsilon_{lmn} \left| \frac{\partial x_r}{\partial y_s} \right| = \varepsilon_{ijk} \frac{\partial x_i}{\partial y_l} \frac{\partial x_j}{\partial y_m} \frac{\partial x_k}{\partial y_n} \tag{10.45}$$

Substituting Eq. 10.44 into Eq. 10.43 we find

$$\frac{\rho^0}{\rho} = \left| \frac{\partial y_i}{\partial x_j} \right| \tag{10.46}$$

However, Eq. 1.45 which is another property of determinants implies that

$$\left|\frac{\partial y_i}{\partial x_j}\right|\left|\frac{\partial x_i}{\partial y_j}\right| = \left|\frac{\partial y_i}{\partial x_j}\frac{\partial x_j}{\partial y_k}\right|,$$

which, due to Eq. 10.12, becomes

$$\left|\frac{\partial y_i}{\partial x_j}\right|\left|\frac{\partial x_i}{\partial y_j}\right| = |\delta_{ij}| = 1 \tag{10.47}$$

Consequently, Eq. 10.46 can be put in an alternative form as follows

$$\frac{\rho}{\rho^0} = \left|\frac{\partial x_i}{\partial y_j}\right| \tag{10.48}$$

Equations 10.46 and 10.48 relate the mass density ratio to a unique transformation from one configuration to another as described in Section 1.13. Moreover, since the mass densities are naturally positive definite, the transformation is admissible and proper.

10.3 STRESS TENSORS IN DIFFERENT DESCRIPTIONS

The equation of equilibrium, Eq. 2.8, obtained in Chapter II corresponds to a deformed configuration and is simply expressed in terms of the *Eulerian stress tensor* σ_{ij}. This is a natural physical concept. However, if strains are referred to the original configuration, it would be more convenient in the course of analysis to define stresses with respect to the same configuration.

As has been established in Eq. 2.6, a stress vector $\overline{X}_i$ on a small area dS in a strained continuum is related to the Eulerian stress tensor σ_{ij} by

$$\overline{X}_i \mathrm{d}S = \sigma_{ji} n_j \mathrm{d}S \tag{10.49}$$

where n_i is the unit outer normal to dS. In the undeformed state, imagine the force $\overline{X}_i$dS acting on the area dS^0 which is the original of dS. Simultaneously, a stress vector $\overline{X}_i^0$ in the same direction as $\overline{X}_i$, and a stress tensor σ_{ij}^0 may be introduced in a relationship similar to Eq. 10.49, i.e.

$$\overline{X}_i^0 \mathrm{d}S^0 = \sigma_{ji}^0 n_j^0 \mathrm{d}S^0 \tag{10.50}$$

where n_i^0 is the unit outer normal to dS^0. Known as the *Lagrangian stress tensor* or the *first Piola-Kirchhoff stress tensor* (G. Piola, 1833 and G. Kirchhoff, 1852), σ_{ij}^0 is a stress component in the x_j-direction on a plane (in the unstrained continuum) perpendicular to the x_i-axis. Equating the last two equations we find

$$\sigma_{ji} n_j \mathrm{d}S = \sigma_{ji}^0 n_j^0 \mathrm{d}S^0 \tag{10.51}$$

In order to find the relationship between σ_{ij} and σ_{ij}^0, we must know the relation between n_jdS and n_j^0dS^0. Let dy_i and δy_i be the sides of the small parallellogram dS. According to the law of a cross product of two vectors (Eqs. 1.6 and 1.32), we can write

$$n_i \mathrm{d}S = \varepsilon_{ijk} \mathrm{d}y_j \delta y_k \tag{10.52}$$

Another equation similar to Eq. 10.52 can be written for the original state as follows

$$n_i^0 \mathrm{d}S^0 = \varepsilon_{ijk} \mathrm{d}x_j \delta x_k \tag{10.53}$$

where $\mathrm{d}x_i$ and δx_i are the originals of $\mathrm{d}y_i$ and δy_i, respectively. Therefore, by Eq. 10.7,

$$n_i^0 \mathrm{d}S^0 = \varepsilon_{ijk} \frac{\partial x_j}{\partial y_l} \frac{\partial x_k}{\partial y_m} \mathrm{d}y_l \delta y_m \tag{10.54}$$

Multiplying the last equation by $\partial x_i/\partial y_n$ and making use of Eqs. 10.45 and 10.48, we get

$$\frac{\partial x_i}{\partial y_n} n_i^0 \mathrm{d}S^0 = \frac{\rho}{\rho^0} \varepsilon_{nlm} \mathrm{d}y_l \delta y_m \tag{10.55}$$

or, due to Eq. 10.52,

$$\frac{\partial x_i}{\partial y_n} n_i^0 \mathrm{d}S^0 = \frac{\rho}{\rho^0} n_n \mathrm{d}S \tag{10.56}$$

Substituting Eq. 10.56 into Eq. 10.51 we find

$$\sigma_{ji} \frac{\rho^0}{\rho} \frac{\partial x_m}{\partial y_j} n_m^0 \mathrm{d}S^0 = \sigma_{ji}^0 n_j^0 \mathrm{d}S^0,$$

or, (since n_j^0 is nonzero),

$$\sigma_{ji}^0 = \frac{\rho^0}{\rho} \frac{\partial x_j}{\partial y_m} \sigma_{mi} \tag{10.57}$$

Equation 10.57 shows that the Lagrangian stress tensor σ_{ij}^0 is not symmetric, though the Eulerian stress tensor σ_{ij} is symmetric, as will be shown formally later. Thus it will be inconvenient to use σ_{ij}^0 in a stress-strain law which is normally in terms of symmetric stress and strain tensors. For this reason, Σ_{ij} known as the *second Piola-Kirchhoff stress tensor* or simply the *Kirchhoff stress tensor* is introduced as

$$\Sigma_{ji} = \frac{\partial x_i}{\partial y_l} \sigma_{jl}^0 \tag{10.58}$$

or, due to Eq. 10.57,

$$\Sigma_{ji} = \frac{\rho^0}{\rho} \frac{\partial x_i}{\partial y_l} \frac{\partial x_j}{\partial y_m} \sigma_{ml} \tag{10.59}$$

From the last equation we can see that Σ_{ij} is a symmetric tensor, i.e. $\Sigma_{ij} = \Sigma_{ji}$. Making use of the identities in Eqs. 10.12, we can obtain the inverses to Eqs. 10.57 to 10.59 as the following

$$\sigma_{ji} = \frac{\rho}{\rho^0} \frac{\partial y_i}{\partial x_m} \sigma_{mj}^0 \tag{10.60a}$$

$$\sigma_{ji}^0 = \frac{\partial y_i}{\partial x_m} \Sigma_{jm} \tag{10.60b}$$

$$\sigma_{ji} = \frac{\rho}{\rho^0}\frac{\partial y_i}{\partial x_l}\frac{\partial y_j}{\partial x_m}\Sigma_{lm} \tag{10.60c}$$

Due to Eqs. 10.6 and 10.9, Σ_{ji} and σ_{ji} can be expressed in terms of displacements

$$\Sigma_{ji} = \frac{\rho^0}{\rho}\left[\sigma_{ji} - \left(\delta_{jm}\frac{\partial u_i}{\partial y_l} + \delta_{il}\frac{\partial u_j}{\partial y_m} - \frac{\partial u_i}{\partial y_l}\frac{\partial u_j}{\partial y_m}\right)\sigma_{lm}\right] \tag{10.61a}$$

$$\sigma_{ji} = \frac{\rho^0}{\rho}\left[\Sigma_{ji} + \left(\delta_{jm}\frac{\partial u_i}{\partial x_l} + \delta_{il}\frac{\partial u_j}{\partial x_m} + \frac{\partial u_i}{\partial x_l}\frac{\partial u_j}{\partial x_m}\right)\Sigma_{lm}\right] \tag{10.61b}$$

At this stage, it is appropriate to consider the directions of the stress vector components on each side of the infinitesimal parallelepipeds. As shown in Fig. 10.5, a stress vector σ^0_{1i} acting on the plane (in the original configuration) normal to x_1-direction has Cartesian components σ^0_{11}, σ^0_{12} and σ^0_{13}. Alternatively, the same stress vector due to Eq. 10.60b, is

$$\sigma^0_{1i} = \Sigma_{11}\frac{\partial y_i}{\partial x_1} + \Sigma_{12}\frac{\partial y_i}{\partial x_2} + \Sigma_{13}\frac{\partial y_i}{\partial x_3} \tag{10.62}$$

It should be noted that each vector on the right-hand side of Eq. 10.62 is parallel to a particular edge of the infinitesimal parallelpiped in the deformed configuration. For example, the vector $\Sigma_{11}\partial y_i/\partial x_1$ is parallel to $dy_i^{(1)}$ (defined in Eq. 10.40), since both vectors are parallel to the same vector $\partial y_i/\partial x_1$. In fact, an original edge in the x_1-direction and with a unit length becomes a vector $\partial y_i/\partial x_1$ in the deformed configuration. Similarly, we can say that the *shear* components $\Sigma_{12}\partial y_i/\partial x_2$ and $\Sigma_{13}\partial y_i/\partial x_3$ are parallel to $dy_i^{(2)}$

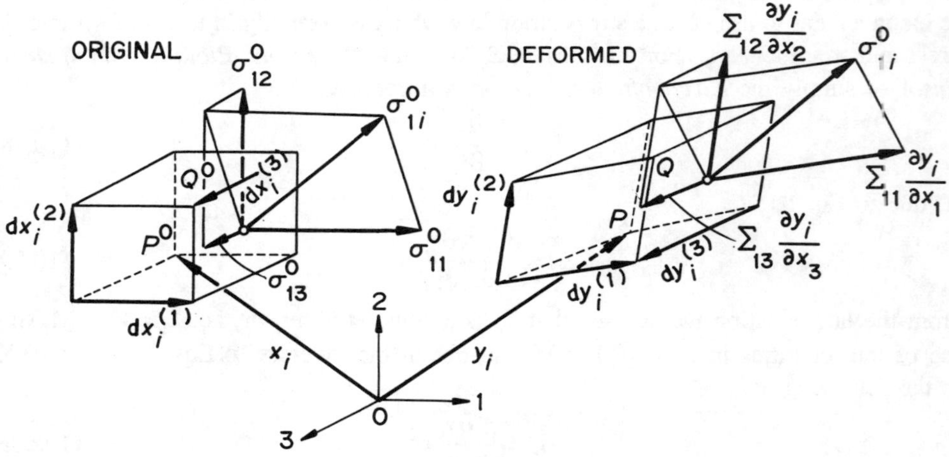

Fig. 10.5 Decomposition of stress vector σ^0_{1i} into three components parallel to edges of each parallelepiped.

and $dy_i^{(3)}$, respectively. Thus the stress vector on the side originally perpendicular to x_1-axis can be decomposed into three components as shown in Fig. 10.5. Repeating the same discussion for other sides of the parallelpiped, we can show that $\Sigma_{22}\partial y_i/\partial x_2$ is parallel to $dy_i^{(2)}$, $\Sigma_{23}\partial y_i/\partial x_3$ to $dy_i^{(3)}$, $\Sigma_{21}\partial y_i/\partial x_1$ to $dy_i^{(1)}$, $\Sigma_{33}\partial y_i/\partial x_3$ to $dy_i^{(3)}$, $\Sigma_{31}\partial y_i/\partial x_1$ to $dy_i^{(1)}$, and $\Sigma_{32}\partial y_i/\partial x_2$ to $dy_i^{(2)}$.

Such knowledge of the directions of the stress vector components is useful in interpreting the phenomena physically and in making rational assumptions in real structural systems. As an example, taking the case in which the original dimension of the structure in the x_3-direction is relatively small, we may assume that Σ_{33}, Σ_{31} and Σ_{32} are zero, since that gives the same effect as the usual assumption that stress vector components in the $dy_i^{(3)}$-direction should vanish.

Instead, we choose the current volume dV as a volume of an infinitesimal rectangular element of the form

$$dV = dy_1 dy_2 dy_3 \tag{10.63}$$

where dy_1, dy_2 and dy_3 are mutually orthogonal line elements in the deformed configuration, and characterized respectively by the following vectors

$$dy_i^{(1)} = dy_1 \delta_{i1} \tag{10.64a}$$

$$dy_i^{(2)} = dy_2 \delta_{i2} \tag{10.64b}$$

$$dy_i^{(3)} = dy_3 \delta_{i3} \tag{10.64c}$$

Substituting Eqs. 10.64 into Eq. 10.7 yields, respectively, the vectors characterizing the same line elements in the original configuration as follows

$$dx_i^{(1)} = \frac{\partial x_i}{\partial y_1} dy_1 \tag{10.65a}$$

$$dx_i^{(2)} = \frac{\partial x_i}{\partial y_2} dy_2 \tag{10.65b}$$

$$dx_i^{(3)} = \frac{\partial x_i}{\partial y_3} dy_3 \tag{10.65c}$$

Two more stress tensors may be introduced as

$$\sigma_{ji}^{00} = \frac{\partial y_i}{\partial x_k} \sigma_{jk} \tag{10.66}$$

and

$$\Sigma_{ji}^{0} = \frac{\partial y_i}{\partial x_l} \frac{\partial y_j}{\partial x_m} \sigma_{lm} \tag{10.67}$$

Note that σ_{ji}^{00} is not symmetric but Σ_{ji}^{0} is. Making use of Eq. 10.12, we can obtain the inverse of Eqs. 10.66 and 10.67 as the following

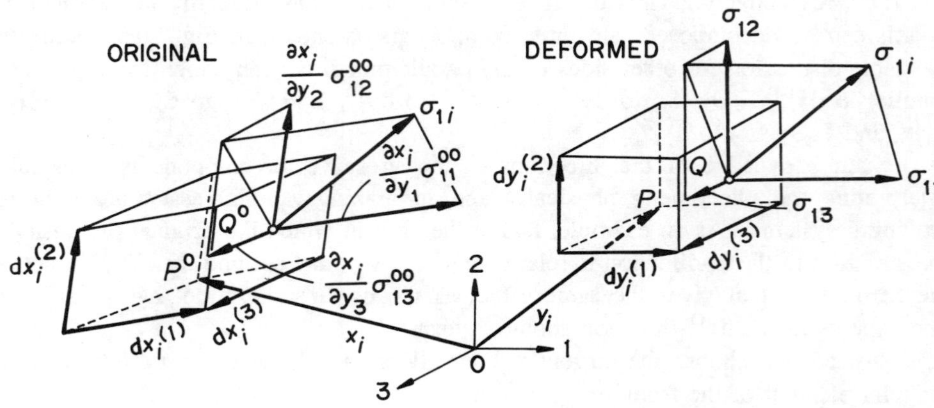

Fig. 10.6 Decomposition of stress vector σ_{1i} into three components parallel to edges of each parallelepiped.

$$\sigma_{ji} = \frac{\partial x_i}{\partial y_k}\sigma_{jk}^{00} \tag{10.68}$$

$$\sigma_{ji} = \frac{\partial x_i}{\partial y_l}\frac{\partial x_j}{\partial y_m}\Sigma_{lm}^{0} \tag{10.69}$$

It should be interesting to see the direction of each component of a stress vector acting on a plane in the deformed configuration which is normal to a Cartesian coordinate axis. As an example, Fig. 10.6 depicts the stress vector σ_{1i} acting on the plane (in the deformed configuration) normal to the y_1-direction. This stress vector has Cartesian components σ_{11}, σ_{12} and σ_{13}, but due to Eq. 10.68,

$$\sigma_{1i} = \frac{\partial x_i}{\partial y_1}\sigma_{11}^{00} + \frac{\partial x_i}{\partial y_2}\sigma_{12}^{00} + \frac{\partial x_i}{\partial y_3}\sigma_{13}^{00} \tag{10.70}$$

of which each vector on the right-hand side is parallel to a particular edge of the infinitesimal parallelepiped in the original configuration, e.g. $\sigma_{11}^{00}\partial x_i/\partial y_1$ is parallel to $dx_i^{(1)}$ (given in Eq. 10.65a).

While σ_{ij} and σ_{ij}^0 are physical by their direct association with the stress vector $\overline{X}_i$; Σ_{ij}, σ_{ij}^{00} and Σ_{ij}^0 are *pseudo* except when we restrict ourselves to small deformation, for which we shall show that

$$\sigma_{ij} = \sigma_{ij}^0 = \Sigma_{ij} = \sigma_{ij}^{00} = \Sigma_{ij}^0, \quad \text{(for small deformation)} \tag{10.71}$$

For an arbitrary function $A(y_i)$, we have, because of Eq. 10.6,

$$\frac{\partial A}{\partial x_i} = \frac{\partial A}{\partial y_j}\frac{\partial y_j}{\partial x_i}$$

$$= \left(\delta_{ij} + \frac{\partial u_j}{\partial x_i} \right) \frac{\partial A}{\partial y_j} \tag{10.72}$$

in which we may neglect the term $\partial u_j / \partial x_i$, and thus arrive at the result

$$\frac{\partial A}{\partial x_i} = \frac{\partial A}{\partial y_i}, \quad \text{(for small deformation)} \tag{10.73}$$

Using Eq. 10.73, we can readily obtain Eq. 10.71 from any of the Eqs. 10.58 to 10.60 and 10.66 to 10.69.

10.4 EQUATIONS OF MOTION IN DIFFERENT DESCRIPTIONS

In Fig. 10.1, a continuum at time t occupies a domain V with a boundary S in the deformed state. The body is subjected to a body force X_i per unit mass and a surface traction $\overline{X}_i$ per unit area of S. Newton's second law of motion states that the rate of change of the *linear momentum* is equal to the applied force, i.e.

$$\frac{\mathrm{d}}{\mathrm{d}t}(\rho v_i \mathrm{d}V) = \mathrm{d}F_i \tag{10.74}$$

where $\mathrm{d}F_i$ is the resultant force acting on $\mathrm{d}V$. Note the identity

$$\varepsilon_{ijk} v_j v_k = 0 \tag{10.75}$$

(since the term represents a cross product of two identical vectors), and the rate of change of the *moment of momentum*

$$\frac{\mathrm{d}}{\mathrm{d}t}(\varepsilon_{ijk} y_j \rho v_k \mathrm{d}V) = \varepsilon_{ijk} y_j \frac{\mathrm{d}}{\mathrm{d}t}(\rho v_k \mathrm{d}V) + \varepsilon_{ijk} v_j v_k \rho \mathrm{d}V \tag{10.76}$$

Substituting Eqs. 10.74 and 10.75 into Eq. 10.76 we find

$$\frac{\mathrm{d}}{\mathrm{d}t}(\varepsilon_{ijk} y_j \rho v_k \mathrm{d}V) = \mathrm{d}T_i \tag{10.77}$$

where $\mathrm{d}T_i$ is the resultant *torque* acting on $\mathrm{d}V$, i.e.

$$\mathrm{d}T_i = \varepsilon_{ijk} y_j \mathrm{d}F_k \tag{10.78}$$

Writing Eqs. 10.74 and 10.77 for V as a whole, we have, respectively,

$$\frac{\mathrm{d}}{\mathrm{d}t} \int_V \rho v_i \mathrm{d}V = F_i \tag{10.79}$$

$$\frac{\mathrm{d}}{\mathrm{d}t} \int_V \varepsilon_{ijk} y_j \rho v_k \mathrm{d}V = T_i \tag{10.80}$$

where

$$F_i = \int_V \rho X_i \mathrm{d}V + \int_S \overline{X}_i \mathrm{d}S \tag{10.81}$$

$$T_i = \int_V \varepsilon_{ijk} \rho y_j X_k \mathrm{d}V + \int_S \varepsilon_{ijk} y_j \overline{X}_k \mathrm{d}S \tag{10.82}$$

Equations 10.79 and 10.80 are *Euler's first and second laws of motion*, respectively, already described in Section 2.4. To derive the equations of motion in terms of the Eulerian stress tensor σ_{ij} in a more formal way using the first law, we substitute Eq. 10.81 into Eq. 10.79 and use Eqs. 10.30 and 10.49 to obtain

$$\int_V \left[\frac{\partial(\rho v_i)}{\partial t} - \rho X_i \right] dV = \int_S (\sigma_{ji} - \rho v_i v_j) n_j dS \tag{10.83}$$

If σ_{ji} and $\rho v_i v_j$ on the right-hand side of the equation above are assumed differentiable with respect to y_j, we can apply Gauss's divergence theorem and obtain

$$\int_V \left[\frac{\partial(\rho v_i)}{\partial t} - \rho X_i \right] dV = \int_V \left[\frac{\partial \sigma_{ji}}{\partial y_j} - \frac{\partial(\rho v_i v_j)}{\partial y_j} \right] dV.$$

Since this equation holds for an arbitrary V, the integrands on both sides must be equal, i.e.

$$\frac{\partial \sigma_{ji}}{\partial y_j} + \rho X_i = \frac{\partial(\rho v_i)}{\partial t} + \frac{\partial}{\partial y_j}(\rho v_i v_j) \tag{10.84}$$

or, due to Eq. 10.13,

$$\frac{\partial \sigma_{ji}}{\partial y_j} + \rho X_i = \rho \frac{\mathrm{d} v_i}{\mathrm{d} t} + v_i \left[\frac{\partial \rho}{\partial t} + \frac{\partial}{\partial y_j}(\rho v_j) \right].$$

However, the last term in this equation must vanish due to an equation of continuity, i.e. Eq. 10.36. Thus

$$\frac{\partial \sigma_{ji}}{\partial y_j} + \rho X_i = \rho \frac{\mathrm{d} v_i}{\mathrm{d} t} \tag{10.85}$$

This is the *Eulerian equation of motion in the differential form*, while Eq. 10.83 is that in an integral form. Alternatively, when the spatial differentiability of ρ and v_i can be assumed, substituting Eq. 10.81 into Eq. 10.79 and making use of Eq. 10.28 yield

$$\int_V \left[\frac{\mathrm{d}(\rho v_i)}{\mathrm{d} t} + \rho v_i \frac{\partial v_j}{\partial y_j} - \rho X_i \right] dV = \int_S \sigma_{ji} n_j dS,$$

or

$$\int_V \left[\rho \frac{\mathrm{d} v_i}{\mathrm{d} t} + v_i \left(\frac{\mathrm{d} \rho}{\mathrm{d} t} + \rho \frac{\partial v_j}{\partial y_j} \right) - \rho X_i \right] dV = \int_S \sigma_{ji} n_j dS,$$

or, due to a continuity condition, Eq. 10.35,

$$\int_V \left(\rho \frac{\mathrm{d} v_i}{\mathrm{d} t} - \rho X_i \right) dV = \int_S \sigma_{ji} n_j dS \tag{10.86}$$

If σ_{ji} is also differentiable with respect to y_j, the equation above can be reduced to Eq. 10.85.

On the other hand, if Euler's second law of motion is used, we will obtain

$$\int_V \varepsilon_{ijk} \left[\rho v_j v_k - y_j \left(\frac{\partial \sigma_{lk}}{\partial y_l} + \rho X_k - \rho \frac{\mathrm{d} v_k}{\mathrm{d} t} \right) \right] dV = \int_V \varepsilon_{ilk} \sigma_{lk} dV,$$

from which, due to Eqs. 10.75 and 10.85, we can obtain

$$\varepsilon_{ilk}\sigma_{lk} = 0.$$

The last equation simply means the symmetry of the Eulerian stress tensor, i.e. $\sigma_{ij} = \sigma_{ji}$.

To adopt Eq. 10.85 as a governing equation of related boundary value problems makes the task difficult, since the variable y_i itself is unknown, and both the velocity and the acceleration defined in Eqs. 10.16 and 10.17 contain nonlinear terms. The problems become simpler when the deformation is considered so small that: these nonlinear terms may be neglected; there is no need to distinguish x_i from y_i; the basic unknown in the problem is u_i.

In the Lagrangian description, the functions involved have x_i and t as independent variables, thus the velocity and the acceleration are

$$v_i(x_j,t) = \frac{\partial u_i(x_j,t)}{\partial t} \tag{10.87}$$

$$\alpha_i(x_j,t) = \frac{\partial v_i}{\partial t} = \frac{\partial^2 u_i}{\partial t^2} \tag{10.88}$$

The linear momentum of the system is

$$I_i = \int_V \rho v_i \mathrm{d}V = \int_{V^0} \rho^0 v_i \mathrm{d}V^0 \tag{10.89}$$

Its material derivative in the Lagrangian description is

$$\frac{\mathrm{d}I_i}{\mathrm{d}t} = \int_{V^0} \rho^0 \frac{\partial v_i}{\partial t} \mathrm{d}V^0 \tag{10.90}$$

Noting the fact that the same body force X_i per unit mass acts in both $\mathrm{d}V$ and $\mathrm{d}V^0$, and that the same surface force (Eq. 10.50) acts on both $\mathrm{d}S$ and $\mathrm{d}S^0$, we can change the resultant force acting on the system (Eq. 10.81) into

$$F_i = \int_{V^0} \rho^0 X_i \mathrm{d}V^0 + \int_{S^0} \sigma^0_{ji} n^0_j \mathrm{d}S^0 \tag{10.91}$$

Equating Eq. 10.90 with Eq. 10.91 we find

$$\int_{V^0} \left[\rho^0 \frac{\partial v_i}{\partial t} - \rho^0 X_i\right] \mathrm{d}V^0 = \int_{S^0} \sigma^0_{ji} n^0_j \mathrm{d}S^0 \tag{10.92}$$

If the spatial differentiability of the stress field can be assumed, we can transform the surface integral in Eq. 10.92 into a volume integral by Gauss's divergence theorem, and finally get the equation of motion for an arbitrary domain V^0 as

$$\frac{\partial \sigma^0_{ji}}{\partial x_j} + \rho^0 X_i = \rho^0 \frac{\partial v_i}{\partial t} \tag{10.93}$$

Alternatively, we can obtain Eq. 10.93 and the symmetry of Σ_{ij} by considering the equilibrium of the infinitesimal parallelepiped after deformation, in the same way that we do for curvilinear coordinates. The equations of motion can be expressed in terms of the Kirchhoff stress tensor by substituting Eq. 10.60b into the last equation:

$$\frac{\partial}{\partial x_j}\left(\frac{\partial y_i}{\partial x_m}\Sigma_{jm}\right)+\rho^0 X_i=\rho^0\frac{\partial v_i}{\partial t} \tag{10.94}$$

or, due to Eq. 10.6,

$$\frac{\partial}{\partial x_j}\left[\Sigma_{jm}\left(\delta_{im}+\frac{\partial u_i}{\partial x_m}\right)\right]+\rho^0 X_i=\rho^0\frac{\partial v_i}{\partial t} \tag{10.95}$$

10.5 FINITE STRAIN TENSORS

In Fig. 10.1, the square of the line element connecting two neighboring points P and Q in the displaced configuration is $dy_i dy_i$, while that in the original configuration is $dx_i dx_i$. The difference between these two squares, because of Eq. 10.4, is

$$dy_i dy_i - dx_i dx_i = 2E_{ij}dx_i dx_j \tag{10.96}$$

or, because of Eq. 10.7, is

$$dy_i dy_i - dx_i dx_i = 2\varepsilon_{ij}dy_i dy_j \tag{10.97}$$

where

$$E_{ij}=\frac{1}{2}\left(\frac{\partial y_k}{\partial x_i}\frac{\partial y_k}{\partial x_j}-\delta_{ij}\right) \tag{10.98}$$

$$\varepsilon_{ij}=\frac{1}{2}\left(\delta_{ij}-\frac{\partial x_k}{\partial y_i}\frac{\partial x_k}{\partial y_j}\right) \tag{10.99}$$

The tensor E_{ij} now known as the *Green's strain tensor* was introduced by G. Green in 1841 and by B. de St.Venant in 1844. The tensor ε_{ij} now known as the *Almansi's strain tensor* was introduced by A. L. Cauchy in 1827 for small strains, and by Almansi (1911) and Hamel (1912) for finite strains. In other words, E_{ij} is a tensor in the Lagrangian description, while ε_{ij} is one in the Eulerian description. Both tensors are symmetric, since

$$E_{ij}=E_{ji},\quad \varepsilon_{ij}=\varepsilon_{ji} \tag{10.100}$$

hence there is at least one set of principal axes for each of them. Another important consequence of Eqs. 10.96 and 10.97 is that: a necessary and sufficient condition that a deformation of a body be a *rigid-body motion* (consisting only of translation and rotation without changing distances between particles) is that all the components of the strain tensor E_{ij} or ε_{ij} be zero throughout the body.

With the use of Eqs. 10.3 and 10.6, the strain tensors can be expressed in terms of the displacement u_i as

$$E_{ij}=\frac{1}{2}\left(\frac{\partial u_i}{\partial x_j}+\frac{\partial u_j}{\partial x_i}+\frac{\partial u_k}{\partial x_i}\frac{\partial u_k}{\partial x_j}\right) \tag{10.101}$$

$$\varepsilon_{ij} = \frac{1}{2}\left(\frac{\partial u_i}{\partial y_j} + \frac{\partial u_j}{\partial y_i} - \frac{\partial u_k}{\partial y_i}\frac{\partial u_k}{\partial y_j}\right) \tag{10.102}$$

Alternatively, we can write E_{ij} and ε_{ij} as

$$E_{ij} = \bar{\varepsilon}_{ij} + \frac{1}{2}(\bar{\varepsilon}_{ik} - \bar{\omega}_{ik})(\bar{\varepsilon}_{jk} - \bar{\omega}_{jk}) \tag{10.103}$$

$$\varepsilon_{ij} = \varepsilon^*_{ij} - \frac{1}{2}(\varepsilon^*_{ik} - \omega^*_{ik})(\varepsilon^*_{jk} - \omega^*_{jk}) \tag{10.104}$$

where

$$\bar{\varepsilon}_{ij} = \frac{1}{2}\left(\frac{\partial u_i}{\partial x_j} + \frac{\partial u_j}{\partial x_i}\right) \tag{10.105}$$

$$\bar{\omega}_{ij} = \frac{1}{2}\left(\frac{\partial u_i}{\partial x_j} - \frac{\partial u_j}{\partial x_i}\right) \tag{10.106}$$

when the displacement u_i is regarded as a function of the Lagrangian coordinates x_i, and

$$\varepsilon^*_{ij} = \frac{1}{2}\left(\frac{\partial u_i}{\partial y_j} + \frac{\partial u_j}{\partial y_i}\right) \tag{10.107}$$

$$\omega^*_{ij} = \frac{1}{2}\left(\frac{\partial u_i}{\partial y_j} - \frac{\partial u_j}{\partial y_i}\right) \tag{10.108}$$

when the displacement u_i is regarded as a function of the Eulerian coordinates y_i.

The *strain energy function* U, *per unit mass*, can be defined as for small deformation (Eqs. 3.1 to 3.7). In the Lagrangian description, the following can be derived

$$\rho^0 \mathrm{d}U = \sigma^0_{ji}\mathrm{d}(\partial u_i/\partial x_j) \tag{10.109}$$

which is analogous to Eq. 3.4. Substituting into the equation above Eq. 10.3 for u_i and Eq. 10.60b for σ^0_{ji}, we get

$$\rho^0 \mathrm{d}U = \Sigma_{jm}\frac{\partial y_i}{\partial x_m}\mathrm{d}\left(\frac{\partial y_i}{\partial x_j} - \delta_{ij}\right)$$

$$= \Sigma_{jm}\frac{\partial y_i}{\partial x_m}\mathrm{d}\left(\frac{\partial y_i}{\partial x_j}\right)$$

$$= \frac{1}{2}\left[\Sigma_{jm}\frac{\partial y_i}{\partial x_m}\mathrm{d}\left(\frac{\partial y_i}{\partial x_j}\right) + \Sigma_{mj}\frac{\partial y_i}{\partial x_j}\mathrm{d}\left(\frac{\partial y_i}{\partial x_m}\right)\right],$$

or, due to the symmetry, $\Sigma_{jm} = \Sigma_{mj}$,

$$\rho^0 \mathrm{d}U = \Sigma_{jm}\frac{1}{2}\left[\frac{\partial y_i}{\partial x_m}\mathrm{d}\left(\frac{\partial y_i}{\partial x_j}\right) + \frac{\partial y_i}{\partial x_j}\mathrm{d}\left(\frac{\partial y_i}{\partial x_m}\right)\right]$$

$$= \Sigma_{jm} \frac{1}{2} \mathrm{d}\left(\frac{\partial y_i}{\partial x_m} \frac{\partial y_i}{\partial x_j} \right)$$

$$= \Sigma_{jm} \frac{1}{2} \mathrm{d}\left(\frac{\partial y_i}{\partial x_m} \frac{\partial y_i}{\partial x_j} - \delta_{jm} \right) \tag{10.110}$$

Replacing the term in the parentheses in the equation above by E_{jm} as defined in Eq. 10.98, we get

$$\rho^0 \mathrm{d}U = \Sigma_{ij} \mathrm{dE}_{ij} \tag{10.111}$$

which is equivalent to Eq. 3.7 for small deformation. If the material is assumed to be elastic, i.e. when the stress components are unique and homogeneous functions of the strain components, there exists the strain energy function U as a scalar homogeneous function of the strain components, i.e.

$$U = U(\mathrm{E}_{ij}) \tag{10.112}$$

Accordingly, its total differential is

$$\mathrm{d}U = \frac{\partial U}{\partial \mathrm{E}_{ij}} \mathrm{dE}_{ij} \tag{10.113}$$

In Eqs. 10.111 and 10.113, each of the nine strain components are considered as an independent variable in the function U. Comparing these two equations, we have

$$\Sigma_{ij} = \rho^0 \frac{\partial U}{\partial \mathrm{E}_{ij}} \tag{10.114}$$

which is equivalent to Eq. 3.10 for small deformation.

Thus we may conclude that a rational form of the constitutive law is the relationship between the Kirchhoff stress Σ_{ij} and the Green's strain E_{ij}.

The small strain tensor as obtained in Chapter II can be specialized from Eq. 10.102 by restricting ourselves to small derivatives of displacements, i.e. $\partial u_i/\partial y_j \ll 1$, and consequently neglecting their products in that equation. Eventually, we get

$$\varepsilon_{ij} = \frac{1}{2}\left(\frac{\partial u_i}{\partial y_j} + \frac{\partial u_j}{\partial y_i} \right), \quad \text{(for small deformation)} \tag{10.115}$$

and Eqs. 10.102 and 10.107 become identical. With the additional help of Eq. 10.73, we can conclude that there is no need to distinguish E_{ij} from ε_{ij}, $\bar{\varepsilon}_{ij}$ or ε_{ij}^* in small deformation.

10.6 REFORMED LAGRANGIAN DESCRIPTION

If a transformation function

$$z_i = z_i(x_j) \tag{10.116}$$

or, reversely,

$$x_i = x_i(z_j) \tag{10.117}$$

is established properly, an undeformed continuum element can be reformed into an element with a simpler shape and more convenient dimensions. So the new description, using z_i and t as independent variables, may be called a *reformed Lagrangian description.* The *reformed Lagrangian coordinates* z_i are sometimes called normal coordinates; they are independent of t just as the x_i. The one-to-one correspondence with the Eulerian description may be written

$$z_i = z_i(y_j, t) \tag{10.118}$$

or, reversely,

$$y_i = y_i(z_j, t) \tag{10.119}$$

Similar to Eqs. 10.10 to 10.12, we can derive the following

$$\frac{\partial}{\partial z_i} A(x_j, t) = \frac{\partial A}{\partial x_k} \frac{\partial x_k}{\partial z_i} \tag{10.120a}$$

$$\frac{\partial}{\partial x_i} A(z_j, t) = \frac{\partial A}{\partial z_k} \frac{\partial z_k}{\partial x_i} \tag{10.120b}$$

$$\frac{\partial}{\partial z_i} A(y_j, t) = \frac{\partial A}{\partial y_k} \frac{\partial y_k}{\partial z_i} \tag{10.120c}$$

$$\frac{\partial}{\partial y_i} A(z_j, t) = \frac{\partial A}{\partial z_k} \frac{\partial z_k}{\partial y_i} \tag{10.120d}$$

$$\frac{\partial x_i}{\partial z_k} \frac{\partial z_k}{\partial x_j} = \delta_{ij} = \frac{\partial z_i}{\partial x_k} \frac{\partial x_k}{\partial z_j} \tag{10.120e}$$

$$\frac{\partial y_i}{\partial z_k} \frac{\partial z_k}{\partial y_j} = \delta_{ij} = \frac{\partial z_i}{\partial y_k} \frac{\partial y_k}{\partial z_j} \tag{10.120f}$$

$$\left[\frac{\partial x_i}{\partial z_j}\right] = \left[\frac{\partial z_i}{\partial x_j}\right]^{-1}, \quad \left[\frac{\partial y_i}{\partial z_j}\right] = \left[\frac{\partial z_i}{\partial y_j}\right]^{-1} \tag{10.120g,h}$$

Denoting the reformed Lagrangian description by the superscript r, and adopting the same procedures as used in the preceding sections to derive various conditions in the correspondence between two different descriptions, we can get the one-to-one mapping of a vector defining an infinitesimal line element as

$$\mathrm{d}x_i = \frac{\partial x_i}{\partial z_j} \mathrm{d}z_j, \quad \mathrm{d}y_i = \frac{\partial y_i}{\partial z_j} \mathrm{d}z_j \tag{10.121a,b}$$

or, reversely,

$$\mathrm{d}z_i = \frac{\partial z_i}{\partial x_j} \mathrm{d}x_j = \frac{\partial z_i}{\partial y_j} \mathrm{d}y_j \tag{10.121c,d}$$

The one-to-one mapping of a volume and the conservation of mass are

$$\frac{\rho^r}{\rho^0}=\frac{dV^0}{dV^r}=\left|\frac{\partial x_i}{\partial z_j}\right|=\left|\frac{\partial z_i}{\partial x_j}\right|^{-1} \tag{10.122a,b,c}$$

$$\frac{\rho^r}{\rho}=\frac{dV}{dV^r}=\left|\frac{\partial y_i}{\partial z_j}\right|=\left|\frac{\partial z_i}{\partial y_j}\right|^{-1} \tag{10.122d,e,f}$$

The one-to-one mapping of a surface is

$$\frac{\partial z_i}{\partial x_j}n_i^r dS^r=\frac{\rho^0}{\rho^r}n_j^0 dS^0 \tag{10.123a}$$

$$\frac{\partial z_i}{\partial y_j}n_i^r dS^r=\frac{\rho}{\rho^r}n_j dS \tag{10.123b}$$

The equality of force vectors on two corresponding surfaces gives

$$\overline{X}_i^r dS^r=\overline{X}_i^0 dS^0=\overline{X}_i dS \tag{10.124a,b}$$

The stress vector formulas are

$$\overline{X}_i^r=\sigma_{ji}^r n_j^r,\quad \overline{X}_i^0=\sigma_{ji}^0 n_j^0,\quad \overline{X}_i=\sigma_{ji}n_j \tag{10.125a,b,c}$$

The *reformed Langrangian stress tensor* is

$$\sigma_{ji}^r=\frac{\rho^r}{\rho^0}\frac{\partial z_j}{\partial x_m}\sigma_{mi}^0 \tag{10.126a}$$

$$\sigma_{ji}^r=\frac{\rho^r}{\rho}\frac{\partial z_j}{\partial y_m}\sigma_{mi} \tag{10.126b}$$

The *reformed Kirchhoff stress tensor* is

$$\Sigma_{ij}^r=\frac{\partial z_j}{\partial y_l}\sigma_{il}^r=\frac{\rho^r}{\rho}\frac{\partial z_i}{\partial y_l}\frac{\partial z_j}{\partial y_m}\sigma_{ml} \tag{10.127a,b}$$

$$\Sigma_{ij}^r=\frac{\rho^r}{\rho^0}\frac{\partial z_i}{\partial x_l}\frac{\partial z_j}{\partial x_m}\Sigma_{ml} \tag{10.127c}$$

The equation of motion is

$$\frac{\partial \sigma_{ji}^r}{\partial z_j}+\rho^r X_i=\rho^r\frac{\partial v_i}{\partial t} \tag{10.128a}$$

or

$$\frac{\partial}{\partial z_j}\left(\frac{\partial y_i}{\partial z_m}\Sigma_{jm}^r\right)+\rho^r X_i=\rho^r\frac{\partial v_i}{\partial t} \tag{10.128b}$$

The strain energy function U is

$$\rho^r dU=\Sigma_{ij}^r dE_{ij}^r \tag{10.129}$$

where E_{ij}^r is the *reformed Green's strain tensor*, i.e.

$$E^r_{ij} = \frac{1}{2}\left(\frac{\partial y_k}{\partial z_i}\frac{\partial y_k}{\partial z_j} - \frac{\partial x_k}{\partial z_i}\frac{\partial x_k}{\partial z_j}\right) \tag{10.130a}$$

$$= \frac{\partial x_k}{\partial z_i}\frac{\partial x_m}{\partial z_j}E_{km} \tag{10.130b}$$

Thus another rational form of the constitutive law is the relationship between the reformed Kirchhoff stress tensor Σ^r_{ij} and the reformed Green's strain E^r_{ij}. For linear elastic solids, Hooke's law applies, and relationships in the Lagrangian and reformed Lagrangian descriptions take the following forms

$$\Sigma_{ij} = C_{ijkl}E_{kl} \tag{10.131a}$$

$$\Sigma^r_{ij} = C^r_{ijkl}E^r_{kl} \tag{10.131b}$$

Making use of Eqs. 10.120e, 10.127c, 10.130b and 10.131, we get the relationship between the stiffness tensors in these two different descriptions as

$$C^r_{ijkl} = \frac{\rho^r}{\rho^0}\frac{\partial z_i}{\partial x_m}\frac{\partial z_j}{\partial x_n}\frac{\partial z_k}{\partial x_o}\frac{\partial z_l}{\partial x_p}C_{mnop} \tag{10.132a}$$

$$C_{ijkl} = \frac{\rho^0}{\rho^r}\frac{\partial x_i}{\partial z_m}\frac{\partial x_j}{\partial z_n}\frac{\partial x_k}{\partial z_o}\frac{\partial x_l}{\partial z_p}C^r_{mnop} \tag{10.132b}$$

An eight-noded three dimensional element, which may not be rectangular, can be transformed into a rectangular element with $2\times2\times2$ dimensions by means of the following transformation

$$x_i = \sum_{a=1}^{8} N^{(a)}(z_j)x_i^{(a)} \tag{10.133}$$

where the superscript a in parentheses denotes a node number, and $N^{(a)}(z_j)$ is an *interpolation function* taking the form

$$N^{(a)}(z_j) = \frac{1}{8}(1+z_1z_1^{(a)})(1+z_2z_2^{(a)})(1+z_3z_3^{(a)}) \tag{10.134}$$

In the equation above, $z_i^{(a)}$ is the position vector of the node a in the reformed Lagrangian description, and can be expressed as

$$z_1^{(a)} = -1, \quad \text{for } a = 1,4,5,8 \tag{10.135a}$$

$$= 1, \quad \text{for } a = 2,3,6,7 \tag{10.135b}$$

$$z_2^{(a)} = -1, \quad \text{for } a = 1,2,5,6 \tag{10.136a}$$

$$= 1, \quad \text{for } a = 3,4,7,8 \tag{10.136b}$$

$$z_3^{(a)} = -1, \quad \text{for } a = 1,2,3,4 \tag{10.137a}$$

$$= 1, \quad \text{for } a = 5,6,7,8 \tag{10.137b}$$

Such a reformed eight-noded three-dimensional element can be put as in Fig. 10.7(a). It can be said that each component of the position vector x_i inside the element can be obtained by interpolation among its nodal values using Eq. 10.133. The same interpolation function can be used for the displacement vector u_i also, i.e.

$$u_i = \sum_{a=1}^{8} N^{(a)}(z_j)u_i^{(a)} \tag{10.138}$$

To avoid some difficulties involved in the treatment of finite rotations in large deformation analysis of shells, Kanok-Nukulchai, Taylor and Hughes (1981) assigned nodes 1 to 4 to the shell midsurface and nodes 5 to 8 to the shell top surface, and proposed the following interpolation function

$$N^{(a)}(z_j) = f(z_j, z_j^{(a)}), \quad \text{for } a = 1 \text{ to } 4 \tag{10.139a}$$

$$= z_3 f(z_j, z_j^{(a)}), \quad \text{for } a = 5 \text{ to } 8 \tag{10.139b}$$

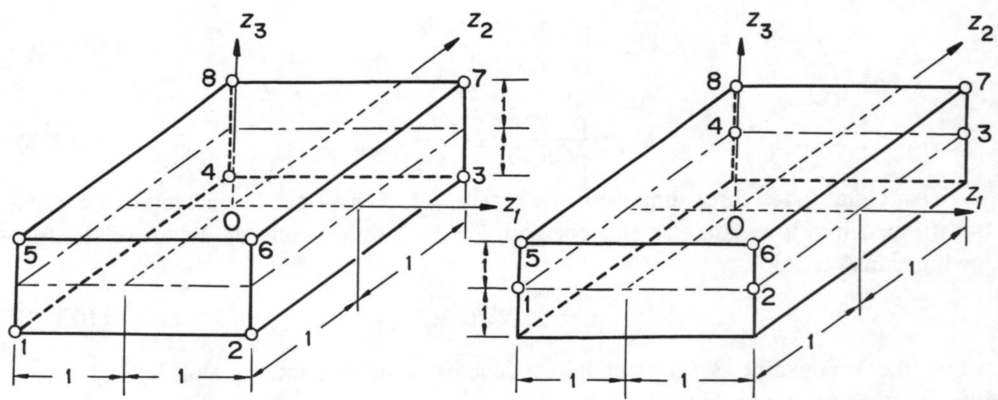

(a) Three dimensional element. (b) Shell element.

Fig. 10.7 Reformed eight-noded continuum elements.

where

$$f(z_j, z_j^{(a)}) = \frac{1}{4}(1 + z_1 z_1^{(a)})(1 + z_2 z_2^{(a)}) \tag{10.140}$$

Moreover, $x_i^{(a)}$ and $u_i^{(a)}$ for $a = 5$ to 8 were assigned relative quantities with respect to nodes $a - 4$; e.g. while $x_1^{(2)}$ is x_1 of the node 2, $x_1^{(6)}$ is equal to x_1 of the node 6 minus that of the node 2. It should be noted that $z_1^{(a)}$ and $z_2^{(a)}$ are as given by Eqs. 10.135 and 10.136 respectively, while Eqs. 10.137 are replaced by the condition $0 \leq z_3 \leq 1$. Figure 10.7(b) depicts such a reformed eight-noded shell element.

10.7 STRAIN TENSORS IN CURVILINEAR COORDINATES

Notations and rules developed in Sections 1.14 and 1.15 will be followed. Figure 10.8 depicts the geometry of an infinitesimal parallelepiped before and after deformation. A point P^0 originally located by $\mathbf{x}$ moves to a new position P located by $\mathbf{y}$, while another point Q^0 located by $\mathbf{x}+d\mathbf{x}$ moves to a position Q located by $\mathbf{y}+d\mathbf{y}$. In terms of curvilinear coordinates, the differential vector $d\mathbf{x}$ is as defined in Eq. 1.173, and the differential vector $d\mathbf{y}$ can be put in a similar form, i.e.

$$d\mathbf{y} = \frac{\partial \mathbf{y}}{\partial \theta^1} d\theta^1 + \frac{\partial \mathbf{y}}{\partial \theta^2} d\theta^2 + \frac{\partial \mathbf{y}}{\partial \theta^3} d\theta^3 \tag{10.141}$$

or

$$d\mathbf{y} = \mathbf{G}_\lambda d\theta^\lambda \tag{10.142}$$

where $\mathbf{G}_\lambda$ is the *covariant base vector after deformation*, i.e.

$$\mathbf{G}_\lambda = \frac{\partial \mathbf{y}}{\partial \theta^\lambda} \equiv \mathbf{y}_{,\lambda} \tag{10.143}$$

Note the direction of each $\mathbf{G}_\lambda$ in Fig. 10.8. In fact, all identities derived in Sections 1.14 and 1.15 in terms of g and x also hold when g and x are changed into G and y, respectively. For examples; the square of the length of the element PQ is

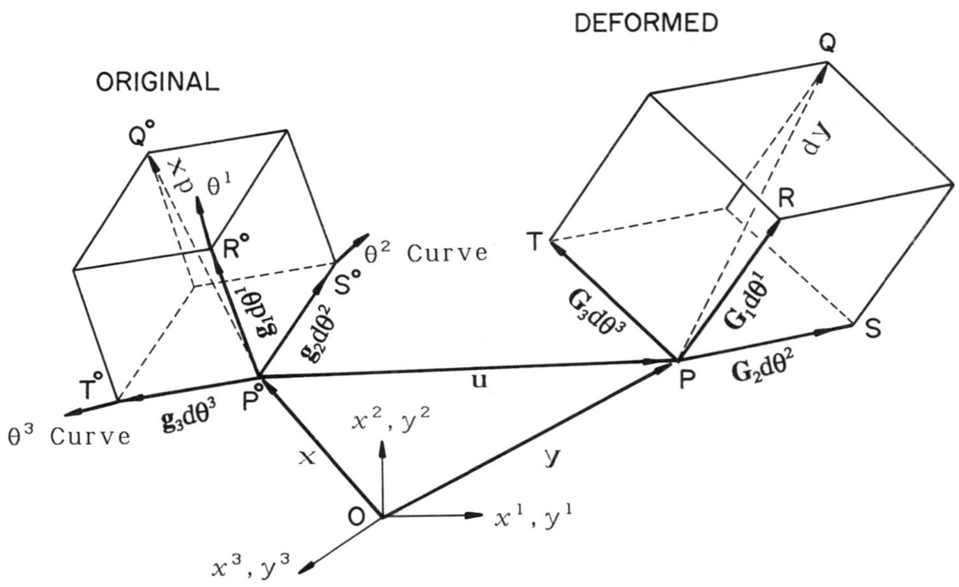

Fig. 10.8 Original and deformed configurations of an infinitesimal parallelepiped.

$$\mathbf{dy} \cdot \mathbf{dy} = G_{\lambda\mu} d\theta^{\lambda} d\theta^{\mu} \tag{10.144}$$

where $G_{\lambda\mu}$ is the *covariant metric tensor after deformation*, the volume of the infinitesimal parallelepiped after deformation is

$$dV = \sqrt{G}\, d\theta^1 d\theta^2 d\theta^3 \tag{10.145}$$

where $G = \det[G_{\lambda\mu}]$, and the derivative of $\mathbf{G}_{\mu}$ with respect to θ^{ν} is

$$\mathbf{G}_{\mu,\nu} = \left\{ \begin{matrix} \lambda \\ \mu\nu \end{matrix} \right\}^{G} \mathbf{G}_{\lambda} \tag{10.146}$$

in which $\left\{ \begin{matrix} \lambda \\ \mu\nu \end{matrix} \right\}^{G}$ is the *Christoffel symbol of the second kind after deformation.*

The difference between Eqs. 10.144 and 1.177 is

$$\mathbf{dy} \cdot \mathbf{dy} - \mathbf{dx} \cdot \mathbf{dx} = (G_{\lambda\mu} - g_{\lambda\mu}) d\theta^{\lambda} d\theta^{\mu} \tag{10.147}$$

Comparing the last equation with Eq. 10.96, we define the *strain tensor in curvilinear coordinates* as follows

$$\Delta_{\lambda\mu} = \frac{1}{2}(G_{\lambda\mu} - g_{\lambda\mu}) = \Delta_{\mu\lambda} \tag{10.148}$$

As shown in Fig. 10.8 the displacement vector $\mathbf{u}(\theta^1, \theta^2, \theta^3)$ is related to the position vectors $\mathbf{x}$ and $\mathbf{y}$ as

$$\mathbf{y} = \mathbf{x} + \mathbf{u} \tag{10.149}$$

The *contravariant* and *covariant components* of $\mathbf{u}$ have been defined in Eqs. 1.199 and 1.202, respectively, i.e.

$$\mathbf{u} = v^{\lambda} \mathbf{g}_{\lambda} = v_{\lambda} \mathbf{g}^{\lambda} \tag{10.150a,b}$$

Substituting Eq. 10.149 into Eq. 10.143, and using Eqs. 1.174, 1.200 and 10.150a, we find

$$\mathbf{G}_{\lambda} = (\delta_{\lambda}^{\kappa} + v^{\kappa}_{;\lambda}) \mathbf{g}_{\kappa} \tag{10.151}$$

Due to the last equation, Eq. 10.148 becomes

$$\Delta_{\lambda\mu} = \frac{1}{2}(g_{\lambda\kappa} v^{\kappa}_{;\mu} + g_{\kappa\mu} v^{\kappa}_{;\lambda} + g_{\kappa\rho} v^{\kappa}_{;\lambda} v^{\rho}_{;\mu}) \tag{10.152}$$

or, making use of Eq. 1.209,

$$\Delta_{\lambda\mu} = \frac{1}{2}(v_{\lambda;\mu} + v_{\mu;\lambda} + v_{\kappa;\lambda} v^{\kappa}_{;\mu}) \tag{10.153}$$

10.8 EQUILIBRIUM EQUATIONS AND STRESS TENSORS IN CURVILINEAR COORDINATES

Consider the equilibrium of an infinitesimal parallelepiped at a point P in the deformed configuration shown in Fig. 10.9. Forces involved are the body force $\mathbf{X}$ (per unit mass) and tractions on the six infinitesimal surfaces. Since the area of the surface θ^λ = constant in the original configuration is, due to Eq. 1.193, proportional to $d\theta^\mu d\theta^\kappa$, it is logical to put the resultant force vector corresponding to this surface in the form

$$d\mathbf{T}^\lambda = \mathbf{t}^\lambda \sqrt{g}\, d\theta^\mu d\theta^\kappa, \quad (\lambda \neq \mu \neq \kappa) \tag{10.154}$$

where g is as defined in Eqs. 1.183b to d. Naturally, the resultant force vector on the side corresponding to the surface $\theta^\lambda + d\theta^\lambda$ = constant in the original configuration is

$$d\mathbf{T}^\lambda + \frac{\partial(d\mathbf{T}^\lambda)}{\partial\theta^\lambda} d\theta^\lambda, \quad (\lambda \text{ not summed}).$$

All resultant force vectors involved are depicted on appropriate sides of the parallelepiped, following the sign convention of stresses (positive if in tension), in Fig. 10.9. Accordingly, the equilibrium of this parallelepiped can be put in the form

$$\frac{\partial(d\mathbf{T}^1)}{\partial\theta^1} d\theta^1 + \frac{\partial(d\mathbf{T}^2)}{\partial\theta^2} d\theta^2 + \frac{\partial(d\mathbf{T}^3)}{\partial\theta^3} d\theta^3 + \rho^0 \mathbf{X} dV^0 = 0 \tag{10.155}$$

Substituting Eqs. 1.186 and 10.154 into the last equation we find

$$(\sqrt{g}\, \mathbf{t}^\kappa)_{,\kappa} + \rho^0 \mathbf{X} \sqrt{g} = 0 \tag{10.156}$$

The vector $\mathbf{t}^\lambda$ may be expressed in the form

$$\mathbf{t}^\lambda = \tau^{\lambda\mu} \mathbf{G}_\mu \tag{10.157}$$

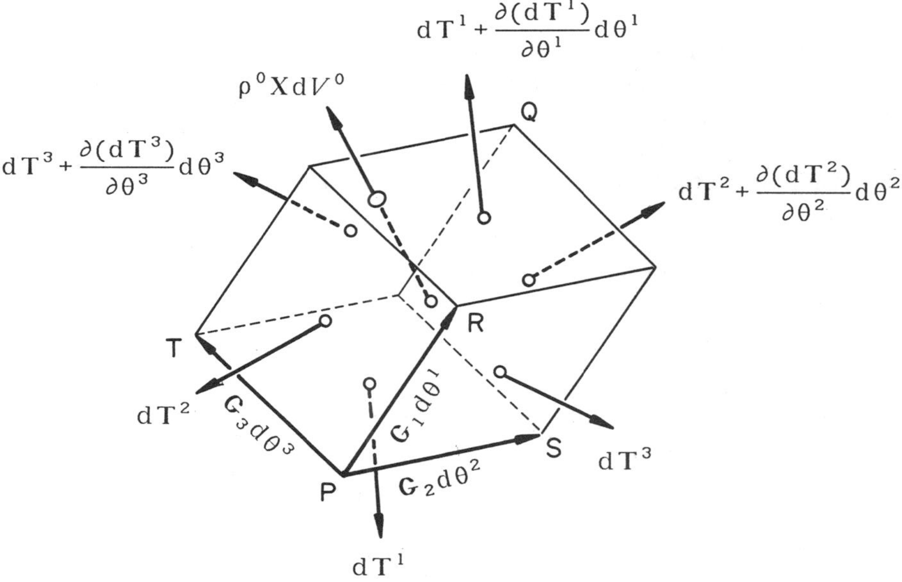

Fig. 10.9 Deformed configuration of an infinitesimal parallelepiped being subjected to forces.

A consideration of the moment equilibrium of the infinitesimal parallelepiped leads to the condition

$$\sum_{\lambda=1}^{3} d\mathbf{T}^{\lambda} \times \mathbf{G}_{\lambda} d\theta^{\lambda} = 0,$$

or, due to Eqs. 10.154 and 10.157,

$$\tau^{\lambda\mu}\mathbf{G}_{\mu} \times \mathbf{G}_{\lambda} = 0,$$

or

$$\frac{1}{2}(\tau^{\lambda\mu} - \tau^{\mu\lambda})\mathbf{G}_{\mu} \times \mathbf{G}_{\lambda} = 0.$$

Since the vector $\mathbf{G}_{\mu} \times \mathbf{G}_{\lambda}$ is nonzero for $\mu \neq \lambda$, the equation above gives

$$\tau^{\lambda\mu} = \tau^{\mu\lambda} \tag{10.158}$$

i.e. the stress tensor $\tau^{\lambda\mu}$ is symmetric.

The body force $\mathbf{X}$ may be expressed in terms of its contravariant components before deformation as follows

$$\mathbf{X} = X_g^{\lambda}\mathbf{g}_{\lambda} \tag{10.159}$$

Substituting Eqs. 10.157 and 10.159 into Eq. 10.156, and using Eqs. 1.194 and 10.151, we find

$$[\sqrt{g}(\delta_{\mu}^{\lambda} + v_{;\mu}^{\lambda})\tau^{\kappa\mu}]_{,\kappa} + \sqrt{g}(\delta_{\mu}^{\rho} + v_{;\mu}^{\rho})\tau^{\kappa\mu}\begin{Bmatrix}\lambda \\ \kappa\rho\end{Bmatrix} + \rho^{0}\sqrt{g}X_g^{\lambda} = 0 \tag{10.160}$$

The equation above holds for dynamic equilibrium as well if $\mathbf{X}$ is considered as the sum of the actual body force vector and the D'Alembert's force vector, which is equal to the negative of the acceleration vector.

If $\tau_0^{1\mu}g_{\mu}$ is the stress vector on the surface where $\theta^1 = \text{constant}$ in the original configuration, the resultant force $d\mathbf{T}^1$ there, because of Eq. 1.193, becomes

$$d\mathbf{T}^1 = \tau_0^{1\mu}\mathbf{g}_{\mu}\sqrt{g\,g^{11}}\,d\theta^2 d\theta^3 \tag{10.161}$$

Comparing the last equation with Eq. 10.154 for $\lambda = 1$, we get

$$\mathbf{t}^1 = \tau_0^{1\mu}\sqrt{g^{11}}\,\mathbf{g}_{\mu} \tag{10.162}$$

or, for any λ,

$$\mathbf{t}^{\lambda} = \tau_0^{\lambda\mu}\sqrt{g^{\lambda\lambda}}\,\mathbf{g}_{\mu}, \quad (\lambda \text{ not summed}) \tag{10.163}$$

On the other hand, Eqs. 10.154 and 10.157 give

$$d\mathbf{T}^1 = \tau^{1\mu}\mathbf{G}_{\mu}\sqrt{g}\,d\theta^2 d\theta^3 \tag{10.164}$$

Equating Eq. 10.161 with Eq. 10.164, we get

$$\tau_0^{1\mu}\mathbf{g}_{\mu} = \frac{1}{\sqrt{g^{11}}}(\tau^{1\mu}\mathbf{G}_{\mu}) \tag{10.165}$$

or

$$\tau_0^{1\mu}\mathbf{g}_\mu = \tau_0^{11}\mathbf{g}_1 + \tau_0^{12}\mathbf{g}_2 + \tau_0^{13}\mathbf{g}_3 = \frac{1}{\sqrt{g^{11}}}(\tau^{11}\mathbf{G}_1 + \tau^{12}\mathbf{G}_2 + \tau^{13}\mathbf{G}_3) \qquad (10.166a,b)$$

or, on the surface where $\theta^\lambda = \text{constant}$,

$$\tau_0^{\lambda\mu}\mathbf{g}_\mu = \tau_0^{\lambda 1}\mathbf{g}_1 + \tau_0^{\lambda 2}\mathbf{g}_2 + \tau_0^{\lambda 3}\mathbf{g}_3 = \frac{1}{\sqrt{g^{\lambda\lambda}}}(\tau^{\lambda 1}\mathbf{G}_1 + \tau^{\lambda 2}\mathbf{G}_2 + \tau^{\lambda 3}\mathbf{G}_3),$$

(λ not asummed) (10.167a,b)

Substituting Eq. 10.151 into Eq. 10.165 then considering the vectorial components of the result in $\mathbf{g}_1$, $\mathbf{g}_2$ and $\mathbf{g}_3$ directions yield, respectively,

$$\tau_0^{11} = \frac{1}{\sqrt{g^{11}}}\tau^{1\mu}(\delta_\mu^1 + v_{;\mu}^1) \qquad (10.168a)$$

$$\tau_0^{12} = \frac{1}{\sqrt{g^{11}}}\tau^{1\mu}(\delta_\mu^2 + v_{;\mu}^2) \qquad (10.168b)$$

$$\tau_0^{13} = \frac{1}{\sqrt{g^{11}}}\tau^{1\mu}(\delta_\mu^3 + v_{;\mu}^3) \qquad (10.168c)$$

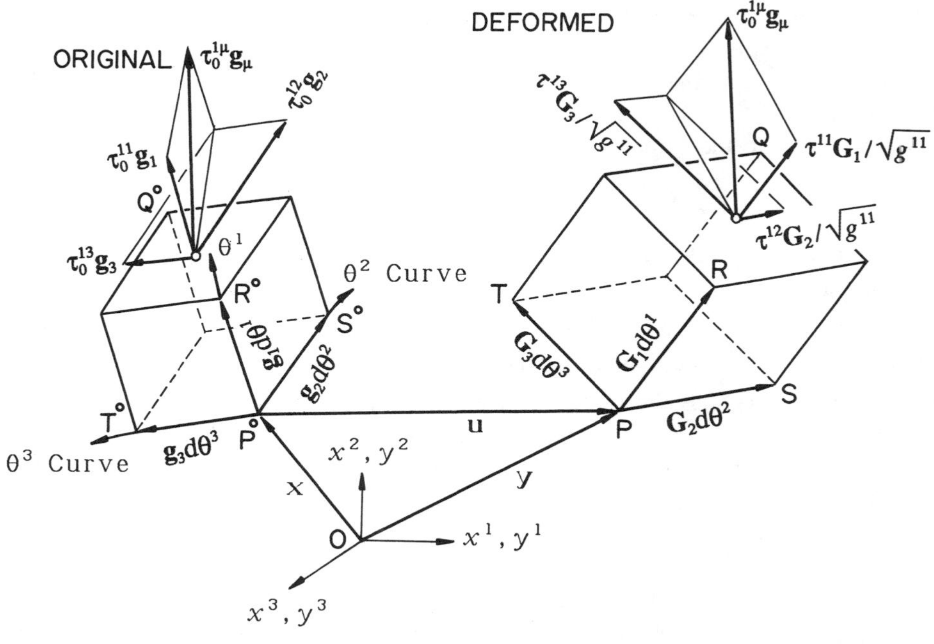

Fig. 10.10 Decomposition of stress vector $\tau_0^{1\mu}\mathbf{g}_\mu$ into three components parallel to edges of each parallelepiped.

Thus we may write a general relationship between $\tau_0^{\lambda\mu}$ and $\tau^{\lambda\mu}$ as the following

$$\tau_0^{\lambda\mu}=\frac{1}{\sqrt{g^{\lambda\lambda}}}\tau^{\lambda\kappa}(\delta_\kappa^\mu+v_{;\kappa}^\mu),\quad (\lambda \text{ not summed}) \tag{10.169}$$

It should be noted that while $\tau^{\lambda\mu}$ carries a meaning similar to the Cartesian Kirchhoff stress tensor Σ_{ij}, $\tau_0^{\lambda\mu}$ carries a meaning similar to the Cartesian Lagrangian stress tensor σ_{ij}^0. Moreover, Eqs. 10.167 imply that the stress vector $\tau_0^{\lambda\mu}\mathbf{g}_\mu$ has three vectorial components in θ^1, θ^2 and θ^3 directions in the original configuration, but in $\mathbf{G}_1$, $\mathbf{G}_2$ and $\mathbf{G}_3$ directions in the deformed configuration. Figure 10.10 depicts, as an example, such vectorial decompositions of the stress vector $\tau_0^{1\mu}\mathbf{g}_\mu$. If the original dimension of the structure in the θ^3-direction is relatively small, we may assume that τ^{33}, τ^{31} and τ^{32} are zero, i.e. the same assumption as of the plane stress conditions with respect to the surface where $\theta^3=\text{constant}$.

10.9 PHYSICAL COMPONENTS OF VECTORS AND TENSORS

When the θ-system is a local rectangular Cartesian coordinate system, mathematical conditions involved are very much simplified as Eqs. 1.211 to 1.213 show, and the physical meaning of the quantities involved can be comprehended more easily. Most obviously, a contravariant or a covariant component of a vector such as in Eqs. 10.150 and 10.159 becomes *indeed a Cartesian component* of that vector. The strain tensor given in Eq. 10.153 carries the same physical meaning as Eq. 10.101, since the former can be reduced to the form

$$\Delta_{\lambda\mu}=\frac{1}{2}(v_{\lambda,\mu}+v_{\mu,\lambda}+v_{\kappa,\lambda}v_{\kappa,\mu}),\quad (\text{for Cartesian system}) \tag{10.170}$$

Next, rewriting Eq. 10.154 under the special conditions mentioned above for $\lambda=1$, as an example, we have

$$d\mathbf{T}^1=\mathbf{t}^1 d\theta^2 d\theta^3,\quad (\text{for Cartesian system}) \tag{10.171}$$

which shows that $\mathbf{t}^1$ has the same physical meaning as σ_{1i}^0 in Eq. 10.62. With an observation that $\mathbf{G}_\lambda$ has the same physical meaning as $\partial y_i/\partial x_1$ for the global Cartesian coordinate system, we can also conclude that the stress tensor $\tau^{\lambda\mu}$ in Eq. 10.157 carries the same meaning as Σ_{ij} in Eq. 10.60b. Further, we can write the equation of equilibrium, Eq. 10.160, in the form

$$[\tau^{\kappa\mu}(\delta_\mu^\lambda+v_{,\mu}^\lambda)]_{,\kappa}+\rho^0 X_g^\lambda=0,\quad (\text{for Cartesian system}) \tag{10.172}$$

which clearly has the same physical meaning as Eq. 10.95.

In general, components of a tensor derived for a curvilinear coordinate system are not mutually orthogonal, and do not have ordinary physical dimensions. Still, it is expedient to formulate some physical problems in terms of such non-physical components, then derive the corresponding physical components at a later stage. The derivation for physical components from the non-physical ones referred to the θ-system can start with setting up a local rectangular Cartesian coordinates system (z-system) at the material point P^0 as shown in Fig. 10.11. Denoting the unit vector in the direction of z^λ-axis by $\mathbf{j}_\lambda$, and noting the identities

$$\frac{\partial(\ )}{\partial\theta^\lambda}=\frac{\partial(\ )}{\partial z^\mu}\frac{\partial z^\mu}{\partial\theta^\lambda} \tag{10.173a}$$

$$\frac{\partial(\ )}{\partial z^\lambda}=\frac{\partial(\ )}{\partial\theta^\mu}\frac{\partial\theta^\mu}{\partial z^\lambda} \tag{10.173b}$$

we have

$$\mathbf{g}_\mu=\frac{\partial\mathbf{x}}{\partial\theta^\mu}=\frac{\partial z^\lambda}{\partial\theta^\mu}\mathbf{j}_\lambda \tag{10.174a}$$

$$\mathbf{j}_\lambda=\frac{\partial\mathbf{x}}{\partial z^\lambda}=\frac{\partial\theta^\mu}{\partial z^\lambda}\mathbf{g}_\mu \tag{10.174b}$$

$$\frac{\partial z^\lambda}{\partial\theta^\mu}=\mathbf{j}_\lambda\cdot\mathbf{g}_\mu \tag{10.174c}$$

$$\frac{\partial\theta^\lambda}{\partial z^\mu}=g^{\lambda\kappa}(\mathbf{j}_\mu\cdot\mathbf{g}_\kappa) \tag{10.174d}$$

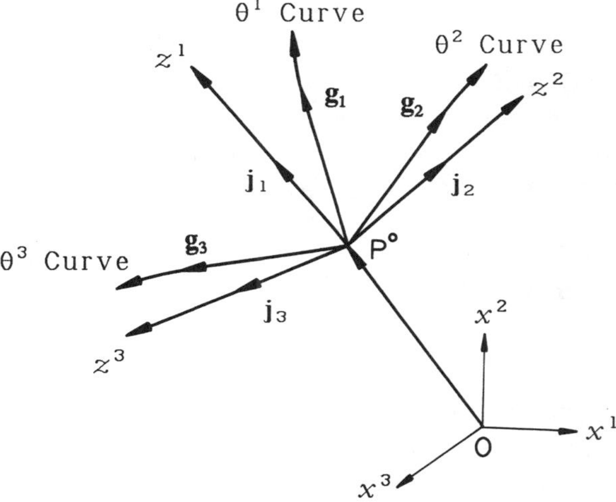

Fig. 10.11 Global Cartesian coordinates x^i, curvilinear coordinates θ^λ, and local Cartesian coordinates z^i of material point P^0.

In addition, the displacement vector **u** can be written as

$$\mathbf{u} = u^{\lambda}\mathbf{j}_{\lambda} \tag{10.175}$$

Equating this equation with Eq. 10.150, and using Eq. 10.174a, we get

$$u^{\lambda} = \frac{\partial z^{\lambda}}{\partial \theta^{\mu}} v^{\mu} \tag{10.176}$$

Similarly, for the body force vector **X**, we can show that

$$X^{\lambda} = \frac{\partial z^{\lambda}}{\partial \theta^{\mu}} X_g^{\mu} \tag{10.177}$$

where X_g^{μ} has been defined in Eq. 10.159, while X^{λ} is the Cartesian component in z^{λ}-direction, i.e.

$$\mathbf{X} = X^{\lambda}\mathbf{j}_{\lambda} \tag{10.178}$$

Being consistent with the derivations of strain tensors of Eqs. 10.98 and 10.148, the physical strain component defined with respect to the z-system is

$$\mathrm{E}_{\lambda\mu} = \frac{1}{2}\left[\frac{\partial \mathbf{y}}{\partial z^{\lambda}} \cdot \frac{\partial \mathbf{y}}{\partial z^{\mu}} - \frac{\partial \mathbf{x}}{\partial z^{\lambda}} \cdot \frac{\partial \mathbf{x}}{\partial z^{\mu}}\right] \tag{10.179}$$

Using Eqs. 10.148, 10.173b and 10.174c, we can rewrite the last equation as

$$\mathrm{E}_{\lambda\mu} = \frac{\partial \theta^{\kappa}}{\partial z^{\lambda}} \cdot \frac{\partial \theta^{\rho}}{\partial z^{\mu}} \Delta_{\kappa\rho} \tag{10.180}$$

or, conversely,

$$\Delta_{\lambda\mu} = \frac{\partial z^{\kappa}}{\partial \theta^{\lambda}} \cdot \frac{\partial z^{\rho}}{\partial \theta^{\mu}} \mathrm{E}_{\kappa\rho} \tag{10.181}$$

In fact, $\mathrm{E}_{\lambda\mu}$ is a *local Cartesian component of the Green's strain tensor.*

For the relationship between stress tensors of the two different curvilinear coordinate systems, the derivation is based on the equilibrium of the system shown in Fig. 10.12 where the infinitesimal tetrahedron *PRST* denotes the deformed configuration of $P^0R^0S^0T^0$ in the original state. In addition to the surface forces on this infinitesimal tetrahedron depicted in Fig. 10.9, there is another force $\overline{\mathbf{X}}dS_0^t$ on the surface *RST*, where dS_0^t is the area of the surface $R^0S^0T^0$. The condition of equilibrium of the forces acting on the tetrahedron can be put in the form

$$\overline{\mathbf{X}}dS_0^t = (d\mathbf{T}^1 + d\mathbf{T}^2 + d\mathbf{T}^3)/2 \tag{10.182}$$

The body (actual and D'Alembert's) forces do not enter into this equation, because they are of higher order of smallness than the surface forces. Substituting Eq. 10.154 into the last equation, and using Eqs. 1.188, we find

$$\overline{\mathbf{X}} = (\mathbf{g}_{\kappa} \cdot \mathbf{n}_0^t)\mathbf{t}^{\kappa} \tag{10.183}$$

where $\mathbf{n}_0^t$ is the unit vector normal to the surface $R^0S^0T^0$. On the other hand, if we write $\overline{\mathbf{X}}$ with respect to the z-system, i.e. for an infinitesimal tetrahedron formed by

$\mathbf{j}_1 dy^1$, $\mathbf{j}_2 dy^2$ and $\mathbf{j}_3 dy^3$, we get

$$\overline{\mathbf{X}} = (\mathbf{j}_\lambda \cdot \mathbf{n}_0^t)\mathbf{S}^\lambda \tag{10.184}$$

where $\mathbf{S}^\lambda$ is introduced in a similar form as $\mathbf{t}^\lambda$ was in Eq. 10.157, i.e.

$$\mathbf{S}^\lambda = \Sigma^{\lambda\mu}\frac{\partial \mathbf{y}}{\partial z^\mu} \tag{10.185}$$

Since $\overline{\mathbf{X}}$, being a physical quantity, does not depend on the choice of coordinate systems, we can equate Eq. 10.184 with Eq. 10.183 and get

$$(\mathbf{j}_\lambda \cdot \mathbf{n}_0^t)\mathbf{S}^\lambda = (\mathbf{g}_\mu \cdot \mathbf{n}_0^t)\mathbf{t}^\mu \tag{10.186}$$

By taking the direction $\mathbf{n}_0^t$ coincident with a z^κ-axis and substituting Eqs. 10.157 and 10.185 into Eq. 10.186, we have

$$\delta_{\lambda\kappa}\Sigma^{\lambda\rho}\frac{\partial \mathbf{y}}{\partial z^\rho} = (\mathbf{g}_\mu \cdot \mathbf{j}_\kappa)\tau^{\mu\lambda}\frac{\partial \mathbf{y}}{\partial \theta^\lambda} \tag{10.187}$$

or, due to Eqs. 10.173b and 10.174d,

$$\Sigma^{\lambda\mu} = \frac{\partial z^\lambda}{\partial \theta^\kappa}\frac{\partial z^\mu}{\partial \theta^\rho}\tau^{\kappa\rho} \tag{10.188a}$$

or, conversely,

$$\tau^{\lambda\mu} = \frac{\partial \theta^\lambda}{\partial z^\kappa}\frac{\partial \theta^\mu}{\partial z^\rho}\Sigma^{\kappa\rho} \tag{10.188b}$$

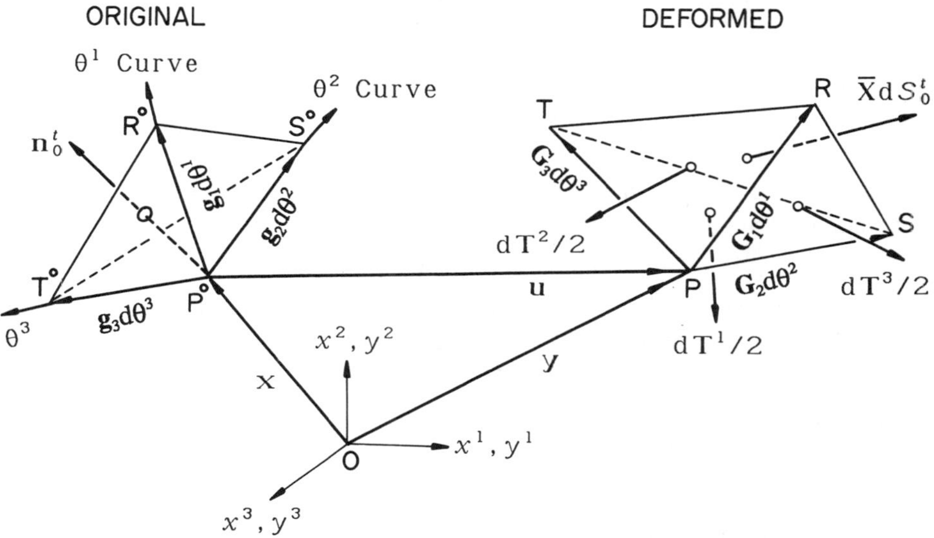

Fig. 10.12 Original and deformed configurations of an infinitesimal tetrahedron being subjected to forces.

In fact, $\Sigma^{\lambda\mu}$ is a *local Cartesian component of the Kirchhoff stress tensor.* Next, we will find the physical components when the θ-system is orthogonal, for which some prevailing conditions are listed in Eqs. 1.210. Setting a local rectangular Cartesian coordinate system (z-system) to have the same orientation as the θ-system, i.e. $\mathbf{j}_1$, $\mathbf{j}_2$ and $\mathbf{j}_3$ are in the same directions as $\mathbf{g}_1$, $\mathbf{g}_2$ and $\mathbf{g}_3$, respectively, we can show that

$$\mathbf{g}_\lambda = \sqrt{g_{\lambda\lambda}}\,\mathbf{j}_\lambda \tag{10.189a}$$

$$\frac{\partial z^\lambda}{\partial \theta^\mu} = \sqrt{g_{\mu\mu}}\,\delta^\lambda_\mu \tag{10.189b}$$

$$\frac{\partial \theta^\lambda}{\partial z^\mu} = \frac{1}{\sqrt{g_{\mu\mu}}}\delta^\lambda_\mu \tag{10.189c}$$

Consequently, *physical components in directions of orthogonal curvilinear coordinates* can be specialized from Eqs. 10.176, 10.177, 10.180 and 10.188 as

$$u^\lambda = \sqrt{g_{\lambda\lambda}}\,v^\lambda \tag{10.190a}$$

$$X^\lambda = \sqrt{g_{\lambda\lambda}}\,X^\lambda_g \tag{10.190b}$$

$$\mathrm{E}_{\lambda\mu} = \frac{1}{\sqrt{g_{\lambda\lambda}g_{\mu\mu}}}\Delta_{\lambda\mu} \tag{10.190c}$$

$$\Sigma^{\lambda\mu} = \sqrt{g_{\lambda\lambda}g_{\mu\mu}}\,\tau^{\lambda\mu} \tag{10.190d}$$

10.10 BOUNDARY CONDITIONS AND CONSTITUTIVE RELATIONSHIP IN CURVILINEAR COORDINATES

At the point P^0 in the original configuration, let $\delta\mathbf{u}$ be a virtual displacement with contravariant components as follows

$$\delta\mathbf{u} = \mathbf{g}_\mu \delta v^\mu \tag{10.191a}$$

dS_0' a small surface passing through that point, $\mathbf{n}_0'$ the unit outward vector normal to dS_0', and $\overline{\mathbf{X}}$ the stress vector on dS_0'. Substituting Eq. 10.163 into Eq. 10.183 we find

$$\overline{\mathbf{X}} = \overline{X}^\mu_g \mathbf{g}_\mu \tag{10.191b}$$

where $\overline{X}^\mu_g$ is a contravariant component of $\overline{\mathbf{X}}$ and has an explicit form as the following

$$\overline{X}^\mu_g = \sum_{\lambda=1}^{3} (\mathbf{g}_\lambda \cdot \mathbf{n}_0')\tau_0^{\lambda\mu}\sqrt{g^{\lambda\lambda}} \tag{10.192}$$

The virtual work due to $\overline{\mathbf{X}}$ through $\delta\mathbf{u}$ is, because of Eqs. 1.178 and 10.191,

$$(\overline{\mathbf{X}} \cdot \delta\mathbf{u})dS_0' = \left[\left(\overline{X}^\mu_g g_{\mu\kappa}\right)\delta v^\kappa\right]dS_0' \tag{10.193}$$

Thus the boundary conditions on the boundary dS_0' involve the prescription of either

$$\overline{X}^{\mu}_{g} g_{\mu\kappa} \quad \text{or} \quad v^{\kappa} \tag{10.194}$$

If dS_0^t is a surface where $\theta^\lambda = \text{constant}$, $\mathbf{n}_0^t$ is in the same direction as $\mathbf{g}^\lambda$, then due to Eqs. 1.192c, 10.192 and 10.194, boundary conditions become either in

$$\tau_0^{\lambda\mu} g_{\mu\kappa} \quad \text{or} \quad v^{\kappa}, \quad \text{(on surface where } \theta^\lambda = \text{constant)} \tag{10.195}$$

At the instant that the position vector $\mathbf{y}$ for the point $P^0(\theta^1, \theta^2, \theta^3)$ in Fig. 10.8 is increased by $d\mathbf{y}$, such vector for the surface where $\theta^\lambda + d\theta^\lambda = \text{constant}$ is increased by

$$d\mathbf{y} + \frac{\partial(d\mathbf{y})}{\partial\theta^\lambda} d\theta^\lambda, \quad (\lambda \text{ not summed})$$

Thus the corresponding increase of the strain energy for the infinitesimal parallelepiped at the point P^0 (Figs. 10.8 and 10.9) is

$$\rho^0 dU dV^0 = \sum_{\lambda=1}^{3} \left\{ \left[d\mathbf{T}^\lambda + \frac{\partial(d\mathbf{T}^\lambda)}{\partial\theta^\lambda} d\theta^\lambda \right] \cdot \left(d\mathbf{y} + \frac{\partial(d\mathbf{y})}{\partial\theta^\lambda} d\theta^\lambda \right) - d\mathbf{T}^\lambda \cdot d\mathbf{y} \right\}$$

$$+ \rho^0 \mathbf{X} \cdot d\mathbf{y} dV^0 \tag{10.196}$$

where U is the strain energy per unit mass. Neglecting the $(d\theta^\lambda)^2$ terms in the equation above yields

$$\rho^0 dU dV^0 = \left[\sum_{\lambda=1}^{3} \frac{\partial(d\mathbf{T}^\lambda)}{\partial\theta^\lambda} d\theta^\lambda + \rho^0 \mathbf{X} dV^0 \right] \cdot d\mathbf{y} + \sum_{\lambda=1}^{3} d\mathbf{T}^\lambda \cdot \frac{\partial(d\mathbf{y})}{\partial\theta^\lambda} d\theta^\lambda \tag{10.197}$$

or, due to the equilibrium given by Eq. 10.155,

$$\rho^0 dU dV^0 = \sum_{\lambda=1}^{3} d\mathbf{T}^\lambda \cdot \frac{\partial(d\mathbf{y})}{\partial\theta^\lambda} d\theta^\lambda,$$

or

$$\rho^0 dU dV^0 = \sum_{\lambda=1}^{3} d\mathbf{T}^\lambda \cdot d\left(\frac{\partial\mathbf{y}}{\partial\theta^\lambda} \right) d\theta^\lambda \tag{10.198}$$

Substituting Eqs. 10.143 and 10.154, in view of Eqs. 1.186 and 10.157, into the equation above, we get

$$\rho^0 dU = \tau^{\lambda\mu} \mathbf{G}_\mu \cdot d\mathbf{G}_\lambda \tag{10.199}$$

or, due to the stress symmetry given by Eq. 10.158,

$$\rho^0 dU = \frac{1}{2} \tau^{\lambda\mu} d(\mathbf{G}_\lambda \cdot \mathbf{G}_\mu)$$

$$= \frac{1}{2} \tau^{\lambda\mu} d(\mathbf{G}_\lambda \cdot \mathbf{G}_\mu - \mathbf{g}_\lambda \cdot \mathbf{g}_\mu),$$

or, due to Eqs. 1.178 and 10.148,

$$\rho^0 dU = \tau^{\lambda\mu} d\Delta_{\lambda\mu} \tag{10.200}$$

which is the same as Eq. 10.111 and equivalent to Eq. 3.7 for small deformation. Thus we may conclude that a rational form of the constitutive law is the relationship between the non-physical stress $\tau^{\lambda\mu}$ and the non-physical strain $\Delta_{\lambda\mu}$.

Following the same scheme as adopted to derive Eqs. 3.11 to 3.15, we obtain the stress-strain relationship for linear elastic solids as

$$\tau^{\lambda\mu} = D^{\lambda\mu\alpha\beta}\Delta_{\alpha\beta} \tag{10.201}$$

and that for the local Cartesian (z^{λ}) system as

$$\Sigma^{\lambda\mu} = C^{\lambda\mu\alpha\beta}\mathrm{E}_{\alpha\beta} \tag{10.202}$$

Recalling Eqs. 10.180 and 10.188a, which relate the Cartesian strain and stress tensors with the non-physical ones, we can get the relationship between stiffness tensors of different coordinate systems as the following

$$D^{\lambda\mu\alpha\beta} = C^{\gamma\delta\varepsilon\zeta}\frac{\partial\theta^{\lambda}}{\partial z^{\gamma}}\frac{\partial\theta^{\mu}}{\partial z^{\delta}}\frac{\partial\theta^{\alpha}}{\partial z^{\varepsilon}}\frac{\partial\theta^{\beta}}{\partial z^{\zeta}} \tag{10.203}$$

or

$$C^{\lambda\mu\alpha\beta} = D^{\gamma\delta\varepsilon\zeta}\frac{\partial z^{\lambda}}{\partial\theta^{\gamma}}\frac{\partial z^{\mu}}{\partial\theta^{\delta}}\frac{\partial z^{\alpha}}{\partial\theta^{\varepsilon}}\frac{\partial z^{\beta}}{\partial\theta^{\zeta}} \tag{10.204}$$

10.11 COMPATIBILITY CONDITIONS

The *Riemann-Christoffel tensor* $R^{\lambda}_{\mu\nu\omega}$ (Spiegel, 1959) is defined in the deformed configuration as

$$R^{\lambda}_{\mu\nu\omega} = \frac{\partial\begin{Bmatrix}\lambda\\ \mu\omega\end{Bmatrix}^{G}}{\partial\theta^{\nu}} - \frac{\partial\begin{Bmatrix}\lambda\\ \mu\nu\end{Bmatrix}^{G}}{\partial\theta^{\omega}} + \begin{Bmatrix}\kappa\\ \mu\omega\end{Bmatrix}^{G}\begin{Bmatrix}\lambda\\ \kappa\nu\end{Bmatrix}^{G} - \begin{Bmatrix}\kappa\\ \mu\nu\end{Bmatrix}^{G}\begin{Bmatrix}\lambda\\ \kappa\omega\end{Bmatrix}^{G} \tag{10.205}$$

and its associated fourth order tensor commonly known as the *Riemann-Christoffel covariant curvature tensor* is

$$R_{\lambda\mu\nu\omega} = G_{\lambda\kappa}R^{\kappa}_{\mu\nu\omega} \tag{10.206}$$

Only 6 of the 81 components of $R_{\lambda\mu\nu\omega}$ are distinct, due to the following symmetry conditions

$$R_{\lambda\mu\nu\omega} = R_{\nu\omega\lambda\mu} = -R_{\mu\lambda\nu\omega} = -R_{\lambda\mu\omega\nu} = -R_{\lambda\omega\nu\mu} \tag{10.207}$$

The 6 distinct covariant components of the curvature tensor may be chosen as

$$R_{2323},\quad R_{3131},\quad R_{1212},\quad R_{1231},\quad R_{2312},\quad R_{3123} \tag{10.208}$$

If we take the strain components $\Delta_{\lambda\mu}$ as arbitrary continuous functions of coordinates of points in the material domain, the deformation of the material body would be insured of continuity only if

$$R_{\lambda\mu\nu\omega} = 0 \tag{10.209}$$

It is easy to prove that these *compatibility conditions* are necessary, but difficult to prove that they are sufficient; the latter will not be given here. The proof of sufficiency is based on the fact that the covariant differentiation in a Euclidean space is commutative, thus the Riemann-Christoffel tensor and curvature tensor are zero in the Eulidean space where the positions of points in the material body are defined after deformation.

Without loss of generality in the application into the compatibility conditions, we can put the curvilinear coordinates identical to the reference Lagrangian (global Cartesian) coordinates, i.e.

$$\theta^1 = x_1, \quad \theta^2 = x_2, \quad \theta^3 = x_3 \tag{10.210}$$

Accordingly, the covariant metric tensor becomes

$$G_{ij} = (\partial y_k/\partial x_i)(\partial y_k/\partial x_j) \tag{10.211}$$

or, due to $\mathbf{y} = \mathbf{x} + \mathbf{u}$,

$$G_{ij} = \delta_{ij} + 2\mathrm{E}_{ij} \tag{10.212}$$

where E_{ij} is the Green's strain tensor. Also, the contravariant metric tensor is given by

$$G^{ij} = \frac{1}{2G}\frac{\partial G}{\partial \mathrm{E}_{ij}}, \quad (i = j) \tag{10.213a}$$

$$= \frac{1}{4G}\frac{\partial G}{\partial \mathrm{E}_{ij}}, \quad (i \neq j) \tag{10.213b}$$

where

$$G = \det[G_{ij}] \tag{10.214a}$$

$$= \begin{vmatrix} 1+2\mathrm{E}_{11} & 2\mathrm{E}_{12} & 2\mathrm{E}_{13} \\ 2\mathrm{E}_{12} & 1+2\mathrm{E}_{22} & 2\mathrm{E}_{23} \\ 2\mathrm{E}_{13} & 2\mathrm{E}_{23} & 1+2\mathrm{E}_{33} \end{vmatrix} \tag{10.214b}$$

$$= (1+2\mathrm{E}_{11})(1+2\mathrm{E}_{22})(1+2\mathrm{E}_{33}) + 16\mathrm{E}_{12}\mathrm{E}_{23}\mathrm{E}_{13}$$

$$-4(1+2\mathrm{E}_{11})\mathrm{E}_{23}^2 - 4(1+2\mathrm{E}_{22})\mathrm{E}_{13}^2 - 4(1+2\mathrm{E}_{33})\mathrm{E}_{12}^2 \tag{10.214c}$$

In fact, $\sqrt{G}$ as indicated by Eq. 10.145 is the ratio between the new and the original volumes; thus, in view of Eqs. 10.37 and 10.46, we can write

$$\sqrt{G} = |\partial y_i/\partial x_j| \tag{10.215}$$

$$= \begin{vmatrix} 1+\dfrac{\partial u_1}{\partial x_1} & \dfrac{\partial u_1}{\partial x_2} & \dfrac{\partial u_1}{\partial x_3} \\ \dfrac{\partial u_2}{\partial x_1} & 1+\dfrac{\partial u_2}{\partial x_2} & \dfrac{\partial u_2}{\partial x_3} \\ \dfrac{\partial u_3}{\partial x_1} & \dfrac{\partial u_3}{\partial x_2} & 1+\dfrac{\partial u_3}{\partial x_3} \end{vmatrix} \tag{10.216}$$

Two compatibility conditions given by Novozhilov (1961) in an explicit form are listed below, while the remaining four conditions can be obtained from these two by cyclic permutation of numerals 1, 2, 3:

$$
\begin{aligned}
&-\frac{\partial^2 E_{11}}{\partial x_2 \partial x_3}+\frac{\partial}{\partial x_1}\left(\frac{\partial E_{12}}{\partial x_3}+\frac{\partial E_{13}}{\partial x_2}-\frac{\partial E_{23}}{\partial x_1}\right)\\
&=G^{11}\left[\frac{\partial E_{11}}{\partial x_1}\left(\frac{\partial E_{12}}{\partial x_3}+\frac{\partial E_{13}}{\partial x_2}-\frac{\partial E_{23}}{\partial x_1}\right)-\frac{\partial E_{11}}{\partial x_2}\frac{\partial E_{11}}{\partial x_3}\right]\\
&\quad+G^{22}\left[\frac{\partial E_{22}}{\partial x_3}\left(2\frac{\partial E_{12}}{\partial x_1}-\frac{\partial E_{11}}{\partial x_2}\right)-\frac{\partial E_{22}}{\partial x_1}\left(\frac{\partial E_{12}}{\partial x_3}+\frac{\partial E_{23}}{\partial x_1}-\frac{\partial E_{13}}{\partial x_2}\right)\right]\\
&\quad+G^{33}\left[\frac{\partial E_{33}}{\partial x_2}\left(2\frac{\partial E_{13}}{\partial x_1}-\frac{\partial E_{11}}{\partial x_3}\right)-\frac{\partial E_{33}}{\partial x_1}\left(\frac{\partial E_{13}}{\partial x_2}+\frac{\partial E_{23}}{\partial x_1}-\frac{\partial E_{12}}{\partial x_3}\right)\right]\\
&\quad+G^{12}\left[2\frac{\partial E_{12}}{\partial x_1}\left(\frac{\partial E_{12}}{\partial x_3}+\frac{\partial E_{13}}{\partial x_2}-\frac{\partial E_{23}}{\partial x_1}\right)+\frac{\partial E_{11}}{\partial x_1}\frac{\partial E_{22}}{\partial x_3}-2\frac{\partial E_{11}}{\partial x_2}\frac{\partial E_{12}}{\partial x_3}-\frac{\partial E_{22}}{\partial x_1}\frac{\partial E_{11}}{\partial x_3}\right]\\
&\quad+G^{13}\left[2\frac{\partial E_{13}}{\partial x_1}\left(\frac{\partial E_{12}}{\partial x_3}+\frac{\partial E_{13}}{\partial x_2}-\frac{\partial E_{23}}{\partial x_1}\right)+\frac{\partial E_{11}}{\partial x_1}\frac{\partial E_{33}}{\partial x_2}-2\frac{\partial E_{11}}{\partial x_3}\frac{\partial E_{13}}{\partial x_2}-\frac{\partial E_{11}}{\partial x_2}\frac{\partial E_{33}}{\partial x_1}\right]\\
&\quad+G^{23}\left[\frac{\partial E_{22}}{\partial x_3}\left(2\frac{\partial E_{13}}{\partial x_1}-\frac{\partial E_{11}}{\partial x_3}\right)+\frac{\partial E_{33}}{\partial x_2}\left(2\frac{\partial E_{12}}{\partial x_1}-\frac{\partial E_{11}}{\partial x_2}\right)-\frac{\partial E_{22}}{\partial x_1}\frac{\partial E_{33}}{\partial x_1}\right.\\
&\qquad\left.-\left(\frac{\partial E_{13}}{\partial x_2}+\frac{\partial E_{23}}{\partial x_1}-\frac{\partial E_{12}}{\partial x_3}\right)\left(\frac{\partial E_{12}}{\partial x_3}+\frac{\partial E_{23}}{\partial x_1}-\frac{\partial E_{13}}{\partial x_2}\right)\right];
\end{aligned}
$$

$$
\begin{aligned}
&2\frac{\partial^2 E_{12}}{\partial x_1 \partial x_2}-\frac{\partial^2 E_{11}}{\partial x_2^2}-\frac{\partial^2 E_{22}}{\partial x_1^2}\\
&=G^{11}\left[\frac{\partial E_{11}}{\partial x_1}\left(2\frac{\partial E_{12}}{\partial x_2}-\frac{\partial E_{22}}{\partial x_1}\right)-\left(\frac{\partial E_{11}}{\partial x_2}\right)^2\right]\\
&\quad+G^{22}\left[\frac{\partial E_{22}}{\partial x_2}\left(2\frac{\partial E_{12}}{\partial x_1}-\frac{\partial E_{11}}{\partial x_2}\right)-\left(\frac{\partial E_{22}}{\partial x_1}\right)^2\right]\\
&\quad+G^{33}\left[\left(2\frac{\partial E_{13}}{\partial x_1}-\frac{\partial E_{11}}{\partial x_3}\right)\left(2\frac{\partial E_{23}}{\partial x_2}-\frac{\partial E_{22}}{\partial x_3}\right)-\left(\frac{\partial E_{13}}{\partial x_2}+\frac{\partial E_{23}}{\partial x_1}-\frac{\partial E_{12}}{\partial x_3}\right)^2\right]\\
&\quad+G^{12}\left[\left(2\frac{\partial E_{12}}{\partial x_1}-\frac{\partial E_{11}}{\partial x_2}\right)\left(2\frac{\partial E_{12}}{\partial x_2}-\frac{\partial E_{22}}{\partial x_1}\right)+\frac{\partial E_{11}}{\partial x_1}\frac{\partial E_{22}}{\partial x_2}-2\frac{\partial E_{11}}{\partial x_2}\frac{\partial E_{22}}{\partial x_1}\right]
\end{aligned}
$$

$$+G^{13}\left[\left(2\frac{\partial E_{13}}{\partial x_1}-\frac{\partial E_{11}}{\partial x_3}\right)\left(2\frac{\partial E_{12}}{\partial x_2}-\frac{\partial E_{22}}{\partial x_1}\right)+\frac{\partial E_{11}}{\partial x_1}\left(2\frac{\partial E_{23}}{\partial x_2}-\frac{\partial E_{22}}{\partial x_3}\right)\right.$$

$$\left.-2\frac{\partial E_{11}}{\partial x_2}\left(\frac{\partial E_{13}}{\partial x_2}+\frac{\partial E_{23}}{\partial x_1}-\frac{\partial E_{12}}{\partial x_3}\right)\right]$$

$$+G^{23}\left[\left(2\frac{\partial E_{13}}{\partial x_1}-\frac{\partial E_{11}}{\partial x_3}\right)\frac{\partial E_{22}}{\partial x_2}+\left(2\frac{\partial E_{12}}{\partial x_1}-\frac{\partial E_{11}}{\partial x_2}\right)\left(2\frac{\partial E_{23}}{\partial x_2}-\frac{\partial E_{22}}{\partial x_3}\right)\right.$$

$$\left.-2\frac{\partial E_{22}}{\partial x_1}\left(\frac{\partial E_{13}}{\partial x_2}+\frac{\partial E_{23}}{\partial x_1}-\frac{\partial E_{12}}{\partial x_3}\right)\right]\;;\qquad \text{etc.} \tag{10.217}$$

If the strain components are sufficiently small compared with unity and can be neglected, then the difference between the covariant and the contravariant components of the Riemann-Christoffel tensor will be slight and consist only of terms of the same order of magnitude as the strain components. In order to incorporate this condition, we must omit all terms in Eqs. 10.217 which have strain components as factors but not their derivatives, since the latters may be considerably greater than strain components themselves. With this modification, we obtain the six distinct compatibility conditions as follows

$$\frac{\partial^2 E_{11}}{\partial x_2^2}+\frac{\partial^2 E_{22}}{\partial x_1^2}-2\frac{\partial^2 E_{12}}{\partial x_1\partial x_2}$$

$$=\left(\frac{\partial E_{11}}{\partial x_2}\right)^2+\left(\frac{\partial E_{22}}{\partial x_1}\right)^2-\left(2\frac{\partial E_{13}}{\partial x_1}-\frac{\partial E_{11}}{\partial x_3}\right)\left(2\frac{\partial E_{23}}{\partial x_2}-\frac{\partial E_{22}}{\partial x_3}\right)$$

$$+\left(\frac{\partial E_{23}}{\partial x_1}+\frac{\partial E_{13}}{\partial x_2}-\frac{\partial E_{12}}{\partial x_3}\right)^2-\frac{\partial E_{22}}{\partial x_2}\left(2\frac{\partial E_{12}}{\partial x_1}-\frac{\partial E_{11}}{\partial x_2}\right)-\frac{\partial E_{11}}{\partial x_1}\left(2\frac{\partial E_{12}}{\partial x_2}-\frac{\partial E_{22}}{\partial x_1}\right);$$

$$\frac{\partial^2 E_{11}}{\partial x_2\partial x_3}-\frac{\partial}{\partial x_1}\left(\frac{\partial E_{12}}{\partial x_3}+\frac{\partial E_{13}}{\partial x_2}-\frac{\partial E_{23}}{\partial x_1}\right)$$

$$=\frac{\partial E_{11}}{\partial x_2}\frac{\partial E_{11}}{\partial x_3}+\frac{\partial E_{22}}{\partial x_1}\left(\frac{\partial E_{12}}{\partial x_3}+\frac{\partial E_{23}}{\partial x_1}-\frac{\partial E_{13}}{\partial x_2}\right)+\frac{\partial E_{33}}{\partial x_1}\left(\frac{\partial E_{13}}{\partial x_2}+\frac{\partial E_{23}}{\partial x_1}-\frac{\partial E_{12}}{\partial x_3}\right)$$

$$-\frac{\partial E_{11}}{\partial x_1}\left(\frac{\partial E_{13}}{\partial x_2}+\frac{\partial E_{12}}{\partial x_3}-\frac{\partial E_{23}}{\partial x_1}\right)-\frac{\partial E_{22}}{\partial x_3}\left(2\frac{\partial E_{12}}{\partial x_1}-\frac{\partial E_{11}}{\partial x_2}\right)-\frac{\partial E_{33}}{\partial x_2}\left(2\frac{\partial E_{13}}{\partial x_1}-\frac{\partial E_{11}}{\partial x_3}\right);$$

$$\text{etc.} \tag{10.218}$$

Finally, if we assume that not only strains but their derivatives are small compared with unity, we can omit all product terms in the previous equations whereupon Eq. 2.62 is obtained.

In Chapter IV, we have illustrated the use of the compatibility conditions in the solution approach, with strains as basic unknowns, to the problems of elastic planes under the small deformation theory. Such an approach under the finite deformation theory is not so practical, because the compatibility conditions become too complicated.

10.12 PROBLEMS

10.12.1 *Constitutive Law in Eulerian Description*

In the Eulerian description, if the strain energy function U is defined as per unit mass, it can be shown with the same procedure as deriving Eq. 3.4 that

$$\rho \mathrm{d}U = \sigma_{ij}\mathrm{d}\left(\frac{\partial u_i}{\partial y_j}\right)$$

$$= \sigma_{ij}\mathrm{d}\varepsilon^*_{ij} \tag{10.219}$$

where ε^*_{ij} is as defined in Eq. 10.107. Substituting Eq. 10.48 into the equation above yields an identity which is similar to Eq. 10.111 of the Lagrangian description, i.e.

$$\rho^0 \mathrm{d}U = \sigma_{ij}\mathrm{d}\tilde{\varepsilon}_{ij} \tag{10.220}$$

where

$$\mathrm{d}\tilde{\varepsilon}_{ij} = \left|\frac{\partial x_m}{\partial y_n}\right|^{-1} \mathrm{d}\varepsilon^*_{ij} \tag{10.221}$$

Note that $\tilde{\varepsilon}_{ij}$ defined in Eq. 10.221 is a symmetric tensor of the second order. If σ_{ij} components are unique and homogeneous functions of the components of the 'new' strain tensor $\tilde{\varepsilon}_{ij}$, we can have the strain energy function as a unique and homogeneous function of such strain tensor components. Thus there can be a rational constitutive law in the Eulerian description relating the Eulerian stress tensor σ_{ij} with the strain tensor $\tilde{\varepsilon}_{ij}$ defined in Eq. 10.221. However, such a law is rather complicated due to the nature of Eq. 10.221 and the type of description.

10.12.2 *Maxwell-Betti Reciprocal Theorem for Finite Deformation*

Adopting superscripts 1 and 2 in parentheses to denote respective systems as in Section 3.4, we can write for a linear elastic solid undergoing a finite deformation the following

$$\rho^0 \int_{V^0} X_i^{(1)} u_i^{(2)} \mathrm{d}V^0 + \int_{S^0} \overline{X}_i^{(1)} u_i^{(2)} \mathrm{d}S^0 = \int_{V^0} \sigma_{ji}^{0(1)} \frac{\partial u_i^{(2)}}{\partial x_j} \mathrm{d}V^0 \tag{10.222}$$

or, due to Eqs. 10.3 and 10.60b,

$$\rho^0 \int_{V^0} X_i^{(1)} u_i^{(2)} \mathrm{d}V^0 + \int_{S^0} \overline{X}_i^{(1)} u_i^{(2)} \mathrm{d}S^0 = \int_{V^0} \Sigma_{jm}^{(1)} \frac{\partial y_i^{(1)}}{\partial x_m}\left[\frac{\partial y_i^{(2)}}{\partial x_j} - \delta_{ij}\right] \mathrm{d}V^0 \tag{10.223}$$

If the generalized Hooke's law, Eq. 3.11, holds relating Σ_{ij} with E_{ij}, the equation above becomes

$$\rho^0 \int_{V^0} X_i^{(1)} u_i^{(2)} dV^0 + \int_{S^0} \overline{X}_i^{(1)} u_i^{(2)} dS^0 = \int_{V^0} C_{jmkl} E_{kl}^{(1)} \frac{\partial y_i^{(1)}}{\partial x_m} \left[\frac{\partial y_i^{(2)}}{\partial x_j} - \delta_{ij} \right] dV^0 \qquad (10.224)$$

Interchanging the superscripts 1 and 2 in the last equation, we get

$$\rho^0 \int_{V^0} X_i^{(2)} u_i^{(1)} dV^0 + \int_{S^0} \overline{X}_i^{(2)} u_i^{(1)} dS^0 = \int_{V^0} C_{jmkl} E_{kl}^{(2)} \frac{\partial y_i^{(2)}}{\partial x_m} \left[\frac{\partial y_i^{(1)}}{\partial x_j} - \delta_{ij} \right] dV^0 \qquad (10.225)$$

Though the stiffness tensor C_{ijkl} is symmetric as in Eq. 3.14, Eqs. 10.224 and 10.225 are not identical. Thus the Maxwell-Betti reciprocal theorem does not exist in finite deformation.

REFERENCES

The article by Truesdell and Toupin (1960) contains an extensive list of references from the years 1678 to 1960. The works cited in this book and published in the twentied century are listed alphabetically below:

1. Abramowitz, M., and Stegun, I.A. (eds.), *Handbook of Mathematical Functions*, Dover Publications, New York, 1965, pp. 355-494.
2. Achenbach, J.D., *Wave Propagation in Elastic Solids*, North-Holland, 1973, pp. 73, 74, 76, 217 and 218.
3. Agabein, M.E., Lee, S.L., and Dundurs, J.,'Bending of an Annular Slabs Supported on Columns', *J. Franklin Inst.*, Vol. 284, No. 5, November 1967, pp. 300-307.
4. Alfrey, T., 'Non-homogeneous Stresses in Viscoelastic Media', *Quart. Appl. Math*, Vol. 2, No. 2, 1944, pp. 113-119.
5. Almansi, E., 'Sulle deformazione finite dei solidi elastici isotropi, I', *Rend. Accad. Naz. Lincei*, (Ser. 5), Vol. 20, 1911, pp. 705-714.
6. Apirathvorakij, V., and Karasudhi, P., 'Quasi-static Bending of a Cylindrical Elastic Bar Partially Embedded in a Saturated Elastic Half-Space', *Int. J. Solids Structures*, Vol. 16, No. 7, 1980, pp. 625-644.
7. Bellman, R., Kalaba, R.E., and Lockett, J., *Numerical Inversion of the Laplace Transform*, American Elsevier, 1966.
8. Bingham, E.C., *Fluidity and Plasticity*, McGraw-Hill, New York, 1922.
9. Biot, M.A., 'General Theory of Three Dimensional Consolidation', *J. Appl. Phys.*, Vol. 12, 1941, pp. 155-164.
10. Boresi, A.P., and Lynn, P.P., *Elasticity in Engineering Mechanics*, Prentic-Hall, Englewood Cliffs, New Jersey, 1974, pp. 310-316.
11. Bridgman, P.W., 'The Compressibility of Thirty Metals as a Function of Pressure and Temperature', *Proc. Am. Acad. Art. Sci.*, Vol. 58, 1923, pp. 163-242.
12. Budiansky, B., 'A Reassessment of Deformation Theories of Plasticity', *J. Appl. Mech.*, Vol. 26, 1959, pp. 259-264.
13. Cagniard, L., *Reflection and Refraction of Progressive Seismic Waves*, E.A. Flinn and C.H. Dix (translated and revised), McGraw-Hill, New York, 1962, pp. 47-49.
14. Cesaro, E., 'Sulle formole del Volterra, fondamentali nella teoria delle ditorsioni etastiche', *Rendiconto dell' scienze fisiche e matematiche (Societa reale di Napoli)*, Vol. 12, 1906, pp. 311-321.
15. Chan, K.S., Karasudhi, P., and Lee, S.L., 'Force at a Point in the Interior of a Layered Elastic Half Space', *Int. J. Solids Structures*, Vol. 10, No. 11, 1974, pp. 1179-1199.
16. Christensen, R.M., *Theory of Viscolasticity-An Introduction*, Academic Press, New York, 1971, pp. 43-51, 63-72, 105-107.

17. Christensen, R.M., amd Schreiner, R.N., 'Response to Pressurization of a Viscoelastic Cylinder with an Eroding Internal Boundary', *AIAA J.*, Vol. 3, 1965, pp. 1451-1455.
18. Churchill, R.V., *Complex Variables and Applications*, 2nd ed. McGraw- Hill, New York, 1960, pp. 34-43, 57, 80-86, 120, 154-173, and 179.
19. Cormeau, I., *Viscoplasticity and Plasticity in the Finite Element Method*, Ph.D. Dissertation, Univ. Coll. Swansea, 1976.
20. Cost, T.L., 'Approximate Laplace Transform Inversions in Vicoelastic Stress Analysis', *AIAA Journal*, Vol. 2, No. 12, 1964, pp. 2157-2166.
21. Crossland, B., 'The Effect of Fluid Pressure on the Shear Properties of Metals', *Proc. Inst. Mech. Engr.*, Vol. 169, 1954, pp. 935-944.
22. Drucker, D.C., 'A More Fundamental Approach to Plastic Stress Strain Relation', *Proc. 1st. U.S. Natl. Congress Appl. Mech.*, Chicago, 1951, pp. 487-491.
23. Drucker, D.C., 'A Definition of Stable Inelastic Material', *J. Appl. Mech.*, Vol. 26, 1959, pp. 101-106.
24. Essenburg, F. 'On Surface Constraints in Plate Problems', *J. Appl. Mech.*, Vol. 29, 1962, pp. 340-344.
25. Ewing, W.M., Jardetzky, W.S., and Press, F., *Elastic Waves in Layered Media*, McGraw-Hill, New York, 1957, pp. 24-31, 136, 211-214.
26. Ferry, J.D., *Viscoelastic Properties of Polymers*, 2nd. ed., Wiley, New York, 1970.
27. Folie, C.M., 'Bending of Clamped Orthotropic Sandwich Plates', *J. Eng. Mech. Div.*, ASCE, Vol. 96, No. EM3, Proc. Paper 7340, June 1970, pp. 243-265.
28. Frazer, R.A., Duncan, W.J., and Collar, A.R., *Elementary Matrices*, Cambridge University Press, 1938.
29. Fung, Y.C., *Foundations of Solid Mechanics*, Prentice-Hall, Englewood Cliffs, 1965, pp. 142, 143, 150-152, 476 and 477.
30. Galerkin, B., 'Contribution à la solution générale du probléme de la théorie de l'élastcité dans le cas trois dimensions', *Comtes Rendus*, Vol. 190, 1930, pp. 1047-1048; *Comtes rendus (Doklady) acad. sci. U.R.S.S.*, Ser. A, Vol. 14, 1930, p. 353.
31. Gilman, J.J., 'Physical Nature of Plastic Flow and Fracture', in *Plasticity*, E.H. Lee, and P.S. Symonds (eds.), Proc. 2nd. Symp. Naval Structural Mechanics, Providence, 1960, pp. 44-49.
32. Gottenberg, W.G., and Christensen, R.M., 'An Experiment for Determination of the Mechanical Property in Shear for a Linear Isotropic Viscoelastic Solid', *Int. J. Eng. Sci.*, Vol. 2, 1964, pp. 45-57.
33. Graham, G.A.C., 'The Contact Problem in the Linear Theory of Viscoelasticity', *Int. J. Eng. Sci.*, Vol. 3, 1965, pp. 27-46.
34. Graham, G.A.C., 'The Contact Problem in the Linear Theory of Viscoelasticity When the Time Dependent Contact Area Has Any Number of Maxima and

Minima', *Int. J. Eng. Sci.*, Vol. 5, 1967, pp. 495-514.

35. Graham, G.A.C., 'The Correspondence Principle of Linear Viscoelasticity Theory for Mixed Boundary Value Problems Involving Time-Dependent Boundary Regions', *Quart. Appl. Math.*, Vol. 26, 1968, pp. 167-174.
36. Haar, D. ter, 'An Easy Approximate Method of Determining the Relaxation Spectrum of a Viscoelastic Material', *J. Polymer Sci.*, Vol. 6, 1951, p. 247.
37. Habip, L.M., 'A Survey of Modern Developments in the Analysis of Sandwich Structures', *Appl. Mech. Review*, Vol. 18, No. 2, February 1965.
38. Hamel, G., *Elementare Mechanik*, Leipzig and Berlin, 1912.
39. Handelman, G.H., Lin, C.C., and Prager, W., 'On the Mechanical Behavior of Metals in the Strain-Hardening Range', *Quart. Appl. Math.*, Vol. 4, 1947, pp. 397-407.
40. Harkrider, D.G., 'Surface Waves in Multilayered Elastic Media, I. Rayleigh and Love Waves from Buried Sources in a Multilayered Elastic Half-Space', *Bull. Seismological Soc. America*, Vol. 54, No. 2, April 1964, pp. 627-679.
41. Hencky, H. 'Zur Theorie plastischer Deformationen und der hierdurch hervorgerufenen Nachspannungen', *Z. angew. Math. Mech.*, Vol. 4, 1924, pp. 323-334.
42. Hill, R., *The Mathematical Theory of Plasticity*, Oxford: Clarenden Press, 1950.
43. Hodge, P.G., Jr., 'Piecewise Linear Plasticity', *Proc. 9th. Int. Congress Appl. Mech.*, Brussel, 1956, and Vol. 8, 1957, pp. 65-72.
44. Hopkins, I.L., and Hamming, R.W., 'On Creep and Relaxations', *J. Appl. Phys.*, Vol. 28, 1957, pp. 906-909.
45. Hovichitr, I., Karasudhi, P., Nishino, F., and Lee, S.L., 'A Rational Analysis of Plates with Eccentric Stiffeners', *PERIODICA*, Int. Assoc. Bridge Struct. Eng., Proceedings, P-9/77, 1977.
46. Hsu, H.P., *Fourier Analysis*, Simon and Schuster, New York, 1970, pp. 107 and 119.
47. Hughes, T.J.R., and Taylor, R.L., 'Unconditionally Stable Alogorithms for Quasi-Static Elasto/Viscoplastic Solid Finite Element Analysis', *Computers and Structures*, Vol. 8, 1978, pp. 169-173.
48. Hunter, S.C., 'The Hertz Problem for a Rigid Spherical Indenter and a Viscoelastic Half-Space', *J. Mech. Phys. Solids*, Vol. 8, 1960, pp. 219-234.
49. Hunter, S.C., 'Tentative Equations for the Propagation of Stress, Strain and Temperature Fields in Viscoelastic Solids', *J. Mech. Phys. Solids*, Vol. 9, 1961, pp. 39-51.
50. Jardetzky, W.S., 'Period Equation for an n-Layered Half Space and Some Related Questions', *Columbia Univ. Lamont Geols. Obs. Tech. Rept. Seismology*, No. 29, 1953.
51. Kalker, J.J., *Threedimensional Elastic Bodies in Rolling Contact*, Kluwer, Dordrecht, the Netherlands, 1990.

52. Kanok-Nukulchai, W., Taylor, R.L., and Hughes, T.J.R., 'A Large-Deformation Formulation for Shell Analysis by the Finite Element Method', *Computers and Structures*, Vol. 13, 1981. pp. 19-27.
53. Kao, J.S., 'Bending of Circular Sandwich Plates', *J. Eng. Mech. Div.*, ASCE, Vol. 91, No. EM1, Proc Paper 4449, August 1965, pp. 168-171.
54. Kao, J.S., 'Circular Sandwich Plates under Eccentric Loads', *J. Eng. Mech. Div.*, ASCE, Vol. 95, No. EM1, Proc. Paper 6413, February 1969, pp. 235-245.
55. Kao, J.S., 'Bending of Circular Sandwich Plates due to Asymmetric Temperature Distribution', *AIAA J.*, Vol. 8, No. 5, May 1970, pp. 951-954.
56 Kao, J.S., 'Differential Equations for Asymmetric Bending of Circular Sandwich Plates', *AIAA J.*, Vol. 11, No. 4, April 1973, pp. 573-575.
57. Karasudhi, P., 'An Infinite Element Algorithm for Elastodynamic Load Transfer Problems', *Computational Mechanics '86 - Theory and Applications, Proc. Int. Conf. Computational Mech.*, G. Yagawa, and S.N. Atluri (eds.), Springer-Verlag, Tokyo, May 1986, pp. IX73-78.
58. Karasudhi, P., Alam, K.M.A., and Lee, S.L., 'Flat Slabs on Three Rows of Columns', *Israel J. Tech.*, Vol. 11, No. 3, 1973, pp. 149-156.
59. Karasudhi, P., and Alvappillai, A., 'An Infinite Element Algorithm for Predicting Land Subsidence', *Computational Mechanics '88-Theory and Applications, Proc. Int. Conf. Computational Engg. Sci.*, S.N. Atluri and G. Yagawa (eds.), Springer-Verlag, Atlanta, Georgia, April 1988, pp. 63.i.1 to 4.
60. Karasudhi, P., Jou, T.S., and Tansirikongkol, V., 'Curved Box Girder Bridges as Sandwich Plates', *J. Struct. Eng.*, Vol. 4, No. 2, 1976, pp. 63-69.
61. Karasudhi, P., Karunasena, W.M., and Puswewala, U.G.A., 'An Infinite Element Algorithm for Multilayered Saturated Half Spaces', *Proc. US-Korea Joint Seminar/Workshop on Critical Engineering Systems*, Seoul, May 1987, Vol. 1, pp. 207-216.
62. Karasudhi, P., Keer, L.M., and Lee, S.L., 'Vibratory Motion of a Body on an Elastic Half Plane', *J. Appl. Mech.*, Vol. 90, 1968, pp. 697-705.
63. Karasudhi, P., Ng, H.K., and Lee, S.L., 'Axisymmetric Bending of Sandwich Plates on Elastic Foundation', *Eng. J. Singapore*, Vol. 4, No. 1, 1977, pp. 17-26.
64. Karasudhi, P., and Prechaverakul, S., 'A Theoretical Prediction of Land Subsidence due to Water Loss from Aquifers', *Proc. US-Asia Conf. Engineering for Mitigating Natural Hazards Damage*, P. Karasudhi, P. Nutalaya, and A.N.L. Chiu (eds.), Bangkok, December 1987, pp. C8-1 to C8-12.
65. Karasudhi, P., Rajapakse, R.K.N.D., and Hwang, B.Y., 'Torsion of a Long Cylindrical Elastic Bar Partially Embedded in a Layered Elastic Half-Space', *Int. J. Solids Structures*, Vol. 20, No. 1, 1984, pp. 1-11.
66. Karasudhi, P., Rajapakse, R.K.N.D., and Liyanage, K.K., 'A Reconsideration of Elastostatic Load Transfer Problems Involving a Half Space', *Trans. Canadian Soc. Mechanical Eng.*, Vol. 1, No. 4, 1984, pp. 218-226.

67. Karman, T. von, 'Festigkeitsversuche unter allseitigem Druck', *Z. VDI*, Vol. 55, 1911, pp. 1749-1757.
68. Keller, H.B., 'Propagation of Stress Discontinuities in Inhomogeneous Media', *SIAM Review*, Vol. 6, No. 4, October 1964, pp. 356-382.
69. Kliushnikov, V.D., 'On Plasticity Laws for Work Hardening Materials', *J. Appl. Math. Mech.* (translated from *Prikl. Mat. i. Mekh.*), Vol. 22, 1958, pp. 129-160.
70. Koiter, W.T., 'Stress-Strain Relations, Uniqueness and Variational Theorems for Elastic-Plastic Materials with a Singular Yield Surface', *Quart. Appl. Math.*, Vol. 11, 1953, pp. 350-354.
71. Koiter, W.T., 'General Theorem for Elastic-Plastic Solids', in *Progress in Solid Mechanics*, I.N. Sneddon, and R. Hill (eds.), North-Holland, 1960, Chapter 4.
72. Koppe, H., 'Über Rayleigh-Wellen an der Oberfläche zweier Medien', *Z. angew. Math. Mech.*, Vol. 28, 1948, pp. 355-360.
73. Kosherick, H.J., Dundurs, J., and Lee, S.L., 'Annular Slabs in Circular Core Buildings', *J. Structural Div.*, ASCE, Vol. 94, No. ST2, February 1968, pp. 471-483.
74. Lamb, H., 'On the Propagation of Tremors over the Surface of an Elastic Solid', *Phil. Trans. Roy. Soc. (London) A*, Vol. 203, 1904, pp. 1-42.
75. Leadermann, H., 'Elastic and Creep Properties of Filamentous Materials and Other High Polymers', *Textile Foundation*, Washington, D.C., 1943, p.175.
76. Lee, E.H., and Radok, J.R.M., 'The Contact Problem for Viscoelastic Bodies', *J. Appl. Mech.*, Vol. 27, 1960, pp. 438-444.
77. Lee, E.H., and Rogers, T.G., 'Solution of Viscoelastic Stress Analysis Problems Using Measured Creep or Relaxation Functions', *J. Appl. Mech.*, Vol. 30, 1963, pp. 127-133.
78. Lee, S.L., and Ballesteros, P., 'Uniformly Loaded Rectangular Plate Supported at the Corners', *Int. J. Mech. Sci.*, Vol. 2, No. 3, 1960, pp. 206-211.
79. Lee, S.L., Karasudhi, P., Zakeria, M., and Chan, K.S., 'Uniformly Loaded Orthotropic Plates Supported at the Corners', *Civil Eng. Trans., Int. Engrs., Australia*, Vol. CE13, No. 2, October 1971, pp. 101-106.
80. Lockett, F.J., 'Interpretation of Mathematical Solutions in Viscoelastic Theory Illustrated by a Dynamic Spherical Cavity Problem', *J. Mech. Phys. Solids*, Vol. 9, 1961, p. 215.
81. Lockett, F.J., and Morland, L.W., 'Thermal Stresses in a Viscoelastic, Thin-Walled Tube with Temperature Dependent Properties', *Int. J. Eng. Sci.*, Vol. 5, 1967, pp. 879-898.
82. Love, A.E.H., *Some Problems of Geodynamics*, Cambridge University Press, London, 1911, 1926.
83. Luco, J.E., 'Torsion of a Cylinder Embedded in an Elastic Half Space', *J. Appl. Mech.*, Vol. 43, No. 3, 1976, pp. 419-423.

84. Ludwik, P., 'Über den Einfluss der Deformationgeschwindigkeit bei bleibenden Deformationen mit besonderer Berücksichtigung der Nachwirkungserscheinungen', *Phys. Zeitschrift*, Vol. 10, 1909, pp. 411-417.
85. Malvern, L.E., 'The Propagation of Longitudinal Waves of Plastic Deformation in a Bar of Material Exhibiting a Strain-Rate Effect', *J. Appl. Mech.*, Vol. 18, 1951, pp. 203-208.
86. Marcus, H., *Die Theorie elastischer Gewebe*, Springer, Berlin, 1932, p. 173.
87. Mase, G.E., *Continuum Mechanics*, Schaum's Outline Series, McGraw-Hill, New York, 1970, pp. 59-61, 64, 67, 68, 74-76, 98, 99, 103, 105-107 and 109.
88. Michell, J.H., *Proc. London Math. Soc.*, Vol. 31, 1900, p. 100.
89. Michell, J.H., *Proc. London Math. Soc.*, Vol. 34, 1902, p. 223.
90. Mindlin, R.D., 'Force at a Point in the Interior of a Semi-Infinite Solid', *Physics*, Vol. 7, 1936, pp. 195-202.
91. Mindlin, R.D., 'Force at a Point in the Interior of a Semi-Infinite Solid', *Proc. 1st. Midwest. Conf. Solid Mech.*, Univ. Ill., Urbana, 1953, pp. 56-59.
92. Mises, R. von, 'Mechanik der festen Körper im plastisch deformablen Zustand', *Göttinger Nachrichten, Math.-Phys.*, Kl. 1913, pp. 582-592.
93. Mises, R. von, 'Mechanik der plastischen Formanderung von Kristallen', *Z. angew. Math. Mech.*, Vol. 8, 1928, pp. 161-185.
94. Morland, L.W., 'A Plane Problem of Rolling Contact in Linear Viscoelasticity Theory', *J. Appl. Mech.*, Vol 29, 1962, pp. 345-352.
95. Morland, L.W., 'Exact Solutions for Rolling Contact between Viscoelastic Cylinders', *Quart. J. Mech. Appl. Math.*, Vol. 20, 1967, pp. 73-106.
96. Morland, L.W., and Lee, E.H., 'Stress Analysis for Linear Viscoelastic Materials with Temperature Variation', *Trans. Soc. Rheology*, Vol. IV, 1960, pp. 233-263.
97. Muki, R., 'Asymmetric Problems of the Theory of Elasticity for a Semi-Infinite Solid and a Thick Plate', *Progress in Solid Mechanics*, Vol. 1, I.N. Sneddon and R. Hill (eds.), North-Holland, Amsterdam, 1960, pp. 399-439.
98. Muki, R., and Dong, S.B., 'Elastostatic Far Field Behavior in a Layered Half Space under Surface Pressure', *J. Appl. Mech.*, Vol. 47, 1980, pp. 69-90.
99. Muki, R., and Sternberg, E., 'On Transient Thermal Stresses in Viscoelastic Materials with Temperature-Dependent Properties', *J. Appl. Mech.*, Vol. 28, 1961, pp. 193-207.
100. Muki, R., and Sternberg, E., 'Elastostatic Load Transfer to a Half Space from a Partially Embedded Axially Loaded Rod', *Int. J. Solids Structures*, Vol. 6, No. 1, 1970, pp. 69-90.
101. Nadai, A., *Forschungsarb.*, Nos. 170 and 171, Berlin, 1915.
102. Nadai, A., *Z. angew. Math. Mech.*, Vol. 2, 1922, p. 1.
103. Nadai, A., *Elastische Platten*, Berlin, 1925.
104. Nadai, A., *Theory of Flow and Fracture of Solids*, McGraw-Hill, New York; Vol. 1, 1950 and Vol. 2, 1963.

105. Naghdi, P.M., 'Stress-Strain Relations in Plasticity and Thermoplasticity', E.H. Lee and P.S. Symond (eds.), *Plasticity*, 1960, pp. 121-169.
106. Neuber, H., 'Ein Neuer Ansatz zur Lösung räumlicher Probleme der Elastizitätstheorie', *Z. angew. Math. Mech.*, Vol.14, 1934, pp. 203-212.
107. Nicholson, D.W., and Philips, A., 'On the Structure of the Theory of Viscoplasticity', *Int. J. Solids Structures*, Vol. 10, 1974, pp. 149-165.
108. Niumpradit, B., and Karasudhi, P., 'Load Transfer from an Elastic Pile to a Saturated Elastic Soil', *Int. J. Num. Anal. Methods Geomech.*, Vol 5, No. 2, April-June 1981, pp. 115-138.
109. Novozhilov, V.V., *Theory of Elasticity*, (translated by J.K. Lusher), Pergamon Press, New York, 1961, pp. 59-61.
110. Osgood, W.R., 'Combined-Stress Tests on 24 ST Aluminum Alloy Tubes', *J. Appl. Mech.*, Vol. 69, 1947, pp. 147-153.
111. Papkovich, P.F., 'Solution générale des équations différentielles fondamentales d'élasticité, exprimée par trois fonctions harmoniques', *Comptes Rendus*, Vol. 195, 1932, pp. 513-515; 'Expressions générale des composantes des tensions, ne renfermant comme fonctions arbitraires que des fonctions harmoniques', *Comptes Rendus*, Vol. 195, 1932, pp. 754-756.
112. Perzyna, P., 'The Constitutive Equations for Rate - Sensitive Plastic Materials', *Quart. Appl. Math.*, Vol. 20, 1963, pp. 321-332.
113. Perzyna, P., 'Fundamental Problems in Viscoplasticity', in *Recent Adv. Appl. Mech.*, Academic Press, New York, 1966.
114. Philips, A., and Wu, H.C., 'A Theory of Viscoplasticity', *Int. J. Solids Structures*, Vol. 9, 1973, pp. 15-30.
115 Plantema, F.J., *Sandwich Construction*, John Wiley and Sons, New York, 1966.
116. Prager, W., 'The Stress-Strain Laws of the Mathematical Theory of Plasticity', *J. Appl. Mech.*, Vol. 15, 1948, pp. 226-233, and Vol. 15, 1949, pp. 215-218.
117. Prager, W., 'The Theory of Plasticity: A Survey of Recent Achievements', *Proc. Inst. Mech. Engrs.*, Vol. 169, 1954, pp. 41-57.
118. Prandtl, L., 'Spannungsverteilung in plastischen Körper', *Proc. 1st. Int. Congress Appl. Mech.*, Delft, 1924, pp. 43-54.
119. Prandtl, L., 'Ein Gedankenmodell zur kinetischen Theorie der festen Körper', *Z. angew. Math. Mech.*, Vol. 8, 1928, pp. 85-106.
120. Press, F., Harkrider, D., and Seafeldt, C.A., 'A Fast, Convenient Program for Computation of Surface-Wave Dispersion Curves in Multilayered Media', *Bull. Seismological Soc. America*, Vol. 51, No. 4, October 1961, pp. 495-502.
121. Quinlan, P.M., 'The Elastic Theory of Soil Dynamics', *ASTM Special Technical Publication No. 156*, Symposium on Dynamic Testing of Soils, Atlanta City, N.J., July 1953, pp. 3-34.
122. Rajapakse, R.K.N.D., and Karasudhi, P., 'Elastostatic Infinite Elements for Layered Half Spaces', *J. Eng. Mech.*, ASCE, Vol. 111, No. 9, September 1985,

pp. 1144-1158.

123. Rajapakse, R.K.N.D., and Karasudhi, P., 'An Efficient Elastodynamic Infinite Element', *Int. J. Solids Structures*, Vol. 22, No. 6, 1986, pp. 643-657.

124. Reissner, E., 'Finite Deflection of Sandwich Plates', *J. Aeronaut. Sci.*, Vol. 15, 1948, pp. 435-440.

125. Reissner, E., 'Small Bending and Stretching of Sandwich Shells', *NACA TN 1832*, 1949.

126. Reuss, E., 'Berücksichtigung der elastischen Formänderungen in der Plastizitätstheorie', *Z. angew. Math. Mech.*, Vol. 10, 1930, pp. 266-274.

127. Robertson, I.A., 'Forced Vertical Vibration of a Rigid Circular Disc on a Semi-Infinite Elastic Solid', *Proc. Camb. Phil. Soc.*, Vol. 62, 1966, pp.547-553.

128. Rongved, L. 'Force at a Point in the Interior of One of the Two Joined Semi-Infinite Solids', *Proc. 2nd. Midwest. Conf. Solid Mech.*, Lafayette, Ind., 1955.

129. Sadowsky, M., 'Zweidimensionale Probleme der Elastizitätstheorie', *Z. angew. Math. Mech.*, Vol. 8, No. 2, April 1928, pp. 107-121.

130. Schapery, R.A., 'Approximate Methods of Transform Inversions in Viscoelastic Stress Analysis', *Proc. 4th. U.S. National Congress Appl. Mech.*, Vol. 2, 1962, pp. 1075-1085.

131. Schiffman, R.L., and Fungaroli, A.A., 'Consolidation due to Tangential Loads', *Proc. 6th. Int. Conf. Soil Mech.*, Montreal, Vol. 2, 1965, pp. 188-192.

132. Scholte, J.G., 'The Range of Existence of Rayleigh and Stoneley Waves', *Monthly Notices Roy. Astron. Soc.: Geophys. Suppl.*, Vol. 5, 1947, pp. 120-126.

133. Schwarzl, F., and Staverman, A.J., 'Time-Temperature Dependence of Linear Viscoelastic Behavior', *J. Appl. Phys.*, Vol. 23, 1952, p. 838.

134. Scordelis, A.C., Pister, K.S., and Lin, T.Y., *J. Amer. Concr. Inst.*, Vol. 28, 1956, p. 241.

135. Sezawa, K., and Kanai, K., 'The Range of Possible Existence of Stoneley Waves and Some Related Problems', *Bull. Earthquake Research Inst. (Tokyo)*, Vol. 17, 1939, pp. 1-8.

136. Sneddon, I.N., *Fourier Transforms*, McGraw-Hill, New York, 1951, pp. 5, 6, 17, 18, 27, 53, 60, 62, 65, 275, 423, 424, 434, 456, 460, 487, 489 and 502.

137. Sneddon, I.N., *Mixed Boundary Value Problems in Potential Theory*, North-Holland, 1966, pp. 30, 40-46, 64, 80-133, 136, 140, 147, 176, 213, 222, 229 and 256.

138. Sokolnikoff, I.S., *Mathematical Theory of Elasticity*, 2nd ed. McGraw-Hill, New York, 1956, pp. 25-29.

139. Sokolovsky, V.V., 'The Propagation of Elastic-Viscoplastic Waves in Bars', *Prik. Mat. Mekh.*, Vol. 12, 1948, pp. 251-280.

140. Spiegel, M.R., *Vector Analysis*, Schaum's Outline Series, McGraw-Hill, New York, 1959, pp. 206, 207 and 212.

141. Spiegel, M.R. *Advanced Calculus*, Schaum's Outline Series, McGraw-Hill, New

York, 1963, pp. 263 and 272.

142. Spiegel, M.R., *Laplace Transforms*, Schaum's Outline Series, McGraw-Hill, New York, 1965, pp. 46, 47, 61, 62, 73, 74, 201, 202, 207 and 210.
143. Spiegel, M.R., *Advanced Mathematics*, Schaum's Outline Series, McGraw-Hill, New York, 1971, pp. 211, 219, 349, 365, 376, 377, 383, 384 and 387.
144. Stakgold, I., *Boundary Value Problems of Mathematical Physics*, Vol. 1, Macmillan, New York, 1967, pp. 6, 7, 18-27, 33, and 273.
145. Sternberg, E., 'On the Analysis of Thermal Stresses in Viscoelastic Solids', *Proc. 3rd. Symp. Nav. Structural Mech.*, Macmillan, New York, 1964, p. 348.
146. Stoneley, R., 'Elastic Waves at the Surface of Separation of Two Solids', *Proc. Roy. Soc. (London), A*, Vol. 106, 1924, pp. 416-428.
147. Sung, T.Y., 'Vibrations in Semi-Infinite Solids due to Periodic Surface Loading', *ASTM Special Technical Publication No. 156*, Symposium on Dynamic Testing of Soils, Atlanta City, N.J., July 1953, pp. 35-68.
148. Suriyamongkol, S., 'Bending of Polygonal Plates', *Japan-Thai Civil Engineering Conf., Proc. Recent Advance Struct. Eng.*, Chulalongkorn Univ., Bangkok, Thailand, March 1985, pp. 85-108.
149. Swanson, S.R., 'Approximate Laplace Transform Inversion in Dynamic Viscoelasticity', *J. Appl. Mech.*, Vol. 47, December 1980, pp. 769-774.
150. Timoshenko, S.P., and Goodier, J.N., *Theory of Elasticity*, 3rd ed. McGraw-Hill, New York, 1970, pp. 122, 124, 146-149.
151. Timoshenko, S., and Woinowsky-Krieger, S., *Theory of Plates and Shells*, McGraw-Hill, New York, 1959, pp. 118-120, 126-135, 139, 140, 143, 161, 187, 191, 192, 194-196, 202, 204, 206, 207, 210-212, 215, 216, 218-220.
152. Ting, T.C.T., 'Contact Problems in the Linear Theory of Viscoelasticity', *J. Appl. Mech.*, Vol. 35, 1968, pp. 248-254.
153. Ting, T.C.T., 'A Mixed Boundary Value Problem in Viscoelasticity with Time-Dependent Boundary Regions', in *Developments in Mechanics*, H.J. Weis, D.F. Young, W.F. Riley, and T.R. Rogge (eds.), Vol. 5, Iowa State Univ. Press, Ames, Iowa, 1969, p. 591.
154. Truesdell, C., 'The Mechnical Foundations of Elasticity and Fluid Dynamics', *J. Rational Mech. Anal.*, Vol. 1, 1952, pp. 125-300; corrected reprint in *The Mechanical Foundations of Elasticity and Fluid Dynamics*, Gordon & Breach, New York, 1966.
155. Truesdell, C., and Toupin, R.A., 'The Classical Field Theories', *Encyclopedia of Physics*, S. Flügge (ed.), Vol. III/1, Springer-Verlag, Berlin, 1960, pp. 226-793.
156. Tsien, H.S., 'A Generalization of Alfrey's Theorem for Viscoelastic Media', *Quart. Appl. Math.*, Vol. 8, 1950, p. 104.
157. Vijakkhana, C., Karasudhi, P., and Lee, S.L., 'Corner Supported Equilateral Triangular Plates', *Int. J. Mech. Sci.*, Vol. 15, 1973, pp. 123-128.
158. Wang, C.-T., *Applied Elasticity*, McGraw-Hill, New York, 1953, pp. 22 and 23.

159. Watson, G.N., *A Treatise on the Theory of Bessel Functions*, Cambridge University Press, 1966.
160. Whittaker, E.T., and Watson, G,N., *A Course on Modern Analysis*, Cambridge University Press, New York, *4th* edition reprinted, 1965, p. 172.
161. Widder, D.V., *The Laplace Transform*, Princeton University Press, Princeton, N.J., 1946.
162. Williams, M.L., 'Stress Singularities Resulting from Various Boundary Conditions in Angular Corners of Plates in Extension', *J. Appl. Mech.*, Vol. 19, 1952, pp. 526-528.
163. Wylie, C.R., *Advanced Engineering Mathematics*, McGraw-Hill Kogakusha, Tokyo, 4th. Edition, 1975, pp. 512, 513, 515 and 528.
164. Ziegler, H., 'A Modification of Prager's Hardening Rules', *Quart. Appl. Math.*, Vol. 17, 1959, pp. 55-65.
165. Zienkiewicz, O.C., and Cormeau, I.C., 'Viscoplasticity-Plasticity and Creep in Elastic Solids, A Unified Numerical Solution Approach', *Int. J. Num. Methods Eng.*, Vol. 8, 1974, pp. 821-845.

AUTHOR INDEX

K

L

M

N

O

P

Q

R

SUBJECT INDEX

Mechanics

From 1990, books on the subject of *mechanics* will be published under two series.

SOLID MECHANICS AND ITS APPLICATIONS

Series Editor: G.M.L. Gladwell

Aims and Scope of the Series

The fundamental questions arising in mechanics are: *Why?*, *How?*, and *How much?* The aim of this series is to provide lucid accounts written by authoritative researchers giving vision and insight in answering these questions on the subject of mechanics as it relates to solids. The scope of the series covers the entire spectrum of solid mechanics. Thus it includes the foundation of mechanics; variational formulations; computational mechanics; statics, kinematics and dynamics of rigid and elastic bodies; vibrations of solids and structures; dynamical systems and chaos; the theories of elasticity, plasticity and viscoelasticity; composite materials; rods, beams, shells and membranes; structural control and stability; soils, rocks and geomechanics; fracture; tribology; experimental mechanics; biomechanics and machine design.

1. R.T. Haftka, Z. Gürdal and M.P. Kamat: *Elements of Structural Optimization*. 2nd rev.ed., 1990 ISBN 0-7923-0608-2
2. J.J. Kalker: *Three-Dimensional Elastic Bodies in Rolling Contact*. 1990 ISBN 0-7923-0712-7
3. P. Karasudhi: *Foundations of Solid Mechanics*. ISBN 0-7923-0772-0

FLUID MECHANICS AND ITS APPLICATIONS

Series Editor: R. Moreau

Aims and Scope of the Series

The purpose of this series is to focus on subjects in which fluid mechanics plays a fundamental role. As well as the more traditional applications of aeronautics, hydraulics, heat and mass transfer etc., books will be published dealing with topics which are currently in a state of rapid development, such as turbulence, suspensions and multiphase fluids, super and hypersonic flows and numerical modelling techniques. It is a widely held view that it is the interdisciplinary subjects that will receive intense scientific attention, bringing them to the forefront of technological advancement. Fluids have the ability to transport matter and its properties as well as transmit force, therefore fluid mechanics is a subject that is particularly open to cross fertilisation with other sciences and disciplines of engineering. The subject of fluid mechanics will be highly relevant in domains such as chemical, metallurgical, biological and ecological engineering. This series is particularly open to such new multidisciplinary domains.

1. M. Lesieur: *Turbulence in Fluids*. 2nd rev. ed., 1990 ISBN 0-7923-0645-7
2. O. Métais and M. Lesieur (eds.): *Turbulence and Coherent Structures*. 1991 ISBN 0-7923-0646-5
3. R. Moreau: *Magnetohydrodynamics*. 1990 ISBN 0-7923-0937-5
4. E. Coustols (ed.): *Turbulence Control by Passive Means*. 1990 ISBN 0-7923-1020-9

Kluwer Academic Publishers – Dordrecht / Boston / London

Mechanics

From 1990, books on the subject of *mechanics* will be published under two series:
FLUID **MECHANICS AND ITS APPLICATIONS**
Series Editor: R.J. Moreau
SOLID **MECHANICS AND ITS APPLICATIONS**
Series Editor: G.M.L. Gladwell

Prior to 1990, the books listed below were published in the respective series indicated below.

MECHANICS: DYNAMICAL SYSTEMS
Editors: L. Meirovitch and G.Æ. Oravas

1. E.H. Dowell: *Aeroelasticity of Plates and Shells.* 1975 ISBN 90-286-0404-9
2. D.G.B. Edelen: *Lagrangian Mechanics of Nonconservative Nonholonomic Systems.* 1977 ISBN 90-286-0077-9
3. J.L. Junkins: *An Introduction to Optimal Estimation of Dynamical Systems.* 1978 ISBN 90-286-0067-1
4. E.H. Dowell (ed.), H.C. Curtiss Jr., R.H. Scanlan and F. Sisto: *A Modern Course in Aeroelasticity. Revised and enlarged edition see under Volume 11*
5. L. Meirovitch: *Computational Methods in Structural Dynamics.* 1980 ISBN 90-286-0580-0
6. B. Skalmierski and A. Tylikowski: *Stochastic Processes in Dynamics.* Revised and enlarged translation. 1982 ISBN 90-247-2686-7
7. P.C. Müller and W.O. Schiehlen: *Linear Vibrations.* A Theoretical Treatment of Multi-degree-of-freedom Vibrating Systems. 1985 ISBN 90-247-2983-1
8. Gh. Buzdugan, E. Mihăilescu and M. Radeş: *Vibration Measurement.* 1986 ISBN 90-247-3111-9
9. G.M.L. Gladwell: *Inverse Problems in Vibration.* 1987 ISBN 90-247-3408-8
10. G.I. Schuëller and M. Shinozuka: *Stochastic Methods in Structural Dynamics.* 1987 ISBN 90-247-3611-0
11. E.H. Dowell (ed.), H.C. Curtiss Jr., R.H. Scanlan and F. Sisto: *A Modern Course in Aeroelasticity.* Second revised and enlarged edition (of Volume 4). 1989 ISBN Hb 0-7923-0062-9; Pb 0-7923-0185-4
12. W. Szemplińska-Stupnicka: *The Behavior of Nonlinear Vibrating Systems.* Volume I: Fundamental Concepts and Methods: Applications to Single-Degree-of-Freedom Systems. 1990 ISBN 0-7923-0368-7
13. W. Szemplińska-Stupnicka: *The Behavior of Nonlinear Vibrating Systems.* Volume II: Advanced Concepts and Applications to Multi-Degree-of-Freedom Systems. 1990 ISBN 0-7923-0369-5
Set ISBN (Vols. 12–13) 0-7923-0370-9

MECHANICS OF STRUCTURAL SYSTEMS
Editors: J.S. Przemieniecki and G.Æ. Oravas

1. L. Frýba: *Vibration of Solids and Structures under Moving Loads.* 1970 ISBN 90-01-32420-2
2. K. Marguerre and K. Wölfel: *Mechanics of Vibration.* 1979 ISBN 90-286-0086-8

Mechanics

3. E.B. Magrab: *Vibrations of Elastic Structural Members.* 1979 ISBN 90-286-0207-0
4. R.T. Haftka and M.P. Kamat: *Elements of Structural Optimization.* 1985
   Revised and enlarged edition see under Solid Mechanics and Its Applications, Volume 1
5. J.R. Vinson and R.L. Sierakowski: *The Behavior of Structures Composed of Composite Materials.* 1986 ISBN Hb 90-247-3125-9; Pb 90-247-3578-5
6. B.E. Gatewood: *Virtual Principles in Aircraft Structures.* Volume 1: Analysis. 1989
   ISBN 90-247-3754-0
7. B.E. Gatewood: *Virtual Principles in Aircraft Structures.* Volume 2: Design, Plates, Finite Elements. 1989 ISBN 90-247-3755-9
   Set (Gatewood 1 + 2) ISBN 90-247-3753-2

MECHANICS OF ELASTIC AND INELASTIC SOLIDS

Editors: S. Nemat-Nasser and G.Æ. Oravas

---

1. G.M.L. Gladwell: *Contact Problems in the Classical Theory of Elasticity.* 1980
   ISBN Hb 90-286-0440-5; Pb 90-286-0760-9
2. G. Wempner: *Mechanics of Solids with Applications to Thin Bodies.* 1981
   ISBN 90-286-0880-X
3. T. Mura: *Micromechanics of Defects in Solids.* 2nd revised edition, 1987
   ISBN 90-247-3343-X
4. R.G. Payton: *Elastic Wave Propagation in Transversely Isotropic Media.* 1983
   ISBN 90-247-2843-6
5. S. Nemat-Nasser, H. Abé and S. Hirakawa (eds.): *Hydraulic Fracturing and Geothermal Energy.* 1983 ISBN 90-247-2855-X
6. S. Nemat-Nasser, R.J. Asaro and G.A. Hegemier (eds.): *Theoretical Foundation for Large-scale Computations of Nonlinear Material Behavior.* 1984 ISBN 90-247-3092-9
7. N. Cristescu: *Rock Rheology.* 1988 ISBN 90-247-3660-9
8. G.I.N. Rozvany: *Structural Design via Optimality Criteria.* The Prager Approach to Structural Optimization. 1989 ISBN 90-247-3613-7

MECHANICS OF SURFACE STRUCTURES

Editors: W.A. Nash and G.Æ. Oravas

---

1. P. Seide: *Small Elastic Deformations of Thin Shells.* 1975 ISBN 90-286-0064-7
2. V. Panc: *Theories of Elastic Plates.* 1975 ISBN 90-286-0104-X
3. J.L. Nowinski: *Theory of Thermoelasticity with Applications.* 1978
   ISBN 90-286-0457-X
4. S. Łukasiewicz: *Local Loads in Plates and Shells.* 1979 ISBN 90-286-0047-7
5. C. Fiřt: *Statics, Formfinding and Dynamics of Air-supported Membrane Structures.* 1983 ISBN 90-247-2672-7
6. Y. Kai-yuan (ed.): *Progress in Applied Mechanics.* The Chien Wei-zang Anniversary Volume. 1987 ISBN 90-247-3249-2
7. R. Negruţiu: *Elastic Analysis of Slab Structures.* 1987 ISBN 90-247-3367-7
8. J.R. Vinson: *The Behavior of Thin Walled Structures.* Beams, Plates, and Shells. 1988
   ISBN Hb 90-247-3663-3; Pb 90-247-3664-1

Mechanics

MECHANICS OF FLUIDS AND TRANSPORT PROCESSES

Editors: R.J. Moreau and G.Æ. Oravas

1. J. Happel and H. Brenner: *Low Reynolds Number Hydrodynamics*. With Special Applications to Particular Media. 1983 ISBN Hb 90-01-37115-9; Pb 90-247-2877-0
2. S. Zahorski: *Mechanics of Viscoelastic Fluids*. 1982 ISBN 90-247-2687-5
3. J.A. Sparenberg: *Elements of Hydrodynamics Propulsion*. 1984 ISBN 90-247-2871-1
4. B.K. Shivamoggi: *Theoretical Fluid Dynamics*. 1984 ISBN 90-247-2999-8
5. R. Timman, A.J. Hermans and G.C. Hsiao: *Water Waves and Ship Hydrodynamics*. An Introduction. 1985 ISBN 90-247-3218-2
6. M. Lesieur: *Turbulence in Fluids*. Stochastic and Numerical Modelling. 1987 ISBN 90-247-3470-3
7. L.A. Lliboutry: *Very Slow Flows of Solids*. Basics of Modeling in Geodynamics and Glaciology. 1987 ISBN 90-247-3482-7
8. B.K. Shivamoggi: *Introduction to Nonlinear Fluid-Plasma Waves*. 1988 ISBN 90-247-3662-5
9. V. Bojarevičs, Ya. Freibergs, E.I. Shilova and E.V. Shcherbinin: *Electrically Induced Vortical Flows*. 1989 ISBN 90-247-3712-5
10. J. Lielpeteris and R. Moreau (eds.): *Liquid Metal Magnetohydrodynamics*. 1989 ISBN 0-7923-0344-X

MECHANICS OF ELASTIC STABILITY

Editors: H. Leipholz and G.Æ. Oravas

1. H. Leipholz: *Theory of Elasticity*. 1974 ISBN 90-286-0193-7
2. L. Librescu: *Elastostatics and Kinetics of Aniosotropic and Heterogeneous Shell-type Structures*. 1975 ISBN 90-286-0035-3
3. C.L. Dym: *Stability Theory and Its Applications to Structural Mechanics*. 1974 ISBN 90-286-0094-9
4. K. Huseyin: *Nonlinear Theory of Elastic Stability*. 1975 ISBN 90-286-0344-1
5. H. Leipholz: *Direct Variational Methods and Eigenvalue Problems in Engineering*. 1977 ISBN 90-286-0106-6
6. K. Huseyin: *Vibrations and Stability of Multiple Parameter Systems*. 1978 ISBN 90-286-0136-8
7. H. Leipholz: *Stability of Elastic Systems*. 1980 ISBN 90-286-0050-7
8. V.V. Bolotin: *Random Vibrations of Elastic Systems*. 1984 ISBN 90-247-2981-5
9. D. Bushnell: *Computerized Buckling Analysis of Shells*. 1985 ISBN 90-247-3099-6
10. L.M. Kachanov: *Introduction to Continuum Damage Mechanics*. 1986 ISBN 90-247-3319-7
11. H.H.E. Leipholz and M. Abdel-Rohman: *Control of Structures*. 1986 ISBN 90-247-3321-9
12. H.E. Lindberg and A.L. Florence: *Dynamic Pulse Buckling*. Theory and Experiment. 1987 ISBN 90-247-3566-1
13. A. Gajewski and M. Zyczkowski: *Optimal Structural Design under Stability Constraints*. 1988 ISBN 90-247-3612-9

Mechanics

MECHANICS: ANALYSIS

Editors: V.J. Mizel and G.Æ. Oravas

1. M.A. Krasnoselskii, P.P. Zabreiko, E.I. Pustylnik and P.E. Sbolevskii: *Integral Operators in Spaces of Summable Functions*. 1976 ISBN 90-286-0294-1
2. V.V. Ivanov: *The Theory of Approximate Methods and Their Application to the Numerical Solution of Singular Integral Equations*. 1976 ISBN 90-286-0036-1
3. A. Kufner, O. John and S. Pučík: *Function Spaces*. 1977 ISBN 90-286-0015-9
4. S.G. Mikhlin: *Approximation on a Rectangular Grid*. With Application to Finite Element Methods and Other Problems. 1979 ISBN 90-286-0008-6
5. D.G.B. Edelen: *Isovector Methods for Equations of Balance*. With Programs for Computer Assistance in Operator Calculations and an Exposition of Practical Topics of the Exterior Calculus. 1980 ISBN 90-286-0420-0
6. R.S. Anderssen, F.R. de Hoog and M.A. Lukas (eds.): *The Application and Numerical Solution of Integral Equations*. 1980 ISBN 90-286-0450-2
7. R.Z. Has'minskiĭ: *Stochastic Stability of Differential Equations*. 1980 ISBN 90-286-0100-7
8. A.I. Vol'pert and S.I. Hudjaev: *Analysis in Classes of Discontinuous Functions and Equations of Mathematical Physics*. 1985 ISBN 90-247-3109-7
9. A. Georgescu: *Hydrodynamic Stability Theory*. 1985 ISBN 90-247-3120-8
10. W. Noll: *Finite-dimensional Spaces*. Algebra, Geometry and Analysis. Volume I. 1987 ISBN Hb 90-247-3581-5; Pb 90-247-3582-3

MECHANICS: COMPUTATIONAL MECHANICS

Editors: M. Stern and G.Æ. Oravas

1. T.A. Cruse: *Boundary Element Analysis in Computational Fracture Mechanics*. 1988 ISBN 90-247-3614-5

MECHANICS: GENESIS AND METHOD

Editor: G.Æ. Oravas

1. P.-M.-M. Duhem: *The Evolution of Mechanics*. 1980 ISBN 90-286-0688-2

MECHANICS OF CONTINUA

Editors: W.O. Williams and G.Æ. Oravas

1. C.-C. Wang and C. Truesdell: *Introduction to Rational Elasticity*. 1973 ISBN 90-01-93710-1
2. P.J. Chen: *Selected Topics in Wave Propagation*. 1976 ISBN 90-286-0515-0
3. P. Villaggio: *Qualitative Methods in Elasticity*. 1977 ISBN 90-286-0007-8

Mechanics

MECHANICS OF FRACTURE

Editors: G.C. Sih

1. G.C. Sih (ed.): *Methods of Analysis and Solutions of Crack Problems*. 1973 ISBN 90-01-79860-8
2. M.K. Kassir and G.C. Sih (eds.): *Three-dimensional Crack Problems*. A New Solution of Crack Solutions in Three-dimensional Elasticity. 1975 ISBN 90-286-0414-6
3. G.C. Sih (ed.): *Plates and Shells with Cracks*. 1977 ISBN 90-286-0146-5
4. G.C. Sih (ed.): *Elastodynamic Crack Problems*. 1977 ISBN 90-286-0156-2
5. G.C. Sih (ed.): *Stress Analysis of Notch Problems*. Stress Solutions to a Variety of Notch Geometries used in Engineering Design. 1978 ISBN 90-286-0166-X
6. G.C. Sih and E.P. Chen (eds.): *Cracks in Composite Materials*. A Compilation of Stress Solutions for Composite System with Cracks. 1981 ISBN 90-247-2559-3
7. G.C. Sih (ed.): *Experimental Evaluation of Stress Concentration and Intensity Factors*. Useful Methods and Solutions to Experimentalists in Fracture Mechanics. 1981 ISBN 90-247-2558-5

MECHANICS OF PLASTIC SOLIDS

Editors: J. Schroeder and G.Æ. Oravas

1. A. Sawczuk (ed.): *Foundations of Plasticity*. 1973 ISBN 90-01-77570-5
2. A. Sawczuk (ed.): *Problems of Plasticity*. 1974 ISBN 90-286-0233-X
3. W. Szczepiński: *Introduction to the Mechanics of Plastic Forming of Metals*. 1979 ISBN 90-286-0126-0
4. D.A. Gokhfeld and O.F. Cherniavsky: *Limit Analysis of Structures at Thermal Cycling*. 1980 ISBN 90-286-0455-3
5. N. Cristescu and I. Suliciu: *Viscoplasticity*. 1982 ISBN 90-247-2777-4

Kluwer Academic Publishers - Dordrecht / Boston / London